Hypercrosslinked Polymeric Networks and Adsorbing Materials:

Synthesis, Properties, Structure, and Applications

Comprehensive Analytical Chemistry

VOLUME 56

Hypercrosslinked Polymeric Networks and Adsorbing Materials:

Synthesis, Properties, Structure, and Applications

VOLUME

**VADIM A. DAVANKOV AND
MARIA P. TSYURUPA**

*A.N. Nesmeyanov-Institute of Organoelement Compounds,
Russian Academy of Sciences, Moscow, Russian Federation*

ELSEVIER

Amsterdam • Boston • Heidelberg • London • New York • Oxford
Paris • San Diego • San Francisco • Singapore • Sydney • Tokyo

Elsevier

The Boulevard, Langford Lane, Kidlington, Oxford OX5 1GB, UK

Radarweg 29, PO Box 211, 1000 AE Amsterdam, The Netherlands

First edition 2011

ISBN: 978-0-444-53700-3
ISSN: 0166-526X

10065192SX

For information on all Elsevier publications
visit our website at www.elsevierdirect.com

Printed and bound in USA

11 12 13   10 9 8 7 6 5 4 3 2 1

CONTENTS

It is my pleasure to introduce this second volume on polymers in the Comprehensive Analytical Chemistry series, following volume 53 on Molecular Characterization and Analysis of Polymers edited by John M. Chalmers and Rob Meier. The book written by Vadim Davankov and Maria Tsyurupa, the authors of the very conception of hypercrosslinked polymeric networks, is a unique multidisciplinary work. The first hypercrosslinked aromatic polymers were introduced by the authors in the early 1970s and were shown to exhibit properties totally incompatible with theoretical expectations for a densely crosslinked polymeric network. The book sums up the results of their 50 years-long experience, providing a platform for the theoretical description of the new generation of polymeric networks and adsorbing materials. While of general interest to specialists in the chemistry and physics of polymers, the book is also of great interest to specialists in analytical chemistry, since the new materials of an exceptional adsorbing potential have become so popular in the field of analytical chemistry over the last 20 years. The book is a valuable reference work that critically reviews the results of analytical applications of the new material, in addition to other application area such as large-scale adsorption technology, medicine, environmental protection, nanotechnology, etc.

This book contains 17 chapters and is divided into 3 parts. Chapters 1–5 deal with the more fundamental aspects of known types of polystyrene networks, like type of networks, the formation of macroporous polymers, the preparation of continuous polymeric beds and the properties of ion-exchange resins, among other examples.

Chapters 7–9 form the second part and describe the hypercrosslinked polystyrene networks. These chapters cover in detail the basic principles of the formation of hypercrosslinked polymers, the synthesis of hypercrosslinked networks and also the properties and morphology of the polystyrene networks. Characterization techniques such as electron microscopy and X ray scattering are also covered. Hypercrosslinked polymers like polysulfone, polyarylates or polyanilines among others are also reported on as a novel class of polymeric materials.

The last part of the book, Chapters 10–17, reports on different applications of polymers in the field of analytical chemistry: sorption of organic compounds, like phenols, pesticides or caffeine, the use of polymeric materials for size-exclusion chromatography, for HPLC packing materials or as solid phase extraction sorbents. The examples cover a broad variety of chemical compounds, like pesticides, pharmaceuticals, organic acids in different matrices, such as food, biological fluids, non-aqueous media, air and the use of hemosorbents in blood purification.

The book is a useful addition to the CAC series since it covers both fundamental and applied aspects of polymeric networks and I am sure that it will attract a broad audience in the field of analytical chemistry, from advanced graduate students to technicians working in different fields of analytical chemistry, like chromatography, environmental chemistry, clinical studies, food safety and others that need polymeric materials to enrich and clean up their samples before final analytical determinations.

I am grateful to the authors and their colleagues from the Department of Stereochemistry of Sorption Processes of the A.N. Nesmeyanov Institute of Organoelement Compounds of the Russian Academy of Sciences for their time and efforts in preparing this comprehensive compilation of developments and research in this area that makes this book on hypercrosslinked polymeric networks and adsorbing materials a unique work.

D Barceló
Barcelona, April 10, 2010

"It is easy to complicate.
It is difficult to simplify."

Science philosophy

INTRODUCTION

The structure and properties of polymeric networks continue to attract the attention of many investigators. This interest is stimulated by the important role these three-dimensional polymers play in developing new materials, including numerous variants of synthetic rubber, adhesives, adsorbents, polymers for medical and hygienic applications, modern nanocomposite materials, etc. On this list, polymeric adsorbents occupy a prominent place as they compete successfully with the traditional adsorbents such as silica gels, zeolites, and activated carbons and open new perspectives for modern adsorption technologies. The majority of synthetic adsorbents are based on copolymers of styrene with divinylbenzene (DVB), because of the simplicity of their synthesis and the possibilities of modifying over a wide range both the sorbent physical structure and chemical nature of their surfaces.

Staudinger [1] obtained in the 1930s the first styrene–DVB copolymers by thermal copolymerization of the monomers. Eleven years later, in 1945, D'Allelio [2] was granted a patent on the application of the copolymers as matrix for the synthesis of ion-exchange resins, though it has become known that ion exchange technology was already intensively used during the Manhattan nuclear weapon project. The successful separation of rare earths and uranium fission products gave a strong impetus to the rapid development of ion exchange science and technology. These early styrene copolymers and ion exchanger were of the *homogeneous gel type* and they comprise "the first generation" of polystyrene networks [3]. Currently, in no country in the world is the process of water demineralization possible without a large-scale application of ion-exchange resins of this generation. The theory of the gel-type copolymers was also developed, mainly by Flory and his followers. It describes the swelling behavior and elastic properties of the homogeneous polymeric gels as a function of the degree of crosslinking of their network and the thermodynamic quality of the solvent.

However, all attempts to use gel-type ion-exchange resins and copolymers that contain more than 5–8% DVB and, consequently, possess a reasonable mechanical strength failed for the sorption of large organic ions or molecules, because of the low permeability of these single-phase polymeric networks. Insofar as the problem of isolation and purification of large organic sorbates became more and more important and neither silica gels nor activated carbons could fit the practical needs, in the 1960s and 1970s numerous publications appeared concerning the synthesis of styrene–DVB copolymers with improved kinetic properties. Starting with the first suggestions of Burney (as stated by Abrams and Millar in their review [3]), the creation of a wide variety of *macroporous* (*macroreticular*) copolymers has been the main achievement of those years. These macroporous copolymers are considered to be sorbents of "the second generation." Here, the success of practical applications, rather than elaborated theories, characterizes the development. The structure of macroporous copolymers is formed under conditions of micro-phase separation during the formation of sorbent beads, resulting in the precipitation of the growing crosslinked and branched polymer chains as nano-sized gel domains and the appearance of channels between the latter. The structure of the final material is thus *heterogeneous*. Being pretty large in diameter, the channels and pores enable large sorbate species to diffuse easily into the interior of the beads. It was soon recognized that, in order to have the polymeric material preserve its heterogeneous macroporous structure, even in solvents causing strong swelling of the polymeric phase, the network of the latter must incorporate a sufficient amount of cross-bridging DVB units, usually more than 6, up to 12%.

A great number of investigations dealing with the influence of various factors on the parameters of macroporous structure led to the development of a number of variants of the second-generation ion-exchange resins and neutral sorbents that are characterized by outstanding adsorption kinetics. Among the latter, the macroporous polystyrene sorbents of Amberlite XAD series manufactured by Rohm and Haas (USA) are the most familiar to the users. The areas of practical application of macroporous materials are very diverse, namely ion exchange, catalysis of many organic reactions, gas and liquid chromatography, adsorption of organic compounds from gaseous and liquid media, etc. At the same time, some fundamental disadvantages of macroporous sorbents have been revealed during the long period of their use. Owing to the high crosslinking degree of the polymeric microphase, adsorption of most organic substances is restricted to the interface only, and, therefore, the adsorption capacity of macroporous copolymers may easily prove insufficient for a commercially acceptable practical process. This shortcoming is built into the very principle of preparation of the heterogeneous copolymers and it cannot be eliminated without the deterioration of other, useful properties of the second-generation macroporous material.

In the 1970s, it became evident that designing of sorbents combining high permeability with high adsorption capacity is possible only by approaching the problem from fundamentally different positions. A new approach proposed by Davankov and Tsyurupa consisted in the formation of a homogeneous, highly expanded network of high rigidity. They first demonstrated [4] the feasibility of the idea by intensive postcrosslinking of preformed polystyrene chains in solution (or in a highly swollen state) by means of a large number of rigid bridges. According to the method suggested and using reactive bifunctional crosslinking agents, it proved to be possible to connect almost all phenyl rings of the initial polystyrene chains to one another, thus producing a *homogeneous*, rigid, and expanded open-network structure. This new type of three-dimensional network provided the polystyrene-type material of "the third generation" with several unusual, even peculiar properties.

Naturally, such a substantially novel and extremely useful material required a new name. Therefore, networks prepared in accordance with the above principles and having a degree of crosslinking above 40% were named "*hypercrosslinked* polystyrene." While being largely homogeneous, these networks do not follow predictions of the classical theory of network elasticity, which is applicable to the homogeneous gel-type copolymers of the first generation. Nevertheless, a phenomenological understanding of the unusual behavior of these networks was fully established: that is the subject of the present book.

Contrary to all traditional expectations, the hypercrosslinked rigid networks demonstrate high mobility. For instance, hypercrosslinked polystyrenes swell in thermodynamically good solvents with a remarkable increase in volume by a factor of 3–5, in spite of the ultimate degree of crosslinking. Obviously, only a cooperative rearrangement of conformations of all interconnected network meshes can account for such a significant change of the entire volume of the material. More importantly, having the aromatic chemical nature similar to that of the polystyrene precursor, the hypercrosslinked polymer swells to the same high extent in liquids that fail to dissolve linear polystyrene. It will be shown later that the reduced energy of polymer–polymer interactions and strong internal stresses within the rigid network of the dry polymer are responsible for the non-trivial ability of the hypercrosslinked network to swell in non-solvents and, in fact, in any kind of liquid and gaseous media, including water or liquid nitrogen.

The analysis of the swelling behavior of hypercrosslinked polystyrenes makes it unavoidable to re-examine and revise the generally accepted notions about the major factors that govern the swelling of three-dimensional polymers. It appears that the mutual arrangement of network meshes in space, their interconnection and interpenetration, that is, the topology of the network, as well as the conformational flexibility of the meshes determine predominantly the mobility of a network. On the other hand, generally considered parameters such as the ratio between inter- and intramolecular crosslinks and even the crosslinking density itself may become totally unimportant under certain conditions. With the topology of a network strongly depending on the conditions of network formation, unexpected relationships can often be observed, such as a significant rise of the swelling ability of the hypercrosslinked polystyrene with an increase in its degree of crosslinking, or with the increasing rate of the chemical crosslinking reaction. Not less surprising is the observation that a water-swollen bead of the hydrophobic porous polymer may exhibit during drying, just before the residual water evaporates from the material, a noticeable decrease in its volume, up to 12% below the final equilibrium volume of the dry polymer. To a similar significant extent the polymer contracts, when immersed into a concentrated solution of LiCl or phosphoric acid.

Another vivid example of the exceptional role of network topology is the unexpectedly high deformation ability of hypercrosslinked polystyrenes under loading, which is usually characteristic of conventional slightly crosslinked networks or linear polymers in the rubber elasticity state. Hypercrosslinked polymers, however, differ from the latter in that they retain their mobility even at very low temperatures. In fact, hypercrosslinked materials do not exhibit typical features of polymeric glasses, nor are they typical elastomers. Their physical state thus cannot be described in terms of generally accepted notions. More likely, the hypercrosslinked networks demonstrate distinctly different, unique deformation and relaxation properties.

When in the dry state, the homogeneous, single-phase hypercrosslinked polystyrenes are characterized by extremely low density of chain packing and exhibit a large free volume. This can be thought of as real porosity of a new type (today, porosity of a material is interpreted only as a result of its heterogeneity, i.e., two-phase nature). No classical interfaces exist in a transparent bead of the new polymer. Nevertheless, the material absorbs large amounts of inert gases, and its (apparent) specific surface area amounts to 1000–2000 m^2/g. The best representatives of conventional macroporous polymers exhibit much smaller specific inner surfaces.

When describing the properties of the hypercrosslinked polystyrenes, we only content ourselves with the phenomenological considerations of the observed facts. The reader can easily understand that any theoretical (i.e., expressed in mathematical language) consideration of the structure–property relationships of the new material would require a basically new approach, starting first of all with a formalized description of network topology, which is totally ignored by the current classical network elasticity theory. Fully recognizing the significance of theory, we believe that the profound understanding and classification of reasons and factors governing the behavior of real polymeric networks are the first and essential steps for developing in the future a more general and more detailed theoretical description of three-dimensional polymers.

From the practical point of view, hypercrosslinked polystyrenes were found to be excellent sorbent materials and were initially referred to as "Styrosorbs." They absorb large amounts of organic substances, both non-polar and polar, from gaseous and liquid media, the adsorption capacity being much higher than that of common gel-type and macroporous polystyrene sorbents. Styrosorbs can be successfully employed in many large-scale adsorption technologies, for solid-phase extraction of contaminants from water and air, as column packing materials in gas and liquid chromatography, and as matrices for preparing various ion-exchange resins, functional polymers, nanocomposites, or catalysts. It would not be an exaggeration to state that Styrosorbs are the first representatives of networks and sorbents of the third generation. Their use substantially enhances the efficiency of many traditional adsorption, concentration, and separation processes. Moreover, fundamentally new phenomena can be observed and exploited with the new materials. Thus, it has become possible to selectively adsorb polar aliphatic and both polar and nonpolar aromatic compounds from nonpolar aliphatic media. Pairs of mineral acids, salts, or bases can be easily separated in a preparative "green" process with an unprecedented spontaneous increase in concentrations of both isolated components. The possibility to partially resolve a salt into parent acid and base constituents using hypercrosslinked polystyrene was theoretically predicted and experimentally confirmed. Adsorption of water vapors from a gaseous phase on the hypercrosslinked polystyrene was reported to result in strong differentiation between the *ortho-* and *para-*spin isomers of water molecules.

In addition to polystyrene, several other polymers have been provided with the hypercrosslinked structure, and, in addition to the postcrosslinking of preformed polymeric chains, other synthetic approaches to hypercros-slinked open networks have been developed, including the direct polymer-ization and polycondensation of appropriate monomers or co-monomers. The recently developed metal-organic frameworks and covalent organic frameworks constitute three-dimensional coordination and element-organic polymers with an unusually high free volume; they fit into the new and rapidly growing class of hypercrosslinked network materials as well.

During the last decade the interest in the hypercrosslinked polymers, both of scientists and industry, increased considerably. This is caused, on the one hand, by their remarkably unusual properties, above all the strong swelling with precipitants for linear polymers, including water, which contradicts all current perceptions. On the other hand, it became quite evident that the exceptional adsorption properties of hypercrosslinked materials open new perspectives for developing new basic adsorption and separation technologies. In the early 1990s, large-scale manufacturing of hypercrosslinked polystyrene-type sorbents was started as a result of coop-eration with Purolite International Ltd. (Wales, UK) under the trademark "Hypersol-Macronet." We express our gratitude to the late Mr. H. Bous-quet, the general manager, Dr. J.A. Dale, director of research & develop-ment unit of the company, and Mr. N.V. Nikitin for their interest in these novel sorbents, kind help, and support of our investigations.

The high adsorption capacity and the ease of regeneration of Styrosorbs result from the high permeability and mobility of their rigid hypercros-slinked network. In order to better understand these fundamental distin-guishing features of the material, it is advisable to start with the description of the structure and properties of traditional copolymers of styrene with DVB of the first two generations. Because of the abundance of information accumulated within half a century of intensive investigation and use of these copolymers, it is neither possible nor necessary to review all related publications. Therefore, Chapters 1–5 (Part I) of the book concentrate on the interconnection between the conditions of network formation and the structure of various single-phase and two-phase polystyrene networks. The concluding parts of these chapters briefly describe the properties of sorbents based on these polystyrene supports.

Part II (Chapters 6–9) considers in detail the new approach leading to the formation of hypercrosslinked-type networks and describes the information accumulated thus far on the synthesis (Chapter 6), structure, and properties (Chapter 7) of hypercrosslinked materials, including polymers other than polystyrene.

An important part of Chapter 7 presents a critical review of existing techniques for the evaluation of the pore size and pore size distribution of porous materials. When applied to hypercrosslinked polymers, many of these commonly used methods fail to give reliable information. Only a combination of a whole palette of techniques permits estimating the major part of "channels" and "pores" in the hypercrosslinked polystyrene as the interconnected system of voids of 2–3 nm in diameter.

Chapter 8 describes the process of introducing numerous crosslinks within the individual coils of dissolved polystyrene molecules. The intramolecularly hypercrosslinked polystyrene thus obtained presents a soluble material consisting of porous rigid macromolecules, called *nanosponge*. These new macromolecular species significantly differ in their structure and properties from the classical coils, globules, helices, as well as hyper-branched molecules and dendrimers. In solution, nanosponges display high values of sedimentation and self-diffusion coefficients and extremely low intrinsic viscosity. A remarkable property of monodisperse nanosponges is their tendency to self-assemble into clusters with a unique regular structure. When in the dry state, nanosponges exhibit a high apparent surface area and the ability to swell in any liquid media, which is typical of hypercrosslinked bulk polymeric materials.

Chapter 9 deals with hypercrosslinked networks as a new class of nanoporous low-density materials. Among them, fully hydrophilic ionic networks, prepared by an elegant process of crosslinking alkylation and spontaneous polymerization of 4-vinylpyridine, as well as metal-organic frameworks and covalent organic networks are of particular interest.

Finally, Part III (Chapters 10–17) is devoted to perspective application areas of hypercrosslinked materials. It gives a critical review of numerous publications describing the adsorption properties and practical applications of experimental Styrosorb materials and commercially available hypercrosslinked sorbents in comparison with conventional gel-type and macroporous polystyrene sorbents. It is appropriate to note here that the terms "sorbent" and "sorption" define Styrosorbs more adequately, because the uptake of various sorbates from gaseous and liquid media results from a superposition of simultaneous adsorption and absorption processes.

In addition to sorption processes from gaseous (Chapter 10) and aqueous media (Chapter 11), the hypercrosslinked materials also permit separations of small molecules according to their size, which is called size-exclusion chromatography (Chapter 12). The nano-sized pores of the rigid hypercrosslinked polystyrene remain fully accessible to water. Though presenting a hydrophobic environment and changing many properties of confined clusters of water, the micropores exhibit an ability to discriminate larger inorganic ions from smaller ones and from protons and hydroxyl anions. This property was utilized in developing a highly productive reagent-free process for the separation of inorganic salts, acids, and bases. The process is unique in that both its productivity and selectivity rise with the concentration of the feed mixture and that it results in a spontaneous self-concentration of the resolved components. This phenomenon, unprecedented for chromatography, leads to the formulation of a conception of "ideal separation process."

Hypercrosslinked polystyrenes proved to be unique HPLC column packing materials (Chapter 13), adsorbents for solid-phase extraction (Chapter 14), hemosorbents (Chapter 15), and a matrix for hypercrosslinked ion-exchange resins (Chapter 16).

The nanoporous hypercrosslinked polystyrene proves to be the material of choice for the preparation of nanocomposites (Chapter 17) by formation of nano-sized clusters of catalytically active metals, for example, platinum or palladium, or of magnetic nano-crystals of iron oxides. Being confined in the nano-cage, the clusters do not agglomerate so that the catalysts remain stable and highly active in any liquid media.

Having been known from the early 1970s and commercially available since the mid-1990s, hypercrosslinked polystyrene cannot be called a new material. However, many important papers describing the unusual structure and properties of these materials appeared in the Russian scientific literature. Also, the early papers in English are nowadays either less available or considered to be superseded by recent treatises. For these reasons, in contrast to gel-type and macroporous networks, the special nature of hypercrosslinked networks is generally poorly understood. Thus, because of the compatibility of hypercrosslinked polystyrene with water, some specialists call them "modified" or "hydroxylated" or even simply "hydrophilic." Moreover, many manufacturers offering numerous polymeric adsorbing materials do not reveal their real chemical nature, so that some users confuse hypercrosslinked polystyrene with styrene–DVB copolymers. Even experienced groups are often forced to classify sorbents according to the values of specific surface area claimed by the manufacturer

[5], rather than by the nature of their network. It is hoped that the first-hand information provided by the authors of the hypercrosslinked networks will prove helpful for both the polymeric scientific community and many users of the new generation of adsorbing materials.

Finally, it can be noted that two earlier published books deal with polymeric networks. However, the book by Stepto [6] is mainly concerned with the theory of rubber elasticity and the preparation and behavior of slightly crosslinked materials and rubbers that belong to the first generation of polymeric networks. The monograph by Osada and Khohklov [7] deals with hydrogels and hydrophilic polymeric systems that convert into gels in response to changing environment conditions. A recently published collective work [8] mainly deals with biocompatible hydrophilic macroporous materials for tissue engineering, cell cultivation, and bioseparations. None of the books describes and discusses hypercrosslinked networks.

The authors thankfully acknowledge the cooperation and contribution of all co-workers at the Department for the Stereochemistry of Sorption Processes of the A.N. Nesmeyanov Institute of Organoelement Compounds of Russian Academy of Sciences, Moscow, who participated in the investigations of the hypercrosslinked networks and sorbents, as well as of numerous partners from industrial and research institutions who showed interest in examining and using the new materials and developing new processes.

We also want to say how thankful we are to the late Dr. Leslie Ettre who passed away on June 1st 2010 for his great help in editing Parts I and II of the manuscript.

REFERENCES

[1] Staudinger, H., Heuer, W. Chem. Ber. 67 (1934) 1164–1172.
[2] D'Alelio, G.F. Patent USA 2366007 (1944); Chem. Abstr. 39 (19) (1945) 4418.
[3] Abrams, I.M., Millar, J.R. React. Funct. Polym. 35 (1997) 7–22.
[4] Davankov, V.A., Rogozhin, S.V., Tsyurupa, M.P. Patent USSR 299165 (1969); Chem. Abstr. 75 (1971) 6841b, US Pat. 3729457.
[5] Fontanals, N., Marcé, R.M., Borrul, F. Trends Anal. Chem. 24 (2005) 394–406.
[6] Stepto, R.F.T. (Ed.), Polymer Networks: Principles of their Formation, Structure, and Properties, Springer, New York, 1998, pp. 343.
[7] Osada, Y., Khokhlov, A. Polymer Gels and Networks, 2001, Marcel Dekker, New York, pp. 400.
[8] Mattiasson, B., Kumar, A., Galaev, I.Y. (Ed.), Macroporous Polymers: Production, Properties and Biotechnological/Biomedical applications, CRC Press, Boca Raton, 2009, pp. 525.

PART One

Known Types of Polystyrene Networks

CHAPTER 1

Gel-Type (Homogeneous) Polystyrene Networks

1. GEL-TYPE STYRENE–DIVINYLBENZENE COPOLYMERS BY FREE RADICAL COPOLYMERIZATION

Many synthetic procedures exist for the preparation of three-dimensional polymers. These are polymerization and polycondensation of bifunctional (polyfunctional) monomers, connection of reactive ends of linear chains into an entire network, or crosslinking of preformed polymeric chains by involving their reactive functional groups or additional cross-agents. Each of these methods imparts a specific topology and, hence, special properties to a network.

Relating to the structure of the final three-dimensional product, all networks can be classified into two main groups. The first group represents the single-phase materials, that is, homogeneous molecular networks, whereas the second group includes the obviously heterogeneous, two-phase materials. As a rule, each of the above-listed methods of network formation results in homogeneous polymers, unless special reasons operate or are intentionally involved in order to induce microphase separation of the system. The latter case will be discussed in Chapter 3, whereas this chapter deals with the main regularities of formation and typical properties of homogeneous styrene–divinylbenzene (DVB) networks. They can be easily prepared by crosslinking bulk (or suspension) free radical or anionic copolymerization of the comonomers. Naturally, copolymerization of other types of comonomers containing one and two (or more) vinyl groups will follow similar regularities, and the polycondensation of corresponding polyfunctional comonomers will also result in homogeneous networks of similar topology.

1.1 Monomer reactivity ratio in crosslinking copolymerization

When copolymerization of two monomers M_1 and M_2 proceeds via a radical mechanism, a chain of irregular structure can be expected to form as follows:

$$M_1 + M_2 = m_1 m_1 m_2 m_1 m_2 m_2 m_2 m_1 m_2 m_1 \ldots$$

In reality, formation of a strictly statistical copolymer seems to be a rare event. The composition of the resulting copolymer is governed by the relative reactive activity of the two monomers and two types of radicals at the ends of the growing polymeric chains. Most frequently, one of the comonomers behaves more actively and predominately binds to the end radicals. Since during the copolymerization of two monomers one type of end group is eventually replaced by another one, the relative activity of the growing macro-radical toward each of the comonomers changes accordingly. A quantitative estimation of the distribution of two monomeric units along the polymeric chain is possible by considering the monomer reactivity ratios (the constants of copolymerization).

During free radical copolymerization of the two comonomers, M_1 and M_2, the polymeric chain grows as the result of the following elementary addition reactions:

$$-m_1^\bullet + M_1 \xrightarrow{k_{11}} -m_1 m_1^\bullet$$
$$-m_1^\bullet + M_2 \xrightarrow{k_{12}} -m_1 m_2^\bullet$$
$$-m_2^\bullet + M_1 \xrightarrow{k_{21}} -m_2 m_1^\bullet$$
$$-m_2^\bullet + M_2 \xrightarrow{k_{22}} -m_2 m_2^\bullet$$

Here, k-values are the corresponding rate constants of the addition of two monomers to two types of macroradicals. The ratio of these constants defines the relative reactivity of the comonomers as

$$r_1 = \frac{k_{11}}{k_{12}} \quad \text{and} \quad r_2 = \frac{k_{22}}{k_{21}} \qquad [1.1]$$

In other words, the constants of copolymerization, r_1 and r_2, show which of the monomers, "own" or "alien," adds faster to a given macroradical.

Let us consider several different situations that possibly arise in the course of the copolymerization. Fig. 1.1 illustrates these situations as the composition of the copolymer that forms from comonomer mixtures of varying compositions related to various values of r_1 and r_2. Plot 1 represents the case when $r_{11} < 1$ and $r_2 > 1$, implying that $k_{11} < k_{12}$ and $k_{22} > k_{21}$. Here, both radicals $-m_1^\bullet$ and $-m_2^\bullet$ react more easily with the more active

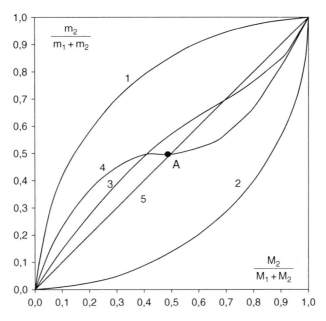

Figure 1.1 Dependence of copolymer composition on that of monomer mixture for various values of r_1 and r_2. $M_2/M_1 + M_2$ is the molar portion of monomer M_2 in the starting monomer mixture and $m_2/m_1 + m_2$ is the molar portion of repeating units m_2 in the copolymer.

monomer M_2; therefore, the copolymer is enriched in units m_2 as compared to the composition of the mixture of monomers. Similarly, if $r_1 > 1$ and $r_2 < 1$, the copolymer will predominantly incorporate units m_1 (plot 2).

In the hypothetical case when $r_1 = r_2 = 1$, none of the two macroradicals differentiates between the monomers, and the composition of the copolymer equals that of the monomer mixture at any degree of conversion (the so-called "azeotropic" mixture of comonomers). Only in this particular case (line 5) is the distribution of the monomers in the macromolecule formed close to statistical. When $r_1 < 1$ and $r_2 < 1$, radical $-m_1^{\bullet}$ tends to react with monomer M_2, whereas radical $-m_2^{\bullet}$ reacts more rapidly with monomer M_1. This tendency to alternation of the monomer units m_1 and m_2 in the growing chain results, among others, in the predominant consumption of the minor component at the beginning of copolymerization (plot 3). The deviation from azeotropic copolymerization is also pronounced in the case of $r_1 = r_2 < 1$ (plot 4). However, if the starting mixture contains both comonomers in equal proportion, a more or less regular alternating copolymer forms with the composition equal to that of the initial mixture, as plot 4 crosses line 5 at point A. A similar "azeotropic" point, though at different comonomer proportion, also exists for the above situation (plot 3) when r_1 is not equal to r_2, both of them being smaller than unity.

When k_{11} is higher than k_{12} and k_{22} is higher than k_{21}, the values of r_1 and r_2 are larger than unity, implying a tendency to the formation of two independent homopolymers.

Finally, a situation is possible when $r_1 < 1$ and $r_2 = 0$. It means that the radical $-m_1^{\bullet}$ reacts more rapidly with monomer M_2, whereas the latter does not react with its own radical (monomer M_2 is not prone to homopolymerization). In this case, in the resulting copolymer the monomer units m_2 will always be surrounded by the monomer units m_1, similar to the situation with the alternating copolymer. It is evident that the content of m_2 units cannot exceed 50 molar percent.

Thus, the monomer reactivity ratios can provide insight into the distribution character of the crosslinks in a network if one of the comonomers is a diolefin. The latter forms a cross-bridge between two chains the moment its second double bond gets involved in polymerization.

For a general case, we have all reasons to assume that if one comonomer is more reactive than the other and binds predominantly to a growing radical chain, then the microstructure of single-phase copolymers must be inhomogeneous and nonuniform. The terms "inhomogeneity" and "heterogeneity" used in this chapter should not be confused, however. The first term refers to the distribution of comonomers along the polymer chain and, consequently, to the distribution of crosslinks in a single-phase network, whereas the term "heterogeneity" refers to macroporous two-phase materials arising from the micro-phase separation and segregation of porogens from the polymer gel. We shall see in the following sections that from the consideration of monomer reactivity ratios of styrene and DVB, one can expect the distribution of crosslinks in the resulting copolymers to be rather inhomogeneous, in both single- and two-phase materials.

1.2 Monomer reactivity ratios of styrene and DVB isomers

Copolymers of styrene with DVB obtained by free radical copolymerization in the absence of any solvent represent the most investigated polymers. These basic networks have often been referred to as standard or conventional, and all the other three-dimensional styrene copolymers, including modified structures, have usually been compared to them. In order to emphasize the single-phase amorphous structure of the conventional styrene–DVB copolymers, they are also called gel-type networks. Although these standard copolymers have been prepared in the 1930s, discussion about their inhomogeneity is still going on, because the results presented by different authors are rather contradictory.

To synthesize the copolymers on an industrial scale, technical-grade DVB has been commonly used as the crosslinking agent. The commercial products currently manufactured by many chemical plants contain 55, 63, or 80% of DVB (mixture of *meta*- and *para*-isomers), the rest being largely a mixture of *meta*- and *para*-isomers of ethylvinylbenzene (ethylstyrene). Concentration of *ortho*-DVB in the technical product is negligibly small, since it is easily converted into naphthalene during production. The latter may be present in amounts up to 5–7% in the 55% technical-grade DVB, while more concentrated DVB brands contain only 1–2% of naphthalene. Besides ethylstyrene, technical-grade DVB used in the early investigations also contained noticeable amounts of diethylbenzene, vinyl toluene, poly-alkyl derivatives of benzene, etc. Since many of these impurities could take part to certain extents in the copolymerization reaction, the quantitative consideration of the process and the determination of monomer reactivity ratios have been carried out with the use of purified individual *meta*- and *para*-isomers of DVB.

A number of publications indicate that the reactivity of *m*-DVB, r_2, is rather close to that of styrene, r_1, both values being nearly equal to unity (Table 1.1). Therefore, a statistical copolymer would have to be obtained as a result of copolymerization. However, according to some other data, during the first stages of the process, the growing polymeric chains are enriched by the *meta*-divinyl monomer, which should inevi-tably result in an inhomogeneous distribution of crosslinks in the final network.

The data concerning the reactivity of styrene and *p*-DVB (Table 1.2) are more unequivocal. The overwhelming majority of the investigations have shown that $r_1 < 1$ and $r_2 > 1$, implying the higher reactivity of *p*-DVB. At 70°C, *p*-DVB is polymerized 2.5 times faster than styrene, whereas the ratio of the copolymerization rates proves to be as high as 3.5 times at 90°C [16]. In the course of copolymerization, the *para*-isomer is consumed first, forming domains with enhanced density of crosslinking, and therefore, the distribu-tion of crosslinks in the network should be extremely inhomogeneous. In 1981, Frick et al. [15] recalculated the published monomer reactivity ratios of

Table 1.1   Monomer reactivity ratios for styrene (r_1) and m-divinylbenzene (r_2)

r_1	r_2	r_1^*	r_2^*	α^a	T^b	References
0.65	0.60	0.62	0.55	3–4	100	1, 2, 15
0.61	0.88	0.74	1.03	2–4	80	3, 15
1.27	1.08	1.26	1.05	3–8	80	3, 15
1.11	1.00	0.59	0.50	4–25	75	4, 5
0.54	0.58	–	–	1–2	70	6
0.54	0.98	–	–	16–29	60	7, 8
–	0.85	–	–	<50	90	9
–	2.0	–	–	>50	90	9
0.86	1.10	–	–	40	60	10
0.30	2.7	–	–	15–30	60	12
0.70	1.0	–	–	15–30	60	11
1.04	0.90	–	–	–	–	12
1.02	0.96	–	–	–	–	12
1.00	1.00	–	–	–	–	12
0.99	1.01	–	–	–	–	12
1.04	0.99	–	–	–	–	12

r_1^* and r_2^* = monomer reactivity ratios recalculated by nonlinear least-squares analysis.[15]
$^a \alpha$ = conversion, %.
$^b T$ = temperature, °C.

styrene and DVB isomers by means of nonlinear least-squares analysis (such an approach to the treatment of experimental data was demonstrated to give a smaller error [17]). The new values of r_1^* and r_2^* are also given in Tables 1.1 and 1.2. The numerical values of the copolymerization constants changed, but the general conclusion remained the same: the distribution of crosslinks is extremely inhomogeneous in styrene–p-DVB copolymers. It is more homogeneous in styrene–m-DVB networks, although m-DVB is still predominantly incorporated into the network. With a probability of 95%, the compositions of styrene–p-DVB and styrene–m-DVB copolymers were found to be described by monomer reactivity ratios of $r_1 = 0.30$ and $r_2 = 1.02$ and $r_1 = 0.62$ and $r_2 = 0.54$, respectively.

Schwachula et al. [18] investigated the copolymerization of styrene with the mixture of two DVB isomers, thus trying to simulate technical-grade DVB. To calculate the r-values, he utilized the following equations, which are known as those of Schwan and Price:

$$log\ r_1 = \left(\frac{1}{T}\right)[6.79(q_1 - q_2) - 3.63\ \varepsilon_1(\varepsilon_1 - \varepsilon_2)] \qquad [1.2]$$

Table 1.2 Monomer reactivity ratios for styrene (r_1) and p-divinylbenzene (r_2)

r_1	r_2	r_1^*	r_2^*	α^a	T^b	References
0.14	0.50	0.20	0.39	3–4	100	2, 15
0.77	2.08	0.91	3.17	0.5–4	80	3, 15
0.77	1.46	0.56	1.33	2–5	70	13, 15,
0.20	1.00	0.17	0.76	2–15	75	4,15
0.15	1.22	–	–	1–2	70	6
0.65	1.10	–	–	3–29	60	7, 8
–	0.43	–	–	<50	90	9
0.44	–	–	–	–	60	14
1.28	1.70	–	–	30	60	10
0.35	2.7	–	–	16–39	60	12
0.70	0.95	–	–	15–30	60	12
0.52	1.69	–	–	–	–	12
0.51	1.92	–	–	–	–	12
0.75	1.5	–	–	–	–	12
0.92	1.09	–	–	–	–	12
0.90	0.99	–	–	–	–	12
0.64	1.87	–	–	–	–	12

$^a \alpha =$ conversion, %.
$^b T =$ temperature, °C.

$$\log r_2 = \left(\frac{1}{T}\right)[6.79(q_1 - q_2) - 3.63\,\varepsilon_2(\varepsilon_1 - \varepsilon_2)] \qquad [1.3]$$

These equations contain two parameters, q and ε. To define their meanings, Schwan and Price [19] considered the reaction of addition of a monomer molecule to the growing macroradical:

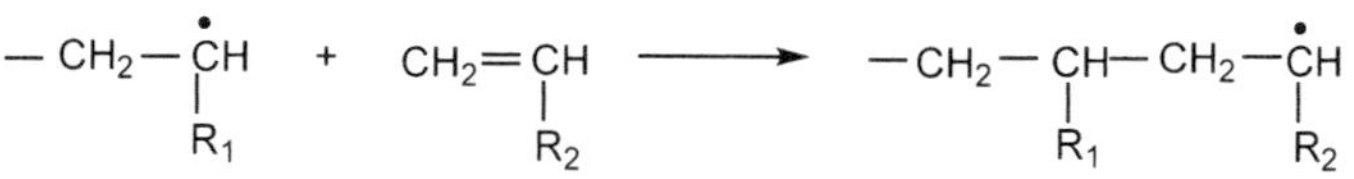

$$-CH_2-\overset{\bullet}{C}H \;+\; CH_2{=}CH \longrightarrow -CH_2-CH-CH_2-\overset{\bullet}{C}H$$
$$\quad\quad\;\; R_1 \qquad\qquad R_2 \qquad\qquad\qquad R_1 \qquad\quad R_2$$

Here, q represents the relative resonance stabilization conferred on the new radical by R_2 and ε may be represented as the charge induced by the radical R_2 or R_1 on either of the carbon atoms forming the new carbon–carbon covalent bond in the transition state. Originating from Eqs. [1.2] and [1.3], the following monomer reactivity ratios were computed for the mixture of styrene(1)–m-DVB(2)–p-DVB(3): $r_{12} = 0.45$, $r_{21} = 0.60$, $r_{13} = 0.14$,

$r_{31} = 0.50$, $r_{23} = 0.42$, and $r_{32} = 1.55$. The molar composition of the ternary copolymer ($m_1 : m_2 : m_3$) may be further calculated from the molar portion of each of the comonomers (M_1, M_2, M_3) and the above-given r-values[20]:

$$m_1 : m_2 : m_3 = M_1 \left[\frac{M_1}{r_{31}r_{21}} + \frac{M_2}{r_{21}r_{32}} + \frac{M_3}{r_{31}r_{23}} \right] \left[M_1 + \frac{M_2}{r_{12}} + \frac{M_3}{r_{13}} \right] :$$

$$M_2 \left[\frac{M_1}{r_{12}r_{31}} + \frac{M_2}{r_{12}r_{32}} + \frac{M_3}{r_{32}r_{13}} \right] \left[M_2 + \frac{M_1}{r_{21}} + \frac{M_3}{r_{23}} \right] :$$

$$M_3 \left[\frac{M_1}{r_{13}r_{21}} + \frac{M_2}{r_{23}r_{12}} + \frac{M_3}{r_{13}r_{23}} \right] \left[M_3 + \frac{M_1}{r_{31}} + \frac{M_2}{r_{32}} \right] \qquad [1.4]$$

The copolymerization diagrams plotted for the mixtures with the molar ratios of p- and m-DVB ranging from 2:1 to 1:4 show that when the p-DVB content in the initial monomer mixture increases, the nonuniformity in crosslinks distribution in the resulting network is also increased (Fig. 1.2).

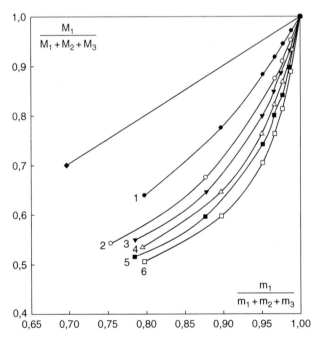

Figure 1.2 Section of copolymer diagram of (1) styrene–*m*-DVB, (2) styrene–*m/p*-DVB 4:1, (3) styrene–*m/p*-DVB 2:1, (4) styrene–*m/p*-DVB 1:1, (5) styrene–*m/p*-DVB 1:2, and (6) styrene–*p*-DVB. *(After [18])*.

The nonuniform distribution of bridges in the networks crosslinked with technical-grade- or *para*-DVB may be reduced by adding a third comonomer, acrylonitrile [21–24]. The copolymerization constants of styrene(1)–*p*-DVB(2)–acrylonitrile(3) are as follows: $r_{12} = 0.14$, $r_{13} = 0.41$, $r_{23} = 4.51$, $r_{21} = 0.50$, $r_{31} = 0.04$, and $r_{32} = 0.204$ [25]. Such calculations indicate that when the acrylonitrile content in the comonomer mixture increases, the composition of the forming copolymer will change toward that corresponding to the azeotropic copolymerization (Fig. 1.3). Indeed, the osmotic stability of ion-exchange resins based on the copolymers containing 6% acrylonitrile was found to improve. However, because of possible hydrolysis of the nitrile groups during further network chemical transformation, the ion exchanger may exhibit properties of a polyfunctional material, which is not always desirable [26].

Networks with a rather homogeneous distribution of crosslinks may probably be obtained by using diisopropenylbenzene (DIPB) as the crosslinking agent [27, 28]. Judging by the monomer reactivity ratios reported by Tevlina et al. [29], *p*-DIPB is not disposed to homopolymerization

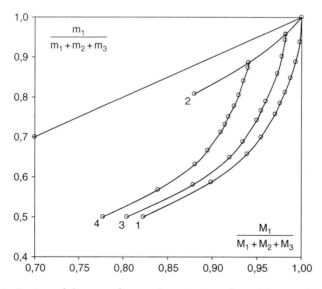

Figure 1.3 Section of diagrams for copolymerization of two binary and two ternary systems: (1) styrene–*p*-DVB, (2) styrene–acrylonitrile, (3) styrene–*p*-DVB-2%mol acrylonitrile, and (4) styrene–*p*-DVB-6%mol acrylonitrile, M_1 is styrene, M_2 is DVB, M_3 is acrylonitrile. *(After [25]).*

($r_2 = 0.0$), and, since $r_1 = 0.61$, in the resulting copolymer the DIPB molecules should always be surrounded with one (or several) styrene repeating units until the complete consumption of the diolefin. Copolymerization of styrene with *m*-DIPB isomer results in the formation of the network relating, to a certain extent, to a statistical copolymer. Interestingly, Hild et al. [10] obtained opposite results. According to their data, the copolymerization of styrene with the *meta*-isomer of DIPB leads to the formation of two homopolymers (or a block copolymer) because $r_1 > 1$ and $r_2 > 1$. On the other hand, the copolymer produced by the reaction with *p*-DIPB should exhibit a structure more similar to that of a statistical copolymer.

1.3 Intramolecular cyclization *vs.* branching during crosslinking copolymerization

The difference in the reactivity of vinyl groups of monovinyl and divinyl comonomers is not the only factor determining the character of crosslink distribution in gel-type networks. Malinsky et al. [30] arrived at a conclusion that the structure of styrene–DVB copolymers very largely depends on the local concentration of unreacted vinyl groups. By using infrared spectroscopy, they studied the composition of styrene–*m*-DVB and styrene–*p*-DVB copolymers at different stages of copolymerization and compared the experimental data with theoretical expectations based on the assumption that the reactivity of the vinyl groups of the monomers and that of the pendent vinyl group (PVG) of a DVB unit (which was already incorporated in a polymeric chain by one of its two double bonds) is independent of the extent of copolymerization. The authors found a lower-than-expected PVG content in the copolymer at the first stages of copolymerization, but their enhanced concentration at higher conversions. By taking into account the enhanced DVB content in the initially formed chains, the authors explained this unexpected finding by suggesting that the probability of consuming the PVGs is proportional to their local concentration in the chain. At low conversions, the initial concentration of PVGs within the domain of the growing macroradical is higher than their average concentration so that their polymerization may proceed locally almost to completion. This is not the case within polymer species formed toward the end, which include less DVB units since most DVB monomer molecules were already used up.

As polymerization starts, DVB molecules are initially incorporated into the growing chain by one of their double bonds. The chain quickly forms a coil and the radical end of the chain may find itself in the vicinity of DVB unreacted double bonds. An attack by the macroradical on one of the PVGs of its own chain results in the incorporation of this bond into the growing chain with closing-up of an intramolecular loop. Monomers present in the swollen coil support further propagation of the macroradical, which, again, may form another intramolecular cycle by reacting with a second PVG of the same chain, and so on. The probability of the intramolecular cyclization obviously increases with the concentration of the divinyl comonomer as well as its reactivity.

As a matter of fact, the intramolecular cyclization (or intramolecular crosslinking) shrinks the coils and gradually converts them into microgel species. The primary microgels remain dissolved in the mixture of comonomers and, at a later point, can grow further by interacting with other macroradicals or initiator molecules. However, these microgel species cannot penetrate into one another for steric reasons so that the intermolecular crosslinking on gelation takes place only through the surface layers of the microgels [31].

As the polymerization proceeds with further consumption of the monomers, viscosity of the reaction mixture dramatically increases, thus additionally restricting the mobility of the growing chain's ends. For this reason, even at a high conversion of monomers, a certain part of pendant vinyl groups miss the chance of meeting a growing macroradical and remain unreacted in the final material. Thus, approximately 30% of DVB units have been found incorporated in a styrene–8% DVB network by one double bond only [32]. Even in the copolymer with 1% p-DVB obtained at 70°C, 40% of DVB vinyl groups remained unreacted. This quantity dropped to 16% when the polymerization was finished at 95°C [33].

Dušek [34] suggested that conducting the copolymerization in a solution has to increase additionally the local network inhomogeneity due to an enhanced consumption of DVB residual double bonds within the macroradical domain. Indeed, although the solvent does not substantially influence the PVG concentration inside a polymeric coil, it reduces the total comonomer concentration in the solution, thus lowering the extent of chain entanglements in the growing network and thereby increasing the mobility of macroradicals. The latter factor has to facilitate the participation of PVGs in the reaction with the macroradical. This thesis was strongly supported by the findings of Kast and Funke [35], who studied thermal homopolymerization of highly reactive p-DVB in a diluted 0.5–0.7 mol/L solution in benzene. Neither linear nor branched polymeric molecules were isolated from the reaction solution prior to total gelation. Only densely crosslinked microgel-like species were the product of the reaction, each of them being formed, as per authors' opinion, as an individual macromolecule.

The intramolecular cyclization during styrene–p–DVB copolymerization was also discussed clearly in the study by Soper et al. [36]. It was demonstrated that at the stages preceding gelation, the intrinsic viscosity of the formed macromolecules, although slightly rising with the conversion, remains considerably lower than that of linear polystyrene obtained under the same conditions. The small radius of gyration and increasing Huggins constants indicate the formation of rather compact species. All these observations indicate that during styrene–p–DVB copolymerization extensive cyclization occurs. The size of the intramolecular cycles appears to be small, approximately one DVB molecule per 7–12 units of styrene when the DVB content in the initial mixture amounts to 13–14 molar percent.

The participation of a noticeable portion of the divinyl comonomer in the formation of intramolecular links was also reported by Dušek and Spevacek [37]. To obtain a three-dimensional styrene copolymer, the authors used ethylene dimethacrylate as the crosslinking agent. It had been found in this case also that at low degrees of conversion, the intramolecular cyclization strongly dominates over intermolecular crosslinking. Yet, the crosslinking agent used here possesses a flexible structure that additionally facilitates the formation of intramolecular cycles [38]. When evaluating the above copolymerization system, Galina and Rupicz [39] concluded that the critical concentration of ethylene dimethacrylate needed to form a gel under the chosen conditions exceeds many times that required to introduce one crosslink per chain. Taken together with the complete absence of unreacted vinyl groups in the resulting network, this fact strongly supports the idea of intense cyclization, which, naturally, is also responsible for the observed delay of gel formation.

Okasha et al. [40] and Hild and Okasha [14] adhere to another opinion. They think that the late gelation (at a conversion of DVB of over 50%) during the copolymerization of styrene with DVB in benzene solution is caused by a reduced reactivity of PVG, as compared with that of the monomer vinyl groups, rather than by the formation of intramolecular cycles. Hild and Okasha [14] found that the styrene–2% DVB macromolecules exhibit a reasonable polydispersity index of approximately 3 when formed at the beginning of polymerization. Polydispersity rises dramatically in the vicinity of the gel point up to a value of approximately 220. The authors assume these species to be highly branched macromolecules, with the DVB units being the branching points. The PVG content of the molecules significantly exceeds the number of consumed PVGs so that the major fraction of inter- and intramolecular crosslinks must form at a later point, that is, after total gelation. These authors found the concentration of free vinyl groups in the copolymers remaining nearly constant up to gelation. The decisive role of the pendant vinyl groups in the network formation was confirmed by a direct experiment [41]. Prior to gelation, all macromolecules were isolated from the reaction solution, washed of the remaining monomers, dried, and redissolved in a new portion of styrene, benzene, and an initiator. On heating the new solution that contained no bifunctional crosslinking comonomer, total gelation of the system was observed to take place soon. On further heating of the gel, the swelling ability of the polymer was found to diminish, while its modulus of elasticity under uniaxial compression grows, indicating the participation of pendant double bonds in strengthening the network and shortening the elastically active chains. Because branched macromolecules would produce a more homogeneous system than microgel species can generate, the authors of the above conception suggested revising the existing opinion on structural inhomogeneity of the gel-type styrene–DVB copolymers and consider them as networks with more or less homogeneous crosslink distribution [10]. (Strangely, the same authors estimate the styrene–p-DVB reactivity ratios as $r_1 = 1.28$ and $r_2 = 1.70$ [10], which would have to imply that styrene and DVB predominantly form two independent homopolymers or a block copolymer during copolymerization. It is not possible to imagine the formation of homogeneous networks from such precursors.)

In this connection, it is worthwhile to mention the report of Matsumoto and Kitaguchi [42] who considered the product of free radical copolymerization of styrene with 0.17% wt of m-DVB as an ideal network (network containing no intramolecular loops or dangling chains). The only reason for this selection is that the regularities of the polymerization process of such a particular comonomer mixture obey the Flory–Stockmayer theory of ideal polymerization. Thus, the conversion of comonomers in gel point (55%) proved to be close to that calculated from the theory (43.5%). As the theory predicts, molecular weight distribution of primary chains formed prior to gelation slightly broadens with conversion because of the gradual formation of high-molecular-weight fractions due to intermolecular crosslinking. Still, the position of the maximum of the molecular weight distribution curve was observed to be independent of the conversion, which was interpreted as formation of a certain preferential distance between adjacent junction points in the final network. Finally, according to the theoretical expectations, the swelling ratio in toluene (>50) of the gel isolated just beyond the gel point was very high, implying no occurrence of microgelation up to the gel point conversion. It should be emphasized, however, that regarding the copolymer with 0.17% *m*-DVB as a perfect network containing no intramolecular loops is a rather fortuitous conclusion. Increasing the degree of crosslinking to 0.5–1% and, particularly, diluting the monomer mixture by one-third with toluene result in a dramatic delay of gelation, compared to the theory. These networks are far from being ideal. The authors traditionally explain this deviation by the interference of intramolecular cyclization.

Shortly after the publication by Okasha et al. [40] appeared, Dušek et al. [43] reexamined the experimental data on the free radical crosslinking copolymerization that have been accumulated by the end of the 1970s and summarized the main features of the process in the following way:

(i)   Theoretically, in a ring-free system, gelation must occur if one repeating unit per each primary chain is involved in crosslinking. This point of gelation must be attained at a certain extent of conversion that is specific for each comonomer composition. As a rule, the experimental critical conversion of the monomers deviates from the theoretical extent toward higher values, sometimes by a factor of 100. This factor increases with increasing content of the bifunctional monomer, increasing length of primary chains, and increasing dilution of the system with a solvent.

(ii) At low conversion, the amount of double-reacted divinyl units is relatively high and exceeds the theoretical value that is needed to attain the gel point in a hypothetical ring-free system. In copolymers with high content of divinyl component, the portion of pendant vinyl groups is high and varies only little with conversion (and dilution).

(iii) With increasing length (flexibility) of the bridges, as in ethylene dimethacrylate, the portion of double-reacted (both intra- and intermolecularly) divinyl units increases. The divinyl component is more involved in intramolecular ring formation than in intermolecular crosslinking.

(iv) The intrinsic viscosity of low-conversion polymers is very low and rather decreases with the content of bifunctional monomer. It is much lower than the value calculated for randomly branched molecules.

Generally, the marked shift of the gel point toward larger conversions may be explained by either extremely low reactivity of the pendant vinyl groups or their strong involvement in cyclization. On the one hand, a relatively high extent of involvement of both groups of the divinyl units at the beginning of polymerization (with only a small portion of them consumed for intermolecular cross-linking) and the extremely low intrinsic viscosity of low-conversion copolymers and their compactness unambiguously suggest cyclization. On the other hand, the independence of the PVG content from the degree of conversion (up to high conversions) and increase of the PVG amount with the crosslinking agent content could testify to the reduced reactive ability of PVGs. Indeed, a lowered rate of PVG polymerization was noted experimentally [9, 44]. In fact, being fixed in a chain and buried in the interior of the coil, a pendant vinyl group can be engaged only in a reaction either with the initiator or with an active radical chain end. However, the latter will definitely try to avoid dense cores of the coil. The apparent lowered reactivity of many buried PVGs is thus logically explained by simple physical factors.

To give an example of reduced activity of PVGs, papers [45, 46] can be mentioned where hydrodynamic properties of the products of bulk copolymerization of styrene with 0.4% DVB (without removing the inhibitor *tert*-butylcatechol) have been examined. If the copolymerization was ideal, one DVB molecule, surrounded by 250 styrene units, would form a tetrafunctional branching point in the macromolecule and the molecular weight of a chain between two branching points should be as high as 25,000 Da. However, the branching starts only at a molecular weight of 175,000 Da, and below this value the molecules behave hydrodynamically as linear polystyrene. Hence, only 1/7th part of the DVB molecules takes part in the formation of branching points prior to gelation, whereas 6/7th part of the crosslinking agent behave passive and remain incorporated into the chain by one vinyl group [46].

With the majority of PVGs buried in the core of the polymeric coil, only shell-exposed PVGs are more likely to participate in the further growth of the polymeric particle through meeting a growing macroradical. Slow dissociation of new initiator molecules and gradual consumption of the remaining monomers, the divinyl component of which is already considerably depleted, first result in combining primary branched molecules or microgels, that is, in total gelation, and then in consolidation of the rather inhomogeneous network.

Nuclear magnetic resonance ^{1}H and [13]C [47], fluorescence probe analysis [48], polarization fluorescence [49] and luminescence [50] techniques confirm the inhomogeneous microgel-like structure of the gel-type polystyrene networks. Elliott and Bowman [51] suggested an analytical solution and numerical integration approaches to theoretically evaluate the influence of various factors on the kinetics of cyclization and the formation of microgels in the network.

With regard to the above discussion on the role of cyclization and branching, it should be noted that, with the complex of methods employed thus far, it is hardly possible to distinguish between a highly branched macromolecule and a species with several intramolecular loops. Both of them exist in solution as particles of enhanced density, compared to a swollen coil of a linear polymer. From the point of view of topology, the main difference between the "internally crosslinked" microgel species and the branched macromolecule is that the former has two end groups, while the latter has more. The former originated from the growth of one single radical chain, while the branching points, that is, DVB units, of the latter were involved in the growth of two distinct radical chains. No doubt, in a real polymerizing system both types of polymeric species will form and, most of all, species of a mixed structure incorporating both intramolecular loops and branching points. It is true that branched molecules have more possibilities to overlap with each other on gelation, but this ability, alone, would not produce a homogeneous network. First, we must accept the preferential rapid consumption of the more-reactive DVB comonomer as an unquestioned fact. Second, we have to take into consideration the extremely uneven character of the growth of polymeric particles. Since radical polymerization is a very fast process and the active radical ends do not live long, each primary polymeric species must form very quickly. Compared to the short life span of an active radical, the many-hours-long duration of the polymerization and curing processes is a long period of time. New radical chains are launched now and then during the whole polymerization period and they do not necessarily combine with the previously formed particles. This implies that the earlier formed species include more DVB units and, therefore, have a higher branching degree or incorporate more intramolecular loops, as compared to species formed at a later stage. (Notably, this time differentiation of the composition of primary polymeric species and the discontinuous, stepwise character of their further growth are largely underestimated, if not at all overseen.) Toward the gel point, neither the branched molecules nor the "microgels" are in position to fully interpenetrate within the gel and form a totally homogeneous network. The latter must consist of domains with both higher and lower crosslinking densities and remain inhomogeneous with respect to the distribution of DVB units till the very end of polymerization. Therefore, only discussions on the extent of the inhomogeneity are appropriate. The problem, however, remains that there are no experimental means to quantitatively measure the inhomogeneity.

Ozol-Kalnin et al. [52] showed theoretically that from the viewpoint of conformational chain statistics, the formation of a heterogeneous structure with a large number of intramolecular cycles dominates the formation of a homogeneously crosslinked polymeric network.

1.4 Suspension polymerization as an additional source of network inhomogeneity

For practical purposes, styrene–DVB copolymers have commonly been obtained by the suspension polymerization method,[53, 54] which is well known to consist of heating and agitating a solution of initiator in monomers with an excess of water containing a stabilizer of the oil-in-water emulsion. Polymerization proceeds in suspended monomer droplets and, in this way, a beaded copolymer is obtained. While looking very simple, this procedure can provide many complications that significantly change the properties of the beaded product as compared to the properties of materials prepared by bulk copolymerization. All parameters of the suspension copolymerization have to be strictly controlled, since even small deviations from optimal conditions of the synthesis can serve as an additional source of heterogeneity in the copolymer beads.[55]

Various polymers, such as poly(vinyl alcohol), polyvinylpyrrolidone, cellulose ethers, gelatin, copolymers of acrylic, methacrylic and maleic acids, and starch, are well known to be used as stabilizers of the suspension. Macromolecules of the polymeric stabilizer sorb on the boundary between the organic and aqueous phases, thus forming a protecting layer, the thickness of which may achieve 100 nm. Both a definite deformational resistance of this layer and its hydrophilicity prevent the sticky oligomer–monomer droplets from coalescence (usually, the sticky step is observed at a 30–50% conversion of the comonomers). However, macromolecules of the stabilizer can be mechanically entrapped in the outer layers of the growing polystyrene network or even grafted chemically to the surface of beads. For these reasons, a complete removal of the polymeric stabilizer is hardly possible. The remaining stabilizer shell does not affect noticeably the permeability of the beads for small inorganic ions, but may cause significant obstacles for the diffusion of larger organic molecules. In order to avoid the complications caused by polymeric stabilizers, phosphates, sulfates, or hydroxides of magnesium or aluminum, silicates, kaolin, etc., have also been used. Fine powders (less than 1 μm in diameter) of the inorganic compounds form a structured layer, preventing mechanically the particles from sticking to each other. After accomplishing copolymerization, these powders must be rigorously washed out, which, however, creates an additional problem of generating large volumes of waste waters.

Water used as the suspension medium must be very pure, because iron or copper ions, even present at trace levels of several part per million, can change the number and the rate of formation of free radicals on the decomposition of peroxide initiators. Obviously, the employed equipment has to be made of an inert material, for example, glass. Gelatin, which is often used as the suspension stabilizer, can facilitate the polymerization, first on the water–organic interface, because amino acid fragments accelerate the decomposition of peroxide initiators. If the polymerization rate of the divinyl comonomer is higher than that of styrene, the outer layers of the forming beads prove to be crosslinked to a larger extent than the central zone of a particle. Subsequently, the application of such a product may be complicated by a low diffusion rate of the substances through the dense shell of the bead. On the other hand, oxygen dissolved in water or captured as air bubbles inhibits copolymerization in the external layers of monomer droplets. In this case, the degree of crosslinking of the bead core turns to be higher than that of the bead surface. Surface inhibition is not always undesirable because the copolymer thus obtained possesses higher osmotic resistance. That is why $Fe(NO_3)_3$ or $Cu(NO_3)_2$ is sometimes added intentionally into the dispersion phase to obtain beads with denser crosslinked cores. Then, at a certain moment, the salts are precipitated by ammonium hydroxide. If this type of beads is not desired, copolymerization needs to be carried out in an inert atmosphere. The heating regime of copolymerization is also very important, especially when it is performed in industrial reactors [56]. As Schwachula et al. [23] have shown, styrene–DVB bulk copolymerization conducted even in an ordinary glass tube results in obtaining a gel rod with zones of different swelling, because of nonuniform warming up of the reaction mass and nonuniform removal of heat that intensively evolves shortly before gelation. Bad heat exchange in a large industrial reactor leads to preparing beads with different properties. In order to avoid uncontrolled heat release and deterioration of batch quality, Zähle et al.

[56] recommend using a reactor with an exterior heat jacket and interior heat exchanger for cold water.

1.5 Formation of popcorn polymer

Already in the beginning of the twentieth century, researches paid attention to an unusual fact that, under certain conditions, the radical bulk polymerization of butadiene or isoprene results in the formation of an insoluble opaque granular mass, along with the usual soluble and transparent product. Since this mass looked like popcorn, it was called appropriately so (sometimes this polymer is also called "α-polymer"). Subsequently, the formation of popcorns was also observed during the polymerization of methyl methacrylate and styrene or the copolymerization of styrene with DVB. Over the years, scant data have been published concerning both the reason for popcorn polymer formation and its properties [57–62]. Popcorn formation and growth are catalyzed by the popcorn itself, rusty iron, and water, especially in the presence of small amounts of peroxide compounds. Exposure to oxygen decreases popcorn activity, while sufficient amounts of nitric oxide, iodine, or $NaNO_2$ present in an aqueous suspension polymerization system completely suppress the growth of popcorn seeds [63, 64].

The reasons for popcorn polymer formation are still not very clear. Abnormally high concentration of free radicals was found to be characteristic of this material [65]. As early as in 1956, Pravednikov and Medvedev [66] explained the accumulation of free radicals by the rapture of stressed chains that fail to tolerate osmotic pressure. However, no new portions of free radicals were observed to appear during additional swelling of the popcorn itself. According to another opinion, popcorn arises as a result of immobilization of a macroradical in a microgel or of chain raptures caused by "chain oriental growing" [67]. Perhaps, a delay, for one reason or another, or even impossibility of chain termination is one of the reasons for the formation of the popcorn polymer.

We were also confronted with the formation of popcorn polymer. In the course of styrene–0.3% tech-DVB copolymerization in bulk, glassy material was obtained with rare incorporations of popcorn seeds. Interestingly, during the storage of the polymer within approximately 20 years at room temperature, almost all glassy part of the copolymer converted into popcorn, though the residual content of monomers in the polymer block must have been small. It testifies, by the way, to the nonequilibrium character of the glassy state and the possibility of conformational

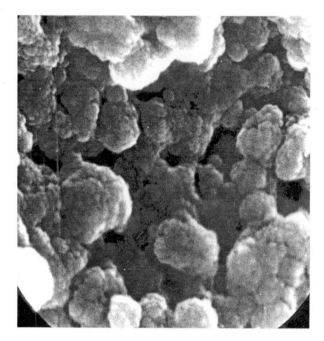

Figure 1.4   Scanning electron micrograph of a popcorn polymer.

rearrangement of chains in the glassy copolymer. In addition, this observation contradicts to the above-mentioned reasons for the formation of this unusual material. Interestingly, although formed in bulk and in the absence of porogens, popcorns obviously represent porous materials. Scanning electron microscopy indicates that their texture is much like that of macroporous copolymers (see Fig. 1.4). However, its specific inner surface area is negligibly small.

1.6 Properties of conventional styrene–DVB copolymers

Free radical copolymerization of styrene with diolefins, although a simple and reliable process, is very sensitive to even insignificant changes in the synthesis conditions. This may well account for many contradictions in the published information on both the regularities of network formation and the properties of the resulting polymers. As an example, we can demonstrate the conflicting data on the reaction of sulfonation of styrene–DVB copolymers. According to Belfer et al. [68], if the degree of crosslinking exceeds 7%, the sulfonation of a copolymer with *m*-DVB is easier than that of a copolymer with *p*-DVB. On the other hand, Wiley [69, 70] reported that at 80°C within 4 h 75% of styrene units in a copolymer with 8% *p*-DVB reacted with sulfuric acid, while under the same conditions only a small amount of functional sulfonic groups could be introduced in a copolymer with 8% *m*-DVB. At the same time, Makarova et al. [71] did not notice any difference in the rates of sulfonation of the copolymers with *m*- and *p*-isomers.

Information about the swelling ability of styrene–DVB networks is less contradictory. The overwhelming majority of research noted [72–75] that, depending on the type of the crosslinking agent, the swelling of networks in good solvents decreases in the sequence tech-DVB > *m*-DVB > *p*-DVB, when benzoyl peroxide or 2,2′-azobis-(isobutyronitrile) is used as the initiators. However, according to Schwachula [76], this regularity changes, when the copolymerization is performed in the presence of lauroyl peroxide: the copolymers with *m*-DVB swell more strongly than those with the technical-grade cross-linking agent. The low swelling of styrene–*p*-DVB copolymers has commonly been explained by the high inhomogeneity in the distribution of its crosslinks [77–80]. However, such conclusions are unfounded and should not be drawn without an analysis of the number of pendant vinyl groups and several other factors.

As for the technical-grade DVB, impurities present in the product take part, to various extents, in polymerization, thus changing the structure of the final network. Among these impurities diethylbenzene, being an active radical transfer reagent (telogen), reduces the length of the radical chains and affects most severely the properties of the final polymers [81, 82]. The present technology of DVB production is such that the crosslinker largely contains the *p*- and *m*-isomers of DVB and *p*- and *m*-isomers of ethylvinylbenzene, while other impurities amount only to about 2%, with diethylbenzene constituting half of them. However, in the 1960s–1970s, technical-grade DVB contained more diethylbenzene, as well as various other impurities, for example, naphthalene. Since the quantitative proportion of components in the technical DVB was rarely reported and considered, the early results obtained with the use of tech-DVB frequently differed from each other.

Telogens are well known as substances, some bonds of which dispose to homolytic cleavage on reacting with a radical. A growing macroradical clips off the mobile hydrogen atom from α-carbon of an aromatic ethyl substituent and binds it, thus terminating its own chain, but simultaneously also starting a new chain:

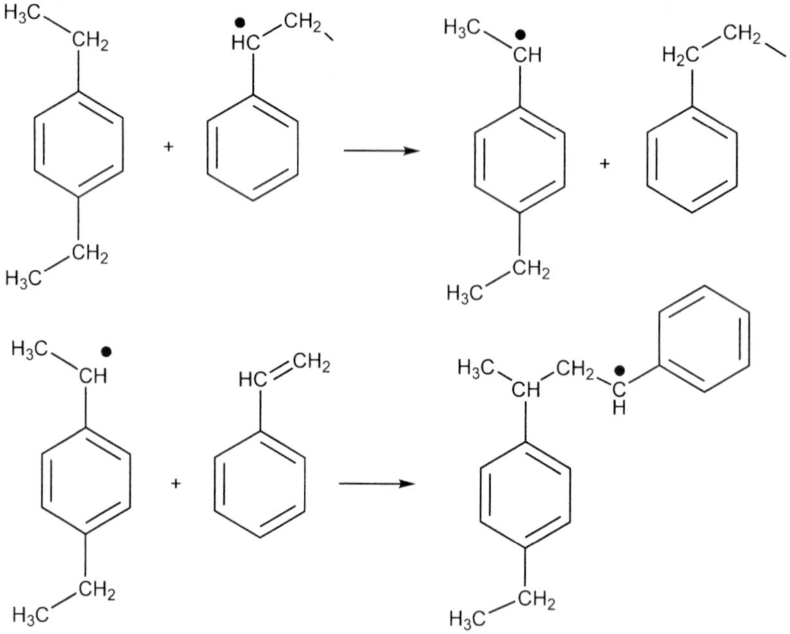

The chain transfer reaction reduces the length of polymeric chains. Networks prepared in the presence of a telogen become looser and swell to a higher extent [83].

Swelling of conventional styrene copolymers with both pure DVB isomers and the technical product is not high when the amount of the divinyl component exceeds 5–10%. Fig. 1.5 demonstrates the toluene uptake by copolymers plotted against their DVB content, as reported by Kressman and Millar [84]. Since the copolymers are mainly used as matrices for the preparation of ion-exchange resins, the problem of their swelling and permeability is of great practical importance. One suggestion was to modulate these properties by conducting copolymerization in the presence of telogens, such as ethyl ester of thioglycolic acid [85] or, more frequently, carbon tetrachloride [86, 87]. The participation of the latter in chain transfer reaction is thought to be proven by the direct dependence of the chlorine content in the final polymer on the carbon tetrachloride concentration in the starting monomer mixture [88, 89], the shift of the monomer conversion at gel point toward higher values [90] and an increase in copolymer swelling [88, 89]. We should, however, remember that carbon tetrachloride acts not only

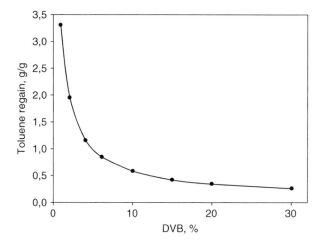

Figure 1.5 Dependence of swelling in toluene of gel-type copolymers on the percentage of divinylbenzene. *(After [84]).*

as a telogen but also as a thermodynamically good solvent for polystyrene, thus additionally enhancing the network swelling and shifting the conversion at gel point due to enhanced intramolecular cyclization. On the other hand, the formation of a loose network in the presence of carbon tetrachloride is confirmed by the fact that the telogen–modified copolymers enter polymer-analogous transformations (for instance, sulfonation) without any preliminary swelling, while in the case of conventional beads only their surface layers can be sulfonated under the same conditions.

Obviously, the gel–type copolymers of styrene with DIPB also belong to the family of telogen–modified networks, since DIPB easily loses a mobile hydrogen from the methyl group, by interacting with a radical. Chain transfer constants for m– and p–DIPB are 8.59×10^{-2} and 8.43×10^{-2}, respectively [91]; for comparison, the chain transfer constant of CCl_4 is 9.2×10^{-3} [92]. DIPB participates in the copolymerization not only as a crosslinking agent but also as a telogen, which explains the unexpectedly high swelling of DIPB copolymers [28].

Gel-type networks with enlarged swelling can also be obtained by the copolymerization of styrene with a relatively small amount of DVB in the presence of thermodynamically good solvents, such as toluene [93–95]. As mentioned above, the solvent would have to assist in increasing network heterogeneity. However, a more powerful factor consists of the fact that a good solvent reduces the concentration of polymeric chains per unit gel volume and, thereby, reduces the degree of chain entanglements. The latter play a role of physical crosslinks. If their number decreases, the network may increase its volume on swelling to a larger degree; this is the reason why Millar et al. [96] suggested considering such networks as networks with expanded structure.

Dry conventional styrene–DVB copolymers are glassy transparent non-porous materials (provided that they contain no popcorn inclusions). By using differential scanning calorimetry, Glans and Turner [97], following Ueberreiter and Kanig [98], found that the glass transition temperature, T_g, of the networks prepared in the absence of any solvent increases with the degree of crosslinking. The elevation ΔT_g (difference between T_g of a network and 103°C, the glass transition temperature of linear polystyrene) is linearly proportional to the DVB content. However, in other reports [97, 99], this simple regularity was not confirmed. Thus, the increase in the amount of p-DVB from 0 to 2.5 and then 10% shifts the T_g of the polymers from 103 to 113 and then to 146°C (Table 1.3). In this table, glass transition temperatures are also given for networks prepared by the copolymerization of styrene with tech-DVB in solution. Both types of copolymers have the same crosslinking density, but the T_g values of the latter are noticeably lower, which is indicative of higher chain mobility in expanded solvent-modified networks.

The difference in properties between various gel-type copolymers described in this chapter, which contain equal amount of cross-bridges, is the logical consequence of the difference in their spatial structure, that is, the difference in *network topology*. Topology deals with most general properties of

Table 1.3 Glass transition temperature, T_g(°C), of conventional styrene–DVB copolymers

Content of DVB, mol-%	Glans et al. [97] p-DVB[a]	Eder et al. [99] tech-DVB[b]	Askadsky [100] DVB, calculated
0 (PS)	103	83	88
2.5	113	–	108.3
5	123	103.5	111.8
10	146	116.5	119.2
20	–	128.5	135.7

PS = linear polystyrene.
[a] copolymers prepared without dilution.
[b] copolymers prepared in heptane–toluene solution 1:9 by volume; the monomer fraction is 0.5; and the polymer has gel structure.

geometrical figures that characterize their form, integrity, mutual arrangement, and the connectivity of their parts. We define structural topology as a combination of paths between two arbitrary selected points of a given system, determined by mutual disposition in space of elements composing this system. As applied to networks, the notion of topology includes several constituents, such as the functionality of the junction points, the ratio of the number of junction points to that of dangling ends, the presence and spatial arrangement of intramolecular loops, and the extent of mutual penetration of network meshes. Here, a junction is a point from which three or more paths can be traced. If one of the paths brings back to the same junction, without including another junction, we deal with an intramolecular loop. In the case a path breaks with a chain end without including another junction, it represents a dangling chain. Dangling chains factually reduce the functionality of the junction from which they originate. The shortest cyclic path from a junction through the network back to the same junction should be considered a network mesh. Each network mesh can embrace several chains belonging to neighboring meshes. They can never get free from the embracing mesh, just like in the case with rings forming a catenane structure. Only dangling chains involved in the physical entanglement with an embracing mesh can theoretically get free by a snake-type (reptation) movement. Some of these topology constituents, as the extremely important extent of mutual interpenetration of meshes, though simple and clear, are hardly susceptible to strict mathematical treatment.

The behavior of a real network is determined by the totality of topological parameters not less than by the density of crosslinking. Neglecting the network topology will result in misleading conclusions. As an example, Table 1.3 demonstrates glass transition temperatures of abstract styrene–DVB copolymers calculated in accordance with the approach developed by Askadskyi [100, 101]. The author depicts the DVB junction point as denoted below (one junction includes four carbon atoms):

Originating from van der Waals volumes of the network junctions and repeating chain units, the packing coefficient for macromolecules in an amorphous solid, increments characterizing the contribution of each atom to weak dispersion (or strong polar) interactions, Askadskyi derived the following simple formulae [1.5]–[1.7]:

$$T_g, \quad K = \frac{65.6 + 109.7\,m}{73.1 + 290.2\,m|} \times 10^3 \qquad [1.5]$$

$$E, \quad MPa = 270\frac{2(m + 0.437)}{4m^2} \qquad [1.6]$$

$$\rho, \quad cm^3 \times g^{-1} = \frac{0.681\,(65 + 104m)}{0.6023\,(65.6 + 109.7m)} \qquad [1.7]$$

for the calculation of glass transition temperatures, the elasticity modulus, and bulk density of gel-type copolymers, respectively. While ignoring topology, the calculations result in T_g values lower than those measured for styrene–p-DVB copolymers, but higher than the T_g of solvent-modified networks. The calculations predict that when the distance between adjacent junction points, m, becomes smaller than five styrene repeating units, both the glass transition temperature and density of the material must increase more sharply, implying, in the author's opinion, a qualitative transition to highly crosslinked networks. Still, if one calculates glass transition temperature for styrene–34% tech-DVB copolymer, while reasonably assuming $m = 2$, one finds $T_g = 163°C$. In fact, such a copolymer does not deform below the temperature of its chemical degradation of ~300°C (Chapter 7, Section 8). A similarly misleading result is obtained on calculating bulk density according to Eq. [1.7]. In every DVB junction, two polystyrene chains are known to cross each other under right angle [102], thus disturbing the packing of polymer chains. That is why with the DVB content rising, the bulk density of a real polymeric material must decrease [103], not increase as Eq. [1.7] predicts. Indeed, the rising permeability of the material for gases implies less-ordered arrangement of the polystyrene chains in the vicinity of a DVB tetrafunctional junction [104].

2. GEL-TYPE STYRENE–DVB COPOLYMERS BY ANIONIC COPOLYMERIZATION

The conclusions about the structure of conventional gel-type styrene–DVB copolymers were drawn largely on the basis of monomer reactivity ratios and the study of polymer precursors formed during the period preceding total gelation. Experimental methods permitting the investigation of the structure of the final insoluble networks are unfortunately not numerous. Many attempts have been made to relate the properties of gel-type styrene copolymers to the predictions of the Flory–Rener theory (or similar theories) of network swelling [105–108]. However, the interpretation of the results has always been ambiguous and inconclusive. This may be explained by the fact that the structure of gel-type networks obtained by free radical copolymerization does not fit the model of an ideal, perfect network that would underlie the theory. The structure of a real three-dimensional polymer has always been complicated, while any theory may operate only with idealized objects and the simplest models, amenable to theoretical description. The postulates and predictions of a theory as such also require to be verified. This work was undertaken by a group of French scientists who developed a new approach to the synthesis of styrene–DVB networks, which, as anticipated, are more uniformly crosslinked and better meet the requirements of an ideal network model. For this reason the networks were called "model" networks, and they have been examined in order to verify the main positions of the swelling theory suggested by Dušek and Prins [109]. Therefore, before considering their synthesis, structure, and properties, it is necessary to define an ideal polymeric network.

The perfect, ideal network has to meet the following requirements:
- it must have a homogeneous single-phase gel-type structure;
- the distribution of crosslinks must be uniform through the entire network volume;
- all chains forming the network must be elastically active, which implies the absence of network defects, such as dangling chains or chains connected to a junction point by two ends (intramolecular loops);
- elastically active chains must obey Gaussian statistics, and hence they must possess a rather high-molecular-weight and, desirably, narrow molecular weight distribution;
- the functionality of the junction points must be known and constant throughout the sample volume; and
- permanent chain entanglements must be absent.

Obviously, because of the difference in the reactivity of styrene and DVB, the networks prepared by free radical copolymerization do not relate to such an ideal system with uniform distribution of DVB units and constant chain lengths between the junction points. Also, it was not possible to eliminate this serious defect by an anionic copolymerization of the comonomers. The anionic copolymerization has often been initiated by n- or sec-butyl lithium [110–112]. Under such conditions, styrene is consumed faster than p-DVB, the monomer reactivity ratios being $r_1 = 1.58$ and $r_2 = 0.32$. Therefore, DVB-enriched domains will form toward the end of the anionic process. On the other hand, the styrene–m-isomer reactivity ratio ($r_1 = 0.65$ and $r_2 = 1.20$) points to the local incorporation of m-DVB crosslinks into the initially formed copolymer [113, 114]. In addition, the anionic process is also accompanied by intramolecular cyclization, similar to radical styrene–DVB copolymerization [115, 116].

Anionic copolymerization of styrene with isomers of DIPB also failed to produce networks with uniformly distributed crosslinks. The copolymerization with the $meta$-isomer may even result in the formation of a block copolymer while the reactivity of the $para$-isomer is negligible [117].

It is amply evident that statistical copolymerization, both free radical and ionic, cannot produce an ideal network because of the unequal reactive ability of the comonomers in their competition for interaction with the active functional end of the growing polymer chain. However, using the so-called living anionic polymerization, it is possible to eliminate the competition between the comonomers by separating the stages of the formation of chain precursors and the formation of network per se, that is, chain crosslinking. Such an approach may be realized in two subsequent stages via anionic stepwise block polymerization of first styrene and then DVB.

2.1 Synthesis of model "ideal" styrene–DVB networks by anionic block copolymerization

It is well known that in pure tetrahydrofuran, alkali metals interact with polycyclic aromatic hydrocarbons, such as naphthalene, by donating one electron and converting the aromatic system into a stable anion–radical. The latter readily transfers the excess electron to an unsaturated monomer, for example, methylstyrene, thus inducing the formation of a new anion radical:

Ph = phenyl

Two such radical species undergo recombination, either immediately or after adding two new monomer units, resulting in the formation of a stable dianion:

$$Na^+ \quad {}^-\underset{\underset{Ph}{|}}{\overset{\overset{CH_3}{|}}{C}}-CH_2-\underset{\underset{Ph}{|}}{\overset{\overset{CH_3}{|}}{C}}-CH_2-CH_2-\underset{\underset{Ph}{|}}{\overset{\overset{CH_3}{|}}{C}}-CH_2-\underset{\underset{Ph}{|}}{\overset{\overset{CH_3}{|}}{C}}{}^- \quad Na^+$$

In this way, a solution containing a desired amount of dianionic α-methylstyrene tetramer was prepared [118] as initiator for the subsequent polymerization of styrene. The polymerization process is very fast even at $-70°C$ and results in linear polystyrene chains with very narrow molecular weight distribution. The polymerization degree of the chains can be easily governed by selecting a desired monomer/initiator ratio, since conversion of the monomers is complete. The outstanding feature of the "living" chains is that, in the absence of air and water, their carbanionic ends maintain their activity and ability to polymerize DVB that is added in the second step of the process. The polymerization of DVB immediately converts the living polystyrene solution into a three-dimensional network.

Not less than three DVB molecules per living end were found to be required for preparing networks with very small amounts of extractable linear polystyrene, provided the molecular weight of the polystyrene precursor was quite small. It was assumed that, under such conditions, the fraction of dangling chains is also negligible. The resulting model network is thus composed of highly crosslinked DVB nodules combined by linear polystyrene chains of nearly equal length.

It should be noted, however, that the networks prepared by this two-step anionic polymerization are still far from ideal. Preparation of polystyrene chains, especially long chains, with very narrow size distribution and two living ends requires a very strict removal of trace moisture and oxygen from all reagents, solvents, and vessels. In addition in the highly crosslinked poly DVB nodule, some pendant double bonds may remain inaccessible for the polystyrene chains; therefore, the functionality f of the nodule is both ambiguous and unknown. Moreover, the addition of two chain ends to the same nodule, that is, closure of the intramolecular loops, cannot be ruled out, and this fact reduces the number of elastically active chains. Finally, a thorough mixing of the added DVB with a rather viscous polystyrene solution under the conditions of fast gelation of the system is problematic (in order to reduce as much as possible the rate of network formation, the polymerization is carried out in a tetrahydrofuran–benzene mixture). Judging by the structure of star-shaped polystyrene molecules obtained under similar conditions but with a monofunctional anionic initiator, the value of f increases with the concentration of the polystyrene solution and, to a lower extent, with the number of DVB molecules taken per living chain end, irrespective of the molecular weight of the polystyrene precursor [119].

In order to produce model networks with a more accurately known functionality, three- and tetrafunctional allyloxy derivatives of triazine (I) and (II),

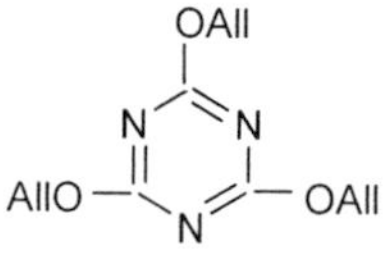

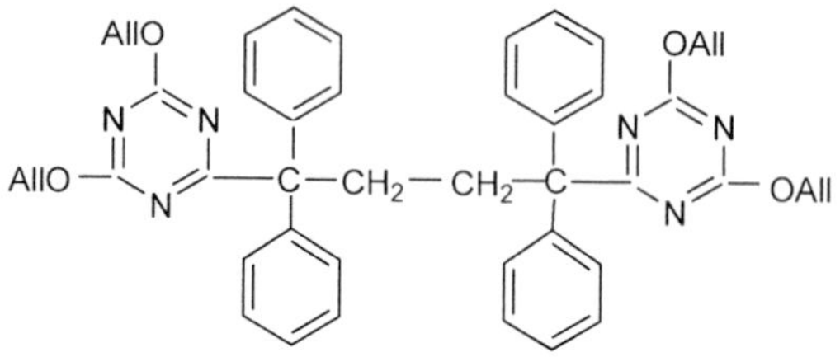

All = -CH$_2$-CH=CH$_2$

respectively, were sometimes used as the crosslinking agents. In these compounds the reactivity of the allyl groups is identical and they react completely with the polystyrene carbanions [120–123]. Probably, these compounds are difficult to prepare and handle, so that DVB remained the preferred crosslinking agent for the preparation of model networks.

It is important that no phase separation accompanied the gel formation through anionic copolymerization [124], that is, no rejection of the solvent from the gel occurred (which generally can result in macro- or microsyneresis).

2.2 Verification of the swelling theory

Correlation between a theoretical description of network properties and the behavior of a real polymer is, undoubtedly, of great interest. According to the theory of Dušek and Prins [109], equilibrium swelling of a gel in a thermodynamically good solvent is given by the following equation:

$$\mu_1 - \mu_{01} = RT\left[\ln(1 - v_3) + v_3 + \chi_{13}v_3^2 + \bar{V}_1 v^*\left(Av_3^{1/3}h_3^{2/3} - Bv_3\right)\right] = 0$$

$$[1.8]$$

where μ_1 is the chemical potential of the solvent in the gel, μ_{01} is the chemical potential of the pure solvent, T is the absolute temperature, R is the gas constant, v_3 is the volume fraction of the polymer in the swollen gel ($v_3 = 1/Q$, Q being the volume swelling, defined as the ratio of volume of a swollen gel to that of the dry polymer), χ_{13} is the thermodynamic parameter of interaction between the solvent and the network (for numerical calculations the authors of the model networks replaced this parameter with that of solvent–precursor interaction), h_3 is the so-called memory term, v^* is the number of elastically active chains (moles per cm^3 volume of dry polymer), $\bar{V}_1$ is the molar volume of the solvent, and A and B are constants.

If the network is ideal, all linear chains of the polystyrene precursor have been converted into a network of elastically active chains and, therefore, their number v^* is

$$v^* = \frac{1}{Mv_3^o} = \frac{\ln(1 - v_3) + v_3 + \chi_{13}v_3^2}{\bar{V}_1\left(Bv_3 - Ah_3^{2/3}v_3^{1/3}\right)} \qquad [1.9]$$

where v_3^o is the specific volume of the dry network and M is the molecular weight of the precursor chain, equal to that of the chain between adjacent junction points. By assuming that A $= 1$, B $= 2/f$, and $h_3 = v_3^c$, where v_3^c is the volume fraction of the polymer at the moment of crosslinking, we can write

$$v^* = \frac{\ln(1 - v_3) + v_3 + \chi_{13}v_3^2}{\bar{V}_1\left(v_3{}^{2}/_f - v_3^{c\,2/3}v_3^{1/3}\right)} \qquad [1.10]$$

If the swelling of the network is sufficiently high and v_3 satisfies the inequalities of $v_3 < 1$ and $v_3 \ll v_3^c(f/2)^{3/2}$, Eq. [1.10] can be written in the following form:

$$v_3^{-1} = Q = \left[\frac{1/2 - \chi_{13}}{\bar{V}_1}\right]^{3/5}\left[\frac{1}{v_3^c}\right]^{2/5}v^{*\,-3/5} \qquad [1.11]$$

Since for a perfect network, according to Eq. [1.9], $v^* = 1/(M \cdot v_3^o)$, it follows that the volume swelling relates to the chain length between junction points, M, as follows:

$$Q = \left[\frac{1/2 - \chi_{13}}{\bar{V}_1}\right]^{3/5}\left[\frac{1}{v_3^c}\right]^{2/5}[v_3^o M]^{3/5} \qquad [1.12]$$

Let us now consider how, in accordance with the concept of model networks by the French authors, the functionality of junctions, the molecular weight of the chains between junction points, the thermodynamic quality of a solvent, and the memory term will affect the swelling of model networks.

Determination of the junction point functionality of model networks turned out to be the most difficult task. It is impossible to experimentally measure f and so the effect of functionality on network swelling can only be estimated by an indirect route. Computer simulation shows [125] that Q decreases with an increase in f and reaches a limited value when $f \geq 10$ (Fig. 1.6). Weiss et al. [126] confirmed experimentally such a profile of $Q = f(f)$ (Fig. 1.7). Swelling of real networks, indeed, proves to be smaller if a higher quantity of DVB molecules per living chain end is introduced into a precursor polystyrene solution; however, more than five DVB molecules do not cause any further decrease in swelling (irrespective of the type of DVB isomers).

According to the prediction of Eq. [1.12], the swelling capacity of model networks must increase linearly with $M^{3/5}$ [126], provided that the

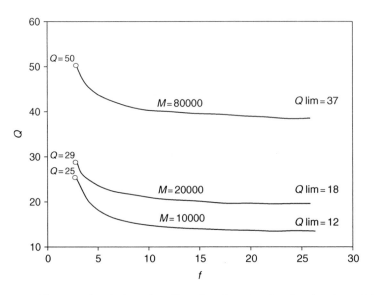

Figure 1.6 Theoretical variation of swelling Q vs. functionality f; molecular weight of polystyrene-precursor is (1) 80,000; (2) 20,000, and (3) 10,000 Da. *(Reprinted from [125] with kind permission of American Chemical Society).*

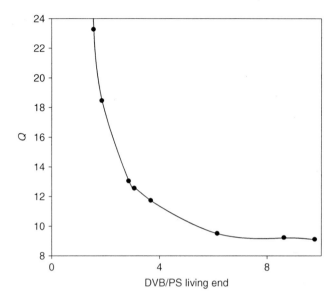

Figure 1.7 Dependence of volume swelling in benzene on the fraction of DVB per living end; $v_3^c = 8.5\%$, $M = 10{,}000$ Da. *(Reprinted from [126] with kind permission of Wiley & Sons Inc.)*

molecular weight of chains between junction points does not exceed 45,000 Da. When the chain length increases significantly, Q increases more rapidly than $M^{3/5}$ (Fig. 1.8). A good correlation between the theoretical prediction and the behavior of real model networks composed of relatively short polystyrene chains would be a serious argument for their ideality. However, if one assumes that $h_3 = v_3^c$ and the solvent–network interaction parameter is equal to that of linear polystyrene with the same solvent (benzene) and inserts all the corresponding numerical values into Eq. [1.12], then the slope of the linear section of the plot of Q *vs.* $M^{3/5}$ will be smaller than the slope observed in practice: 2.67×10^{-2} instead of 4.85×10^{-2} [127]. The authors tend to attribute this discrepancy to either incorrect estimation of the values of A, B, χ_{13}, and h_3, or to nonideality of the network structure. Indeed, on the basis of a relatively high-molecular-weight polystyrene precursor, the model networks contain a marked fraction of soluble macromolecules. This fact implies that a certain portion of the linear precursor lost both of its active ends or failed to find any DVB junction partners to attach, because of steric reasons. In this situation it is logical to expect an even larger portion of chains to be attached to a

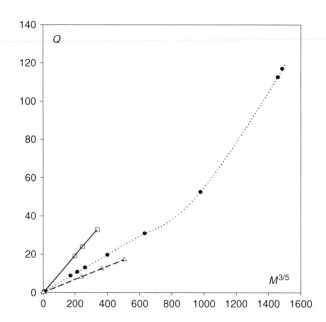

Figure 1.8 Dependence of volume swelling in benzene of model networks *vs.* molecular weight *M* of chain precursors for three sets of networks; each set prepared at a constant concentration and constant portion of DVB per living end. *(After [126]).*

network by only one end, that is, that the probability of formation of dangling chains is rather high.

The classic Flory–Dušek–Prins approach to the swelling of networks predicts that the solvent regain increases with the fraction p of dangling chains [128, 129]:

$$C_e(p) \propto \varphi_e(p) < (1-p)^{3/5} \varphi_e(p=0) \qquad [1.13]$$

Here, C_e is the concentration of polymeric chains in the swollen network (i.e., ν_3), φ_e is the volume fraction of polymer at swelling equilibrium, and $p = \nu_{mf}/(\nu_{mf} + \nu_{bf})$ is the fraction of monofunctional chains, ν_{mf}, in their mixture with bifunctional ones, ν_{bf}.

The above prediction can be examined experimentally, since the method of anionic polymerization allows preparation of networks with an accurately known proportion of dangling chains. A polystyrene living precursor with a molecular weight of about 30,000 Da and two active ends was mixed in the desired proportion with polystyrene having the same molecular weight, but only one living end (the latter polymer was obtained with potassium cumyl as the initiator). Five DVB molecules per living end were then added to the mixture to form entire networks. Fig. 1.9 shows that swelling of the latter in benzene is independent of the fraction of dangling chains until it approaches 0.3, and only a larger portion of pendent chains will slightly increase the network swelling [128, 129]. One can also see from Fig. 1.9 that the experimental points locate far above the theoretical line calculated according to Eq. [1.13]. The classic theory fails to explain these experimental results [129]. On the contrary, the analogy between a swollen network and a semidiluted polymer solution proposed by De Gennes describes much better the above finding. This approach is based on an assumption that swelling of a chain between two adjacent junctions is equal to swelling of an equivalent chain between two neighboring entanglements in a semidiluted solution [129].

Interestingly, a dissimilar dependence is observed when the model network with dangling chains stands at equilibrium with cyclohexane. Here, the overall network swelling at room temperature is much smaller than swelling in benzene. Unexpectedly, it slightly decreases with an

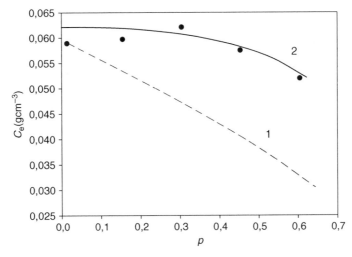

Figure 1.9 Plot of concentration of polystyrene chains, C_e, vs fraction p of monofunctional chains in networks swollen with benzene; $p = \nu_{mf}/(\nu_{mf} + \nu_{bf})$, ν_{mf} and ν_{bf} denote the number of monofunctional and bifunctional chains; (1) experimental results, (2) upper limit calculated from Eq. [1.13]. *(Reprinted from [129] with kind permission of Marcel Dekker, Inc.)*

increasing fraction of pending chains (Fig. 1.10). Neither the classic theory nor analogy with a semidiluted solution provides an explanation for this fact.

According to the latter approach, the equilibrium concentration of polymeric chains in the swollen gel, C_e, may also be related to the critical concentration C^* of a solution of equivalent linear macromolecules at which the macromolecular coils begin to overlap [129]:

$$C \propto C_e \propto C^* \propto \frac{N}{R_g^3} \qquad [1.14]$$

Here, N is the number of links at chain ends and R_g is the radius of gyration of the macromolecular chain. One can expect then that, for a network in equilibrium with benzene (BZ) and cyclohexane (CH), the following relationship has to be valid:

$$\frac{C_e(\mathrm{CH})}{C_e(\mathrm{BZ})} = \left[\frac{R_e(\mathrm{BZ})}{R_e(\mathrm{CH})} \right]^3 \qquad [1.15]$$

Small-angle neutron scattering (SANS) technique seems to show (see below) that the R_g values of swollen network strands are of approximately

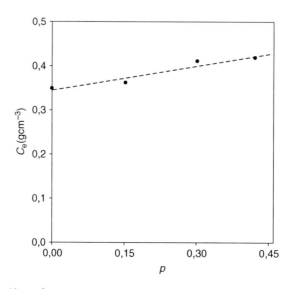

Figure 1.10 Plot of concentration of polystyrene chains, C_e, vs fraction p of monofunctional chains in networks swollen with cyclohexane (for denotes see Fig. 1.9). *(Reprinted from [129] with kind permission of Marcel Dekker, Inc.)*

the same order of magnitude as the radius of gyration of linear macromolecules with identical molecular weight in corresponding dilute solutions. From solution data for polystyrene chains of ~30,000 Da molecular weight, one finds the ratio $[R_g(BZ)/R_e(CH)]^3$ to be equal to 2.3, and one may expect a similar value for a model network prepared by crosslinking the above precursor with five DVB molecules per living end. It turned out, however, that the ratio of polymer concentrations in swollen gels $C_e(CH)/C_e(BZ)$ is much higher and amounts to about 6. Consequently, there is no direct analogy between the size of the individual polymeric coils in solution and the swelling of chains combined in a network.

Contrary to the extent of swelling, the deformation of swollen model networks under uniaxial compression (or elongation) strongly depends on the quantity of dangling chains (Fig. 1.11). The elastic modules E_p of swollen samples decrease linearly with increasing portion of dangling chains up to about $p = 0.45$ and then decrease even more rapidly [129]. Indeed, the elastic modulus is proportional to the number of elastically active chains

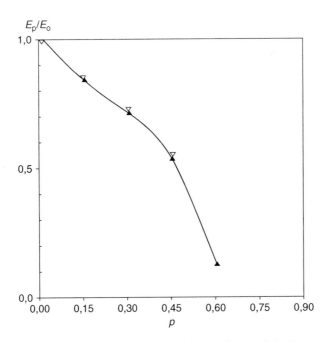

Figure 1.11 Plot of E_p/E_o for model networks swollen (▲) in benzene and (Δ) cyclohexane *vs* fraction *p* of monofunctional chains. The straight line has a slope equal to −1. *(Reprinted from [129] with kind permission of Marcel Dekker, Inc.)*

per unit volume of the sample, with this number decreasing on dilution of the material by ballast chains. When the fraction of pendent chains exceeds half of the polymer, the deformation increases more sharply, since the probability of formation of "bifunctional junctions," which means no crosslink at the DVB nodule, is also increasing.

The deformation ability of networks strongly swollen with benzene and those slightly swollen in cyclohexane was unexpectedly found to be the same. What is surprising here is the absence of any correlation between the volume increase of model networks on swelling and their deformation under compression or elongation [130], as it would have to follow from the classic theory of rubber elasticity. This theory does not predict any difference between the extensional modulus and the shear modulus that controls the swelling. Nevertheless, the experimental ratio of $C_e(CH)/C_e(BZ) = 6$ is twice as large as the ratio of $E(CH)/E(BZ) = 3$ (irrespective of p) [123].

Strikingly, besides the dangling chains, free polymeric chains entangled inside a network, and even short chains, considerably reduce the extensional modulus of polystyrene model networks, but nearly have no effect on their swelling [130]. As regards this finding, the authors wrote: "It is quite striking that only the presence of chains inside the networks modifies the response to the deformation so much. At this stage, we can only acknowledge this difference." According to our opinion, there is no major contradiction in the different impact of dangling or free polymeric chains on swelling and mechanical deformation of networks. A mechanical stress can only be opposed by elastically active chains. Therefore, reducing the volume concentration of the chains in the sample by diluting them with elastically inactive dangling or with free polymeric chains must unavoidably reduce the elasticity modulus of the material. On the other hand, swelling is the result of a (rather weak) interaction of the solvent molecules with all polymeric chains. Solvent molecules cannot distinguish between repeating units of elastically active, dangling, and free chains and solvate them to equal extent. It is true that only elastically active chains set a limit to the swelling and their volume concentration is reduced by the ballast polymer. However, we cannot expect swelling to rise proportionally to the reduction of the concentration of elastically active chains. We have to take into consideration the fact that the ballast has already expanded the elastically active network and has strained the chains bonded to junction points with their two ends. For the final expansion of the network on swelling, it is rather irrelevant whether the interior of the network is filled with polymer chains or with solvent molecules. In order to be correct in the estimation of the "true" extent of network expansion on swelling, we should add the volume of the ballast to the determined solvent uptake and then divide the sum by the weight of the sample less the weight of the ballast. If we do not introduce these corrections, we should not be surprised by the fact that the swelling value, estimated and calculated in the traditional way, proves rather insensitive to the presence of ballast polymer.

Returning back to the analysis of Eq. [1.12], it can be seen that swelling of model networks has to be dependent linearly on $(v_3^c)^{-2/5}$, provided that the earlier made assumption is valid that $h_3 = v_3^c$. Yet, the function obtained in practice proved to be nonlinear.

Now, it is appropriate to define the memory term h_3 more precisely. Dušek and Prins relate the memory term to such concentration of polymeric segments at which no forces are exerted on crosslinks; in other words, the chains are in the reference state:

$$h_3^{2/3} = \frac{\langle r^2 \rangle_d}{\langle r^2 \rangle_{os}} = \frac{\langle r^2 \rangle_d}{\langle r^2 \rangle_c} \times \frac{\langle r^2 \rangle_c}{\langle r^2 \rangle_{os}} = \frac{\langle r^2 \rangle_c}{\langle r^2 \rangle_{os}} \ (v_3^c)^{2/3} \qquad [1.16]$$

where $\langle r^2 \rangle_{os}$ is the mean square end-to-end chain distance of the free polymeric chain precursor in solution, $\langle r^2 \rangle_d$ is the mean square end-to-end chain distance in dry network, and $\langle r^2 \rangle_c$ is the mean square dimension of the elastically active chain just after the introduction of crosslinks. Thus, h_3 depends on the estimation of $\langle r^2 \rangle_c / \langle r^2 \rangle_{os}$. If the end-to-end-linking process does not result in significant extension or shrinkage of the polymeric coils, this ratio should be close to unity and, therefore, the memory term should be determined by the polymer concentration at which crosslinks are introduced. This assumption seems to be rather reasonable as the crosslinking does not cause any syneresis of the initial solvent from the final gel. On the other hand, a different opinion exists [131, 132], namely that polymeric coils shrink by a factor of about two on crosslinking, and then h_3 should be defined as

$$h_3 = \left[\frac{\langle r^2 \rangle_c}{\langle r^2 \rangle_{os}} \right]^{3/2} v_3^c = 0.354 v_3^c \qquad [1.17]$$

Obviously, the reference state of chains in the model networks obtained in the presence of solvent can be associated neither with completely dried chains nor with chains swollen to maximum. The most logical assumption would be that the reference state relates to the swelling degree of the gel at the moment of its formation: $h_3 = v_3^c$. Computer simulation showed, however, that with this concept the entire calculated curve $Q = f(f)$ locates below the swelling $Q = 20$ of the network at the moment of its formation (Fig. 1.12). Such result implies that the macrosyneresis of the excess solvent would have to be observed; however, this is not the case. Hence, the hypothesis of $h_3 = v_3^c$ is erroneous [125]. On the contrary, the calculations testify more reasonably to enlarged network swelling, $Q > 20$, if one assumes $h_3 = 0.354v_3^c$ and the network functionality to be less than 10.

The same conclusion on the invalidity of $h_3 = v_3^c$ idea may be drawn from examining the following relation between the network swelling and the molecular weight of polystyrene precursor [133]:

$$M = \frac{\bar{V}_1 \left(\frac{2}{f} Q^{-1} - h_3^{2/3} Q^{-1/3} \right)}{v_3^o \left[\ln(1 - Q^{-1}) + Q^{-1} + \chi_{13} Q^{-2} \right]} \qquad [1.18]$$

As shown in Fig. 1.13, the best agreement was achieved between experimental and predicted results only with $h_3 = 0.354v_3^c$, provided that the Flory–Huggins χ_{13} parameter depends on the polystyrene concentration as $\chi_{13} = 0.45 + 0.9v_3^c$, $v_3^c = 0.1$, and $f = 4$ if three DVB molecules were added per living chain end.

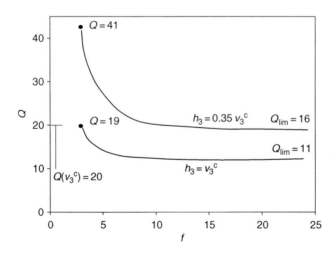

Figure 1.12 Theoretical variation of Q with f for model networks. Calculations were performed for $M = 10,000$, $v_3^c = 0.05$, $\chi = 0.456$. Two curves correspond to the hypothesis of $h_3 = v_3^c$ and $h_3 = 0.35v_3^c$. (Reprinted from [125] with kind permission of American Chemical Society.)

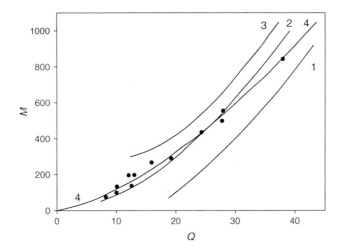

Figure 1.13 Calculated relation between molecular weight of polystyrene precursor, M, and swelling in benzene of model networks, assuming $\chi_{13} = f(C)$, $h_3 = 0.354v_3^c$ and functionality of junction points of (1) 3, (2) 4, and (3) 5; (4) experimental curve. *(After [125].)*

The functionality $f = 4$ of the DVB nodules appears to be close to reality, insofar as no difference is observed between the swelling capacity of networks obtained in the presence of three DVB molecules per living chain end and of networks crosslinked with the tetrafunctional cross-agent (allyloxy)triazine derivative (II) [134].

We can summarize that the swelling behavior of model networks is generally inconsistent with the predictions of the theory of Dušek and Prins. If one still assumes that Eq. [1.18] may adequately describe the network swelling, as in the above case [133], and accepts that $h_3 = 0.354v_3^c$, then the dubious suggestion must be also accepted that the linear dimensions of polymeric chains decrease by a factor of two on crosslinking, without causing any phase separation or syneresis in this process. Actually, the situation is markedly more complicated.

2.3 Characterization of model networks by SANS technique

If the polymers under consideration are regarded as model networks in which f chains connect a given junction point with f nearest topological neighbor junction points, one can assume a certain ordering in the network structure [134]. If now the junction is placed into a cell of a crystalline lattice and different types of crystals with different numbers of closest neighbors (i.e., functionality) are considered, then f affects appreciably the distance, d, between two junction points. For an individual polystyrene chain of 25,000 Da dissolved in benzene at a volume fraction of 0.1, the root mean square end-to-end distance calculated from the Mark–Kuhn–Hauwink and Flory–Fox equations amounts to 115 Å. When all ends of such chains are combined by tetrafunctional junctions (diamond lattice), d decreases to 81 Å in swollen state. By assuming $f = 12$ (model of a face-centered lattice), the distance between neighboring junctions increases to 152 Å. When regular networks of these functionalities form, the cross-linking must result in a considerable shrinkage or stretching of the polymer strands. In this case, for the networks under consideration, h_3 should depend not only on the molecular weight of the strands but also on the functionality of the junction points.

To determine the distance between junction points in a real polymer, Benoit et al. [135] synthesized polystyrene chains of 21,000–50,000 Da molecular weights containing 5–6 deuterated styrene units on both living ends of the macromolecules. Therefore, in networks obtained by cross-linking these precursor chains with 10 DVB molecules per living end, each DVB nodule proved to be surrounded by labeled styrene units. The networks were then examined by neutron scattering under small angles with the expectation that the deuterated zones will scatter neutrons. Indeed, the intensities of coherent scattering plotted against the angles show a clearly cut and quite narrow peak in the region of small angles. It gives evidence to the expectation that the model networks are characterized by a preferred distance between neighboring junction points, that is, they exhibit a rather uniform distribution of crosslinks throughout the entire sample.

The distance between neighboring junctions in a dry network, calculated from neutron scattering by Benoit and coauthors [135] increases with the molecular weight of the polystyrene precursor (Table 1.4). This distance increases also on swelling of the network in good solvents. At that, the product of d and $Q^{-1/3}$ was found to remain constant (at least, up to the very high swelling of about 10). This relation points to the "affinity" of network deformation, that is, to the expected proportionality between the network volume and the distance between its junction points.

However, the values of distances, d, between two junction points calculated from SANS measurements seem to be unreal. Thus, according to SANS data presented in Table 1.4, the network with strands of

Table 1.4 Experimental values of swelling ratio, Q, and mean junction points separation distance, d, Å, calculated from SANS data for model networks with strains of different M-values crosslinked with 10 DVB molecules per living end

Solvent	21,000		30,000		44,000		50,000	
	Q	d	Q	d	Q	d	Q	d
Non (dry)	1	95.4	1	112.2	1	130.5	1	136.6
C_6H_{12}	2.34	136.3	2.67	162.3	2.9	192.8	3.01	217.2
THFA	5.07	178	6.3	227	8.1	272.1	9.01	307.4
C_6H_6	9.8	209.8	12.6	252.5	17.5	298.9	19.5	329.4

THFA = tetrahydrofuryl alcohol; C_6H_{12} = cyclohexane.
Source: After [135].

21,000 Da (and prepared at a dilution of approximately $Q = 10$) should have an average distance d of ~210 Å, when swollen with benzene. At the same time, according to both estimations from the Mark–Kuhn–Hauwink relation and from SANS measurements, for a solution of polystyrene 21,000 Da in benzene, the size of an individual free polystyrene precursor coil is only 100 Å. Hence, being crosslinked, the polymeric coil stretches rather than shrinks [134], which seems quite improbable. Indeed, at the rather high concentration of the initial solution ($Q = 10$), polymeric coils must either strongly overlap and mutually interpenetrate or adopt a rather compressed conformation. In this situation, either several junction nodes must emerge within the volume of a single coil (at a realistic value of functionality of junction points) or the crosslinking must freeze the compressed conformations of coils. In no case the distance between the neighbor junctions can become as large as 210 Å. A similar obvious discrepancy is also characteristic of polystyrene chains having other molecular weights.

Another important statement of French authors from their neutron scattering experiments was that the size of polystyrene strands in dry model networks is equal to the size of the precursor chains dissolved under Θ-conditions (see Table 1.5). At first sight, this result confirms the idea [136, 137], expressed by Flory [138] in the middle of the last century and later picked up by De Gennes [139], that in bulk polymer the chain retains its unperturbed dimension characteristic of its solution in a Θ-solvent. Flory believed to have experimentally confirmed his idea [140]. By analyzing elastic neutron scattering from a mixture of linear protonated and deuterated polyisobutylene with a molecular weight of $48,000 \pm 3,000$ Da, he found that the gyration radii of the macromolecules

Table 1.5   Network chain dimensions according to small angle neutron scattering

Polymer precursor	$M \times 10^{-3}$ Dry network	Chain dimension in dry network d, Å	Root square end-to-end distance of free chain, Å	
			Good solvent	Θ-Solvent
PS[a]	21	80	100	80[b]
PS	27	90	114	87[b]
PS	42	109	142	109[b]
PDMS[c]	4.5	24	48	40
PDMS	8.7	30	70	56
PDMS	17.5	39	105	79

Source: After [141].
[a] Crosslinked with 10 DVB molecules per living end; good solvent is benzene, Θ-solvent is cyclohexane.
[b] Mean end-to-end distance.
[c] $f = 4$, good solvent is *n*-heptane, Θ-solvent is toluene.

are 75 ± 5 and 77 ± 5 in bulk and Θ-solvent, respectively. If this were a general rule, any polymer would have to exhibit the same property.

Similar to the above polystyrene-based systems, any dry polymeric network would preserve dimensions of its strands on swelling in a Θ-solvent. Nevertheless, this is not the case. In addition to polystyrene that, according to SANS data presented in Table 1.5, perfectly follows the above expectation, polydimethylsiloxane (PDMS) networks have been also examined by this technique. They were prepared by crosslinking dissolved linear PDMS chains, fitted on both ends with silane functions, with tetraallyloxyethane

$$(CH_2 = CH - CH_2 - O)_2 CHCH(O - CH_2 - CH = CH_2)_2.$$

In the presence of chloroplatinic acid as the catalyst, silane adds to the double bonds of the tetrafunctional crosslinking agent with the formation of stable Si-C bonds:

$$--\rangle SiH + CH_2 = CH - CH_2 - O-- \rightarrow --\rangle Si - CH_2 - CH_2 - CH_2 - O--.$$

As can be seen from Table 1.5, chain dimensions in the dry PDMS networks are found to be much smaller than those of the unperturbed macromolecules in the Θ-solvent, toluene.

Generally speaking, the idea of unperturbed chain dimensions in dry polymers is based on a rather simple reasoning. If a solvent is sufficiently poor for a given macromolecule, such as cyclohexane for polystyrene, its energy of interaction with the polymeric chain at a certain temperature

(Θ-temperature) is such that the mutual attraction of chain segments completely compensates their mutual repulsion caused by their own chain volume. Then, the chain does not experience any perturbing effects from the surrounding polymeric segments or solvent molecules and its dimension is determined entirely by the bond lengths and valence angles. In other words, the chain is unperturbed. In the unperturbed polymeric coil, the energies of the polymer–polymer, polymer–solvent, and solvent–solvent interactions are identical. In terms of interactions under Θ-conditions for a given coil, it is all the same whether molecules of the Θ-solvent occupy the interior of the coil or segments belonging to another coil replace some portion of the solvent by penetrating into the coil. On this account, Flory wrote that unperturbed macromolecules behave as hypothetical point molecules, which may freely distribute themselves over the volume. According to him, the same situation must also arise in a bulk polymer. Here, the interactions between the polymeric segments inside one macromolecular coil equal those between segments of different coils, and therefore, the polymeric chains must attain unperturbed dimensions in bulk material.

Undoubtedly, in the preparation of model networks from rather long polymeric precursors by the end-linking reactions, the strands between the junction points will maintain the conformation of coils. In accordance with classical ideas, the coils in the dry model network should acquire unperturbed dimensions, similar to coils in a Θ-solvent. However, on contacting a model polystyrene network with cyclohexane at room temperature, which is far below the Θ-point for linear polystyrene (34.5°C), its volume increases by a factor of ~3. Thus, the solvent breaks some polymer–polymer interactions, stimulates swelling of the network, and may result in an increase of chain dimensions. Hence, even at room temperature, the polystyrene chains prefer replacing the alien chain segments by cyclohexane molecules.

This swelling would be plausible if the Θ-temperature for the polystyrene network falls below the Θ-temperature for the linear polymer. For star-shaped polystyrene macromolecules having 9.4 branching arms, each arm having a molecular weight of 50,000 Da, the Θ-temperature in cyclohexane was found to be as low as 22°C [142]. To a certain extent, the model networks might be considered as a collection of similar highly branched macromolecules. Therefore, if in the model polystyrene networks the Θ-temperature decreases to room temperature, it could explain the network swelling in cyclohexane.

The thermodynamic affinity of cyclohexane to polystyrene is known to increase with temperature and, naturally, increasing the temperature must further raise the volume of the polystyrene networks in cyclohexane. There is, however, an additional point we should consider. The plot of Q *vs.* temperature exhibits a steplike discontinuity at around 30°C (Fig. 1.14). This discontinuity, resembling very much a Θ-transition, is located 3–5°C below the Θ-temperature for linear polystyrene in cyclohexane and about 8°C above the Θ-point for star-shaped polystyrene macromolecules. This phenomenon is outside the scope of the questions discussed here, but, naturally, the first assumption of the authors [143] seems to be very logical, according to which the discontinuity reflects a transition from Gaussian coil to a "supercoiled" compact structure on cooling the swollen gel below that temperature zone.

A reversible transition from a swollen to unperturbed coil and further to a more compact globule-like conformation of polystyrene on cooling its diluted solutions in cyclohexane below the Θ-temperature is well documented by viscosimetric measurements [142, 144], determination of

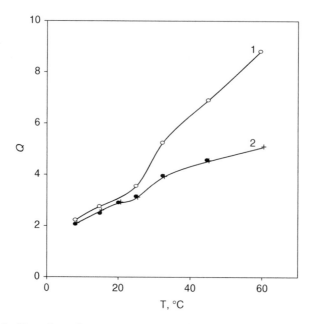

Figure 1.14 Plot of swelling *vs.* temperature for two model networks based on polystyrene precursor with molecular weight of (1) 31,500 and (2) 14,700 Da. *(After [143]).*

diffusion constants, [145] as well as light scattering technique [146–148]. It is very doubtful that in a model network, polymeric strands would not experience the same tendency for shrinkage. We have all reasons to expect the network strands to shrink first with the temperature coming down below the Θ-point and then with the removal of the solvent, that is, on transition to dry bulk material. From this point of view, too, the finding of unperturbed chain dimensions for polystyrene strands in the dry model network mentioned in Table 1.5 is rather dubious.

In order to obtain information about the gyration radii of polystyrene chains between adjacent junction points in a model network, another approach was also examined by the same French group [135]. A small portion of deuterium-labeled linear polystyrene chains was added to an identical hydrogenated precursor during network formation. Then the gyration radius, R_g, was calculated from SANS of the network, both in dry and in swollen state. The results proved to be rather surprising. While the distance d between two junctions was found to increase with network swelling, as illustrated above in Table 1.4, the calculated gyration radius remained nearly constant (Table 1.6). For example, a network prepared by crosslinking a polystyrene precursor having a molecular weight of 26,000 Da with six DVB molecules per living end will increase its volume by a factor of 15 on swelling with benzene, whereas R_g only increases by 25%. According to the principle of affine deformation, the gyration radius would have to increase by a factor of 2.5 in order to correlate with the strong swelling.

Thus, the above-detailed investigation of dry and swollen model networks by using SANS technique also failed to provide any correlation between the increasing junction-to-junction distance on swelling the

Table 1.6 Swelling ratio, Q, and radius of gyration, R_g, Å, from SANS experiments for model networks with M = 26,000 crosslinked with different amounts of DVB

Solvent	3 DVB		6 DVB		10 DVB	
	Q	R_g	Q	R_g	Q	R_g
Non (dry)	1	45	1	44	1	43
C_6H_{12}	3.02	49	2.85	46	2.64	40
THFA	7.55	57	7.15	51	6.2	45
C_6H_6	17.4	67	14.9	57	12.5	50

THFA = tetrahydrofuryl alcohol; C_6H_{12} = cyclohexane.
Source: After [135].

network and rather constant gyration radii of chains between the junctions. Still, to explain the above disproportionality, the following mechanism could be suggested for network swelling [131, 129].

2.4 Presumptive mechanism of swelling of model networks

Networks obtained by anionic block copolymerization consist of coils, the ends of which are combined by junctions. A given junction has two types of neighboring junctions, namely, topologically nearest junctions, *t*, which are connected with the given junction by polystyrene chains (hence their number is equal to junction functionality), and the nearest spatial neighbors, *s* (Fig. 1.15). The ratio of the total number of neighbors for a given junction to the topologically nearest neighbors depends on the molecular weight of the polystyrene precursor and amounts to 15–18 for the usually used polystyrenes with $M = 20,000$–$30,000 \, \mathrm{Da}$ [137]. It implies that the shortest topological path connecting two spatial neighbors is much longer

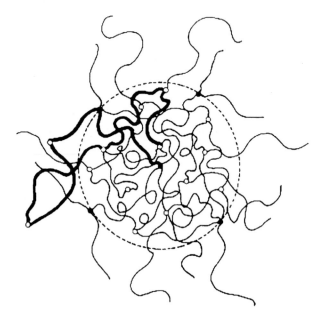

Figure 1.15 Schematic presentation of a dry model network. The sphere has a radius equal to the end-to-end distance of a chain. The filled and the open circles represent the topological and the spatial neighbors of the central junction, respectively. The dark line represents the shortest path from the central junction to its nearest spatial neighbor. *(Reprinted from [129] with kind permission of Marcel Dekker, Inc.)*

than the length of the starting polystyrene molecules and, therefore, it incorporates several crosslink points. In the presence of a solvent, the spatial neighbors, due to longer chains between them, can move more freely than t-neighbors, the mobility of which is more restricted. Hence on swelling, the distance between the s-neighbors increases more than the distance between the t-neighbors, thus leading to the total mutual rearrangement of junctions and an increase in the network volume. This mechanism of swelling can be illustrated by a simple model construction, composed of rigid metallic rods connected by flexible joints (Fig. 1.16). Such a system can deform along three axes, although the dimensions of the rod chains remain unchanged. French authors [129] named this mechanism of network expansion as "desinterspersion of junction points."

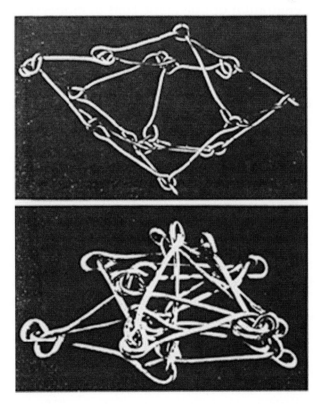

Figure 1.16 Illustration of the desinterspersion effect on a model network formed of rigid strands. *(Reprinted from [129] with kind permission of Marcel Dekker, Inc.)*

In principle, the suggested model may explain the disproportional change in d and R_g on network swelling, at least for small swelling values. On the other hand, the augmentation of the network volume by a factor of 15 on its swelling with only 25% alteration in chain gyration radius would imply a predominant solvation of the junction points and their mutual strong repulsion, rather than a preferred solvation of polystyrene chains. However, it is unlikely that this can happen. During the process of removing the solvent from the initially swollen gel, the free "desinterspersion" model implies that the spatial-neighbor junctions were allowed to easily penetrate into zones occupied by coils belonging to topological-neighbor junctions. This, again, is difficult to imagine. Also, the model provides no explanation for many other interesting features of network deformation both on swelling and on mechanical impacts.

2.5 Interpenetration of polymeric coils in model networks

It seems appropriate to discuss here the probability of interpenetration of polystyrene coils in the model networks. As already mentioned, according to the theoretical considerations of Flory [138] and De Gennes [139], polymeric coils in an amorphous solid state retain unperturbed dimensions. Since the volume fraction of the polymer in an unperturbed coil under Θ-conditions is well known to be very small, only about 2%, the transition from swollen coils to solid state has to be accompanied by the replacement of all solvent molecules with fragments of other polymeric molecules. In other words, theoretical notions predict extremely high mutual interpenetration of the polymeric chains in bulk state. Indeed, in order to maintain the coil dimension that is characteristic for a Θ-solution, the coil must accommodate, on removing the solvent, a 50- to 100-fold amount of alien polymeric matter. In the 1970s this problem was discussed in full [149–165]. The authors of the "tailor-made" networks also took part in the discussion.

Herz et al. [141] suggested distinguishing between the chemical functionality (f_{ch}) of a crosslink, which is the average number of elastically active chains protruding from a junction, and the geometric functionality (f_{geom}), which is defined as the number of its nearest neighboring junctions. If $f_{ch} > f_{geom}$, some chains must attain stretched conformation and bind more remote DVB nodules. The character of chain connection can be estimated by means of the so-called connectivity factor X^2 [166], the ratio of the mean-square end-to-end distance of elastic chains to the square of the average distance between the closest neighbor crosslinking points.

If elastically active chains connect only the nearest neighbors, then $X^2 = 1$; however, when $f_{ch} > f_{geom}$ then $X^2 >> 1$. In this way the parameter X^2 reflects the extent of chain entanglements.

Theoretically the modulus of elasticity, E, at uniaxial compression of a network swollen in benzene is related [167] to the molecular weight M of the polystyrene chains and the connectivity factor X^2 as

$$E = 3.18 \times 10^3 \; C^2 X^2 M^{-3/2} \qquad [1.19]$$

C is a structural parameter reflecting the "dislocation geometry of crosslinks surrounding a junction" and its value is close to unity (it varies between 0.8 and 1.2). For networks crosslinked with sym-triallyloxytriazine (I), the dependence obtained experimentally meets the case of

$$E = 3.20 \times 10^3 \; M^{-3/2} \qquad [1.20]$$

Here $X^2 = 1$ and, therefore, the permanent entanglements are negligible in such a network crosslinked with the trifunctional cross-agent. In this respect, it meets quite well the requirements of an ideal structure. For model networks crosslinked with DVB, the slope of the straight line E vs. $M^{-3/2}$ proved to be steeper than that for the above-discussed case of $f = 3$. The more DVB molecules per living end are used to tie a network, the steeper will be the slope, which reflects an increase in the functionality of the nodes formed. The functionality of the latter can be estimated by using the ratio of the elasticity modulus of a given system to that characteristic of the above trifunctional network. However, such estimation may lead to unrealistically high values of functionality, up to 55, when 10 DVB molecules are used per living end. Obviously, a part of the chains connects junctions that are not the first neighbors, and the probability of permanent chain entanglements increases with an increase in the amount of the introduced divinyl cross-linking agent. This conclusion is further confirmed by comparing the distance between the neighboring ferrocene-labeled junctions measured by X-ray scattering, with mean square end-to-end distance of elastic chains obtained from uniaxial compression data. All measurements were made on networks swollen in benzene. With increasing DVB amount taken per living chain end, the swelling of networks drops. Then, the inter-junction distance drops faster than the coils' dimensions, which is interpreted as an evidence of growing chain entanglements.

Rempp and Herz [168] conclude that an especially intensive mutual interpenetration of network strands begins on drying the networks that are initially prepared in the swollen state. Yet, this conclusion follows automatically from the notion on unperturbed chain dimensions in dry networks, which the authors uncritically accept. It is appropriate to add here that the French authors of the model networks do not question the results of data processing that is used for the SANS technique. As a typical example, let us consider some results reported by Benoit [150]. By mixing solutions of hydrogenated and deuterated polystyrenes, followed by a complete evaporation of the solvent, a solid mixed polymer was obtained. Examining this sample by the SANS technique points to unperturbed dimensions of individual polystyrene chains. It would be logical to expect that the complete interpenetration of individual coils, with each of them forced to incorporate up to 100-fold amount of alien chain loops, in order to preserve its size and conformation, may require much more time than that available for the evaporation procedure. On drying the highly viscous solution, the polymeric coils are more likely to attain a more compact packing of their segments than that corresponding to unperturbed dimensions of the coils. We can expect the coils to overlap one another only in their outer periphery. The interpenetration of the strands of a network should be even more restricted, since they are devoid of flexible chain ends able to penetrate the neighbors by "reptation" movement. The finding of unperturbed dimensions of macromolecules in crystalline polyethylene [155], derived from SANS data, are similarly surprising and doubtful. Formation of a crystal would require a highly ordered arrangement of segments, whereas an unperturbed coil is a totally random structure with Gaussian distribution of segments.

We may stress here that neutron scattering, like the X-ray scattering technique, cannot be considered as a direct means of measuring object dimensions. It only provides information on the spatial distribution of scattering heterogeneities or density fluctuations in the sample. Mathematical transformation of these data into information on the size of the heterogeneities is always based on certain model assumptions, the validity of which for real polymers still has to be proven.

The SANS technique requires perfect mixing of deuterated and protonated macromolecular species in the mixed sample. Hereby it is assumed that these macromolecules are fully compatible and do not recognize the nature of the partner. Meanwhile, sufficient information has accumulated, showing a substantial difference in the properties of deuterated and protonated molecules. Substitution of hydrogen by deuterium gives shorter bond lengths, decreases molar volumes, and reduces bond polarizabilities for the deuterated compounds. Wang et al. [169] found that deuterated polystyrene had a smaller mean segment size and formed more compact coils than a protonated polystyrene chain of the same length. Thin-layer chromatography [170] can be used for the separation of deuterated and protonated polystyrenes. In reversed-phase chromatography [171] on C_{18}-silica or carbonaceous HPLC column packings, deuterated styrene $n = 10$ and 14 oligomers have approximately 9 and 13% shorter retention times than protonated analogues. More importantly, Hong et al. [172] found a significant surface enrichment of deuterated polystyrene in its mixtures with conventional polystyrene, when the surface of solid samples was examined by surface-enhanced Raman scattering. Even the SANS showed that 50:50 mixtures of protonated/deuterated polymers do not form ideal solutions and will phase separate at temperatures below a certain upper critical solution temperature [173, 174]. Thus, being mixed together, deuterated and protonated macromolecular coils are most likely to change their conformation and size, reducing significantly the value of information provided by the SANS technique.

2.6 Study of model networks by other methods

Investigation of the model networks by other physicochemical methods failed to produce additional useful information about their structure. To evaluate the distances between polymeric chains in swollen networks, Hild et al. [175] analyzed the phase distribution of linear polystyrene samples between the gel and the surrounding solution. As could be expected, the static distribution coefficients decrease with increase in size of the polystyrene standards, but no reliable conclusions could be made concerning the pore dimensions, since the diffusion of polystyrene into the gel was accompanied with a considerable network deswelling (it is interesting that the observed deswelling is larger than that predicted by Flory's lattice model; Bastide et al. [176] found the phenomenon to be better accounted for as part of the scaling theory). Size-exclusion chromatography was also helpless for bringing out mesh dimensions, because no equilibrium was achieved under dynamic conditions. For the same reason, the static and dynamic distribution coefficients were found to be quite different [177].

The glass transition temperature of the model networks decreases noticeably with an increase in the chain length between junction points and a decrease in the functionality of the latter (Table 1.7) [178].

Table 1.7 Glass transition temperature, T_g, °C, for linear polystyrene and model networks having joints of different functionality f. [178]

M_n	T_g, °C Linear PS	T_g, °C f 5–7	M_n	T_g, °C f 3	M_n	T_g, °C f 10–12
7,500	77.7		9,600	98.5	8,000	109.5
10,450	83	103	12,100	98	20,000	103.5
10,660		102.5	27,550	97.7	36,000	102
13,200	84.9	101.7				
19,100	90					
19,900	92	100.7				
30,000	93.5	99				
45,600	94.5	98.6				
75,000	95.3	97.7				
110,000	97.4					

T_g was measured by differential scanning calorimetry technique.

As a conclusion, we may state that the very intensive studies of model networks prepared by anionic block polymerization provided much interesting experimental data. However, they failed to prove the validity of the current network elasticity theories and provided rather ambiguous explanations of the swelling and mechanical properties of model networks. Moreover, the results obtained raised more questions regarding the evaluation and interpretation of the neutron scattering technique data.

3. GEL-TYPE ION-EXCHANGE RESINS

Gel-type styrene–DVB copolymers are well known to be the traditional matrices for manifold ion-exchange resins. Two types of inhomogeneities are characteristic of these supports. The first type is the inherent shortcoming of all styrene–DVB copolymers consisting of the nonuniform distribution of crosslinks caused by a large difference in the reactivity of the two comonomers. Addition of a third comonomer, such as acrylonitrile, results in functional heterogeneity of the final material, although it makes the distribution of crosslinks slightly more uniform. The second type of structural inhomogeneities including popcorns, shell-core differentiation, and formation of zones with different swelling along the radius of a bead arises from the method of bead preparation. These shortcomings may be largely eliminated by the optimization of the synthesis conditions, such as the used temperature, and the type and quantity of the initiator, and also by the selection of the proper type of suspension stabilizer, by the introduction or complete removal of the polymerization inhibitor, and by conducting the copolymerization in an inert atmosphere. Each of these factors and, even more, their combinations significantly affect the structure of the beaded copolymer and, thereby, the properties of the ion exchanger.

Currently, the manufacturers managed to considerably improve the properties of gel-type resins by optimizing their production protocols. However, because of their high commercial value, this knowledge is kept as undisclosed know-how. For this reason, scientific evaluation of the conditions of preparation, the structure of the matrix, and properties of commercial ion exchanger as well as their comparison with published data became practically impossible. This particularly concerns the publications of the last decades. Nevertheless, the most general features of ion-exchange resins based on conventional gel-type styrene–DVB supports are well known, since this subject was debated in all aspects in the scientific literature of the 1960s–1970s. Detailed description of this information may be found in the fundamental books of Helfferich [179], Tremillon [180], Soldatov [181] Samuelson [182], the series edited by Marinsky and Marcus [183], in "Ion Exchangers" edited by Dorfner [184], etc. Here, we would like to discuss briefly the influence of the gel-type network structure on the permeability of ion exchanger for organic molecules and ions.

The crosslinking density of the matrix influences the ability of the gel-type styrene copolymers to incorporate various polar functional groups by well-known polymer-analogous reactions. A large amount of functional groups can only be introduced into the copolymer beads if their DVB content does not exceed 10–12%. At any higher degree of crosslinking, only the beads' surface layers interact with the sulfonating or chloromethylating reagents [185, 186]. Copolymers of styrene with technical-grade DVB can be sulfonated under milder conditions than beads crosslinked with pure DVB isomers [73], the highest rate of sulfonation being reported for the copolymers prepared with a technical-grade crosslinking agent containing 30% p-DVB and 70% m-isomer [187].

With the advent of ion exchange technologies, gel-type ion-exchange resins are the workhorses of the water-softening and demineralization processes, largely due to their low cost. The problem of permeability of the gel matrix for small inorganic ions was thought to be of minor importance, compared to the osmotic stability of the beads. Strong local inner stresses, resulting from a considerable difference in swelling between slightly and highly crosslinked domains[188, 189], cause bead breakdown when regenerating the resin. Destruction is especially severe at low contents of crosslinks (copolymers with technical DVB were reported to possess the lowest osmotic stability [190]). As a rule, an increase in the crosslinking density reduces the total swelling and its fluctuations, and partially preserves the resin beads from negative consequences of an osmotic shock, but it simultaneously facilitates fouling of the resins with high molecular weight organics that are common in surface waters. Gradually diffusing into the beads of an ion exchanger, these will shield the functional sites, preventing a complete regeneration of the resin. This circumstance is associated with the very low diffusion rate of large species within the gel matrix.

Recently, it was suggested [191] to control the depth of functionalization of the beads in order to reduce the diffusion path within a bead and facilitate the ion exchange and regeneration processes. The core of the so-called shallow-shell resins has no functional sites at all. Such cation exchangers exhibit improved osmotic resistance and good mechanical strength, their regeneration proceeds faster and the consumption of regenerants is smaller.

Contemporary optimization of large-scale applications of standard resins pursues the object of enhancing column performance, that is, increasing the flow rate and the operating capacity of a given resins bed until the breakthrough of sorbates. One of the important technological improvements is using smaller and monodisperse beads. The usual suspension polymerization procedure always provides a rather broad, Gaussian-type bead size distribution. Two particularly important approaches have been suggested to produce monosized beads.

In the first method, the so-called seed technique, a slightly crosslinked styrene–DVB copolymer, having a narrow bead size distribution, is subjected to swelling in the mixture of styrene, DVB, and the initiator. Next, the mixture is polymerized inside the seeds [192–194] to result in larger beads, with the bead size distribution slightly wider than that of the seeds. Strictly speaking, the beads obtained in this way cannot be regarded as typical gel resins, and they belong to the family of interpenetrating networks. The company Bayer A.G. managed to produce interpenetrating strong acidic and strong basic ion-exchange resins with really narrow bead size distribution: the diameter of more than 90% of "Lewatit MonoPlus" beads deviate by no more than 0.05 mm from the mean value [195]. The product exhibits good osmotic and mechanical stability and improved kinetic properties.

The second method is based on the use of the vibration technique permitting the dispersing of the liquid monomer mixture into uniformly sized droplets.[196, 197] However, during the subsequent polymerization step using the common suspension technique, care should be taken to prevent further coalescence and splitting of the initial droplets.

Attempts to apply gel-type ion-exchange resins to solving numerous problems of isolation, purification, and analysis of organic substances have not been very encouraging. All functional sites were found to be accessible even for the relatively small tetramethyl- and tetraethylammonium ions only under the condition that the supporting copolymer contains no more than 2% of crosslinks (Table 1.8). The ion exchange capacity markedly decreases with an increase in the support's crosslinking density. Hydrated tetrabutylammonium ions of 9.8 Å diameter [198] can substitute only 70% of the counter ions in a cation exchanger at its degree of crosslinking of 2%. In this resin, only one-third of the HSO_3 groups can interact with trimethylcetyl ammonium ions; these sites are certainly located on the periphery of the beads.

Regarding the difference between copolymers with pure p- and m-isomers of DVB, no reliable data are available. One could expect large organic ions to diffuse faster into sulfonated resins crosslinked with p-DVB, since the more irregular structure of the latter should offer a larger number of relatively long and flexile polystyrene chains between the dense microgel domains. Resins crosslinked with p-DVB were found [199] to offer smaller diffusion rates for tetrabutylammonium ions (but larger equilibrium exchange capacity) than resins incorporating m-DVB and having

Table 1.8 Ion exchange capacity of styrene–p-DVB copolymer resins toward organic ions (percent of Na^+ capacity)

Organic ion	Equilibrium absorption^a (%) at p-DVB content (%)		
	2	5	10
$(CH_3)_4N^+$	100	86	70
$(C_2H_5)_4N^+$	100	83	70
$(C_4H_9)_4N^+$	71	66	51.5
$(CH_3)_3C_6H_5N^+$	80	67.5	65.5
$(CH_3)_3C_6H_5CH_2N^+$	82.5	75.5	76
$(CH_3)_3C_{16}H_{33}N^+$	28.6	26.0	20

Source: After [87].
^{a} Sorption from 0.02 N solutions.

comparable swelling values (Table 1.9). However, the same group [200] expressed later a different opinion, namely, that the permeability does not depend on the type of crosslinking agent and is only governed by resin swelling. There is, however, no question that different conditions in the preparation of the copolymers and their sulfonation may easily lead to different results.

Ion exchange resins based on copolymers with technical-grade DVB exhibit larger swelling capacity with water, but their permeability, too, rapidly drops with an increase in the DVB content [201–203]. An increase of the degree of crosslinking from 2 to 4% reduces the diffusion rate of antibiotics by one or even two orders of magnitude [204–206]. The strong basic anion exchanger AV-17 (Russia) with a crosslinking density as high as 12% does not absorb antibiotic novobiocyn at all [207]. Generally, sorption of organic ions is always accompanied by the dehydration of ion exchanger [208], and even irreversible changes have been reported in the network structure [209].

Copolymerization of styrene and DVB in the presence of a good solvent such as toluene leads to expanded network structures with a reduced degree of chain entanglements and higher swelling values. By adding varying amounts of toluene to the comonomer mixture (Table 1.10), Millar et al. [210] prepared a set of sulfonated resins with different degrees of cross-linkings, but equal water-regaining properties. In these experiments, a sample with 7% DVB represented the conventional resin. A sample prepared in the presence of 38% toluene and crosslinked with 15% DVB had a typical expanded structure, while a third sample with 43% toluene and 27% DVB showed an intermediate between expanded and macroporous

Table 1.9 Permeability of sulfonated resins based on copolymers with pure DVB isomers for tetrabutylammonium ions

DVB (%)	m-DVB			p-DVB		
	K_{sw}	F	$\bar{D} \times 10^8$	K_{sw}	F	$\bar{D} \times 10^8$
4	0.353	70	6.7	0.350	88	5.9
6	0.216	64	3.8	0.217	83	1.7
9	0.154	61	1.0	–	–	–
12	–	–	–	0.149	52	0.57

K_{sw} = swelling in water, g/meq.
F = capacity toward tetrabutylammonium ions, percent of Na^+ capacity.
$\bar{D}$ = interdiffusion coefficient, cm^2/s.
Source: After [199]

Table 1.10   Permeability of toluene-modified sulfonated resins for organic ions

DVB %	F_m	K_{sw}	Me_4N^+		Et_4N^+		Pr_4N^+		Bu_4N^+	
			F_m	$\bar{D} \times 10^6$	F_m	$\bar{D} \times 10^6$	F_m	$\bar{D} \times 10^6$	F_m	$\bar{D} \times 10^6$
7	1.00	0.20	100	1.08	100	0.26	0.95	0.10	92	0.015
15	0.62	0.20	–	–	100	0.29	–	–	82	0.0332
27	0.57	0.22	–	–	87	0.23	–	–	67	0.088

F_m = volume fractions of monomers in their mixture with toluene;
K_{sw} = specific water regain, mL/meq;
F = exchange capacity for the organic ion, percent of Na^+ capacity;
$\bar{D}$ = interdiffusion coefficient, $cm^2\ s^{-1}$.
Source: After [210].

structures. Naturally, at higher DVB contents of the starting comonomer mixtures, larger amounts of toluene are needed in order to obtain resins with equal swelling capacities. Whereas the increase in the degree of crosslinking reduces the network's mesh size, dilution with toluene reduces the degree of chain entanglements in the final copolymer. The two factors, affecting the permeability of the matrix in opposite directions, actually change the ratio between physical entanglements and chemical crosslinks. The above series of sulfonated cation exchangers was tested in their ability to tetraalkylammonium ion sorption. The rate of diffusion of the ions dropped by nearly two orders of magnitude with increasing the ion diameter from ~4.5 to 12.2 Å. Interestingly, the diffusion rate of the relatively small tetraethylammonium ions (diameter: 7.8 Å) was found to be identical in these three ion exchanger, while the sorption of tetrabutylammonium ions (diameter: 12.2 Å) decreased in materials with gradually decreasing mesh size.

Generally, at the beginning, the diffusion of ions predominantly proceeds into network domains with reduced crosslinking density, whereas the highly crosslinked domains are less accessible and mostly influence the total exchange capacity at final equilibrium. The latter drops with increase in both the crosslinking density and the size of ions.

Addition to the initial comonomer mixture of telogens, such as carbon tetrachloride or diethylbenzene, which reduce the average polymer chain length via chain transfer reactions, enhances swelling and permeability of the ion-exchange resins at any given DVB content [80, 88, 211]. Even at equal swelling in water, the diffusion rates of tetrabutyl- and trimethylcetylammonium ions or tetracyclin into telogenated resins were found to be higher than the diffusion rates in conventional exchangers [81, 82, 212].

It is also useful to mention that the mechanical stability of sulfonated resins with the matrix prepared in the presence of diethylbenzene is markedly better [82, 190].

Strong basic and strong acidic ion-exchange resins are known to swell only with polar organic solvents, such as methanol, ethanol, or dioxane, which restricts their application in catalytic processes carried out in less-polar media. For this reason an attempt has been made to render the standard ion exchanger swellable also in nonpolar media, by subjecting the styrene–DVB copolymers to acylation with C_3-C_{12} aliphatic fatty acids, followed by the common procedures of functionalization [213–219]. Such type of ion exchanger is referred to as oleophilic resins. Although the oleophilic exchangers swell in nonpolar organic solvents, the large amount of ballast aliphatic chains significantly reduces their ion exchange capacity and considerably hinders the diffusion of organic ions or molecules into the beads.

Networks obtained by anionic block copolymerization were found to be unsuitable for the preparation of ion-exchange resins. Although the model networks can be easily sulfonated [220], they find no any practical application because of extremely high swelling with water. Popov et al. [221] managed to reduce the volume swelling of sulfonated resins to 3–7 (v/v) by end linking the polystyrene precursor of very low molecular weight, 3,000–7,000 Da. In order to retard the total gelation and allow better mixing of the reaction mixture with DVB, one diisopropenylbenzene (DIPB) molecule per living end had to be previously added. Various publications [113, 222–224] report good osmotic stability and kinetic properties of the sulfonated exchangers obtained in this way, but this method of synthesis is too complicated for any practical use.

CHAPTER 2

Interpenetrating Polystyrene Networks

1. INTERPENETRATING STYRENE–DIVINYLBENZENE NETWORKS

Interpenetrating networks (IPNs) composed of two or several chemically independent crosslinked polymers present an important family of polymeric materials. Over nearly half a century of their existence, IPNs have been studied in detail and currently used in practice as shockproof plastics, adhesives, soft contact lenses, matrices for ion-exchange resins, tooth fillings, and so on. As a rule, IPNs have been obtained on the basis of polymers with limited thermodynamic compatibility. Styrene–divinylbenzene (DVB) copolymers are often used as one of the IPN components, while polyurethanes, polyacrylates, epoxy resins, butadiene–styrene rubbers, castor oil, and so on, represent the other constituent network. Different thermodynamic compatibility of two polymers composing the IPN materials and different ways of their synthesis permit one to obtain a wide variety of homogeneous and heterogeneous copolymers having interesting physicochemical properties for practical applications.

A comprehensive review of the synthesis and properties of IPNs is beyond the scope of this book, and in this section we will deal only with the properties of IPN networks formed by styrene–DVB copolymers. These IPNs were named "Millar IPNs" after the author who was the first to prepare this novel type of networks as early as in 1960 [225]. These materials are also named "homo-IPNs," either keeping in mind the fact that they are formed by chemically identical polymers or sometimes assuming a homogeneous character of their physical structure.

If one allows a preformed copolymer of styrene with, for example, 2% of DVB, to swell with a styrene–2% DVB monomer mixture containing a polymerization initiator, removes the excess liquid, and then polymerizes the retained monomers inside the swollen primary particles, the final product will be composed of two chemically identical but

Comprehensive Analytical Chemistry, Volume 56
ISSN 0166-526X, DOI 10.1016/S0166-526X(10)56002-2

independent networks. In this example each network is crosslinked with 2% DVB and, therefore, the crosslinking density of the 2 × 2 IPN (degree of crosslinking of the first and second networks, respectively) is also 2%. Nevertheless, the properties of the individual constituent networks and the IPN are markedly different. Millar [225] has noted that, as the second and then third networks are introduced, the density of the resulting product slightly increases, whereas its swelling ability noticeably decreases in comparison with those of the simple network (Figs. 2.1 and 2.2). Considering IPNs as a homogeneous structure, Millar explained these findings by the effect of internetwork chain entanglements acting as additional crosslinks. He emphasized, however, that there is no simple quantitative correlation between the swelling of IPNs and that of the constituent individual networks. Reasoning from the observation that 2 × 10 or 4 × 10 IPNs take up smaller amounts of toluene than 2 × 2 and 4 × 4 networks (Table 2.1), he concluded that the solvent regain of IPNs is governed by the lower swelling ability of the network with the highest crosslinking density. The less crosslinked network proves to be only partially extended after swelling and so the whole system exhibits a complex behavior. Millar also concluded that network chain entanglements are partially reversible, because the

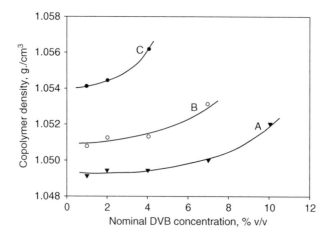

Figure 2.1 Copolymer density at 25°C: (A) primary copolymer, (B) secondary intermeshed copolymers (DVB content $n = n_1$), (C) ternary intermeshed copolymers ($n = n_1 = n_2$). After [225].

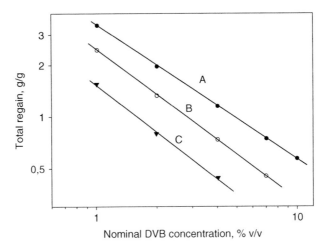

Figure 2.2 Copolymer toluene regains: (A) primary copolymer, (B) secondary intermeshed copolymers $(n = n_1)$, (C) ternary intermeshed copolymers $(n = n_1 = n_2)$. After [225].

apparent DVB content calculated from the swelling of sulfonated IPNs in water is always smaller than that calculated from the toluene regain of the starting IPN. He explained this discrepancy by a suggestion that the considerable swelling pressure, arising from swelling of the sulfonate in water, causes essentially larger extension of the IPN compared to weaker van der Waals interactions of toluene with the initial hydrocarbon matrix (Table 2.7).

Shibayama and Suzuki [226] also assumed that internetwork chain entanglements play an important role in the behavior of IPNs. According to their opinion, it is the entanglements that result in increasing rubber modulus of IPNs in comparison with that of the primary network; for this reason, the crosslinking density of IPNs calculated from the viscoelastic modulus is almost twice as large as the crosslinking density calculated from the total monomer composition. They also believe that the entanglements noticeably reduce the swelling of IPNs.

Interestingly, Thiele and Cohen [227] arrived at the opposite conclusion. They state that the physical chain entanglements in IPNs do not affect their swelling at all. Using the approach developed by Hermans for the study of swelling properties of a network originally polymerized in the presence of a diluent, and assuming that the first and second networks are

Table 2.1 Swelling and apparent DVB content for various IPNs. *(Reprinted from Millar [225] with permission of Royal Society of Chemistry.)*

Hydrocarbon copolymer			Sulfonated copolymer				Apparent DVB content, % v/v	
Nominal DVB (n, n_1, n_2), % v/v			TR, g/g	WR, g/g	EC, meq/g	WR_{sp}, g/meq	From, WR_{sp}	From, TR
1	–	–	3.38	10.5	5.66	1.85	1.0	1.0
1	1	–	2.44	5.24	5.27	0.995	1.3	1.5
1	1	1	1.57	3.64	5.25	0.694	1.7	2.7
2	–	–	1.93	2.67	5.00	0.534	2.1	2.1
2	1	–	1.30	2.13	5.29	0.403	3.2	3.5
2	2	–	1.33	2.19	5.22	0.419	3.0	3.3
2	10	–	0.76	1.15	4.96	0.232	6.0	6.9
2	2	2	0.79	1.75	5.31	0.329	4.1	6.5
4	–	–	1.17	1.74	5.17	0.337	4.0	4.0
4	1	–	0.68	1.15	5.08	0.216	6.6	8.0
4	4	–	0.73	1.19	5.06	0.235	6.0	7.2
4	10	–	0.66	0.91	4.24	0.215	6.6	8.2
4	4	4	0.44	0.94	5.17	0.182	7.3	14.0
7	–	–	0.74	0.91	5.01	0.182	7.3	7.3
7	1	–	0.44	0.65	4.99	0.130	10.7	14.0
7	7	–	0.44	0.66	5.00	0.132	10.2	14.0

1% v/v = 1.01%; w/w = 0.82 mol% n, n_1, n_2 = DVB content in primary, second, and third networks, respectively; DVB = divinylbenzene; EC = exchange capacity; TR = toluene regain; WR = water regain.

elastically independent, they derived an equation describing the swelling of IPNs as follows:

$$
\mu_1 - \mu_1^0 = RT\left[\ln\{1 - (v_2 + v_3)\} + (v_2 + v_3) + \chi(v_2 + v_3)^2 \right.
$$

$$
+\frac{\bar{V}_1}{\bar{v}_2 M_{c(2)}}\left(1 - \frac{2M_{c(2)}}{M}\right)\left(v_2^{1/3} - \frac{v_2}{2}\right)
$$

$$
\left. +\frac{\bar{V}_1}{\bar{v}_3 M_{c(3)}}\left(1 - \frac{2M_{c(3)}}{M}\right)\left(v_3^{0\,2/3}\, v_3^{1/3} - \frac{v_3}{2}\right)\right] \qquad [2.1]
$$

Here, subscript 1 refers to the solvent, 2 to the primary network, and 3 to the secondary network; on the left-hand side of the equation superscript 0 refers to the pure solvent, while on the right-hand side it refers to unswollen IPNs, μ is the chemical potential, v the volume fraction, χ

the polymer-solvent interaction parameter, $\bar{V}_1$, $\bar{v}_2$, $\bar{v}_3$ the molar volumes (of a solvent or a monomer repeating unit), M_c the molecular weight of the polymeric chain between crosslinks, and M the number average molecular weight of an uncrosslinked polymer prepared under identical conditions.

In order to predict equilibrium swelling of IPNs, the $M_{c(2)}$ and $M_{c(3)}$ values first need to be calculated from the swelling of simple styrene–DVB networks, according to the well-known Flory–Rener equation [138], and then inserted into Eq. [2.6]. Comparison of the thus expected swelling with experimental results showed that Eq. [2.6] predicts the swelling behavior of IPNs rather well, provided that none of the constituent networks is densely crosslinked. Since Eq. [2.6] means the complete independence of the networks involved in IPN formation and relates the swelling of IPNs only to the degree of crosslinking and the volume fraction of each component, it was concluded that physical chain entanglements do not influence the swelling of IPNs. However, even moderately crosslinked IPNs were found to swell less than it was expected. The authors, not being logical in their argumentations, explained this finding by the appearance of additional physical crosslinks in the IPNs with enlarged DVB content, due to some trapped entanglements. Also, they did not rule out a partial grafting of the second network to the primary one.

Sperling and coworkers supported the viewpoint of Thiele and Cohen. Furthermore, they came out with the suggestion [228] that the behavior of IPNs is largely governed by the first network and found the corroboration of this idea in the results of the following experiments. A set of 0.22×2.2 Millar IPNs was synthesized in such a manner that the volume fractions of both networks were varied from 75:25 to 25:75, and for all the IPNs, Youngs modulus, measured in the experiment, was compared with the modulus, E, calculated according to Eq. [2.2] (see ref. [229]) as follows:

$$E = 3\left(v_1^{1/3}\zeta_1 + v_2\zeta_2\right)RT \qquad [2.2]$$

where ζ_1 and ζ_2 are the number of moles of polymeric chains comprising the first and the second networks in $1\ cm^3$, v_1 and v_2 are volume fractions of each of the networks, and R and T correspond to the gas constant and the absolute temperature. It was found that the modulus of real IPNs does not depend on the volume fraction of the second network, although theoretical Eq. [2.2] predicts an increase of E with this fraction increasing.

This provided a basis for the conclusion on the predominant role of the first network.

If the primary network dominates the mechanical properties (as well as the swelling behavior) of IPNs, the second network should be composed of more or less separated fragments. Indeed, electron microscopy of 0.4×4.0 IPNs showed that the second network forms domains of 50–100 Å in diameter (depending on the portion of the second network) (see Fig. 2.3). At first sight, this result is rather surprising because these IPNs are formed by the copolymers that should be completely compatible. Siegfried and coworkers [230] believe that when the first network swells with the monomer mixture of the second network, the liquid predominantly locates in small regions of space with accidentally lower crosslinking density, because in these regions the elastic compressive pressure should be markedly lower.

Certainly, the regions with lower crosslinking density contain larger proportions of monomers, but undoubtedly the liquid styrene–DVB mixture forms a continuous phase inside the primary swollen network. The segregation of the second network must take place during copolymerization. Strictly speaking, the copolymers with 0.4 and 4% DVB are not chemically identical: they contain different numbers of pendent double

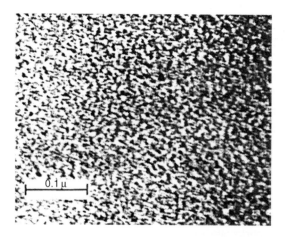

Figure 2.3 Morphology of $0.4 \times 4\%$ DVB Millar's IPN's. The ratio of two networks is 75 : 25 by weight. The second network contains 1% of isoprene fragments to contrast the second phase by exposure to osmium tetraoxide vapors. Light fragments are the first network. Reprinted from [230] with kind permission of Wiley & Sons, Inc.

bonds and a larger amount of impurities (such as ethylstyrene) introduced with a larger quantity of technical-grade DVB. We can see here a definite analogy, with the phase separation occurring during styrene-DVB copolymerization in the presence of linear polystyrene, which will be treated in more detail in Chapter 3, Section 1.5. (At the same time, one cannot but note that 1% isoprene added to the monomer mixture for the second network to intensify the contrast on the micrographs may additionally stimulate phase segregation.) Possibly, even in the case of IPNs composed of two chemically identical final networks, for example, 2 × 2 IPNs, the tendency for microseparation of the second network should become apparent because of the local chemical nonidentity of the preformed primary network and growing second network. Nevertheless, the second network cannot form its phase completely independent of the primary network. Polymeric chains of both phases get unavoidably entangled and interlaced, at least partially, and this influences the swelling and mechanical properties of the IPNs. The above occasional coincidence of the theoretically expected swelling with the real solvent regain does not yet imply the absence of chain entanglements in IPNs.

What is more interesting here is to ask ourselves: Why do IPNs swell? The first conventional network is assumed to be fully expanded when swollen to the maximum limit in the monomer mixture for the second network and, therefore, strong internal stresses arise in the first network. These stresses get fixed in the final IPN and they always act against any further increase of the polymer volume, that is, swelling. It may seem that the matrix network should not expand additionally and so the IPNs would not have to swell at all. However, this is actually not the case. Strong tendency for solvation of the fragments of the second phase, which was formed in an unstressed state and has an equally great affinity to the thermodynamically good solvent, represents the driving force causing additional expansion of the first already expanded network. Consequently, a conventional network, even swollen to a "maximum," will realize only a part of its potential capability to swell and expand. (In Chapter 4, we shall see other examples confirming this statement.) Therefore, a theoretical consideration of an IPN has to be based on the assumption that solvation of the second, "not-strained" network represents the major driving force for swelling, whereas the first, "pre-strained" network mainly sets limits to the volume increase. Without taking into account these opposite tendencies, no mathematical combination of the properties of the individual "unstrained" networks could ever adequately describe the behavior of IPN materials.

2. INTERPENETRATING ION-EXCHANGE RESINS

When talking about the properties of materials based on IPNs, we should take into account the fact that the second network forms a more or less discrete phase inside the primary continuous phase of the first network. This permits the consideration of styrene-DVB IPNs as rather heterogeneous materials, and although similar to gel-type homogeneous supports, they exhibit no real porosity.

Among other application examples, IPNs have found widespread use as matrices for ion-exchange resins. Millar et al. [231] compared the permeability of the industrial gel-type styrene–DVB sulfonate Zeo-Carb 225 with that of a laboratory sample of sulfonated styrene–DVB 4.5 × 4.5 IPN. In spite of the same swelling in water (the apparent DVB content was 7% in both cases), the interdiffusion coefficients of $(C_2H_5)_4N^+-H^+$ exchange were found to be higher for IPNs and equal to that of a gel-type cation exchanger with 6% DVB. Undoubtedly, the slightly enhanced permeability of the IPN sample is caused by the higher flexibility of its network having larger linear polymer segments between the network joints. This effect, however, is no more perceptible on exchanging kinetics of smaller proton–metal ions; here, both resins exhibit the same permeability. On the other hand, the sulfonate 4.5 × 4.5 displays a higher selectivity of Na^+-H^+ exchange, which is similar, in this respect, to that of a standard resin with 15% DVB. Moreover, the selectivity of $(C_2H_5)_4N^+-H^+$ ion exchange for the 4.5 × 4.5 resin is higher than that of both Zeo-Carb 225 and a standard ion exchanger with 15% DVB.

CHAPTER 3

Macroporous (Heterogeneous) Polystyrene Networks

1. MACROPOROUS STYRENE–DIVINYLBENZENE COPOLYMERS

As far back as in the late 1940s to the early 1950s, it was noticed that if the suspension copolymerization of mono- and divinyl monomers is carried out in the presence of an inert solvent, it will yield beads that are opaque in appearance and much more resistant to osmotic shock compared with the known gel-type copolymers. It soon became evident that the improved physical properties of the new resins are caused by their special internal structure, which, in its turn, results from the phase separation of the initially homogeneous comonomer–diluent solution. This finding opened the door to a new generation of polymeric adsorbents, the so-called macroporous resins that exhibit stable porosity in both the dry and the solvated states.

Since the pioneering reports of the early 1960s (the first scientific publication on porous styrene–divinylbenzene (DVB) ion-exchange resins appeared in 1962 [232]), various methods for the preparation of macroporous resins have been extensively pursued in a number of countries. By today, the main principles of porosity formation have been established and the influence of synthesis conditions on the parameters of porous polymer structure investigated in detail. This information had been summarized in a number of reviews (see [74, 232–237]); still it is important to discuss here the main principles of porosity formation in styrene–DVB copolymers, particularly emphasizing the reasons for phase separation and presenting a critical analysis of the physicochemical methods commonly used for the determination of the parameters of the porous polymeric structure. Understanding the phase separation process is very important for the synthesis of adsorbent resins with predetermined porous structure, and taking into account all possible pitfalls in the interpretation of data provided by the various physicochemical measurements can give a feeling of the reliability of the estimated porosity parameters.

Comprehensive Analytical Chemistry, Volume 56
ISSN 0166-526X, DOI 10.1016/S0166-526X(10)56003-4

1.1 Determination of porous structure parameters

Precipitation of the growing polymer from the initial solution of styrene and DVB in an inert diluent during crosslinking copolymerization results in the formation of a two-phase heterogeneous network, in which one phase is presented by the highly crosslinked and rigid polymer, while the rejected diluent forms another phase. After removing the diluent, permanent voids remain in the copolymer beads. The total pore volume, W_o, and the inner surface area, S, are the major characteristics of the porous structure; these are intimately related to pore size and pore size distribution. These parameters determine the practical application fields of the polymeric adsorbent resins; therefore, a precise quantitative characterization of resin porosity becomes an important task.

According to the IUPAC recommendations [238], the term "micropores" should be applied to voids smaller than $20\,\text{Å}$ in diameter, "mesopores" are those with diameters between 20 and $500\,\text{Å}$, and "macropores" have diameters larger than $500\,\text{Å}$. The micropores and thin mesopores mostly contribute to the value of the inner surface area, whereas the wide mesopores and, in particular, macropores make up most of the total pore volume of porous materials. It should be noted that the overwhelming majority of porous styrene–DVB copolymers are typically mesoporous materials; however, the initially introduced name "macroporous polymers" became generally accepted and we will not deviate from this traditional scientific jargon.

A whole series of methods has been developed for the investigation of porous solids such as activated carbons, porous glasses, silica gels, and zeolites. Together with some new suggestions, they all have been applied to the characterization of the porosity of polymeric adsorbents. We will briefly review these methods related to the porosity parameters and emphasize the problems that may arise on their application to polymers, when the specific properties of the polymeric materials are not taken into account.

Specific inner surface area: The most widespread method to determine the inner surface area is measuring the adsorption isotherms of inert gases, largely nitrogen or argon, at low temperatures [239, 240]. Sometimes adsorption of vapors of organic solvents, for example, *n*-perfluorooctane, at room temperature [241] is employed. Calculation of the surface area from adsorption isotherms involves the theory of polymolecular adsorption suggested by Brunauer, Emmett and Teller (BET) [242] or sometimes the procedure of Barrett, Joyner and Halenda (BJH) [243]. The BET method provides a reliable value of inner surface area of a porous solid only if the adsorption isotherm has an S-shaped form, in other words belonging to the IInd or IVth types (Fig. 3.1) in accordance with the classification of Brunauer et al. (BDDT) [244]. In the range of low relative pressures, usually up to 0.35, the adsorption isotherm must produce a straight line in the coordinates of the semiempirical BET equation:

$$\frac{p/p_s}{a\left(1 - p/p_s\right)} = \frac{1}{a_m C} + \frac{C - 1}{a_m C} \times \frac{p}{p_s} \qquad [3.1]$$

where p and p_s are the current pressure and the pressure of saturated vapor at the temperature of measurement, respectively; a and a_m are the current adsorption and the capacity of the monolayer, respectively; and C is a constant related to the adsorption heat and the temperature. Additional essential conditions for the applicability of the BET method are that the B-point (the point corresponding to the completion of a monolayer and the onset of straight line part of the isotherm (Fig. 3.1)) should fit the BET straight line dependence and the C constant should be neither very small nor very high, somewhat between 50 and 150 [240]. If the constant is much higher, one cannot rule out that the investigated sorbent contains a noticeable portion of micropores; these fill up with sorbate molecules (condensate) even at very low relative pressures, and therefore, the inner surface area proves to be overestimated. On the other hand, a too low value of constant C implies that multiple adsorption layer formation starts before the completion of the monolayer and the surface area may be underestimated.

   Sometimes mercury porosimetry and adsorption of p-nitrophenol from an aqueous solution [245, 246] are also used to determine the surface area. The disadvantages of mercury intrusion technique will be discussed below. Regarding the method of adsorption from solution, its drawback is associated, first of all, with the uncertainty in the determination of the cross-sectional area of the p-nitrophenol molecules, in which the benzene rings may adsorb either transversely or parallel to the surface. Also some sorbents swell in water, and this may lead to a misrepresentation of the surface area of the adsorbing resin in dry state.

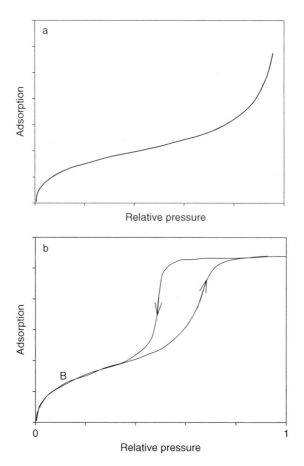

Figure 3.1   Schematic representation of adsorption isotherms of (a) II and (b) IV types according to BDDT classification.

Pore volume: The pore volume is frequently also determined from the gas adsorption isotherm. The value of low-temperature nitrogen adsorption at the relative pressure $p/p_s \to 1$ is generally accepted to be equal to the pore volume of the sorbent. However, this is valid only for type IV isotherms (Fig. 3.1.b), which clearly level off in the vicinity of $p/p_s = 1$. Leveling of the isotherm indicates that all pores in the sorbent are already filled with the liquid sorbate. If the isotherm has a different shape in this ultimate range of relative pressures and adsorption continues to rise at $p/p_s \to 1$ as in Fig. 3.1.a, the estimation of the pore volume becomes inconclusive because capillary condensation in the largest pores is not yet complete. The same uncertainty exists in the estimation of the pore volume when the upper range of isotherm measurements is limited by a relative pressure p/p_s value as low as 0.95–0.98, while the unaccounted large pores could significantly contribute to the total pore volume.

Another method to determine pore volume, which is generally believed to be sufficiently precise for macroporous materials, is based on measuring the true (ρ_{tr}) and apparent (ρ_{app}) densities of the material and calculating W_o in accordance with the following simple equation [247]:

$$W_o = \frac{1}{\rho_{app}} - \frac{1}{\rho_{tr}} \qquad [3.2]$$

(Some important critical remarks to the very definitions of these terms, as well as to their experimental measurements, will be given in Chapter 7, Section 5.1.) The true density of macroporous polymers is usually estimated by helium or nitrogen densitometry, while the apparent density can be determined by measuring the diameters of a sufficient number of spherical beads and weighing them.

Mercury intrusion porosimetry may also yield sufficiently realistic values for both the apparent density and the total pore volume of macroporous polymers. However, it should be taken into account that mercury does not wet glass walls and polymer beads; therefore, small air bubbles may remain in the interstitial space and/or at the walls of the densitometer, thus mimicking a reduced apparent density and providing an overestimated total volume of porous space. The danger of overestimating the pore volume by mercury porosimetry is particularly high in the case of small beads or irregular particles. This is because the wedge-shaped space near the contact points between polymer particles cannot be correctly distinguished from large pores, since the distance between the two surfaces gradually approaches zero at the contact point.

If the polymer does not swell with certain solvents (such as methanol or octane in combination with macroporous styrene–DVB copolymers), the pore volume may be estimated from the uptake of the solvent by simply weighing the dry and wet polymer samples [248].

As can be expected from Eq. [3.2], the total pore volume was experimentally found to be linearly proportional to the (apparent) specific volume

$(V_a, \mathrm{cm}^3/\mathrm{g})$ of dry beaded porous material, even having pretty wide bead size distribution [249]:

$$W_o = aV_a - b \qquad\qquad [3.3]$$

For a large number of macroporous styrene–DVB copolymers, sulfonated cation exchangers KU-23, and weak basic anion exchangers AN-221, the specific volumes and the true and apparent densities were measured independently and the values of W_o and V_a were correlated in accordance with Eq. [3.3]. The coefficients a and b of the above equation, obtained after statistical treatment of all the results [250], are given in Table 3.1.

Pore size and pore size distribution. Particularly two methods have been used to determine the pore size in adsorbing materials of both organic and inorganic nature, namely, the gas adsorption technique and the mercury intrusion porosimetry. On the basis of information provided by these methods, a number of serious conclusions have been drawn on the porous structure of macroporous styrene–DVB copolymers. This necessitates a more critical analysis of the possible errors in the interpretation of the results of measuring adsorption isotherms and mercury intrusion.

All estimations of the pore size of a material from its *gas adsorption isotherm* measurements are based on the well-known Kelvin equation suggested more than 100 years ago. It considers an equilibrium between the vapor phase and the bulk liquid at a constant temperature and relates the relative vapor pressure p/p_s to the radius r_m of the convex (plus) or concave (minus) spherical meniscus of the liquid placed in a capillary:

$$\ln\left(\frac{p}{p_s}\right) = \pm\frac{2\sigma V_m}{r_m RT} \qquad\qquad [3.4]$$

Table 3.1 Coefficients a and b and correlation coefficient, r_c, for the Eq. [3.3]

Coefficients	Styrene–DVB copolymer	KU-23	AN-221
a	0.62	0.55	0.51
b	0.88	0.60	0.77
r_c	0.92	0.96	0.94

Source: After [250]

where σ and V_m are the gas–liquid surface tension and molar volume of the liquid, respectively ($\sigma = 8.72$ mN/m and $V_m = 34.68$ cm^3/mol for nitrogen at 77.4 K [240]), and R and T have their usual meanings. It is believed that vapors of any adsorbate may form a condensed liquid in the pores at a certain relative pressure. The filling starts in the micropores and the finest mesopores and then continues in the larger pores where the radius of the meniscus is larger.

Capillary condensation is indicated by the fact that the desorption branch of the isotherm deviates from the adsorption branch, thus forming a hysteresis loop. However, hysteresis is not an essential attribute of capillary condensation. Let us assume that all pores have a form of closed cylindrical capillaries, similar to a test tube. When condensation starts during the sorption experiment, a spherical meniscus will form at the bottom of the pore and move along the capillary toward its opening, having a radius r_m equal to that of the capillary. During desorption, the condensate will evaporate with the same meniscus moving toward the bottom of the pore. Since the radius of the meniscus during the sorption and desorption processes remains constant, the sorption and desorption isotherms will not deviate from each other. Hysteresis is also absent at condensation in conical pores with their large opening directed toward the surface, as well as in the case of nonparallel slits.

On the other hand, a regular cylindrical pore, provided both its ends are open, would generate a hysteresis loop. Such a pore does not have a spherical bottom of the curvature $1/r_m$ that would initiate condensation at the p/p_s value determined by the Kelvin equation. Indeed, curvature of an open cylindrical through-pore is smaller by a factor of 2 than that of the bottom of a test tube-like pore, since it is $1/r_m$ at right angle to the pore axis but $1/r_\infty$ parallel to the axis, that is $1/2(1/r_m + 1/r_\infty) = 0.5/r_m$. Therefore, condensation in an open cylindrical capillary will start at a higher p/p_s value, corresponding to a smaller curvature of $0.5/r_m$. The capillary will get spontaneously filled at that higher pressure. However, the filled cylindrical capillary will then expose a single spherical meniscus with the constant sharp curvature of $1/r_m$ during the whole desorption process. Desorption from that filled cylinder will start and proceed at a smaller pressure (corresponding to the curvature of $1/r_m$), thus giving rise to a distinct hysteresis loop. Still another situation will characterize the sorption–desorption relations in slit-type pores with parallel walls, which also generate hysteresis. These simple examples demonstrate that the form and size of the hysteresis loop strongly depend on the form of the pores. Pore geometry is largely unknown for the majority of polymeric adsorbents. One can only be certain that most of them are interconnected and have openings on both ends. For this reason, it is believed that mesoporous polymers must exhibit hysteresis loops.

Many strategies have been developed for processing the adsorption data. They try to account for several additional phenomena that accompany capillary condensation. Most important of them is the fact that filling of pores with the condensate is preceded by the formation of a film of one or several adsorption layers on the walls of (open) pores. On the one hand, it justifies the assumptions that the wetting angle of the film is equal to zero regardless of the nature of the support, and, therefore, the meniscus formed during the condensation in a cylindrical capillary has spherical shape. On the other hand, it dictates involvement of the film thickness into the calculations, since, now, the pore radius is a sum of the film thickness and the radius of the liquid "core" (or its meniscus). The film thickness increases with the rise in relative pressure. At a given relative pressure, it depends on the local curvature of the surface and on the adsorption power (i.e., chemistry) of the solid surface. The latter factor is not considered, though it may change not only the thickness but also the density of the adsorbed film. Thus, in the first adsorption layer, a nitrogen molecule would occupy only $0.13\,nm^2$ on the regular surface of corundum that displays a high adsorption potential, but as much as $0.28\,nm^2$ on Teflon, the surface of which exposes less or less-active adsorption sites [251]. For carbon-type materials, silica, and macroporous styrene–DVB copolymers, the cross-sectional area of a nitrogen molecule is usually assumed to be $0.162\,nm^2$ and that of an argon atom $0.176\,nm^2$. Actually, the real sizes of these adsorbing species are smaller: an ellipsoid with the axes of 0.41×0.3 nm for N_2 and a sphere with a diameter of 0.21 nm for Ar. The argon atom would thus require a spot smaller than $0.044\,nm^2$. However, as shown above, the adsorbate species occupy on the surface much more space than needed, since the adsorption monolayer is rather loose, with densities depending on the chemistry of the solid support.

It is quite impossible to account correctly for all these factors. The uncertainty introduced by different modes of calculating the data of the same adsorption isotherm can be illustrated by the following example. For a porous aluminosilicate, classical procession of the nitrogen adsorption data results in a pore–size distribution curve with a sharp maximum at a pore radius of 1.0 nm. It should be considered that the strength of the surface adsorption field will shift this maximum to 1.5 nm, while also considering the form and curvature of the pores will position the maximum at 2.7 nm [251].

Another interesting and important observation is that for mesoporous adsorbents, each adsorbate is characterized by a specific minimal relative pressure where the hysteresis loop closes in the adsorption isotherms, irrespective of the nature of the material. Thus, for nitrogen adsorption isotherms, hysteresis is not observed below $p/p_s = 0.44 \pm 0.02$ (see Fig. 3.1b). This relative pressure value corresponds to condensation–evaporation equilibrium in pores as small as 1.6–1.7 nm in diameter. Obviously, filling and evaporation of liquid nitrogen from micropores smaller than this value proceed without hysteresis. Considering that all the artifacts particularly increase the uncertainty of the evaluation of the smallest pores, one usually stops calculations at this boundary of the p/p_s axis. Therefore, it is assumed that nitrogen adsorption isotherms may be useful for the evaluation of pores in the range from 1.5–2.0 to ~50 nm in diameter only, although other sorbates may slightly shift the applicability range of the calculations according to the Kelvin equation.

In spite of all problems and the conditional character of calculations, evaluating the porous structure of materials from gas adsorption isotherms remains the most widely accepted technique. When applied to a series of similar materials under identical experimental conditions and modes of calculation, the method has reproducibility better than 10% and can provide relative data that are useful for making predictive conclusions within the examined series.

With respect to the absolute values of porosity parameters, the data from adsorption isotherms are much less reliable. When considering these data, one has to remember that they actually characterize an equivalent model system composed of an ensemble of open-ended, independent cylindrical capillaries of constant width. This model is far from the real structure of a polymeric adsorbent. Another serious drawback is the rather arbitrary choice between the adsorption and the desorption branches of the hysteresis loop for the calculations. If, indeed, open-ended channels are anticipated in the material, the desorption branch should give more representative results. On the other hand, if closed ink-bottle-type pores are present, the adsorption branch could be used. Filling of a bottle-type pore starts at a low p/p_s value corresponding to the diameter of the neck and ends at a higher relative pressure corresponding to the size of the bottle's interior, whereas evaporation proceeds at a single p/p_s value determined by the meniscus in the neck. Partially for this reason, the adsorption branch is generally known to provide a much wider pore–size distribution curve, compared with that provided by the corresponding desorption branch.

A recently suggested simple method to evaluate the desorption isotherm of a solvent, without measuring the vapor pressure of the latter [252], is based on the precise measurement of the weight loss of a solvent-filled (pre-swollen) sorbent sample on drying under quasi-equilibrium conditions. In this case, the rate of solvent evaporation from the sample container through a small channel is considered to be proportional to the equilibrium vapor pressure in the container, which permits the transformation of the evapora-tion rate curve into a desorption isotherm and the calculation of pore size distribution in accordance with the Kelvin equation. However, this method cannot constitute an independent alternative to the gas adsorption technique. Although the weight measurements are simple and precise, all the above-discussed inherent shortcomings of using the Kelvin equation to calculate the pore size distribution also characterize this technique. A similar method of "reference drying" was also reported [253, 254], which com-pares the drying rate of a solvent-impregnated sorbent with that of a reference porous material under identical conditions. The pore size dis-tribution of the reference material is assumed to be known, for example, established by gas adsorption or mercury intrusion porosimetry. Subsequent simple calculations permit one to obtain integral or differential curves of pore volume distribution versus pore size of the sample, but the trust-worthiness of such calculation totally depends on the reliability of the data for the calibrating reference sample.

In conclusion, it should be also said that the origin of the hysteresis loop of the adsorption–desorption isotherms of porous polymers is still debated and can be interpreted in different ways. For example, there exists an opinion that hysteresis is not related to traditional capillary condensation in the pores, but may be a consequence of the out-of-equilibrium character of phase transitions in real disordered mesoporous polymers [255]. A failure to reach equilibrium under the given experimental conditions may be caused by the slow diffusion rate of the sorbate [256] or slow swelling of the polymeric sorbent on adsorption and slow relaxation of its swollen structure on desorption. Quite often, a subsequent adsorption on the same material results in larger adsorption capacity values. It is the so-called conditioning effect [256] that may imply a nonequilibrium character of the process. Even the reproducibility of the shape and location of a hysteresis loop of the isotherms may indicate the establishment of fast local equilibriums, while equilibration on a global, macroscopic scale may require much more time. This is generally the case with glassy polymers below their glass transition temperature. Here, the determination of pore size distribution in terms of classical capillary condensation may be rather meaningless and scientifically incorrect, though, still, being widely used in practice.

Mercury porosimetry is another popular method to determine pore size distribution. It capitalizes on the relationship between the pore size and the pressure needed to force mercury into the pore.

$$r = 2\sigma \cos\theta \ P^{-1} \qquad\qquad [3.5]$$

The surface tension σ is considered as 484 mN/m, the value for pure mercury surface, in spite of the likelihood of its contamination by hydraulic liquid. Because of the high surface tension, mercury does not wet the material and it is assumed that it forms a convex meniscus with a contact angle of 141° with the material surface. Actually, this value has been measured for silica, although even for this material, contact angles from 125° to 152° can be found in the literature. These variations alone can introduce large uncertainty in the final results. Thus, variations of the pore diameter from 1.79 μm (125°) to 2.65 μm (152°) for gigapores and from 7.5 nm (125°) to 11.9 nm (152°) for mesopores of a silica material have been reported [257], just because the real value of the material/mercury contact angle is unknown.

Despite the above-mentioned problems, mercury porosimetry is successfully used for the characterization of rigid macroporous materials, but it has also been applied to polymers having smaller mesopores [258, 259]. With relatively soft polymeric materials, there is always a danger that the porous sample may be susceptible to compression. When applying pressure to mercury with polymeric beads immersed in it, there is no way to distinguish between the intrusion of mercury into the pores at a certain pressure and the simple compression of the beads. With polymers we never know what we measure, and so an easy compression of the soft porous polymer may simulate intrusion of mercury and result in a dramatic overestimation of the pore size. Still another uncertainty arises again from the fact that the real geometry of the pore is totally ignored, while the relationship (Eq. [3.5]) is actually valid for cylindrical pores only.

Generally, mercury porosimetry is believed to give acceptable results for pores within a diameter range of 2.5–30,000 nm. The range from ~2.5 to

~50 nm thus appears to be amenable to both gas adsorption and mercury intrusion techniques, giving a possibility to compare results generated by these two most popular methods for pore characterization. It is not surprising that the agreement between the values obtained is often found to be poor.

Still another method to examine the porous structure of polymeric sorbents is *inversed size-exclusion chromatography* [260]. The material first has to be prepared in the form of small particles, even better in the beaded form, suitable to pack an efficient chromatographic column. One then measures the retention volumes of a series of solutes of known sizes in the column. Small solutes are able to explore all pores of the polymer filled with the mobile phase and they reside longer in the column. On the other hand, large solutes spend most of the time in the mobile phase moving in the interstitial space, visiting briefly only the largest pores, and they emerge rather early at the exit of the column. One prepares a size versus retention plot for a series of solutes and analyzes it in terms of pore size distribution of the material. The method can cover a very wide pore size range, starting from very thin micropores that are accessible to ions or to the small molecules of acetone or toluene, and up to those accessible to macromolecular coils of several thousand angstroms in diameter.

A crucial precondition of the method is that all attraction interactions between the sorbent surface and the test molecules be suppressed, which is usually met when both partners interact better with the solvent molecules than with each other. This immediately introduces a certain solvation shell around both partners, the thickness of which is difficult to estimate. Solvation can significantly alter size relations between small solutes and small pores. Its influence gradually becomes negligible on moving toward large pores and macromolecular solutes. Inhibiting adsorption phenomena by using good solvents has another consequence, namely, the polymer matrix of the sorbent swells, which generally affects the porous structure of the material. Thus, the method can easily provide information about dimensions of the space between chains in swollen gel-type polymers that are nonporous at all, when in dry state.

Besides the solvation and swelling phenomena, a serious inherent drawback of inversed size exclusion method is that the general correlation between pore size and size of molecules able to penetrate the pores during the chromatographic experiment is ambiguous. On the one hand, for an unhindered diffusion the pore diameter must be larger than that of the macromolecular coil; however, on the other hand, giving sufficient time, a macromolecule can diffuse into a smaller pore, since the polymeric coil is flexible, can acquire an elongated conformation, and even diffuse via "reptation"-type movements. Moreover, the results of the chromatographic porosimetry measurement are markedly dependent on the structural and mathematical models used for calculations, which is especially problematic for polymers containing small pores.

Thermoporosimetry is a term suggested for a technique providing insight into pore dimensions of ion-exchange resins swollen with water. It is based on calorimetric measurements of the depression of water freezing/melting temperature that, in its turn, depends on pore diameters [261–263]. However, there is no clear or generally accepted theory behind this phenomenon. Similarly, relationship has been found between the pore size of unmodified styrene–DVB copolymers and the freezing point depression of benzene, by using differential scanning calorimetry [262] and ^{1}H-NMR spectra of benzene [263]. A shift toward higher magnetic field of the ^{1}H-NMR signal of chloroform placed into "aromatic environment" of a styrene–DVB matrix was also analyzed in terms of crosslinking density and pore size of the matrix [264].

Finally, *electron microscopy* is sometimes employed to visualize the microstructure of porous materials. This is the only technique permitting direct observation of the pores and measurement of their dimensions, while all the other methods measure properties of a system consisting of the porous material and a test species (nitrogen gas, mercury, polymeric coils, water, chloroform, etc.) placed into the pores. Calculating the pore size from all such measurements requires formalization of pore geometry and developing a rather complex mathematical model of the system. However, microscopy also has some shortcomings. A two-dimensional cut is not really suitable for the quantitative characterization of an irregular three-dimensional structure. By preparing a microtome cut for transmission microscopy or a fresh chip for scanning electron microscopy, we definitely distort and tear the initial structure. Thus, we have to recognize that no methods exist that can provide unambiguous trustworthy information about the pore size distribution of adsorbing materials.

Naturally the list of methods discussed above is not complete. Other ways such as *X-ray scattering* or *positronium annihilation* may also be used. It is very important to understand that each method, besides showing certain advantages over other methods, has definite limitations, and only a combination of several methods, followed by a critical analysis of the whole body of results, can give us the most reliable estimate of the porous structure of a polymer. Such a comprehensive approach, however, is quite rare in occurrence [102, 257]. Nevertheless, abundant information accumulated for macroporous styrene–DVB copolymers by means of various methods permits us to reveal major regularities of porosity formation and its dependence on the conditions of synthesis.

1.2 Phase separation during the crosslinking copolymerization in the presence of diluents

Macroporosity of styrene–DVB copolymers arises from the presence of certain diluents in the initial homogeneous solution of the comonomers; these do not participate in free radical polymerization but phase-separate from the growing polymer. Three types of diluents have generally been recognized: (i) precipitating media for linear polystyrene, (ii) good solvents for polystyrene, and (iii) linear polystyrene or other polymers.

Generally, phase separation during crosslinking copolymerization of mono- and divinyl monomers starts in the vicinity of the gel point and can proceed in the form of either macrosyneresis when the liquid and swollen gel produce two continuous phases or microsyneresis when microdroplets of the separated liquid phase remain dispersed in the polymeric gel phase. Sometimes macro- and microsyneresis may proceed simultaneously. In both cases, phase separation is governed predominantly by thermodynamic factors, namely, by the inability of the growing network to incorporate all the diluent. Combining the general concepts of two theories, that of polymer solutions and rubber elasticity of networks [138], Dušek [265, 266] has attempted to formulate conditions describing the incipience of phase separation occurring on introducing crosslinks into a polymeric chain of infinite molecular weight dissolved in a diluent:

$$\ln\left(1 - v_3^o\right) + v_3^o + \chi_{13}\left(v_3^o\right)^2 + v^*\overline{V}_1 v_3^o/2 = 0 \qquad [3.6]$$

In this equation, $\left(1 - v_3^o\right)$ is the volume fraction of the diluent in the network phase at the moment of phase separation (or, which is the same, in the solution prior to phase separation), v_3^o the volume fraction of the network phase, v^* (mol/cm^3) the concentration of the network active chains in unit dry volume of the gel (which is proportional to the concentration of crosslinks), χ_{13} the Flory–Huggins parameter, $\overline{V}_1$ the partial molar volume of the diluent, and subscripts 1 and 3 relate to the diluent and polymer, respectively. Notably, Eq. [3.6] incorporates four independent parameters: the thermodynamic quality of the diluent χ_{13}, its partial molar volume $\overline{V}_1$, the dilution of the system $1 - v_3^o$, and the varying crosslinking density v^*.

Crosslinking polymerization of comonomers in a diluent differs from the above model in that the monomers themselves represent a good solvent for the emerging polymer, but their concentration, as well as that of the polymer and of the crosslinks, steadily changes with monomer conversion and remains unknown. These factors dramatically limit the practical usefulness of Eq. [3.6]. Still, it is worthwhile to analyze which of these four parameters are really critical for the phase separation to occur during crosslinking copolymerization.

In the case of copolymerization proceeding in a *polymer-precipitating liquid*, phase separation occurs when the thermodynamic parameter of interaction between the growing polymer and the reaction medium becomes larger than 0.5. Evidently, achieving this Θ-point with $\chi = 0.5$ happens soon in a system with a high proportion of the diluent to the comonomers. In such system, phase separation (which manifests itself by emerging turbidity of the reaction mixture) may even occur before total gelation. With a small amount of the non-solvent in the starting mixture, the Θ-point will be reached at very late stages of the process, only long after gel formation. In such a system, the portion of the separated liquid phase is small and the porosity of the final material is low. Thus, it is the ratio of volumes of the monomers and the non-solvent and, to a smaller degree, the precipitating power of the latter that determine the moment of phase separation and govern the porous structure of the product. Obviously, the crosslinking density of the network is of smaller importance.

It is worthwhile to make an additional point on the polymer–solvent interaction parameter χ_{13}. In reality it may be a function of the reaction temperature: thus, isooctane, being a strong precipitating media for polystyrene at room temperature, may represent a better solvent at 70–80°C. Consequently, the start of phase separation and the resulting porous structure may well depend on the reaction conditions [267]. In the case of acrylic-type copolymers, the influence of temperature on the Flory–Huggins parameter was found to be even more important [268].

If crosslinking copolymerization proceeds in the presence of a *thermodynamically good solvent*, phase separation starts when the swelling ability of the forming network drops below the amount of the solvent taken initially for the reaction. In this case, the comonomers/solvent volume ratio and the degree of crosslinking seem to play a decisive role in phase separation and porous structure formation. With relatively small amounts of the solvent and the crosslinking agent, one may expect the formation of a nonporous polymer that would be able to accommodate additional amounts of the solvent, if available. The resulting network will have the so-called expanded structure with reduced number of chain entanglements. At higher crosslinking densities and dilution, phase separation will take place rather early and the final material will exhibit macropores between domains of expanded polymeric phase. Thus, solvents and polymer-precipitating diluents provide different structures of the polymeric phase.

In accordance with these considerations it is logical to expect that it should be easier to induce phase separation and implement macroporosity using precipitating diluents. Indeed, solid lines in Fig. 2.3 demonstrate that larger DVB contents and diluent concentrations are needed to observe turbidity (a sign of phase separation) in polymerizing systems with toluene (a good solvent) as the diluent compared with systems with isooctane. Fig. 3.2 also illustrates the important fact that observation of the phase separation phenomenon does not necessarily imply the formation of a final porous product. Obviously, any porous structure has a tendency to eliminate voids when the diluent is removed from the voids during drying of the product. In order to obtain the material with a permanent porosity in

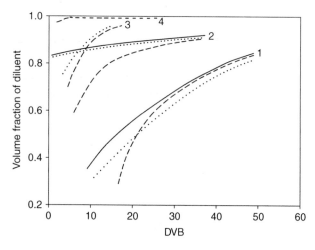

Figure 3.2   Volume fraction of diluent and DVB concentration (wt%) for the incipience of phase separation in copolymerization of styrene and DVB. Diluent: 1) toluene (a quasi binary system, $\chi = 0.46$); 2) isooctane χ(styrene–polymer) $= 0.46$, χ(isooctane–polymer) $= 1.40$, χ(styrene–isooctane) $= 0.40$; 3) polystyrene mol.wt. 50,000 χ(copolymer–polystyrene–porogen) $= 0$; 4) polydimethylsiloxane 50cP; —— development of turbidity, — — formation of network porosity in dry state, ... theory. Reprinted from [266] with kind permission of Plenum Press.

dry state, sufficiently larger amounts of the diluent and/or crosslinking agent are needed than predicted by the theory.

Regarding the role of the partial molar volume of the diluent, which is considered by Dušek as one of the important factors in phase separation (see Eq. [3.6]), we may note that there is no reliable experimental support for this suggestion. Any new solvent taken as diluent will mainly influence the situation through its Flory–Huggins parameter χ_{13}. Still, large molar volume has been suggested as a major factor for micro phase separation in systems where *linear polystyrene* is taken *as the porogen* (curve 3 in Fig. 3.2). Trying to describe these systems with Eq. [3.6], Dušek [266] considers polystyrene as the "best solvent" for the polystyrene network and sets $\chi_{13} = 0$ "because of complete similarity of their segments." However, in the capacity of a diluent, linear polystyrene should be more correctly ascribed as a Θ-solvent for polystyrene network and χ_{13} would have to be about 0.5, rather than 0. (According to the definition of a Θ-solvent, there is no difference in the interaction among the polymer segments themselves and the interaction between segments and solvent; $\chi_{13} = 0.5$ for the Θ-solvent.) Indeed, toluene, being similar in structure and chemical nature to polystyrene repeating units and being well known as one of the thermodynamically best solvents for the polymer, is characterized by $\chi_{13} = 0.46$. This value, one of the lowest known, is still much closer to 0.5 than to 0. Therefore, any statement that $\chi_{13} = 0$ for polystyrene as the diluent seems to be unjustified.

Polymeric porogens have been found to generate the formation of very large pores, up to several microns in diameter. The size of a single swollen polystyrene coil with molecular weight of 3×10^5 Da would amount only to 300–400 Å at infinite dilution in good solvents and would fail to make such big holes in the network. Wide macropores may be formed only by aggregates of polystyrene coils. On the other hand, aggregates may appear only as a result of an intense phase separation, which again contradicts the assumption of $\chi_{13} = 0$. However, an entirely different situation arises if we assume that $\chi_{13} \approx 0.5$: a system close to its Θ-point must be very sensitive and prone to phase separation.

We would like to point out that DVB exhibits higher reactivity and so the growing polymer chains must get enriched in DVB fragments. If they form crosslinks, they dramatically change conformation, vibration frequencies, and other properties of neighboring chain segments. If not involved in polymerization, the second vinyl groups of the DVB units remain pending, and they distinguish the chain from pure polystyrene. If technical-grade DVB is used, the forming polymer may also incorporate isomers of ethylstyrene. In any case, the growing styrene–DVB network is not pure polystyrene. As a rule, polymers of different chemical natures are fundamentally incompatible, except in highly diluted solutions. Hence, at certain threshold values of DVB concentration in the monomer mixture and of the molecular weight of the polystyrene porogen, an intensive phase separation will take place when the conversion of the monomers produces sufficient amounts of the polymeric network.

We may find confirmation for our statement about the thermodynamic incompatibility of linear polystyrene with the styrene–DVB copolymer in experiments by Wong et al. [269]. These authors reported similarities in the phase separation power of linear polystyrene and that of linear styrene–methylmethacrylate copolymer, the latter being *a priori* incompatible with the styrene–DVB network. Complete incompatibility of polystyrene with linear polydimethylsiloxane facilitates phase separation and results in the formation of a porous styrene–DVB network on adding as little as 0.5–1% of the above porogen to the initial comonomer mixture (Fig. 3.2, curves 4). It is also not surprising that the porosity of copolymers induced by linear polystyrene and linear polydimethylsiloxane is almost the same when the DVB content exceeds 10%. At a DVB content that high, the network formed differs fundamentally from linear polystyrene, as from any alien polymer.

1.3 Formation of macroporous copolymers in the presence of precipitating diluents

1.3.1 Experimental findings

Copolymerization of styrene with DVB is carried out most often in the presence of solvents that are miscible with the comonomers but precipitate the forming polymer. These are usually aliphatic hydrocarbons or alcohols of different chain lengths. In compliance with theory, true porosity appears only at certain threshold concentrations of the crosslinking agent and precipitant in the reaction mixture. When hydrocarbons such as, for example, *n*-heptane are used, the DVB proportion must be higher than 5–10% and the precipitant content must exceed 10–15% [233, 270]. The total porosity of the polymer produced under these conditions is still rather low, only 5–7%, when in dry state. Obtaining products with higher and stable porosity requires larger amounts of DVB and/or porogen [271]. As a rule, with increasing DVB concentration (and constant dilution), not only the total pore volume but also the inner surface area of the resulting materials will increase (Table 3.2). At the same time, an increase in the

Table 3.2   Parameters of the porous structure of macroporous styrene–tech.-DVB copolymers prepared in the presence of n-heptane

DVB (%)	Porogen (vol.%)	S (m^2/g)	W_o (cm^3/g)	R_p^{eff} (nm)*
4	100	8.4	0.15	35.8
8	100	32	0.50	31.3
12	100	75	0.67	17.9
15	50	3.7	0.12	64.9
15	70	60	0.60	20.0
15	100	76	0.84	22.1
20	50	8.5	0.27	63.5
20	70	120	0.84	14.0
20	100	91	1.12	24.6
30	50	123	0.60	9.8
30	70	215	0.94	8.7
30	100	163	1.60	19.6
40	50	208	0.67	6.4
40	70	308	1.19	7.4
40	100	300	1.60	10.7
60	50	295	0.71	4.8
60	70	378	1.33	7.1
60	100	375	1.70	9.1
100**	70	419	1.33	6.4
100**	100	457	1.82	8.9

$^*R_p^{eff}$ = average effective radius of pores.
**Poly(tech.-DVB).
Source: After [273]

volume fraction of the diluent to around 0.3 increases the surface area, but further dilution tends to reduce it again (Fig. 3.3), while the pore volume continues to increase [271–277]. Obviously, the ratio of narrow and wide pores is changing in favor of the latter.

Thus, the pore size distribution of polymers strongly depends on the conditions of polymerization. Higher dilution of monomer mixtures with porogens always increases the average pore diameter [278]. At a constant and high dilution with n-heptane, an increase in DVB concentration from 8 to 30% shifts the maximum of pore size distribution toward larger pores. For poly(tech.-DVB) containing at least 55% DVB, rather wide pore size distribution was obtained (Fig. 3.4).

The influence of the precipitating power of porogens on the polymeric porous structure can be illustrated by the following example. At the reaction temperature of 80°C, n-butyl alcohol was found to be a better precipitant for styrene–DVB copolymers than isooctane: the χ_{13} parameters are 2.31 and 1.27, respectively [267]. The better precipitant favors the

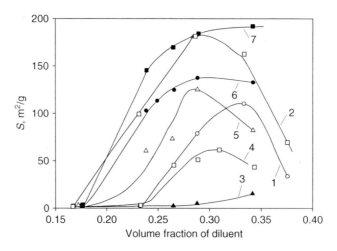

Figure 3.3 Dependence of inner surface area on the volume fraction of diluent for the macroporous copolymers of styrene with *tech*-DVB prepared in the presence of (3-7) *n*-heptane and (1, 2) isooctane, the concentration of DVB, %: (1, 4) 20, (2, 5) 30, (3) 8, (6) 40, (7) 60. After [271].

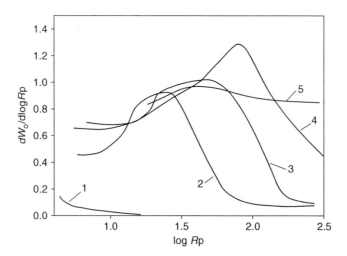

Figure 3.4 Pore size distribution for the macroporous copolymers of styrene with *tech*-DVB prepared in the presence of 134 vol.% *n*-heptane; concentration of DVB, %: (1) 8, (2) 15, (3) 20, (4) 30, (5) 100, poly(tech-DVB). After [278].

formation of larger pores and larger total pore volume of the dry copolymer, because large pores are more resistant to the drying process [279]. Still
larger macropores form in the presence of C_5–C_8 normal alcohols (Fig. 3.5)
and particularly branched alcohols. Some authors attribute this fact to the
inclination of alcohols to form associates [280–283], in addition to their
high χ_{13} values.

As for the assumed influence of the molar volume of porogens on
network porosity, no unambiguous and convincing experimental data can

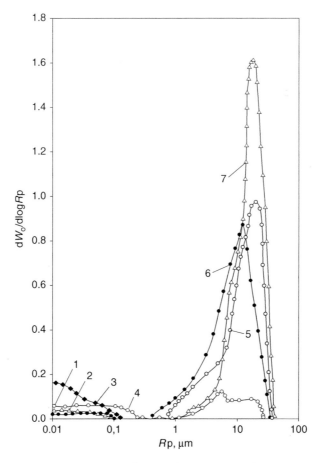

Figure 3.5 Pore size distribution for (1) dry conventional styrene–6% *tech*-DVB
copolymer and macroporous copolymers prepared in the presence of (2) *n*-heptane,
(3) isooctane, and (4–7) buthyl alcohol; the concentration of DVB %: (4) 2, (1–3, 6) 6, (7)
20. After [282].

be found in the literature that would confirm this effect. Wolf et al. [284] indicated that in order to obtain a porous material in the presence of high molecular weight hydrocarbons (with boiling temperatures from 120 to 300°C), the required amounts of DVB or diluent could be reduced when the diluent has larger molecular dimensions. However, with increasing chain length and molecular volume of aliphatic hydrocarbons, their thermodynamic affinity to polystyrene networks may gradually drop, strongly affecting phase separation. For example, the strong precipitating power of n-decane is known to result in the formation of macroporous copolymers even with the DVB content as low as 4% [285].

It appears that the precipitating power of aliphatic hydrocarbons also influences the character of pore size distribution and the total porosity. Indeed, the copolymerization of styrene with 20% DVB in the presence of equal amounts of n-heptane, n-octane, n-decane, and n-hexadecane at a volume fraction of 0.5 of the monomers was reported to result in obtaining polymers with pore volumes of 0.41, 0.83, 1.19, and 1.25 cm^3/g, respectively [286]. However, the last two values seem to be too high, as they exceed the volume of the porogens taken for the synthesis. Assuming the density of comonomers mixture to be 0.9 g/cm^3, we can expect the mixture of equal volumes of monomer and diluent to result in a polymer with a maximal porosity of about 1.1 cm^3/g, provided that the conversion of the comonomers was complete. Unfortunately, values unreasonably exceeding the expected limit are not uncommon in publications; they can also be found in Table 3.2.

Naturally the total pore volume of dry network can be smaller than the volume of porogens added to the mixture of the comonomers. If the network is insufficiently rigid, thin pores may disappear and large pores may shrink on removal of the porogen, sharply reducing both the pore volume and the inner surface area of the dry material [287, 288]. Interestingly, the porosity of the dried copolymers was found to depend on the type of solvent that filled the pores before drying [289, 290]. The smallest pore volume has always resulted from evaporating good solvents. Obviously, good solvents function as plasticizers and enhance the conformational rearrangement of the network during its shrinkage on removal of the solvent. (More information about this factor will be given later in Chapter 7, Section.5.2.) Importantly, the shrinkage of networks on solvent removal is reversible. If the dry copolymer (or resin) with reduced porosity is placed again in a good solvent and the latter is then replaced by washing with a precipitant before drying, the initial high porosity of the material is restored [291].

In addition to the thermodynamic quality of the evaporating solvent, shrinkage of the network must also strongly depend on the elasticity of the chains and cross-bridges and on the strength of the mutual attraction of polymeric segments. For example, it is well known that after drying from water, sulfonated resins shrink, resulting in materials having significantly less pores and surface area than the initial macroporous styrene–DVB copolymers had [292, 293]. Generally speaking, three reasons may be responsible for the marked reduction of the porosity of dry ion exchanger. First, the introduction of highly polar groups increases the mutual attraction of polymeric segments on drying. Second, spacious functional groups occupy a significant portion of the previously vacant porous volume in the copolymer. Finally, the HSO_3 groups are rather heavy and increase the weight of the material by a factor of 1.8. (In the above-cited publications this factor has been taken into account.) That the first factor is really important follows from the fact that weak basic anion-exchange resins do not demonstrate a significant loss of porosity of the initial copolymer [294], since the tertiary amino groups are almost nonpolar.

As indicated above, the pore volume of highly crosslinked polymers should be equal to the initial volume of the precipitating diluent, especially if the pores are large [295, 296], but it cannot exceed the diluent's volume. Higher porosity values generally imply incorrect measurements of the pore volume (mercury porosimetry often generates overestimated values) or incomplete conversion of the monomers. In the latter case, the unreacted monomers just function as an additional diluent. Interestingly, when trying to account for excess porosity, Seidl et al. [233] suggested diffusion of the unreacted monomers into the network phase shortly after phase separation ("inversed syneresis"), leading to expansion of the network, which would leave voids that are occupied neither by the polymer nor by the diluent liquid. However, nature abhors a vacuum and no migration of the components within the volume of the organic droplet-bead could ever increase its total volume. Undoubtedly, "inversed syneresis" provides no explanation for the formation of excessive porous space as large as that presented for some products listed in Table 3.1. One can note that inexplicably high porosity is mostly reported for products obtained with hydrocarbons as diluents. If such cases are reliable, they can be explained only by the partial capture of water in the interior of the droplet beads of the organic phase during polymerization. In this respect we can mention only a possibility of formation of double emulsions, "oil in water" and "water in oil" [297].

The parameters of the porous structure of copolymers with tech.-DVB may differ from those of copolymers crosslinked with the pure p-isomer, although literature data are not very consistent. At the same dilution of monomeric mixtures, larger total pore volume and pore size are characteristic of the copolymer with p-DVB, whereas the specific surface areas differ insignificantly [298]. Yet, Tsilipotkina et al. [299] found both W_o and S to be smaller for copolymers with the tech.-DVB. When using tech.-DVB as the crosslinking agent, reduced porosity was also found for its copolymers with vinylpyridines [300]. Compared to copolymers with tech.-DVB, the surface area of styrene–p-DVB copolymers achieves maximum values at slightly lower dilution with n-heptane (0.5 and 0.4, respectively), while the maximum of pore volumes is observed at approximately the same degree of crosslinking of about 20% (Figs. 3.6 and 3.7).

Wojaczynska et al. [301] studied the porous structure of ternary styrene–DVB–acrylonitrile copolymers with varying content of the third comonomer, from 18 to 62%. As found, the addition of acrylonitrile results in decreasing surface area and pore volume, as compared with binary styrene–DVB or acrylonitrile–DVB copolymers; however, the pore size distribution for the three types of copolymers was found to be approximately the same (Table 3.3).

Characterizing the texture of products, that is, the average size of primary microparticles and the average distance between them, originating from measured total pore volume and specific surface area values as was suggested by Hradil [302], Wojaczynska and Kolarz [301] arrived at the conclusion that the primary particles are packed more densely in the binary copolymers. It was found that the porosity of the acrylonitrile copolymers, too, depends on the type of solvent that is being removed on drying the polymer beads.

As was mentioned above, phase separation (induced by precipitants) results in the formation of solvent dispersion within the gel. To reveal the reasons why microsyneresis has priority over macrosyneresis during the crosslinking copolymerization, Dušek and Sedláček [268] investigated how gel-type copolymers of 2-hydroxyethyl methacrylate, containing small amounts of ethylene glycol dimethacrylate as the crosslinking agent, swell in n-butanol. At an elevated temperature of 80°C n-butanol represents a thermodynamically good solvent for the copolymer and causes

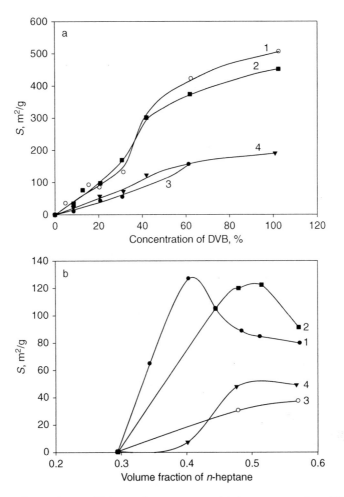

Figure 3.6 Dependence of inner surface area upon (a) the concentration of DVB and (b) the degree of dilution with n-heptane for (1, 2) macroporous styrene–DVB copolymers and (3, 4) sulfonates based on the copolymers with (1, 3) p-DVB and (2, 4) tech-DVB; (a) the volume fraction of n-heptane: 0.57, (b) the concentration of DVB: 20%. After [298].

swelling of the gel. However, with decreasing temperature, the solvent–polymer affinity decreases, and so the swollen transparent beads gradually become opaque when cooling to room temperature. Turbidity is particularly strong in the case of slightly crosslinked copolymers that incorporate more solvent on swelling. Remarkably, the volume of the beads continues to decrease at room temperature and, eventually, opaque beads again

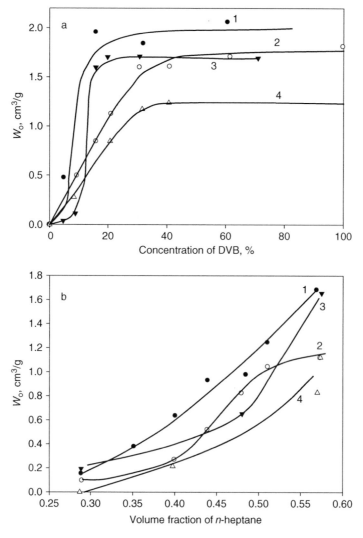

Figure 3.7 Dependence of total pore volume on (a) the concentration of DVB and (b) the degree of dilution with *n*-heptane for (1, 2) macroporous styrene–DVB copolymers and (3, 4) sulfonates based on the copolymers with (1, 3) *p*-DVB and (2, 4) *tech*-DVB; (a) the volume fraction of *n*-heptane: 0.57, (b) the concentration of DVB: 20%. After [298].

became transparent. Consequently, the microsyneresis slowly transforms into the macroform of phase separation. On the other hand, a relatively highly crosslinked copolymer with 4% ethylene glycol dimethacrylate maintains transparency when the temperature drops from 80 to 25°C,

Table 3.3 Morphological characteristics of macroporous styrene (S)–DVB, styrene–acrylonitrile (AN)–DVB, and acrylonitrile–DVB copolymers

Sample	$\phi_{AN}{}^a$, %	$V_{sp}{}^b$, cm³/g	S, m²/g	Pore radius, %		
				10–100 nm	100–1000 nm	1000–7500 nm
S/DVB	0	0.81	492	87.6	9.3	3.1
S/3AN/DVB	18.75	0.52	292	90.7	8.3	1.0
S/5AN/DVB	31.25	0.72	310	88.2	8.1	3.7
S/10AN/DVB	62.5	0.72	281	90.5	8.5	1.0
AN/DVB	100	0.79	465	88.0	10.4	1.6

[a] ϕ_{AN} = acrylonitrile content in mixtures with styrene.
[b] V_{sp} = specific volume.
Source: After [301]

and excess of the solvent separates from the beads only in the form of macrosyneresis.

Generally, a decrease in the thermodynamic affinity of the solvent (χ-induced syneresis) during cooling would have to result in decreasing copolymer swelling. However, if much solvent has to be removed from the beads, it first separates as droplets inside the gel, because the network relaxation is fairly slow [266]. Excess solvent is then slowly pushed out of the beads having a rather flexible gel-type matrix that is homogeneous under equilibrium conditions. Consequently, the microsyneresis is a nonequilibrium state of the particular gel-type polymer–solvent system.

During the preparation of macroporous materials by crosslinking copolymerization in the presence of precipitants, phase separation takes place within a relatively short period of time. This fast phase separation naturally results in the formation of microdroplets of the rejected porogen. Since the conversion of comonomers at that moment is very low, the rapidly growing polymeric network fixes in the gel the emerging liquid droplets. The nonequilibrium microsyneresis thus transforms into the stable form of phase separation within a heterogeneous system. Thus, the fast arrival at the unstable local polymer–solvent relationship, as compared with the slow rates of solvent macrosyneresis and of network relaxation, leads to the formation of gel-included microdroplets and, finally, to a permanent macroporous structure of copolymers.

1.3.2 Formation of macroporous texture in the presence of precipitating diluents

It is important to understand how the macroporous texture of polymers is formed. Electron microscopy has always been useful in exploring the formation of porosity in styrene–DVB copolymers [303–308]. This method shows that in the case of pretty large DVB concentrations and moderate dilutions, the morphology of macroporous copolymers is characterized by a three-level organization. The phase separation during crosslinking copolymerization initially starts with the precipitation of microgel particles (or nuclei). Their spherical shape is stimulated by the interfacial tension on the growing polymer–monomer solution border. These primary particles, 50 to 100–200 Å in diameter, aggregate and form microspheres with diameters of about 1000–2000 Å. The microspheres, being also aggregated, produce the internal texture of polymeric beads, which is very much like a cauliflower. As an example, Fig. 3.8 demonstrates such a kind of internal structure of highly crosslinked styrene–DVB bulk

Figure 3.8   Cauliflower texture of macroporous styrene–DVB bulk copolymer. After [309].

copolymer prepared in the presence of a mixed porogen composed of water, low molecular weight alcohols, or their mixture with formamide [309]. In this case, however, the dimensions of microspheres and their aggregates proved to be markedly larger.

The pores of such macroporous polymers may be classified into three families [235]. Pores between the primary nuclei are smaller than 150–200 Å, and they are mainly responsible for the high internal surface area of around 400–500 m^2/g. The pores, 200–500 Å in size, between the microspheres can provide a moderate contribution of about 100 m^2/g to the total surface area. Finally, the very irregular pores between microsphere agglomerates are, on the average, 500–1000 Å, and they largely add to the total pore volume that may have a high value, up to 3 cm^3/g [296], but do not practically contribute to surface area. When this type of macropores predominates, the surface area of copolymers is only several square meters per gram or even less. Jacobelli et al. [310] suggest considering only those polymers as truly macroporous the pore volume of which is mainly formed by the space between microspheres.

Such a porosity formation via nucleation and subsequent aggregation of primary nuclei was described for the first time by Kun and Kunin [311] at the end of 1960s. However, later it was found [312–314] that, being prepared under slightly different conditions, macroporous resins may acquire an entirely different texture. Fig. 3.9a shows a schematic diagram initially suggested by Häupke and Pientka [312], in which, as a function of the DVB and diluent concentrations, two border lines confine domain II where all the above three levels of porosity may be observed. Below the lower boundary (domain I), where the concentrations of DVB and the diluent are smaller

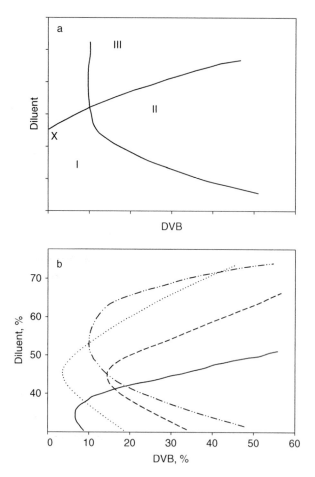

Figure 3.9 Styrene–DVB copolymerization in the presence of precipitants. (a) Schematic representation of boundaries separating (I) gel-type domain; (II) macroporous domain; (II) domain where network texture is formed only by large fused species; (x) incipient precipitation of polystyrene in diluent–styrene mixture; (b) boundary lines for macroporous domains of styrene–DVB copolymers obtained with non-solvents as porogenic agents: (....) 2-ethyl-1-hexanoic acid, (·········) benzyl alcohol, (- - -) heptane, and (———) pentanol. After [314].

than the threshold values, gel-type polymers are obtained. Above the upper boundary, in domain III at excessively high dilution, macroporous copolymers are obtained with a texture strongly differing from the cauliflower texture of domain II. In the process of copolymerization at the very high dilution, the primary particles, rather well separated at the beginning, completely lose their individuality at a certain point in time of further polymerization, and only large fused spherical species characterize the final morphology [313].

Figure 3.9b shows the same boundaries confining the macroporous domain II for real styrene–DVB copolymers prepared in the presence of 2-ethyl-1-hexanoic acid, benzyl alcohol, heptane, or pentanol [314]. Judging from the shape of the plots, 2-ethyl-1-hexanoic acid and benzyl alcohol generate macroporous copolymers with cauliflower texture and high surface area in a wider range of crosslinking densities and dilutions compared with heptane or pentanol diluents.

Neither the microsyneresis theory of Dušek [265], nor the mechanism of nucleation and agglomeration of the nuclei can explain the above dramatic variation in network morphology with rising dilution. The concept developed by Brutskus et al. [315, 316] also fails to provide insight into the situation. These authors have tried to characterize the polymer-precipitant system from the positions of classic colloid chemistry by placing primary emphasis on the "degree of oversaturation," that is, the extent of deviation of the reaction system from the thermodynamically stable state. Their reasoning is as follows: The higher the degree of oversaturation and the longer the metastable state maintained, the larger the number of seeds of a new phase would form, which should enhance the dispersity of condensed structures and increase the surface area of the final dry polymer. Generally, an increase in DVB and diluent concentrations reduces polymer solubility, thus favoring oversaturation. On the other hand, dilution of the monomers reduces their concentration and thereby the rate of copolymerization, which leads to the decrease in oversaturation. The first factor is assumed to have priority at a low degree of dilution, whereas the reduced rate of copolymerization dominates at high dilution. Brutskus et al. believe that an interplay of these two factors results in the extreme profile of surface area-dilution dependence (Fig. 3.3). Still, the mechanism of "oversaturation" and "seed formation" prior to transition from a nonequilibrium system to phase separation [317] seems to be rather artificial and cannot explain the disappearance of the cauliflower texture with increased dilution.

Chung et al. [313] tried to find an explanation for such variations in network morphology by considering the sequence of three "critical events" taking place during copolymerization: (i) formation of an insoluble gel that is affected largely by the DVB content and, to a lesser extent, by the amount of the diluent; (ii) phase separation occurring when the polymer–environment interaction parameter becomes higher than $\chi_{13} = 0.5$, as affected by the amounts of diluent and DVB; and (iii) overlap of branched and partially crosslinked macromolecules occurring when the volume occupied by the swollen molecules becomes equal to the available volume. However, no transpositions of the critical events could provide a simple answer to the question why excessive dilution of the comonomer mixture should change network morphology.

We believe that the reason for the enigmatic phenomenon is quite simple. We only have to pay attention to the self-evident fact that the copolymer will grow where the radical-forming initiator is located. Commonly used initiators such as 2,2′-azobis(isobutyronitrile) and benzoyl peroxide are soluble in styrene–DVB mixtures, but insoluble in most conventional polystyrene-precipitating diluents. Naturally, the starting solution is homogeneous, even with a high diluent content. Precipitation of the initial primary nodules and their coalescence to microspheres with further agglomeration to a cauliflower texture create a second, mostly aromatic phase. Meanwhile, the monomer content in the liquid phase drops to a level that forces the initiator molecules to concentrate in (or on the surface of) the polymeric phase. Now, the polymerization will proceed locally, from the surface of the initially formed cauliflower structure into the bulk solution, rather than statistically throughout the whole volume of the droplet/bead. No doubt, the initial fine structure of the polymeric phase will soon be lost and only agglomerates of large spherical units will remain distinguishable. On the other hand, with a moderate content of the diluent in the comonomer mixture, the initiators will stay in the solution and continue generating radicals there, thus giving birth to new primary nodules and microgel particles, which will later merge with the polymeric phase and provide it with fine structure and high surface area.

Certainly, the boundaries between domains I, II, and III cannot be as sharp as shown in Fig. 3.9b. Depending on the ratio of comonomers to diluent and the nature of the latter, the relocation of the initiator and, hence, the change in the mode of growth of the polymeric phase (and in the evolution of its morphology) will take place sooner or later during the long polymerization process. Hence, the plot of surface area values versus dilution exhibits a markedly extreme but smooth character (Fig. 3.3). Even inside the macroporous domain II, there are many possibilities for fine-tuning the morphology, parameters of the porous structure, and target properties of macroporous styrene–DVB copolymers.

1.4 Formation of macroporous copolymers in the presence of solvating diluents

When styrene–DVB copolymerization proceeds in the absence of any diluent, the network will grow through the whole volume of the homogeneous system. Individual microgel species formed at the beginning of the free radical crosslinking copolymerization merge to form the entire network at later stages of the process by interacting with each other or with new polymer chains via their pending double bonds. The same is valid for a system where a small amount of a thermodynamically good diluent, such as toluene, is added to the monomer mixture, provided that the latter contains moderate amounts of divinyl comonomer. In this case, the resulting network displays an expanded structure characterized by a reduced degree of chain entanglement, because the dilution of comonomers has reduced the chain concentration in unit volume of the final network. Expanded networks are homogeneous, too, and only display enlarged swelling ability.

However, with the proportion of DVB increasing to about 25–30%, the polymerization product obtained in the presence of two parts of toluene becomes opaque and its apparent density drops dramatically (Fig. 3.10), indicating that phase separation has taken place [318]. Obviously, the highly crosslinked primary microgels were unable to accommodate through swelling all the amounts of the good solvent that were added to the initial

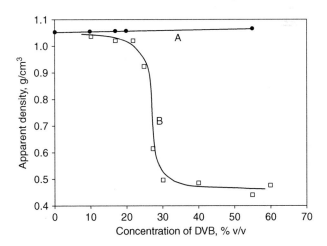

Figure 3.10 Apparent density of styrene–tech-DVB copolymers prepared in the presence of toluene at volume fraction of the polymer: (A) 1.0 and (B) 0.33. After [318].

mixture (ν-induced syneresis). The excess solvent will remain as a separate liquid phase between the swollen microspheres and their agglomerates. Obviously, this type of phase separation requires certain, relatively large threshold amounts of the solvent and crosslinking agent in the initial system. The high rigidity of the final network will prevent its complete shrinking upon drying, resulting in a material with permanent porosity and cauliflower texture.

As an alternative to networks prepared with n-decane, the toluene-modified copolymers exhibit narrower pore size distribution and smaller pore volume. A substantial part of the porous volume is provided by small voids [319] located within the initially swollen primary microgels. These pores are poorly accessible even to the small molecules of methanol. This is the reason why the surface area of the toluene-modified copolymers calculated from adsorption of methanol vapors proves to be markedly smaller than that determined by the conventional nitrogen adsorption technique [320]. In addition to micropores with a diameter below 15 Å, the polymer exhibits mesopores with diameters up to 300 Å, representing the space between the microgels and their aggregates. Owing to the increased portion of small pores, the surface area of the toluene-modified networks can achieve large values and exceed that of the polymers prepared with precipitants [321, 315]. In order to raise both the pore volume and the pore size of solvent-modified materials, a precipitant needs to be added to toluene [322–326].

It had been assumed that the molecular dimensions of solvating diluents influence polymer porosity. In the series of methyl-, ethyl-, propyl-, dimethyl-, and diethylbenzene used as the diluent (all other conditions being the same) both the size and the volume of pores increase, while the surface area changes randomly, in the range between 300 and 500 m^2/g [327]. On the other hand, the solvation power of these liquids has not been analyzed, and the increasing capability of compounds to participate in chain transfer reaction was not considered. Radical chain transfer agents (or telogens) reduce the average lengths of each polymeric chain and may affect the porous structure of the copolymer. It was noted that they shift the moment of phase separation toward increasing conversion of the comonomers. Once again, it is not easy to give preference to the molecular volume of the diluent, since many factors influence the porous structure.

Whatever the DVB content and dilution of the comonomers may be, the toluene uptake ability of toluene-modified networks meets the following dependence [318]:

$$K_{sw}^f = K_{sw}^g + N, \text{mL/g} \qquad [3.7]$$

where K_{sw}^f is the swelling in toluene of the final network, K_{sw}^g is the swelling in toluene of gel-type styrene copolymer incorporating the same amount of DVB, and N is the volume of toluene diluent calculated per gram of initial monomer mixture. Densely crosslinked copolymers swell poorly, even with toluene, and, therefore, with increasing crosslinking density the swelling of toluene-modified copolymers becomes closer to the value of N.

It is noteworthy to mention that the densely crosslinked toluene-modified resins swell by taking up almost the same amounts of toluene, cyclohexane, n-heptane, and nitromethane. For the network containing 55% DVB and prepared at the monomer volume fraction of 0.33, these quantities are 2.21, 2.10, 2.09, and 2.01 mL/g, respectively [318]. It was also stated in this publication that water fills out only the internal space of the permanent pores in the copolymer, 1.10 mL/g, causing no increase of its volume. As a matter of fact, highly crosslinked styrene–DVB copolymers, when synthesized in toluene, swell even in water. The fact of swelling of toluene-modified networks in non-solvents and water was reported for the first time in 1962 by Lloyd and Alfrey [328]. They did not comment on the increase in network volume when taking up these solvents; however, one can see this happening from their data presented in Fig. 3.11. Forty years later [329], the swelling in water was confirmed by direct experiments with macroporous resins containing 55% DVB, prepared in a toluene–isooctane mixture (10:2, v/v), at a volume fraction of the monomers equaling 0.33. Millar et al. [318] explained the unusual swelling behavior of toluene-modified copolymers by the ability of organic non-solvents to solvate the polystyrene chains connecting the microgels. At the same time they demonstrated in the same publication that neither cyclohexane nor n-heptane can solvate polystyrene chains in a conventional styrene–7% DVB copolymer. The latter does not swell in the non-solvents for polystyrene.

With respect to swelling in non-solvents, toluene-modified styrene–DVB copolymers have much in common with hypercrosslinked polystyrenes [330]. Both are prepared in accordance with the same basic principle, the formation of rigid networks in strongly solvated state. It will be shown in detail in Chapter 7 that rigid expanded networks possess a relaxed favorable conformation only in their swollen state and, therefore, exhibit a marked tendency to acquire this state by swelling and incorporating any liquid, even non-solvating one.

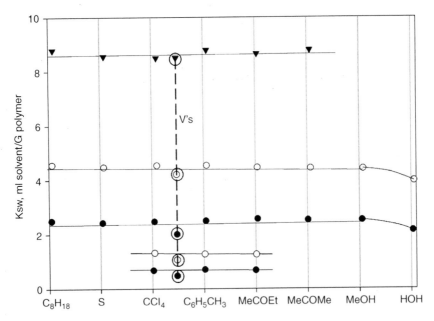

Figure 3.11 Equilibrium swelling in different solvents of 74.7 % divinylbenzene polymers prepared at varying degree of initial dilution, V's. Reprinted from [328] with kind permission of Wiley & Sons, Inc.

The morphology of toluene-modified macroporous copolymers is similar to that of networks obtained in precipitants; however, the aggregates of microparticles are said to be packed more loosely [331, 332]. The morphology changes on resin swelling, "the agglomerates of microgels disintegrate [332, 333]," the dimensions of the microgels and the average pore sizes increase [319], and the pore size distribution becomes wider compared with that of a dry network [334]. Perhaps because of swelling in water of the toluene-modified nonionic vinyl-aromatic sorbents (as well as other porous adsorbing materials listed in Table 3.4), the inner surface area of resins measured by the adsorption of p-nitrophenol from aqueous solutions appears larger than that measured from the dry material by standard nitrogen adsorption at low temperature. Millar [335] estimated the increase in surface area on swelling to be roughly proportional to 2/3 power of the volume increase (Table 3.4). It should be kept in mind, however, that, just like the nitrogen adsorption technique, the method of determining surface area by adsorption of p-nitrophenol has serious limitations. First of all, the

Table 3.4   Comparison of solvated and dry surface areas

Sorbent	Type[a]	Surface area, m^2/g	
		Solvated (PNP)	Dry (BET/N_2)
Duolite S-861	VA	985	500
Resin A[b]	VA	859	652
Resin B[c]	VA	509	420
Duolite ES-863	VA	460	390
Amberlite XAD-2	VA	420	300
Resin C[d]	VA	474	14
Resin D[e]	VA	333	114
Duolite S-761	P	450	295
Amberlite XAD-7	A	520	450
Duolite C-464	A*	300	30
Duolite ES-771	N	650	400
Duolite S-37	N	390	168
Duolite A-7	N*	340	60
Duolite A-561	N*	280	123

"Dry" surface was measured by nitrogen adsorption; "solvated" surface was measured by adsorption of p-nitrophenol from aqueous solution.
[a] VA = vinylaromatic; P = phenolic; A= acrylic; N = nitrogen containing.
[b] 60% DVB, volume fraction of comonomers, F_m 0.33 with toluene.
[c] 37% DVB, F_m 0.37 with toluene.
[d] 37% DVB, F_m 0.37 with heptane.
[e] 40% DVB, F_m 0.67 with toluene.
* Ion-exchange resin.
Source: After [335]

effective cross-section area of this molecule depends on the mode of adsorption, which may be unknown; also chemosorption on adsorbents with functional groups is not ruled out.

Completing the considerations of macroporosity formation in the presence of precipitants and solvating diluents, it should be said that well-defined macroporous beads can be also synthesized by using supercritical carbon dioxide as the porogen [336]. It was illustrated by an example of free radical suspension polymerization of trimethylolpropane trimethacrylate that a simple change in the pressure of carbon dioxide permits a fine control over its thermodynamic quality and the resulting bead porosity: with pressure rising from 200 to 400 bar, the surface area of the beaded product increases from 220 to 470 m^2/g, while the pore volume drops from 1.05 to 0.69 cm^3/g, and the median pore diameter decreases from 2000 to 400 Å. This indicates that with an increase of pressure and density of the supercritical fluid its quality becomes better with respect to the polymer. It is hoped that in future supercritical carbon dioxide will occupy a fitting place as a porogenic medium because it is inexpensive, nontoxic, and inflammable.

1.5 Formation of macroporous copolymers in the presence of linear polystyrene

The third approach to the formation of macroporous structure of styrene–DVB copolymers consists of adding a linear polymer, largely polystyrene, to the monomer mixture.

The molecular weight and quantity of polystyrene have to exceed certain threshold values to produce a mutually connected pore system in the copolymer beads after extraction of the porogen. This precondition of phase separation directly follows from the above-developed notion (see Chapter 3, Section 1.2) of the incompatibility of polystyrene and styrene–DVB copolymer. It is well known that only oligomeric molecules, rather than macromolecules, differing in chemical nature may be completely compatible. Hence, linear polystyrene having, for example, a molecular weight up to 5000 Da does not provoke phase separation at the early stages of styrene–DVB copolymerization. Therefore, low molecular weight polystyrene acts similar to a good solvent (toluene), boosting the swelling ability of the styrene–DVB network [321]. Dušek et al. [337] estimate that 20% of polystyrene with a molecular weight of at least 15,000 Da is needed to introduce macroporosity into a copolymer containing 8–15% of the crosslinking agent. The nonporosity of polymers obtained at lower polystyrene loadings and/or low molecular size of polymeric porogens can be accounted for several reasons, such as absence of phase separation, low connectivity of rare pores in the polymer matrix, and contraction of the network after extraction of the diluent. Still, polystyrene is a more efficient porogen than, for example, n-heptane. The former requires a lower crosslinking density or smaller dilution of the monomers to implement macroporosity. The porosity of the final networks increases with rising polystyrene chain length and reaches a constant maximum value at the degree of polymerization of the additive of about 500–600 (Fig. 3.12). At a constant fraction of polystyrene in the monomer mixture, the porosity rises sharply in the range of crosslinking density from 5 to 15%, and then levels off (Fig. 3.13) [338]. Evidently, 5–15% of DVB are sufficient to make the chemical structure of the growing network different from that of linear polystyrene, thus provoking phase separation, and also to provide a final polymer with reasonable rigidity. In the extreme case of high DVB content (a 1:1 mixture of

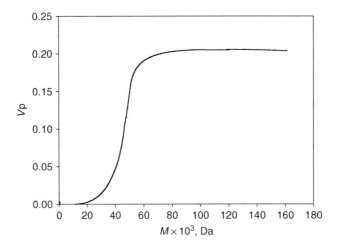

Figure 3.12   Volume fraction of pores, V_p, in dry copolymers of styrene with 10 % DVB prepared in the presence of 20 % of linear polystyrene of different molecular weight, M. Reprinted from [266] with kind permission of Elsevier.

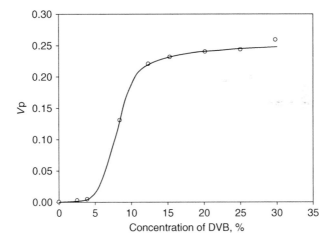

Figure 3.13   Dependence of volume fraction of pores, V_p, in dry styrene–DVB copolymers prepared in the presence of 20 % linear polystyrene ($M = 87000$) on the concentration of divinylbenzene, wt%. Reprinted from [266] with kind permission of Elsevier.

styrene with 80% DVB) and high porogen concentrations (about 50%), polystyrene with a molecular weight as low as 4100 will also generate the formation of many macropores of about 55 nm in size [269].

Copolymerization of styrene with DVB in the presence of ^{14}C-labeled polystyrene with a molecular weight of ~37,000 Da demonstrated that high molecular weight fractions of the polystyrene-diluent are partially incorporated into the network, most probably due to chain transfer reaction (although an alternative possibility could be the incomplete extraction of these fractions). At the same time, a small amount of newly formed inactive polymer with molecular weights of 10,000–20,000 Da was found in the extract. Interestingly, the amount of this leachable polymer is much larger than the soluble fraction formed during the copolymerization of the same monomers without the porogen [339]. Probably the above binding of some porogen chains to the network is the reason for the constancy of porosity when more than 20–30% polystyrene was added to the monomers. Above this range the pore volume of the polystyrene-modified copolymer becomes smaller than the volume of the added porogen [338].

The extracted polystyrene is reusable; however, the progressive loss of high molecular weight fractions and contamination with newly formed low molecular weight fractions result in the drop of the average molecular weight and broadening of the molecular weight distribution of the porogen, deteriorating the efficiency of the reused polystyrene [340].

The specific inner surface area of polystyrene-modified networks is low, usually 5–10 m^2/g, and the total pore volume is about 0.4–0.7 cm^3/g, similar to that obtained with a good solvent [341] (Table 3.5). The average pore radius depends on the amount of the diluent, increasing from 800–1200 Å to 5000 Å with the polymer fraction rising from 15 to 25 wt% [342]. The pore size and the width of pore size distribution are also reported to be proportional to porogen polydispersity [343]. Still, formation of extremely large macropores, amounting to 5000–10,000 Å [343–346] and up to several microns [343], must be caused by aggregation of the polystyrene molecules during phase separation, and this fact should eliminate correlation with the polydispersity of the porogen.

Kolarz [347] synthesized macroporous copolymers in the presence of a mixture of polystyrene with xylene as the solvating diluent (Tables 3.5 and 3.6). As found, at a constant dilution, the increase of DVB concentration from 8 to 12% involves a noticeable increase in the porosity of the copolymers, while further increase of crosslinking density does not change anymore the parameters of the porous structure. The highest porosity was characteristic of samples prepared in the presence of approximately equal amounts of polystyrene and xylene. Similar to the copolymers prepared with polystyrene alone, it was observed that pore size

Table 3.5 Characteristics of porous styrene–DVB copolymers prepared with polystyrene and mixtures of polystyrene with xylene as the porogens

DVB, % wt	Polystyrene $M_w \times 10^3$	% wt	Xylene, % wt	φ	P, %	W_o, cm³/g	S, m²/g	Radius of pores forming 80% of pore volume, μm
12.5	27.1	24.0	—	—	17.6	0.167	5.6	—
12.5	49.7[a]	24.0	—	—	23.0	0.44	15.5	0.06–2.0
12.5	61.2	24.0	—	—	16.6	0.416	13.2	0.3–2.50
12.5	193.6[b]	24.0	—	—	29.5	0.416	9.8	1.6–3.2
15.0	27.1	28.2	27.4	1.2	50.1	1.17	22.8	0.1–1.5
12.7	49.7[a]	25.6	25.1	1.0	46.5	0.867	18.2	0.2–0.8
15.0	193.6[b]	28.2	27.4	1.2	40.8	0.77	12.4	0.2–0.5
15.0	250	28.2	27.4	1.2	26.4	0.46	8.36	0.3–0.6

φ = volume ratio of diluent(s) to comonomer.
S = specific surface area measured in the pore range of 0.0075–0.5 mm.
P = porosity.
[a] $\overline{M}_w/\overline{M}_n = 2.08$.
[b] $\overline{M}_w/\overline{M}_n = 3.14$.
Source: After [343]

Table 3.6 Characteristics of the porous structure of styrene–DVB copolymers prepared in the presence of polystyrene ($\bar{M}_w = (193, 000)$ and xylene

DVB, % wt	PS, % wt	Xylene, % wt	φ	P, %	W_o, cm³/g	S, m²/g	R_p^{eff}, µm	R, 80% µm
8.55	28.6	26.6	1.2	21.3	0.314	10.2	0.059	0.15–0.3
10.5	28.6	26.9	1.2	32.3	0.494	9.0	0.110	0.2–0.5
12.7	28.2	27.1	1.2	40.2	0.69	11.7	0.118	0.2–0.5
15.0	28.2	27.4	1.2	40.8	0.77	12.4	0.108	0.2–0.5
18.0	27.7	27.7	1.2	42.1	0.68	11.3	0.119	0.2–0.5
15.0	25.6	25.1	1.0	39.6	0.654	12.5	0.104	0.07–0.7
15.0	29.2	28.2	1.3	47.4	1.3	24.7	0.308	0.1–0.6
15.0	–	53.0	1.3	–	–	–	–	–
15.0	–	51.0	1.2	–	0.04	3.6	0.022	0.01–0.7
15.0	11.5	39.6	1.2	38.8	0.666	8.8	0.160	0.5–0.12
15.0	22.6	29.5	1.2	49.5	–	–	–	–
15.0	33.4	19.4	1.2	39.6	–	–	–	–

Source: After [347]

distribution of the product obtained with the mixed diluent becomes wider with an increase of the polydispersity of the polymeric porogen.

Styrene–15% DVB copolymers obtained in the presence of polystyrene and C_5–C_{10} aliphatic alcohols as the precipitating agents are characterized by smaller values of porosity, pore volume, and specific surface area compared with those of copolymers prepared with alcohols only. Kolarz [348] found that copolymerization in the presence of *n*-decyl alcohol and polystyrene at the diluents to comonomers volume ratio of 1.2 results in a copolymer having the porosity of 44%, surface area of 16.5 m²/g, and total pore volume of 0.88 cm³/g, 80% of pore volume being formed by voids with radii ranging from 2,000 to 4,000 Å. In comparison, the pore volume of 1.94 cm³/g and surface area of 43.6 m²/g are characteristic of a copolymer synthesized with isodecyl alcohol under the same conditions.

Not much information is available about employing other linear polymers for the preparation of porous styrene–DVB resins. A review paper [233] mentions poly(vinyl acetate) and poly(methyl methacrylate) to induce formation of porous styrene–10% DVB beads. Poinescu et al. [349] managed to synthesize highly porous materials by carrying out styrene–DVB copolymerization in the presence of poly(vinyl acetate) and C_{10}–C_{11} aliphatic alcohols as co-porogens. More recently, Macintyre and Sherrington [350] suggested using oligomeric poly(propylene glycol) 1000, poly(propylene glycol) 4000, and poly(dimethyl siloxane) and their mixtures with toluene to generate porous polydivinylbenzene. The oligomeric porogens taken in a 1:1 volume ratio with DVB provide formation of sufficiently large pore volumes, 0.8–1.2 cm^3/g, but a relatively small surface area of 15–60 m^2/g. Addition of toluene to the oligomeric porogen reduces, although not too strongly, the porosity of the resulting product and increases its specific surface area by one order of magnitude. The authors believed that when using 80 vol% toluene and 20 vol% poly(dimethyl siloxane), they were able to prepare a product with bimodal pore size distribution, one pore population being micropores and the other macropores.

2. MACROPOROUS ION-EXCHANGE RESINS

Knowledge of the factors that govern the porosity of styrene–DVB copolymers permitted the preparation of numerous adsorbing materials with a wide variety of pore dimensions, pore volume, and specific surface area. This in turn provided ways to solve many practically important tasks and stimulated the rapid evolution of adsorption technologies.

Properly designed macroporous styrene–DVB copolymers have been employed as matrices for various ion-exchange resins. Presently macroporous ion exchanger are in general use for the demineralization of drinking and waste waters, in hydrometallurgy, in the nuclear industry, in the production of pharmaceuticals, and in biochemistry and biotechnology, and medicine, just to name a few fields. Already the first publications [232, 351] describing the properties of macroporous ion-exchange resins have demonstrated the basic advantages of these materials over conventional gel-type ion exchanger, namely, that porous polystyrene sulfonates absorb relatively large organic substances and continue functioning even in low polar and nonpolar media. For example, the functional sulfonic groups were found to remain accessible and capable of sorbing triethylamine from butanol or n-heptane. This finding immediately offered new opportunities to employ macroporous ion exchanger in nonaqueous media for heterogeneous catalysts of organic reactions, not only in the laboratory practice but also on an industrial scale [184, 352].

Neutral nonfunctionalized macroporous adsorbents were suggested for the removal of organic pollutants from all types of waste and drinking waters and industrial gas emissions. On an analytical scale macroporous polystyrene adsorbents, both functionalized and nonfunctionalized, gained wide acceptance as stationary phases in gas and liquid chromatography, as well as sorbents for solid-phase extraction of organic components from aqueous environment. Macroporous polystyrene resins also represent a useful starting material for the preparation of various polymeric reagents, catalysts, or scavengers.

Early studies have also demonstrated that macroporous ion exchanger and adsorbents are mechanically stable and rigid and, in contrast to conventional gel-type exchangers, do not suffer from osmotic shock. Macroporous ion-exchange resins based on relatively low crosslinked matrices were found to absorb organic ions from water more rapidly than gel-type resins having the same degree of crosslinking (Table 3.7).

Soon, however, it has become evident that only those functional sites are easily substituted with large organic ions, such as methylene blue or alkyl ammonium ions, that are located on the pore walls, while at least half of the total number of functional groups remain inaccessible within the highly crosslinked polymer phase, because of insufficient swelling of this phase. Both the high diffusion rate of organic ions into macroporous beads at the early stage of adsorption and its dramatic decrease at a certain point in time, resulting in the relatively low total adsorption capacity, have been well documented in a number of publications [355, 356].

In general, the accessibility of functional sites is determined by the relationship between the dimensions of the pores and of solute molecules. The peculiar behavior of macroporous resins logically follows from the bimodal character of their pore size distribution. Space between the primary nodules and their aggregates presents transport channels with easily accessible functional groups on their walls. When the porous support was formed in the presence of a solvating diluent, such as toluene, its transport

Table 3.7 Permeability of macroporous and conventional polystyrene sulfonates for methylene blue [356] (the macroporous copolymers were prepared in the presence of 100 %-wt n-heptane)

DVB, %	Macroporous resins			Conventional resins	
	S, m^2/g	F^*, %	$\bar{D}$, cm^2/min	F^*, %	$\bar{D}$, cm^2/min
4	7	73	5.81×10^{-9}	60	2.21×10^{-9}
8	12	70	4.19×10^{-9}	48	3.37×10^{-10}
15	–	–	–	31	1.0×10^{-10}
20	65	70	2.81×10^{-9}	30	4.71×10^{-11}
30	100	55	2.35×10^{-9}	12	5.67×10^{-12}
40	148	54	5.54×10^{-9}	–	–

$^*F$ = portion of accessible sulfonic groups.
Source: After [356]

pores are only reasonably large, but the polymer phase has an expanded structure that does not totally shield the majority of functional groups located inside the swollen polymeric phase. Naturally, only small organic ions can migrate through all the pores to those functional groups. For this reason toluene-modified polymer sulfonates take up large quantities of small $CH_3NH_3^+$ or $(CH_3)_4N^+$ ions, much larger than resins prepared in the presence of aliphatic hydrocarbons as porogens. The latter resins have larger transport pores, but the polymeric phase comprising their walls is scarcely swellable. They display higher adsorption capacity for large $(C_2H_5)_3(C_6H_5)N^+$ ions [356, 357], but it still remains only a small portion of their theoretical capacity value.

Since macroporous resins are mainly designed for sorption of organic molecules and ions and sorption of these species mainly proceeds on wall surfaces, the values of inner surface area of commercial materials always attract the attention of customers. Unfortunately, it is often overseen that there is no direct correlation between the reported surface area and the real adsorption capacity of the material, for the simple reason that the surface area is measured by adsorption of nitrogen, whereas the target sorbates are usually much larger molecules.

Real macropores and large mesopores contribute insignificantly to the total surface area of a porous matrix. Surface area is mostly provided by pores having a diameter of less than 200 Å, typically 50–100 Å. If they are present in significant proportions, the surface area of a macroporous sorbent may amount to 400–500 m^2/g. The well-known mesoporous-adsorbing materials of the type of Amberlite XAD-4 may display a surface of up to 1100–1200 m^2/g. However, the diffusion into the fine pores may be much hindered, especially if large organic molecules are concerned. The main problem in selecting the optimal macroporous support in such a case arises from the impossibility to combine simultaneously the accessibility of the most active sorption sites (high sorption capacity) with the high diffusion rate of solutes to these sites. A user has always been forced to look for a reasonable compromise between these two important features of macroporous sorbents. In many contemporary processes, such as catalysis or chromatographic separations that involve large molecules, rapid diffusion to active sites is considered the dominating factor; it is the fast diffusion that provides a rapid turnover and high efficiency of these processes.

In flow reactors and chromatographic columns, the total rate of mass exchange would further rise considerably, if pores could become large enough to permit the mobile phase to flow through the sorbent particles, not just around them. However, suspension crosslinking copolymerization in the presence of commonly used diluents failed to produce supports with interconnected pores larger than 5000–8000 Å (being measured by mercury porosimetry, the pore dimensions can be easily overestimated). Therefore, in the last decade new approaches have been developed to overcome the limitations to the transport of liquids through the polymeric material.

CHAPTER 4

Gigaporous Polymeric Separating Media

1. FORMATION OF GIGAPOROUS TEXTURE IN THE PRESENCE OF SOLID POROGENS

An obvious approach to obtain porous polymers with very large pores (gigapores) consists of admixing fine ground inert inorganic salts, for instance, sodium chloride [358] or sodium sulfate [359, 360], to the initial monomer mixture. After the completion of network formation, the salts are easily washed out with water, leaving behind wide pores in the bulk material. Naturally, these gigapores can be useful only if they are interconnected. To secure this, large amounts of mineral filling must be involved. The remaining bulk polymer can be provided with gel-type, macroporous, or any other type of network structure.

Thus, copolymerization of glycidyl methacrylate, triallyl isocyanurate, and divinylbenzene (DVB) in the presence of porogenic solvents, toluene and n-heptane, and 40 vol of 0.7–1.5 μm sodium sulfate granules per 100 vol of monomers in a glass tube of 100×10 mm inner diameter (ID) resulted in obtaining a monolithic polymer with a biporous structure [359]. If one trusts the data of mercury porosimetry, the polymer contains both smaller pores of 100–1500 Å in diameter and very large transport pores ranging from 0.2 μm (2000 Å) to 3 μm. After introducing anion-exchange groups by opening the epoxy rings with diethylamine, particles of the crushed monolith were tested as chromatographic separation medium for proteins (lysozyme and bovine serum albumin). In contrast to another polymer obtained according to the same protocol but without salt, the new material showed low back pressure at high mobile phase velocities, up to 1260 cm/h, as well as high column efficiency (small theoretical plate heights). It is important to note that the dynamic sorption capacity of the biporous resin toward albumin was found to be very close to the static capacity, even at a high mobile phase flow rate, indicating the absence of diffusion limitations in the new material and the availability of flow-through pores.

Comprehensive Analytical Chemistry, Volume 56
ISSN 0166-526X, DOI 10.1016/S0166-526X(10)56004-6

By using powdered sodium chloride as the large-pore-forming agent, hypercrosslinked polystyrene with wide pore size distribution was prepared for the immobilization of penicillin amidohydrolase with the purpose of subsequent enantioselective hydrolysis of N-phenacetyl-D, L-phenylglycine [361–363]. The preparation of hypercrosslinked polystyrene by intensive crosslinking of the dissolved polystyrene chains with bifunctional crosslinking agents is usually accompanied by macrosyneresis of the solvent and substantial shrinking of the swollen monolithic product. Interestingly, the introduction of 40–50 vol% of the inert salt prevents the monolith from shrinking and changes the process of phase separation and pore formation. Almost no macrosyneresis takes place any more, and the rejected part of the solvent separates in the form of micro-droplets, thus forming additional fairly large pores inside the polymeric phase. The crushed particles of the final salt-modified hypercrosslinked material possess pores in a wide range from 2–3 nm, characteristic of the hypercrosslinked polystyrene, to several microns in size. The largest pores originating from the sodium chloride powder are very convenient for positioning large enzyme molecules. Immobilization of penicillin amidohydrolase in these pores was performed via functionalization of the hypercrosslinked matrix with ethylenediamine or triethylenetetramine and formation of a mixed complex between these attached ligands, transition metal ions (Zn^{2+}, Cu^{2+}), and the enzyme. Mesopores and smaller macropores of the polymeric matrix were thought to provide fast diffusion of the acylamino acid substrate molecules to the enzyme and removal of the reaction products.

Shi and Sun [364] suggested a simple way for preparing spherical polymer particles having a biporous structure. A mixture of glycidyl methacrylate, triallyl isocyanurate, DVB, toluene, n-heptane, initiator, and insoluble granules of calcium carbonate was shaken at 40°C until the viscosity of the mixture became high enough to prevent the inert salt from settling. Afterward the mixture was dispersed in aqueous media and beaded material was obtained by the usual suspension technique. Removing the carbonate provided the beads with gigapores. Via epoxy ring opening with diethylamine, the polymer was converted into an anion exchanger. Good separation of lysozyme, myoglobin, and bovine serum albumin was achieved with this chromatographic material even at the high mobile phase velocity of 1535 cm/h.

2. FORMATION OF GIGAPOROUS TEXTURE BY POLYMERIZATION OF REVERSED EMULSIONS

Menger et al. [365] suggested another interesting way for providing a styrene–DVB copolymer with extra large pores and thus increasing the outer surface of relatively large (150–250 US mesh) particles of an otherwise nonporous material. It was found that the surfactant Aerosol OT, sodium 1,4-bis(2-ethylhexyl)sulfosuccinate (AOT), being dissolved in a styrene–DVB mixture, solubilizes large amounts of water in this organic phase. At water concentration of $\sim$2–12% in the comonomers, a stable water-in-oil microemulsion forms, in which fine water droplets are dispersed and do not touch each other. Bulk polymerization of the monomers, initiated by a free radical initiator and ultraviolet (UV) irradiation at ambient temperature, fixes the droplets in the resulting matrix and produces a monolith with closed pores. After the monolith is ground, a certain portion of water micelles forms open cavities (pores) on the surface of the irregularly shaped particles. Even more pores remain closed and inaccessible for the solutes, allowing the consideration of this material as a polymer with surface porosity. The investigation of the resulting product carried out by the authors of the above publication and, subsequently in more detail, by Zhu et al. [366, 367] showed that the idea to synthesize porous polymers with reversed micellar templates is simple only at first sight. In fact, the formation of outer porosity is sufficiently complicated and not all observations can find simple and obvious explanations.

First of all, it should be noted that the surface area and pore volume formed by the open micelles on the surface of the crushed monolith are small and do not exceed $40\,m^2/g$ and $0.1\,cm^3/g$, respectively. On the other hand, the pore size may be very large and achieve 1500 and even 2000 Å in diameter. These dimensions are much larger than those of micelles formed by water prior to polymerization ($<100\,$Å) and thus, in a number of cases, the droplets must coalesce during polymerization. The coalescence leads not only to an increase of the pore diameter but also to the formation of partially interconnected pores.

The size of water droplets dispersed in the starting organic monomer liquid and the stability of the emulsion depend on the water-to-AOT molar ratio W. The smallest droplets and the most stable emulsion are formed when $W = 1$. This ratio may change from 1 to 18, but, beyond this value, the initial system becomes unstable and macro phase separation takes place. In principle, W determines the surface area, pore volume, and pore size distribution in the resulting monolith, though the correlation between W and these parameters of the monolithic porous structure is ambiguous. As a whole, the following factors govern the porosity of the materials under consideration:

(i)   When the monoliths are obtained at a constant water-to-AOT ratio and at low concentration of water, the droplets are small and located far from one another. The probability of their coalescence is low and the surface area of the final particles is also small. An increase in the water amount results in an increasing surface area, due to the increasing number of cavities. However, when the amount of introduced micelles becomes very large, the small droplets coalesce much easier, thus forming large droplets, and the surface area of the final products becomes small again. In other words, with the increasing amount of water and constant W, the surface area and pore volume increase until maximum at a certain point and then drop dramatically (Fig. 4.1).

(ii)   When the water-to-AOT molar ratio is small, for example, equal to unity, the surface area of the monolithic crushed particles is $\sim 1\,m^2/g$, because it is provided by small micelles. If now the concentration of AOT is kept constant but W is increased, then the size of droplets gradually increases while their stability decreases due to the reduced concentration of the surfactant relative to water. In addition, the number of micelles in the system can also change. With increasing W from 1 to 14, the pore volume and surface area of the particles also increase, and the maximum of pore size distribution of the styrene–40% DVB copolymer shifts from small pores of less than $100\,\text{Å}$ in diameter toward $400\,\text{Å}$, while the distribution itself becomes considerably wider. When W exceeds 14, the surface area and pore volume become independent of W. Under these conditions, the contribution of droplet size and droplet number to porosity appears to compensate each other (Figs. 4.2 and 4.3).

(iii) In the comparison of styrene–6.4% DVB copolymers obtained at a constant water concentration of 2.78 M, but with increasing concentration of the surfactant from 0.2 to 0.5 M (W is decreasing from 9.2 to 5.6), the surface area reduces from 24 to $2.2\,m^2/g$, that is, by 1 order of magnitude. This suggests that smaller number of larger microemulsion droplets may generate larger surface area than larger number of independent small droplets.

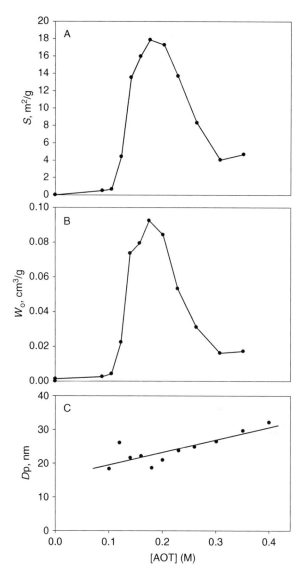

Figure 4.1 Inner surface area (a), pore volume (b), and average pore diameter (c) plotted as a function of AOT concentration. For all samples the volume ratio of styrene to DVB is 1:1, $W = 12$. *(After [367]).*

(iv) Quite surprising is the fact that the porosity of the discussed polymers depends on the degree of crosslinking. Both the surface area and the pore volume will rise more or less linearly with the DVB content

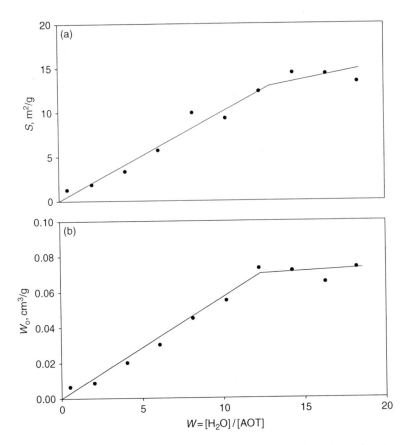

Figure 4.2 Inner surface area (a) and total pore volume (b) plotted as a function of water concentration in reversed micelles. Sample with varying W: styrene : DVB volume ratio is 1:1, [AOT] = 0.2 M. *(After [366]).*

(Fig. 4.4) (other conditions of the synthesis being the same) [367]. Most probably, rigidity of the matrix formed is responsible for the phenomenon observed.

A styrene–DVB copolymer with reversed micellar imprints was used for preparing a porous material with catalytically active sites, $(CH_3)_2N-(CH_2)_2N(CH_3)(CH_2)_{12}OCH_2-X$ (X is the styrene unit). The idea of employing the imprinted support consisted in that a marked portion of the polar functional groups attached to the styrene units are likely to be exposed to the water–polystyrene interface after

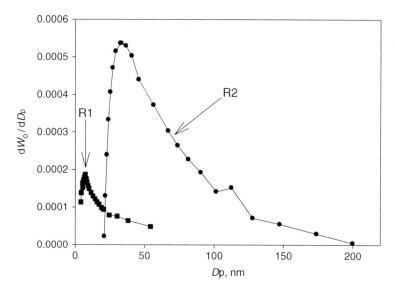

Figure 4.3 Pore size distribution curves for two polymer resins with different pore diameters. For both samples [AOT] = 0.2 M, styrene:DVB = 3:2 (v/v). R1, $W = 1$; R2, $W = 14$. *(Reprinted from [366] with permission of the American Chemical Society).*

crushing the monolith. Indeed, by using the copper complex of this diamine, the hydrolysis of *p*-nitrophenyl diphenyl phosphate was found to be catalytically accelerated and accomplished with high yield in a short period of time [368].

Powdered neutral imprinted polystyrene obtained at [H_2O] = 2.78 M and [AOT] = 0.2 M, and having a surface area of 19.4 m^2/g, was used for the preparation of a thin-layer chromatography plate and the separation of nitrobenzene, phenol, aniline, benzoic acid, and nitrophenol positional isomers. The plate showed fairly different R_f values for all these compounds, while the selectivity of a commercial silica plate was found to be much worse [365].

The above-discussed synthesis of reversed micellar templates is based on employing low-concentration emulsions, in which water forms a discrete phase while styrene and DVB form a continuous phase. Water droplets remain separated from each other and only form cavities on the outer surface of the disintegrated material. Only when the volume concentration of water is as high as 74% will its droplets start to be tangential to one another and lose their spherical form. Interestingly, the viscosity of such

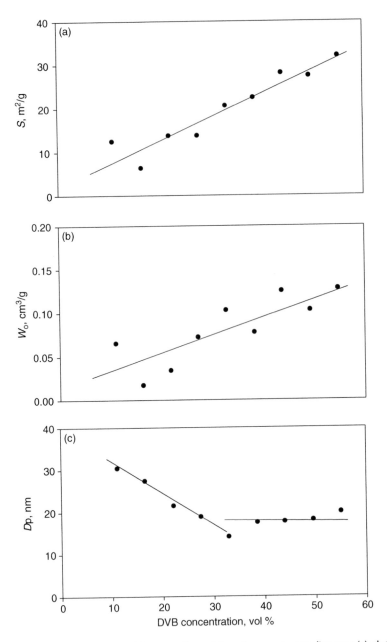

Figure 4.4 Inner surface area (a), pore volume (b), and average pore diameter (c) plotted as a function of volume percentage of pure DVB in the styrene–DVB mixture. [AOT] = 0.3 N, W = 12. *(Reprinted from [367] with permission of the American Chemical Society).*

concentrated emulsions becomes very high. During polymerization of the comonomers of an 85–95% emulsion, its water droplets merge under the formation of 1- to 3-μm openings between holes of 5–10 μm. The final very light polymer block has an open-cell structure with interconnected pores [369]. In this way solid foam samples with crosslinking densities ranging from 2 to 55% of DVB were obtained. The structural parameters of the resulting supports, referred to as PolyHIPEs (high internal phase emulsions), are given in Table 4.1.

When using the nonionic polymeric surfactant Hypermer 1070, it is possible to prepare more rigid PolyHIPEs at a water content as low as 60%; its pores are still interconnected in a united system [370]. However, judging from a large difference between the pore volume measured by mercury intrusion and that by the methanol uptake (Table 4.1), a noticeable fraction of the pores remains isolated. The specific surface area of PolyHIPEs does not exceed $20 \, m^2/g$, but it can be significantly enhanced by using a pre-cipitating solvent or toluene as co-porogens for the styrene phase, which would generate a secondary pore structure in the pore walls [371]. The technique of preparing HIPEs makes it also possible to easily introduce chloromethyl groups into the final product by using vinylbenzyl chloride as the staring comonomer. Depending on the further application of the material, the chloromethyl groups can then be subjected to various chemical transformations.

A PolyHIPE was employed in the form of a disk as support for the immobilization of flavine for further catalysis of the oxidation of benzyldi-hydronicotinamide in a flow system, demonstrating very good results [369]. PolyHIPE polymers also proved to be a useful support for solid–phase peptide synthesis [372]. In this case, large cells of the monolith were filled

Table 4.1   Physical characteristics of some poly(styrene-co-DVB) PolyHIPEs [372]

PolyHIPE	Crosslink ratio	Phase volume H_2O, %	Pore volume, mL/g		Surface area, m^2/g	
			Hg intrusion	CH_3OH imbibition	Hg intrusion	N_2 adsorption
A	20	90	5.7	5.0	18.0	21.0
B	20	95	10.4	6.4	51.9	22.9
C[a]	20	90	3.9	3.3	3.9	17.7
D[b]	55	90	9.6	3.5	9.6	137

[a] 50 % organic phase volume (chloromethylstyrene).
[b] The porogen was toluene (monomer:porogen = 1:0.5).

with a secondary soft, solvent-swollen polyamide gel support. When employed in the continuous flow peptide synthesis, such a composite provided higher yields of the peptides and higher purity of the products compared with conventional supports.

Employing an HIPE technique, poly(aryl ether sulfone) monoliths were obtained by the copolymerization of maleimide-terminated aryl ether sulfone macromonomer with styrene, DVB, or bis-vinyl ether in a solution in which petroleum ether (80% by volume) was dispersed [373]. The resulting product possessed an open-cell structure with porous cell walls and enlarged thermostability compared with poly(styrene-co-DVB) monoliths. Unfortunately the utilization of the material as a possible medium for chromatographic separation has not been reported.

The main advantage of sponge-like supports such as PolyHIPE involves the flow of the mobile phase all the way through polymer particles, permitting convective mass transfer; this proceeds much faster than conventional diffusion mass transport through the stagnant zones of a liquid phase in the pores. Fast mass transport is especially important in such analytical techniques as high-performance liquid chromatography (HPLC) or capillary electrochromatography (CEC) that are irreplaceable in contemporary science and technology. Accordingly, PerSeptive Biosystems Inc. has suggested the use of 20-μm irregular-shaped flow-through particles for the HPLC separation of proteins [374, 375]. The particles have a biporous structure, where large 6000–8000 Å pores transect the whole particle and permit convective transport of the mobile phase, while the system of interconnected smaller pores of 500–1500 Å in diameter provides diffusive mass transfer. Importantly, in these materials, the diffusion path length through the polymeric phase is short, because the distance between two neighboring transport pores is smaller than 0.5–1 μm. Consequently the materials named POROS possess excellent permeability for biomolecules. The breakthrough curve for bovine serum albumin is very sharp, almost vertical, and practically independent of the flow rate of the solution in the range between 0.58 and 2.0 mL/min, through a column of only 2.1 mm ID (Fig. 4.5). The manufacturer states that it is easy to produce and pack 20-μm particles, and the columns operate at low back pressure. A new term, perfusion chromatography, has been coined for this kind of fast chromatography. For HPLC separations, however, beaded supports are more desirable, rather than irregular particles.

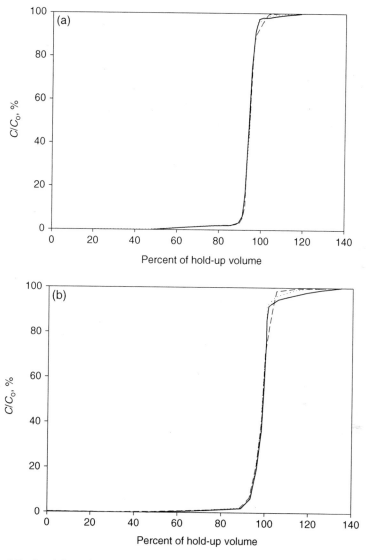

Figure 4.5 Breakthrough curves for (a) POROS Q/M and (b) POROS Q/H; 0.1 mg/mL bovine serum albumin dissolved in 20 mM Tris–HCl buffer, pH 8.0; 30 × 2.1 mm ID column; detection at 280 nm; —— = 0.58 mL/min = 1000 cm/h; ⋯ = 1.2 mL/min = 2000 cm/h; ———— = 20 mL/min = 3500 cm/h. (*Reprinted from [374] with permission of Elsevier*).

3. POROUS POLYMERIC MONOLITHS

3.1 *In situ* preparation of porous continuous polymeric beds

Fast mass transfer determining the efficiency of chromatographic columns and the rapidity of separations, high surface area of the stationary phases determining the loadability of the column, and low-column back pressure influencing the technical parameters of the instrument are the dominating targets in the development of separating systems for HPLC. Unfortunately, these requirements conflict with each other. The use of nonporous spherical beads eliminates the slow diffusion of solutes within the beads and may even provide sufficient surface area if the particles are very small, but the flow resistance of the packing becomes unacceptably high. Nonporous monosized silica particles as small as 2–4 μm were tested in the high-performance chromatography of biomolecules. Although fast separations of peptides, proteins, and DNA fragments [376–380] were indeed demonstrated, this approach holds not much promise for smaller solutes, because of high back pressure and still low binding capacity of the surface. In addition, the packing of particles, even monosized spheres, is never perfect and presents particularly high technical problems with very small beads.

Generally, a column packed with nonporous particles presents a body in which interstitial space plays a role of flow-through channels. Ideally packed monosized beads occupy up to 76% of the column volume, while in the case of irregularly shaped particles the packing density drops to 60–70%. Thus, the total flow-through volume of a column packed with a nonporous sorbent particulate cannot be increased above 25–40% of the column volume. Besides, this interstitial volume is composed of compartments having a highly complex shape and channels with strongly varying cross-sections between these compartments. Theoretically, it is desirable to increase the total flow-through volume and improve the shape of channels, without increasing their sizes above 1–2 μm. This would reduce the back pressure of the packing without increasing the diffusion paths for the solutes to the sorption sites.

In 1989, Hjertén at al. [381] suggested a promising approach to such "ideal" chromatographic separation medium by preparing a continuous gel plug immediately in the chromatographic column. They conducted *in situ* copolymerization of acrylic acid and N,N'-methylene bis-acrylamide under precipitating conditions to arrive at a continuous macroporous polymer plug in a column of 0.6 cm ID. The solvent-swollen copolymer sponge was sufficiently soft and could be strongly compressed in the column to a height of 3 cm. In this compressed state the plug preserved pores large enough to permit a flow of the mobile phase through the plug with the flow rate of up to 0.5 mL/min at a moderate pressure of 15 bar. The separation of proteins, alcohol hydrogenase, horse skeletal muscle myoglobin, whale myoglobin, ribonuclease A and cytochrome C, through ion exchange mechanism on the compressed plug of 3 × 0.6 cm ID, was very good, nearly independent of the flow rate ranging from 0.05 to 0.5 mL/min, and was performed fast, within 15 min. Authors suggested that the polymer phase is nonporous and impermeable to proteins, while the channels between the sponge walls in the compressed plug are small enough to reduce the diffusion paths for the solute molecules from the mobile phase to the surface functional groups. The method of *in situ* polymerization is simple and does not require tedious preparation of beads, their fractionation, and packing.

Three years later, Svec and Fréchet [382] showed that the *in situ* prepared macroporous continuous polymeric phase, or monolith, does not need to be compressed in the column to a dense plug, if the precipitating porogen generates an appropriate porous structure of the polymer. The main principles of the synthesis of macroporous copolymers have been already fully described in the literature, before the macroporous monolithic columns were introduced; however, the preparation of polymers with large flow-through pores required optimization of all conditions. Naturally, generation of large pore volume requires higher dilution of the comonomers with a diluent. Usually, one adds about 1 vol of the diluent per volume of the monomer mixture. Besides the type and amount of the porogen, several other variables were found to govern the formation of the flow through pores. These are the degree of cross-linking, the type and amount of the initiator, and the temperature of polymerization. Up to then the importance of the latter parameter had never been recognized in the synthesis of conventional macroporous copolymers.

The influence of temperature on the porosity of a molded monolith prepared with a precipitating porogen is described by a simple rule: the higher the temperature of polymerization, the smaller are the pores (Fig. 4.6) [383]. This tendency may be understood in terms of the "classical" mechanism of porosity formation via nucleation and subsequent aggregation of the primary nuclei. Phase separation occurs when the growing polymer becomes insoluble in the monomer–diluent mixture.

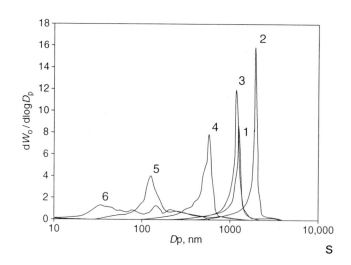

S

Figure 4.6 Differential pore size distribution curves for the sheets prepared by polymerization of 24% glycidyl methacrylate and 16% ethylene dimethacrylate in 54% cyclohexanol and 6% dodecanol at temperatures of (1) 55, (2) 60, (3) 65, (4) 70, (5) 80, and (6) 90°C. *(Reprinted from [383] with permission of the American Chemical Society).*

The precipitated polymer, mostly highly crosslinked or branched, forms primary small nonporous nuclei. Their number in the polymerization system depends, first of all, on the number of initiating free radicals, which, in its turn, depends on both the type of initiator used in the process and the temperature of polymerization: the higher the temperature, the larger is the number of free radicals formed in the initial solution. At high temperature the polymerization proceeds quickly and, probably, the primary nuclei have not enough time to form large clusters. Therefore, such a polymer has smaller pores formed by many nuclei and larger surface area than a polymer prepared at a lower temperature when favorable conditions exist for the formation of large aggregates. It is also obvious that at the same temperature, different initiators produce different amounts of free radicals. For example, the decomposition of benzoyl peroxide is slower than that of 2,2′-azo-bis(isobutyronitrile) by a factor of about 4. That is why benzoyl peroxide induces the formation of polymers with larger pores between large aggregates (Fig. 4.7).

Primary nuclei that precipitate at early polymerization stages are formed mostly by a divinyl monomer the reactivity of which is known

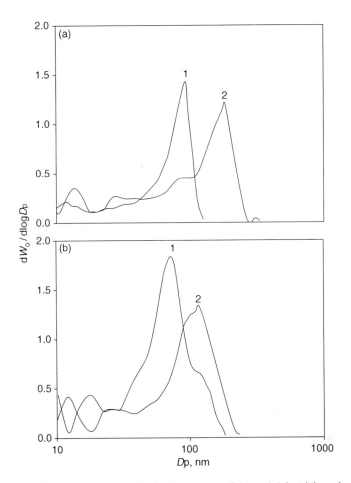

Figure 4.7 Differential pore size distribution curves of the poly(glycidyl methacrylate-co-diethylene glycol dimethacrylate) prepared by suspension polymerization at the temperatures (1) 90 and (2) 70°C initiated by (a) benzoyl peroxide and (b) 2,2'-azo(bisisobutyronitrile). *(Reprinted from [383] with permission of the American Chemical Society).*

to be higher than that of a monovinyl monomer. Therefore, swelling of the nuclei in surrounding liquid is fairly small, and the probability to grow via further polymerization in the interior of the nuclei is rather low. The primary seeds may maintain, more or less, their individuality, if further polymerization mostly proceeds in the liquid phase through the formation of new radicals, new seeds, and new nuclei. On the other hand, if the polymerization of the remaining monomers predominantly proceeds on the surface of the precipitates, aggregation of the nuclei may result in a partial or even complete loss of their individuality. It is clear that in the latter case a decrease in the micropore fraction and, hence, a decrease in the surface area can be expected with increasing conversion of the monomers. This last tendency can be illustrated by the example of copolymerization of 24% glycidyl methacrylate with 16% methylene dimethacrylate, proceeding in the presence of two precipitants, 54% cyclohexanol and 6% dodecanol, at 55 °C [384]. With an increasing conversion of the comonomers from 20 to 100%, the specific surface area of the product drops by a factor of 2, and the diameter of its largest pores shifts toward larger values. The distribution itself becomes narrower, also due to the disappearance of very large pores of 1–10 μm, which are present in the low-conversion dry rod after the extraction of both the porogens and the unreacted monomers. According to mercury intrusion porosimetry, the final product of 100% conversion has two types of pores: large pores of around 1 μm and smaller pores of 100–500 Å. Interestingly, some additional structural rearrangements still proceed in this polymer if it is heated longer than 10 h, required for achieving 100% conversion of the monomers. This observation confirms the nonequilibrium character of microphase separation and the tendency for further compactization, with a possible involvement of unreacted double bonds.

The dimensions of pores in the poly(glycidyl methacrylate-co-ethylene dimethacrylate) mold strongly depend on the concentration of dodecanol in its mixture with cyclohexanol: the larger the proportion of dodecanol, the greater the pores [384]. An analogous effect of porogen composition was also observed for the copolymer of butyl methacrylate, ethylene dimethacrylate, and 2-acrylamido-2-methyl-1-propanesulfonic acid prepared in a mixture of water, 1-propanol, and 1,4-butanediol. As shown in Fig. 4.8, it is possible to accomplish a fine control over the pore size distribution by simply changing the ratio of 1-propanol to 1,4-butanediol. Adjustment of rod porosity presents a tool for optimizing both the chromatographic efficiency and the electroosmotic flow through the monolith used in CEC [385]. Optimization of the monomer composition and the ratio of 1-propanol to 1,4-butanediol was also adopted by Grafnetter et al. [386] to prepare a poly(butylmethacrylate-co-ethylene dimethacrylate) monolith for the capillary liquid chromatography of low molecular weight organic molecules. The monolithic column exhibited good performance, and its separating properties did not change in a wide range of flow rates and applied pressures, up to 30 MPa.

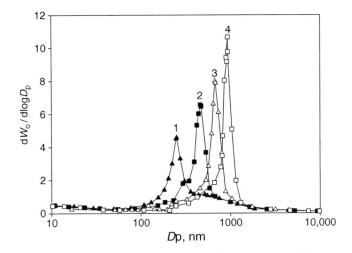

Figure 4.8   Differential pore size distribution curves of porous polymer of monolithic capillary columns obtained by *in situ* polymerization of 40% ethylene dimethacrylate with 60% mixture of butyl methacrylate and 2-acrylamido-2-methyl-1-propanesulfonic acid in 10% water and 90% mixture of 1-propanol and 1,4-butanediol taken in various ratios; mode pore diameter: (1) 255, (2) 465, (3) 690, and (4) 1000 nm. (*Reprinted from [385] with permission of the American Chemical Society*).

A glycerol dimethacrylate (GDMA)-based polymeric monolith with a very favorable three-dimensional (3D) continuous skeletal structure was reported to form by applying a 1–5% solution of monodisperse ultra-high molecular weight polystyrene (M up to 3,840,000 Da) in chlorobenzene as the porogen [157]. At the GDMA/porogenic solution ratio of 35/65 (v/v), the monolith exhibits two modes of pores, namely, with a narrow maximum at 3–4 nm in the mesopore region and 1–2 μm in the macropore region.

The fact that alcohols, too, induce the formation of large pores in molded rods is well consistent with the same observation for beaded polymers prepared by suspension polymerization. In general, suspension polymerization is considered to be bulk polymerization on a mini scale. Hence, all dependences that are characteristic of bulk polymerization would have to be valid for both the molded rods and the beads. Indeed, poly(glycidyl methacrylate-co-ethylene dimethacrylate) obtained under identical conditions in the form of rod or beads have similar surface area and pore volume values. Both polymers can be expected to have more or less identical pore size. Meanwhile, measured by mercury intrusion porosimetry, the mean diameter of pores was found to be much larger for the rod. Svec and Fréchet [384] adhere to the opinion that in the suspension variant of polymerization, the aqueous phase–organic phase interfacial tension causes shrinking of the beads, thus forcing the nuclei to aggregate closer. At the same time they suggest that it is hardly the only reason for this observation. Indeed, this explanation would contradict the fact that the pore volume is independent of the polymerization mode. The basic "reason" rather lies in the method of mercury intrusion used by the authors for the pore size determination.

The mercury intrusion method has often given inadequate results with porous polymers, as the resulting pore size distribution curve more likely reflects the pressure under which the porous polymer gets compressed. Thus, when beaded styrene–6% DVB copolymer obtained in the presence of n-butyl alcohol is subjected to mercury porosimetry, it will show, due to its inadequate rigidity, simulated pores as large as 4–5 μm [387]. Irregularly shaped particles of a crushed rod are especially liable to compression. Spherical beads having the same porous structure are more resistant. Because of the need of higher pressures, the pores in spherical beads may seem to be smaller. To answer the question of whether a molded rod and suspension beads have the same structure, all one needs to do is to compare the properties of crushed molded rod and crushed beads. The inadequacy of the mercury intrusion porosimetry may also explain why this technique and inversed size-exclusion chromatography give different information about the structure of monolithic rods [388].

It might be pointed out here that sometimes, depending on the conditions of suspension polymerization, the structure of the beads' outer layers may strongly differ from that of the beads' core. The core–shell structure of beads may contribute to their pressure resistance. Also, the pore openings on the shell may be smaller than the pore diameters within the bead. It seems plausible that the dense structure of the beads' outer surface may also be responsible for the apparent absence of large flow-through pores in highly porous beads obtained in the presence of alcohols.

The porosity of molded rods depends on the crosslinking density. With increasing percentage of DVB in the styrene–DVB monolith, the pore diameter gradually decreases, while the surface area increases [389, 390]. Accordingly, the highest surface area is characteristic of a rod based on 80% $tech$-DVB, especially when toluene was employed as the diluent. Toluene and long-chain aliphatic alcohols are typical porogens, widely used for the synthesis of beaded macroporous polymers, and their influence on the porosity is well known. As an added bonus, the technique of molded rod preparation permits the use of those porogens that are water miscible and, therefore, have never been used in the suspension variant of copolymerization. Thus, polymerization of 80% *tech*-DVB in the presence of tetrahydrofuran results in obtaining of rods with fine-grained texture, with a surface area of $820 \, m^2/g$ and narrow pore-size distribution with the maximum located at $\sim$50 Å. Acetonitrile provides a very wide pore size distribution and a moderate surface area of $325 \, m^2/g$. In contrast to these two good solvents, pure methanol is not favorable for the preparation of porous materials. Styrene–DVB copolymerization occurring in methanol at 65°C results in monoliths exhibiting a surface area of less than $1 \, m^2/g$ and a negligible pore volume of $0.01 \, cm^3/g$. Since the monomer-to-porogen volume ratio in the synthesis was fairly large, 1:1, these pore volume values suggest that macrosyneresis takes place predominantly during the phase separation. Scanning electron micrographs document that the texture of this polymer is composed of globular species of several microns in diameter, these species being packed fairly tightly [390]. Nevertheless, methanol proved to be a useful additive to other porogens such as the mixture of water and ethanol; all the three co-porogens provide the formation of porous monoliths acceptable for chromatographic separations [309].

Porous molded monolithic rods were designed as separating media for HPLC and CEC. From this standpoint, the radial and axial homogeneity of the porous structure and quite probably the contraction of rods during polymerization are the questions of primary importance.

The homogeneity of the rod's porous structure was found to be strongly dependent on the conditions of polymerization and, above all, the temperature that represents a general powerful tool controlling the porosity of rods. When a rod is small, for example, $50 \times 8 \, mm$ ID, the copolymerization of glycidyl methacrylate with ethylene dimethacrylate in a mixture of cyclohexanol and dodecanol at 55 or 70°C results in homogeneous rods, the porous structure of which is identical at the bottom and at the top of the rod. However, the structure of the rod changes when the polymerization is carried out at 55°C for 1 h followed by heating at 70°C for additional 14 h. The top and the bottom parts of the rod acquire the same large pores of $\sim$1.2 μm, but the top of the rod also gains markedly larger portion of smaller pores. This observation indicates that it is possible to set up a porosity gradient along the column bed by using a complex temperature profile. Pore gradient might prove to be useful in a number of separations [384].

The preparation of homogeneous rods in narrow tubes is simple, because the heat resulting from polymerization readily dissipates. However, the evolving heat represents a serious problem when preparing large-scale rods. Copolymerization of styrene with 50% DVB in a tube of 20×5 cm ID initiated by 1% benzoyl peroxide at 70°C leads to a 39°C jump in the temperature in the center of the rod. The rod will crack, due to exceeding the boiling point of toluene used as co-porogen with 1-dodecanol. Reducing the temperature of polymerization to 60°C will diminish the exothermic effect; in this case it amounts only to 3°C, but the polymerization proceeds much slower, providing pores with diameters as large as 20 μm [391].

The character of pore size distribution over the entire volume of the 20×5 cm ID styrene–DVB macroporous rod may be improved by conducting radical living polymerization in dodecanol in the presence of 2,2,6,6-tetramethyl-1-piperidyloxy-based initiating system conjugated with benzoyl peroxide and acetic anhydride [392]. The living radical polymerization proceeds slowly even at 130°C and, therefore, the exothermic effect is small. The polymeric product has no radial or axial heterogeneity. The rod derived from this reaction is characterized by a wide pore size distribution and enlarged surface area, but by reduced fraction of flow-through pores, compared with rods prepared by conventional free radical polymerization.

Initiation of the living polymerization with 2,2,5-trimethyl-3-(1-phenylethoxy)-4-phenyl-3-azahexane allows reducing the reaction temperature to 110°C. However, the desired flow-through pores with diameters around 1 μm can be obtained only when octadecanol or a 5% solution of polystyrene of 2.8×10^5 Da molecular weight in toluene is employed as the porogen [393]. Another efficient porogen is a combination of poly(ethylene glycol) 400 with n-decanol [394].

The capped radical-generating centers of living polymerization, located on the surface of monolith pores, can be further used for functionalization, by grafting various monomers, for example, vinylbenzyl chloride, tert-butyl methacrylate, or vinylpyridine [393], as well as 2-hydroxyethyl methacrylate and 3-sulfopropyl methacrylate [394]. Another possibility for changing the surface chemistry is the involvement of pendent double bonds that remain on the pore surface of highly crosslinked styrene–DVB rods, in a variety of chemical reactions; these have been reviewed by Hubbard et al. [395].

Thus, by varying the composition of the porogenic solvents, the temperature of polymerization, or the type of initiator, the structure of molded macroporous monolithic rods can be adjusted for the particular tasks of chromatographic separation. Many findings discussed in the above-mentioned publications are also valid for suspension processes, and, thus, they are useful additions to our basic knowledge about the formation of macroporosity. Obviously, some aspects of the polymerization are more important for the *in situ* bulk process than for the suspension procedure. For example, the density of solid polymers is well known to be higher than that of liquid monomers and, therefore, the volume of the polymer will unavoidably be smaller than the initial volume of the monomers. While some reduction in the bead volume is irrelevant for the further application of the beaded sorbent, the contraction of a rod could result in the appearance of free space between the rod and the walls of the mold; nevertheless, Svec and Fréchet [384] did not observe this phenomenon. According to their reports, the rod grows slowly from the bottom to the top of the mold standing vertically in a bath, and, thus, the unreacted liquid monomers at the top have enough time to fill the space arising from gel contraction at the bottom. The resulting rod exhibits no radial shrinkage; rather, its length will become somewhat shorter than the initial height of the liquid monomers in the tube.

Peters et al. [385, 396] believe that it is not necessary to bind the continuous column bed to the walls of the fused-silica capillary; even with having an unattached bed, durable columns can be prepared and excellent and fast separations can be achieved (Fig. 4.9). However, other researchers, as was mentioned in the review of Zou et al. [397], prefer chemical binding of the bed, in order to avoid its possible displacement under electroosmotic flow or the applied pressure. In addition, the chemical modification of the walls followed by binding of the continuous bed reduces the possibility of interactions of the analytes and the mobile phase with the fused-silica surface. Such modification may be done by preliminary vinylization of the capillary tube's inner surface, for example, with γ-methacryloyl-oxypropyl-trimethoxysilane, and subsequent involvement of introduced double bonds in the copolymerization with styrene–DVB monomers forming the rod [398]. It is interesting to note that boiling of a glass capillary tube in deionized water followed by simple silylation was also found to enhance the strength of monolith attachment to capillary walls [399].

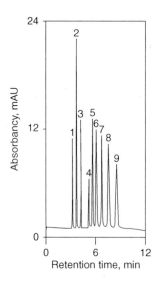

Figure 4.9 Electrochromatographic separation of benzene derivatives on the selected monolithic capillary. Conditions: capillary column 100 μμm ID × 30 cm active length; stationary phase: 0.3 mol% of 2-acrylamido-2-methyl-1-propanesulfonic acid in monomer mixture; pore size: 750 nm; (1) thiourea, (2) benzyl alcohol, (3) benzaldehyde, (4) benzene, (5) toluene, (6) ethylbenzene, (7) propylbenzene, (8) butylbenzene, (9) amylbenzene. *(Reprinted from [396] with permission of the American Chemical Society).*

3.2 Polymeric monoliths in chromatography and electrochromatography

Continuous bed columns of various diameters are obtained by *in situ* crosslinking polymerization of the comonomers in the column, and, consequently, this technique eliminates the tedious procedures of preparing polymeric beads, their fractionation, and packing the column, as well as any manipulations with frits. Highly important is that the monoliths can be even more reproducible and stable than the packed HPLC columns. Still, the most outstanding property of continuous bed packings is that their chromatographic performance estimated by the number of theoretical plates per unit length is practically independent of the flow rate.

Since continuous rod columns have no interstitial space, all mobile phase is forced to flow through the large transport pores. As experiments have shown, with a proper porous structure the back pressure of continuous bed columns can be kept sufficiently low. For example, it amounts to only 6.5 MPa on percolating an aqueous–organic mobile phase with a flow rate of 23 mL/min through a 50×8 mm ID styrene–DVB rod having gigapores as large as 19.5 μm [400]. Certainly, the back pressure further drops with decreasing flow rate, while the efficiency of reversed-phase chromatography on the column increases. At a flow rate of 1 mL/min, this short rod can completely separate two peptides, bradykinin and a similar peptide differing only in the seventh amino acid residue (D-phenylalanine instead of L-proline).

A molded styrene–DVB rod prepared in the presence of dodecyl alcohol, having largely pores of 10 μm in diameter, proved to be unsuitable for conventional size-exclusion chromatography of polymers. It did not separate at all the polystyrene standards within the molecular weight range from 10^4 to 10^6 Da, while a difference in the retention volumes of oligomers with molecular weights of 10^2–10^4 was noticeable, although very small. This difference is indicative of the presence of relatively small pores in the rod, which mercury porosimetry did not show. Still, an excellent separation of polystyrene standards proved to be possible in the precipitation–redissolution mode (Fig. 4.10) [401]. The principle of this chromatographic technique involves injection of a solution of the macromolecules to be separated, for example, polystyrene standards of different molecular weights, into the stream of aqueous methanol or aqueous acetonitrile used as the mobile phase that is a poor solvent for these molecules. The injected polymers precipitate and adsorb onto the surface of the porous support in the vicinity of column inlet. Then, the solvation power of the mobile phase is progressively raised by the addition of a good solvent, for example, tetrahydrofuran. In this way at a certain moment, the low molecular weight polystyrene molecules start to redissolve. They move along the large pores with the mobile phase, but the latter has a longer diffusion path, since it can penetrate into all available pores, while the diffusion path of the dissolved polystyrene coils is limited to larger pores. Because of this, the dissolved molecules tend to move ahead of the mobile phase gradient; they soon find themselves in an unfriendly environment and precipitate again. The redissolution takes place later, when the local thermodynamic quality of the mobile phase again exceeds the required threshold value. This process repeats itself many times, until all molecules leave the column in accordance with their increasing molecular weight. Interestingly, Fig. 4.10 shows no difference in the separation of polystyrene standards carried out at low and high flow rates, as well as at slow and fast gradients. Indeed, the precipitation–redissolution proceeds on the surface of the large pores in the rod and is a fast process. This separation mode of macromolecules had been

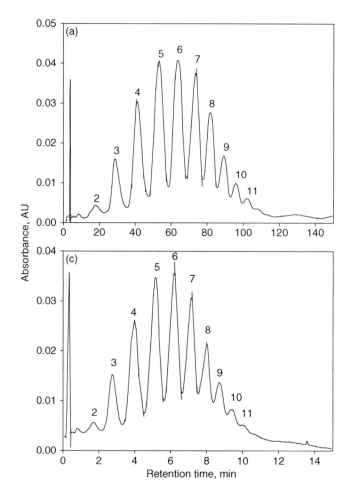

Figure 4.10 Effect of flow rate and gradient time on the separation of styrene oligomers in molded poly(styrene-co-divinylbenzene) rod. Conditions: column, 50 × 8 mm ID; (a) mobile phase, linear gradient from 60 to 30% water in tetrahydrofuran within 200 min; flow rate, 1 mL/min; (b) mobile phase, linear gradient from 60 to 30% water in tetrahydrofuran within 20 min; flow rate, 10 mL/min; UV detection, 254 nm; peak numbers correspond to the number of styrene unit in the oligomer. *(Reprinted from [401] with permission of the American Chemical Society).*

elaborated in detail by Glökner [402], and may be an efficient alternative to conventional size-exclusion chromatography of some polymers. Gigaporous polymeric molds are the best possible separating media for this kind of chromatography.

Nowadays, porous monoliths have found an extensive use in CEC of organic compounds, which represents a powerful separation tool, complementary to HPLC. CEC is a hybrid method in which the separation is performed through the phase distribution mechanisms of traditional HPLC (reversed-phase, ion exchange, etc.), while the flow of the mobile phase through column packing is affected by electroosmotic forces, as in electrophoresis. The coexistence of a stationary phase and an electric field permits separation not only of ions but also of neutral compounds, due to their different electrophoretic mobility and different distribution between the mobile and the stationary phases.

In order to generate an electroosmotic flow of the mobile phase in the applied electric field, the surface of the pores of a monolithic polymer used in CEC must have charges. To introduce a charge, the monomer mixture is usually provided with a certain portion of a comonomer capable of electrolytic dissociation. Such an example is 2-acrylamido-2-methyl-1-propanesulfonic acid, which was used for preparing butyl methacrylate-based cation-exchanging monoliths [385, 396]. Another approach to charged monoliths consists of involving functional sites located on the polymer surface in chemical transformation. For example, positively charged ammonium groups have been introduced into the copolymer of vinylbenzyl chloride and DVB by treatment with N,N-dimethyloctylamine [403] or diaminoethane [404]. With increasing amounts of the charged sites, the mobile phase flow rate through the continuous bed under the action of electric field significantly increases. However, when the concentration of the polar groups is high, the surface of the polymer becomes hydrophilic, and the separation of nonpolar alkylbenzenes via reversed-phase mechanism becomes poor. The porous structure of the monolith can be compromised by the introduction of polar comonomers, as was the case with poly (butyl methacrylate-co-ethylene dimethacrylate) rod containing certain amounts of the above propanesulfonic acid. To resolve these contradictions, it was suggested to adjust the pore structure by increasing the proportion of 1-propanol in the mixed porogenic solvent, water, 1-propanol, and 1,4-butanediol. An optimized monolith exhibited excellent separation of thiourea, benzene, and alkylbenzene within 5–7 min, even at a high concentration of the sulfonic acid functions [385].

Fig. 4.11 demonstrates a twofold increase in the mobile phase flow rate through monolithic capillary columns when increasing their pore diameter from 250 nm to 1.3 μm. This finding contradicts the generally accepted perception that in traditional CEC on packed columns, the rate of

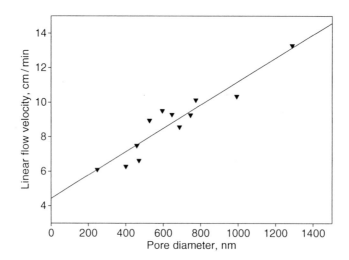

Figure 4.11 Effect of mode pore diameter on flow velocity of mobile phase through monolithic capillary columns. Conditions: capillary column, 100 μm ID × 30 cm active length; stationary phase with 0.3 wt% 2-acrylamido-2-methyl-1-propanesulfonic acid; mobile phase, 80:20 v/v mixture of acetonitrile and 5 mmol phosphate buffer, pH = 7. *(Reprinted from [416] with permission of Wiley & Sons Inc).*

electroosmotic flow is independent of the particle size. Indeed, insofar as the size of the beads is usually small, 2–4 μm, and the dimensions of the interstitial voids are comparable with the channels in monolithic rods, one may assume that the pore size–electroosmotic flow rate relationship in CEC separations would have to be similar on both monoliths and packed columns. Peters et al. [385] believe that a decrease in the bead diameters results in increasing number of charges exposed to their surface, and the increasing elecroosmotic driving force compensates for the increase in column resistance. In the case of monolithic rods, the dependence of the electroosmotic flow on pore size simply reflects the decrease in bed resistance with increasing dimensions of the large channels.

The very good performance of molded polymeric stationary phases in HPLC and CEC at high flow rates is especially important when sensitive biomolecules are separated. Thus, the reversed-phase separation of five proteins was easily achieved in less than 20 s on a styrene–DVB rod with an optimized porous structure [405]. The extensive use of molded polymeric monoliths as stationary phases in HPLC and CEC demonstrated within the last decade their wide applicability and many inherent advantages over traditional beaded supports. Glycidyl methacrylate rods, after reacting their epoxy rings with amines, were successfully used in the anion exchange chromatography of peptides, proteins [406], and oligonucleotides [407]. The 300×8 mm ID rod of this polymer permits the preparative separation of a mixture consisting of 66% chicken egg albumin and 33% bovine serum albumin: 300 mg of the mixture could be baseline separated within 60 min [408]. Glycidyl methacrylate monolithic disks were used for fast processing and in-process control of biopolymers [409]. Such disks with immobilized ligands were also employed in the affinity chromatography of biomolecules and enzymatic conversion of substrates [410]. The copolymer of hydroxymethyl methacrylate with ethylene glycol dimethacrylate and N-methacryloyl-(L)-histidine methyl ester was suggested for the isolation and purification of immunoglobulin IgG from human plasma [411]. A representative of another class of porous polymeric materials, urea–formaldehyde monolith obtained by the condensation of the monomers in water, can be used after its modification with Cibacron blue F3GA dye as the separating medium for human serum albumin, ovalbumin, and lysozyme in affinity chromatography [412]. A porous acrylic template–imprinted monolith was employed for the molecular recognition of positional isomers of diaminonaphthalene and the enantiomeric resolution of racemic phenylalanine amide [413]. The use of a poly(methacrylic acid-co-ethyleneglycol dimethacrylate) rod, imprinted with pentamidine, was recommended for the analysis of this medication [414]. Pentamidine, a drug used for the treatment of AIDS-related pneumonia, possesses undesirable side effects, and, therefore, the control of its concentration in physiological fluids is very important. Since the concentration of pentamidine in urine is very low, at the nanomole level, its quantitative measurement requires the preliminary enrichment of the probe. This was carried out by selective on-column adsorption of the analyte on a 150×3 mm ID column rod. The same column with an appropriate eluent was then used as the separating medium in the subsequent chromatographic analysis of the drug.

A small, 20×1 mm ID, monolithic rod, obtained by the polymerization of 80% *tech*-DVB in a mixture of dodecanol and toluene, was employed for the solid-phase extraction of polar organic compounds from aqueous solutions [415]. It was demonstrated again that by maintaining constant dilution of the monomer and changing only the weight percent ratio of toluene to dodecanol from 2:58 to 10:50, it is possible to increase the surface area of the sorbent from 188 to 348 m²/g and to reduce the mean pore diameter from 7.59 to 0.39 μm; the total pore volume remains constant and is equal to 1.7 cm^3/g. This material proved to be poorly suited for the extraction of the polar phenol, with the recovery of only 58%; at the same time the recoveries for nitro and chloro derivatives of phenol were quite acceptable, between 77 and 88%. To enhance the adsorption of polar compounds and the wettability of the internal surface of the poly-DVB monolith, Xie et al. [415] suggested adding some amount of 2-hydroxyethyl methacrylate to the mixture of 91% tech-DVB, tetradecanol, and a small quantity of toluene. The resulting monolithic rod showed excellent recoveries for phenol, nitrophenols, and chlorophenols, ranging from 91 to 97%. Many other applications of the styrene–DVB-, glycidyl methacrylate-, or acrylamide-type rods in HPLC and CEC separations have been reviewed in the literature [397, 416–420].

More exotic porous monolithic rods have also been prepared and tested in chromatographic analyses, for example, a rod obtained by ring-opening metathesis polymerization of norborn-2-ene and 1,4,4a,5,8,8a-hexahydro-2,4,5,8-exo,endo-dimethanonaphthalene in different solvents via the mechanism of living polymerization [421]. The polymers were found to have a regular microstructure composed of porous micro-globules, the size of which may be adjusted between 2 and 30 μm. The unfunctionalized rods demonstrate good chromatographic separation of proteins in reversed-phase mode. Functionalization of the polymer by in situ derivatization results in rods with a wide variety of surface chemistry, as illustrated by the example of the enantioselective separation of proglumide (β-blocker) on a β-cyclodextrin-modified continuous rod.

Another interesting approach to polymeric monoliths consists of obtaining block copolymers of styrene with 40% D,L-lactide [422]. In this copolymer lactide forms cylindrical fragments randomly embedded in the polystyrene matrix. Applying shear to the melted copolymer results in the alignment of the cylinders. After removing the lactide fragments by hydrolysis under mild conditions, a porous monolith is obtained with vertically oriented pores of 16 μm in diameter.

It should be mentioned that silica-based monolithic columns are even more widely used in CEC and HPLC compared with various polymeric rods (see, e.g., *Journal of High Resolution Chromatography, Special issue on monolithic columns*, vol. 23, N1, 2000 or [423]). In addition, even carbon-type monoliths have started to be commercially available. They are manufactured (MAST Carbon Ltd., Guildford, UK) by carbonization and subsequent activation of extruded phenolic resins [424]. The MAST Carbon square channel monolith has a peculiar honeycomb structure with cell density of $\sim$89 cells/cm^3, the size of the cells being 0.6×0.4 mm. The approximately 300- to 500-μm thick cell walls represent a porous material having a BET surface area of 800 m^2/g, and micropores and macropores of 4 Å and 5 μm diameter, respectively. Curiously, the carbon has no mesopores at all. This material has not been tried yet as a separating medium for liquid chromatography, but large columns packed with short monoliths have been successfully used for the adsorption of volatile organic compounds.

Finally, it is interesting to mention that capillary monolithic columns have also started to be used in gas adsorption chromatography [425, 426]. Poly-DVB monolith obtained in the presence of 1.5 vol of dodecanol–toluene mixtures possesses good separating power; however, its efficiency (the theoretical plate height) still yields by a factor of 3–10 to that of traditional open capillary columns. On the other hand, the theoretical plate height for a similar monolith prepared for use in liquid chromatography proved to be comparable with that of conventional capillary silica-packed column [427].

This chapter demonstrated the success we can achieve in the separation of small and large organic molecules by means of continuous bed columns. Certainly, the monolithic columns are not an absolute alternative to the conventional beaded packing materials but Guiochon finds reasons to say: "The recent invention and development of monolithic columns is a major technological change in column technology, indeed the first original break-through to have occurred in this area since Tswett invented chromatography, a century ago" [428].

CHAPTER 5

Isoporous Anion-Exchange Resins

Styrene–divinylbenzene (DVB) copolymers show a remarkable advantageous property: they can readily enter into many chemical transformations. For that reason they have enjoyed well-deserved popularity as the matrices in the synthesis of various catalysts, polymeric reagents, chromatographic separating media, and so on. Above all, styrene–DVB copolymers present an indispensable starting material for functionalization aiming at the preparation of numerous types of ion-exchange resins. Thus, sulfonation proceeds when treating the copolymers swollen in ethylene dichloride with concentrated sulfuric acid under rather mild conditions. As a conclusion of the reaction, strong cation exchangers are obtained with a reproducible capacity of up to 5 meq/g, with well-controlled swelling in water, selectivity, and high stability. The preparation of anion-exchange resins is more complicated. At first chloromethyl groups need to be introduced, the reactive chlorine atoms of which are then substituted by any appropriate amino group in the second reaction step, the so-called amination process. At present, this protocol is accepted by all manufacturers of the anion-exchange resins. However, the properties of resins produced on industrial scale by different manufacturers were found to differ, the difference being sometimes significant.

As early as 1964, Anderson [429] paid attention to the fact that some standard commercial quaternary ammonium anion-exchange resins exhibit essentially enhanced osmotic stability compared with other resins of the same type. He found links between the resistance of the resins to osmotic shock and the conditions of the chloromethylation reaction of the starting styrene–DVB copolymers. The analysis of numerous data led him to conclude that chloromethylation is accompanied to a larger or smaller extent by the formation of additional crosslinks in the polymeric network. Such links are formed by a side reaction of the chloromethyl groups with the unsubstituted phenyl rings of the polystyrene chains, which occurs in the presence of Lewis acids and results in the introduction of methylene bridges between phenyl ring pairs. The new crosslinking bridges are introduced into the swollen copolymer, and they drastically change the network topology and, hence, the properties of the product. At the same ion exchange capacity and swelling in water, the anion exchangers having a larger proportion of methylene-type to DVB-type bridges distinguish favorably by the higher mechanical strength of the beads and lower "breathing" the change in the bed volume during resin regeneration. As these are the properties that significantly extend the lifetime of commercially used anion exchangers [430], some manufacturers adopted the simultaneous conducting of the two processes of chloromethylation and additional methylene bridging of the styrene–DVB matrix; this process serves as the basis for the industrial production of the so-called isoporous anion-exchange resins [431, 432]. The term "isoporous" comes from the notion that methylene bridges arise statistically throughout the volume of a swollen bead, contrary to the very nonuniformly distributed DVB-type bridges. For this reason, the network meshes ("pores") of the methylene-bridged copolymer must have a narrower size distribution.

As to the laboratory procedures of preparing isoporous chloromethylated copolymers, these are just little different from the findings and suggestions of Anderson. Most often, one follows an obvious path of heating a swollen, slightly crosslinked styrene–DVB copolymer with excess monochlorodimethyl ether, CH_3OCH_2Cl, in the presence of a Friedel–Crafts catalyst [433–435]. Hauptmann and Schwachula [436] carried out the chloromethylation and bridging of beaded styrene–1% DVB copolymers with a mixture of methylal, $(CH_3O)_2CH_2$, chlorosulfonic acid, $HOSO_2Cl$, and sulfuryl chloride, SO_2Cl_2. The authors also noted that this mixture is appropriate for the crosslinking and comprehensive chloromethylating of a linear polystyrene, a beaded isoporous product being obtained when the solution of the linear polystyrene in methylal and sulfuryl chloride is suspended in chlorosulfonic acid. With the concept of simultaneous introduction of crosslinks and functional groups into a swollen matrix, the preparation of isoporous sulfonates also became possible by treating beads of slightly crosslinked styrene–DVB copolymers with the mixture of paraformaldehyde, chlorosulfonic acid, and sulfuric acid [437]. The resulting cation-exchange resins demonstrated high exchange capacity, but, contrary to expectations, the mechanical strength of beads was found to be insufficient for the practical application of the resins. However, it may be slightly improved by introducing a third comonomer, acrylonitrile, into the initial styrene–DVB copolymers.

Surprisingly, one cannot find any interesting or useful information from the publications concerning the structure and properties of isoporous resins other than their water regain and exchange capacity. All researchers have noted that the degree of the additional methylene crosslinking of styrene–DVB copolymers strongly depends on both the conditions of the chloromethylation reaction and the swelling extent of the initial copolymers, that is, their DVB link content. However, no reliable experimental methods exist for the determination of the degree of methylene bridging in isoporous resins. As a measure of methylene bridging, Laskorin et al. [438] suggested to use the ratio of optical densities of the band of out-of-plane deformational vibrations at $831\,cm^{-1}$ (C–H groups in *para*-disubstituted benzene rings) and of the polystyrene band at $865\,cm^{-1}$. This method, however, may be considered useful only if the concentration of the DVB links in the initial copolymer is negligible, so that DVB itself does not contribute to the infrared (IR) spectrum. Trushin et al. [439–441] also attempted to qualitatively estimate the degree of methylene bridging in the products of post-crosslinking of swollen chloromethylated styrene copolymers with 0.5–6% DVB via Friedel–Crafts reaction. They measured both the amount of hydrogen chloride released during the crosslinking reaction and the change in swelling ability of the beads. The authors were confused by the unexpected finding that the appreciable amount of the released hydrogen chloride does not correlate with the very insignificant drop of the swelling ability of the polymers. When judging by the swelling properties, they thought of a very low efficiency of the post-crosslinking process. On the other hand, intensive HCl evolution could have no explanation other than the formation of many methylene bridges.

Belfer et al. [442–444] studied the structure of polymers prepared by a similar procedure. The quantity of chloromethylating agent (equimolar mixture of methylal and chlorosulfonic acid) [445], time and temperature of the reaction, as well as the type of solvent were found to strongly influence the swelling and morphology of isoporous products. With regard to the resin texture, pores of several hundred angstroms to several microns in diameter and even "clusters of polystyrene crystals" were found when using scanning electron microscopy. The formation of a "heterogeneous supramolecular structure" was accounted for by phase separation induced by fast and nonuniform appearing of new links during the "cationic post-polymerization." These pretentious conclusions, however, were based on typical artifacts of electron microscopy. Grebenyuk et al. [446], who studied the electroconductivity of isoporous anion-exchange resins, also consider the structure of the isoporous matrix to be more heterogeneous than that of conventional resins.

Interestingly, having discovered the post-crosslinking effect, Anderson [429] was not sure at all about the statistical distribution of the new links. He even wrote "... the methylene crosslinking is not introduced randomly. Since the methylene bridging reaction is very dependent on the proximity of aromatic rings, it is probable that the reaction will take place where two polymer chains are intertwined. Once such a new crosslink formed, it increases the probability that another link will occur in the same vicinity. This would lead to accelerated, localized crosslinking which would contract the polymer." Followers of this opinion may be found even today [447]. Without going into a debate on this viewpoint (we shall deal with this in Chapter 6), we only note that these conclusions appealingly damped the scientific interest in isoporous networks, although the industrial production of isoporous anion-exchange resins never ceased.

In spite of the negative opinion of many investigators regarding the synthesis of isoporous ion-exchange resins, we assume that the post-crosslinking approach can be considered as an important first step in the desired direction of improving the network structure.

REFERENCES TO PART ONE

[1] Wiley, R.H., Sale, E.E. J. Polym. Sci. 42 (1960) 491–500.
[2] Wiley, R.H., Davis, B. J. Polym. Sci. B: Polymer Lett. 1 (1963) 463–464.
[3] Wiley, R.H., Mathews, W.K., O'Driscoll, K.F. J. Macromol. Sci., A: Chem. 1 (1967) 503–516.
[4] Wiley, R.H., Prabhakara, S., Jin, J.-I., Kim, K.S. J. Macromol. Sci. A4 (1970) 1453–1462.
[5] Wiley, R.H. Pure Appl. Chem. 43 (1975) 57–75.
[6] Malinsky, J., Klaban, J., Dušek, K. Coll. Czech. Chem. Commun. 34 (1969) 711–715.
[7] Popov, G., Stiebitz, V., Schwachula, G. Plaste Kautsch. 27 (1980) 671–675.
[8] Popov, G., Stiebitz, V., Schwachula, G. Plaste Kautsch. 28 (1981) 10–11.
[9] Kwant, P.W. J. Polym. Sci.: Polym. Chem. Ed. 17 (1979) 1331–1338.
[10] Hild, G., Rempp, P. Pure Appl. Chem. 53 (1981) 1541–1556.
[11] Popov, G., Stiebitz, V., Schwachula, G. Acta Polym. 30 (1979) 412–415.
[12] Popov, G., Schwachula, G. Plaste Kautsch. 26 (1979) 374–377.
[13] Wiley, R.H., Tae-Oan-Ahn, J. Polym. Sci. A6: Polym. Chem. (1968) 1293–1298.
[14] Hild, G., Okasha, R. Makromol. Chem. 186 (1985) 93–110.
[15] Frick, C.D., Rudin, A., Wiley, R.H. J. Macromol. Sci. A, 16 (1981) 1275–1282.
[16] Storey, B.T. J. Polym. Sci. A3 (1965) 265–283.
[17] McFarlane, R.C., Reilly, P.M., O'Driscoll, K.F. J. Polym. Sci.: Polym. Chem. Ed. 18 (1980) 251–257.
[18] Schwachula, G. J. Polym. Sci.: Polym. Symp. N53 (1975) 107–112.
[19] Schwan, T.C., Price, C.C. J. Polym. Sci. 40 (1959) 457–468.
[20] Alfrey, T., Jr., Goldfinger, G. J. Chem. Phys. 12 (1944) 322.
[21] Schwachula, G., Wolf, F., Kathe, H. Plaste Kautsch. 19 (1972) 731–734.
[22] Schwachula, G., Schmidt, H. Plaste Kautsch. 18 (1971) 577–579.
[23] Schwachula, G., Wolf, F. Plaste Kautsch. 15 (1968) 33–35.
[24] Schwachula, G., Wolf, F. Plaste Kautsch. 14 (1967) 879–880.
[25] Schwachula, G., Wolf, F., Gatzmanga, H. Plaste Kautsch. 17 (1970) 255–258.
[26] Hoffmann, H., Schwachula, G. Plaste Kautsch. 18 (1971) 98–99.
[27] Hering, R. Z. Chemie 5 (1965) 149.
[28] Kolesnikov, G.S., Tevlina, A.S., Frumin, L.E., Kirilin, A.I. Vysokomol. Soedin. A13 (1971) 549–454.
[29] Tevlina, A.S., Korshak, V.V., Frumin, L.E., Kamnev, Yu.V., All-Union Conference on Application of Ion Exchange Materials, NIIPM, Moscow, 1973, p. 8.
[30] Malinsky, J., Klaban, J., Dušek, K. J. Macromol. Sci.: Chem. A5 (1971) 1071–1085.
[31] Dušek, K. in: Composition Polymeric Materials, Naukova Dumka, Kiev, 1975, pp. 14–28.
[32] Dušek, K. Coll. Czech. Chem. Commun. 32 (1967) 1182–1190.
[33] Periyasamy, M., Ford, W.T., McEnroe, F.J. J. Polym. Sci. Part A: Polym. Chem. 27 (1989) 2357–2366.
[34] Dušek, K. Makromol. Chem. Suppl. 2 (1979) 35–49.
[35] Kast, H., Funke, W. Makromol. Chem. 180 (1979) 1335–1338.
[36] Soper, B., Haward, R.N., White, E.F.T. J. Polym. Sci. A10 (1972) 2545–2564.
[37] Dušek, K. Spevacek, J. Polymer 21 (1980) 750–756.
[38] Elias, H.G. Chimia 22 (1968) 101–110.
[39] Galina, H., Rupicz, K. Polym. Bull. 3 (1980) 473–480.
[40] Okasha, R., Hild, G., Rempp, P. Eur. Polym. J. 15 (1979) 975–982.
[41] Hild, G., Okasha, R., Rempp, P. Makromol. Chem. 186 (1985) 407–422.
[42] Matsumoto, A., Kitaguchi, Y. Macromolecules 32 (1999) 8336–8339.

[43] Dušek, K., Galina, H., Mikeš, J. Polym. Bull. 3 (1980) 19–25.
[44] Fink, J.K. J. Polym. Sci.: Polym. Chem. Ed. 19 (1981) 195–202.
[45] Ambler, M.R., McIntyre, D. J. Appl. Polym. Sci. 21 (1977) 3237–3250.
[46] Ambler, M.R., McIntyre, D. J. Appl. Polym. Sci. 21 (1977) 2269–2282.
[47] Speváček, J., Dušek, K. J. Polym. Sci.: Polym. Phys. Ed. 18 (1980) 2027–2035.
[48] Shea, K.J., Stoddard, G. J. Macromolecules 24 (1991) 1207–1209.
[49] Leicht, R., Fuhrmann, J. Polym. Bull. 4 (1981) 141–148.
[50] Fuhrmann, J., Leicht, R. Colloid Polym. Sci. 258 (1980) 631–637.
[51] Eliott, J.E., Bowman, C.N. Macromolecules 32 (1999) 8621–8628.
[52] Ozol-Kalnin, B.G., Gravitis, Ya.A. Vysokomolek. Soedin. B24 (1982) 329–332.
[53] Yuan, H.G., Kalfas, G., Ray, W.H. J. Macromol. Sci. – Rev. Macromol. Chem. Phys. C31 (1991) 215–299.
[54] Vivaldo, E., Wood, P.E., Hamielec, A.E. Ind. Eng. Chem. Res. 36 (1997) 939–965.
[55] Patterson, J.A., in: G.R. Stark (Ed.), Biochemical Aspects of Reactions on Solid Supports, Academic Press, New York, 1971, pp. 189–213.
[56] Zähle, F., Häupke, K., Schwachula, G. Chem. Tech. (DDR) 33 (1981) 462.
[57] Letsinger, R.L., Hamilton, S.B. J. Am. Chem. Soc. 81 (1959) 3009–3012.
[58] Letsinger, R.L., Kornet, M.J., Mahadevon, V., Jerina, D.M. J. Am. Chem. Soc. 86 (1964) 5163–5165.
[59] Letsinger, R.L., Kornet, M.J. J. Am. Chem. Soc. 85 (1963) 3045–3046.
[60] Tatsumi, M., Yamamoto, S. Technol. Repts. Kansai Univ. N23 (1982) 105–114.
[61] Schwachula, G., Wolf, F., Schmidt, H. Plaste Kautsch. 14 (1967) 802–805.
[62] Kraus, M.A., Patchornik, A. J. Polym. Sci.: Polym. Symp. N47 (1974) 11–18.
[63] Immergut, E.H. Makromol. Chem. 10 (1953) 93–106.
[64] Tatsumi, M., Yamamoto, S. Nippon kagaku kaishi N8 (1982) 1386–1393.
[65] Breitenbach, J.W., Goldenberg, H., Olaj, O.F. J. Polym. Sci. B10: Polym. Lett. Ed. (1972) 911–913.
[66] Pravednikov, A.N., Medvedev, S.S. Dokl. Akad. Nauk SSSR 109(3) (1956) 579–581.
[67] Breitenbach, J.W., Campbell, G., Schindler, A. J. Polym. Sci. B: Polym. Lett. 3 (1965) 1017–1019.
[68] Belfer, S.I., Saldadze, K.M., Vitsnudel, M.B. Plastmassy N7 (1967) 8–10.
[69] Wiley, R.H. Chim. et Ind. 94 (1965) 602–610.
[70] Wiley, R.H. J. Polym. Sci. A4 (1966) 1892–1894.
[71] Makarova, S.B., Aptova, T.A., Vinogradova, N.M., Chernyavskaya, T.M., Egorov, E.V. Plastmassy N10 (1970) 5.
[72] Belfer, S.I., Saldadze, G.K., Kudryavtsev, E.A., Stepanyan, S.A., Shabadash, A.N. Plastmassy N8 (1970) 24–26.
[73] Wiley, R.H., Venkatachalam T.K J. Polym. Sci. A3 (1965) 1063–1067.
[74] Samsonov, G.V., Trostyanskaya, E.B., Elkin, G.E., Ion Exchange. Sorption of Organic Compounds, Nauka, Leningrad, 1969.
[75] Wiley, R.H., Allen, J.K., Chang, S.P., Musselman, K.E., Venkatachalam, T.K. J. Phys. Chem. 68 (1964) 1776–1779.
[76] Schwachula, G. Zeitschr. Chemie 18 (1978) 242–251.
[77] Wiley, R.H., Smithson, L.H. J. Macromol. Sci. A2 (1968) 589–594.
[78] Wiley, R.H., Jin, J.-I., Reich, E. J. Macromol. Sci. A4, (1970) 341–348.
[79] Braun, D., Kim, U.Y., Kolloid, Z. 216/217 (1967) 321–325.
[80] Trostyanskaya, E.B., Makarova, S.B., Aptova, G.A., Murashko, I.N. Vysokomol. Soedin. A7 (1965) 2083–2088.
[81] Denisenko, E.I., Trushin, B.N., Khimia i Khimicheskaya Technologia, Sb. Trudov N36, Kuzbasskii Politechnicheskii Institut, Kemerovo, 1972, pp. 141–144.
[82] Saldadze, K.M., Belfer, S.I., Tkatchuk, S.M., Kurtskhalia, Ts.S., Plastmassy N10 (1968) 6–7.

[83]   Saldadze, K.M., Kurtskhalia, Ts. S., Plastmassy N6 (1968) 5–7.
[84]   Kressman, T.R.E., Millar, J.R. Chem. Ind. N45 (1961) 1833–1834.
[85]   Aptova, T.A., Babushkin, Yu.Ya., Gukasova, E.A., Egorov, E.V., Korolev, G.V., Makarova, S.B. et al., Vysokomol. Soedin. A12 (1970) 1246–1253.
[86]   Tevlina, A.S., Korshak, V.V., Geoklenov, B., Kirillova, A.P. Plastmassy N3 (1973) 13–14.
[87]   Trostyanskaya, E.B., Makarova, S.B., Aptova, T.A., Vinogradova, N.M. Vysokomol. Soedin. A9 (5) (1967) 1066–1070.
[88]   Tevlina, A.S., Kolesnikov, G.S., Samsonov, G.V., Dmitrenko, L.N., Chuchin, A.E. Chemically Active Polymers and their Application, Khimia, Leningrad, 1969, pp. 24–28.
[89]   Tevlina, A.S., Sadova, S.F. Zh. Priklad. Khimii 38 (1965) 1643–1646.
[90]   Trostyanskaya, E.B., Makarova, S.B., Aptova, T.A., Murashko, I.N., Egorov, E.V. Synthesis and Properties of Ion Exchange Materials, Nauka, Moskva, 1968, pp. 17–22.
[91]   Tevlina, A.S., Korshak, V.V., Frumin, L.E., Kamnev, Yu.V., Vysokomol. Soedin. B15 (1973) 903–907.
[92]   Losev, I.P., Trostyanskaya, E.B., Chemistry of Synthetic Polymers, Khimia, Moskva, 1965, p.143.
[93]   Dušek, K., Malinsky, J. Chem. Prumysl. 16/41, (1966) 219–224.
[94]   Mikeš, J.A., J. Polym. Sci. 30 (1958) 615–623.
[95]   Wiley, R.H., Badcett, J.T. J. Macromol. Sci. A2: Chem. (1968) 103–110.
[96]   Millar, J.R., Smith, D.G., Marr, W.E., Kressman, T.R.E. J. Chem. Soc. (1963) 218–225.
[97]   Glans, J.H., Turner, D.T. Polymer 22 (1981) 1540–1543.
[98]   Ueberreiter, K., Kanig, G. J. Chem. Phys. 18 (1950) 399–406.
[99]   Eder, M., Wlochowicz, A., Kolarz, B.N. Angew. Makromol. Chem. 126 (1984) 81–88.
[100] Askadskyi, A.A. Vysokomol. Soedin. A32 (1990) 2149–2156.
[101] Askadskyi, A.A. Uspekhi Khimii 67 (1998) 755–787.
[102] Tsyurupa, M.P., Davankov, V.A. React. Funct. Polym. 66 (2006) 768–779.
[103] Grubhofer, N. Makromol. Chem.30 (1959) 96–108.
[104] Artamonov, V.A., Soldatov, V.S. Izvestia Akad. Nauk BSSR, Ser. Khim. Nauk N3 (1974) 34–37.
[105] Trostyanskaya, E.B., Babaevskii, B.G. Uspekhi Khimii 40 (1971) 117–141.
[106] Tevlina, A.S., Frumin, L.E., Korshak, V.V., Kamnev, Yu.V. Vysokomol. Soedin. A15 (1973) 1187–1190.
[107] Saldadze, K.M., Belfer, S.I. Plastmassy N3 (1967) 10–12.
[108] Barr-Howell, B.D., Peppas, N. J. Appl. Polym. Sci. 30 (1985) 4583–4589.
[109] Dušek, K., Prins, W. Adv. Polym. Sci. 6 (1969) 1–102.
[110] Bates, F.S., Cohen, R.E. J. Am. Chem. Soc., Polym. Prepr. 22 (1981) 159–160.
[111] Schmidt, M., Burchard, W. Macromolecules 14 (1981) 370–376.
[112] Popov, G., Schwachula, G. Plaste Kautsch. 27 (1980) 245–246.
[113] Schwachula, G., Popov, G. Pure Appl. Chem. 54 (1982) 2103–2114.
[114] Popov, G., Schwachula, G. Plaste Kautsch. 28 (1981) 372–374.
[115] Eschwey, H., Burchard, W. J. Polym. Sci.: Polym. Symposia N53 (1975) 1–9.
[116] Worsfold, D.J. Macromolecules 3 (1970) 514–517.
[117] Lutz, P., Beinert, G. Makromol. Chem. 183 (1982) 2787–2797.
[118] Weiss, P., Hild, G., Herz, J., Rempp, P. Makromol. Chem. 135 (1970) 249–261.
[119] Worsfold, D.J., Zilloix, J.-G., Rempp, P. Canad. J. Chem. 47 (1969) 3379–3385.
[120] Herz, J., Strazielle, C. C. R. Acad. Sc. (Fr.) C272 (1971) 747–749.
[121] Herz, J., Hert, M., Strazielle, C. Makromol. Chem. 160 (1972) 213–225.
[122] Hert, M., Strazielle, C., Herz, J. C. R. Acad. Sc. (Fr.) C276 (1973) 395–398.

[123] Hild, G., Herz, J.E., Rempp, P. J. Am. Chem. Soc.: Polym. Prepr. 14 (1973) 601–605.

[124] Hild, G., Rempp, P. C. R. Acad. Sc. (Fr.) C269 (1969) 1622–1624.

[125] Hild, G., Haeringer, A., Rempp, P., Benoit, H. J. Am. Chem. Soc.: Polym. Prepr. 14 (1973) 352–357.

[126] Weiss, P., Herz, J., Rempp, P. Makromol. Chem. 141 (1971) 145–159.

[127] Hild, G., Rempp, P., C. R. Acad. Sc. Paris, (Fr.) C271 (1970) 1432–1435.

[128] Bastide, J., Picot, C. J. Polym. Sci.: Polym. Phys. Ed. 17 (1979) 1441–1456.

[129] Bastide, J., Picot, C., Candau, S. J. Macromol. Sci.: Phys. B19 (1981) 13–34.

[130] Ramzi, A., Hakiki, A., Bastide, J., Boue, F. Macromolecules 30 (1997) 2963–2977.

[131] Froelich, D., Crawford, D., Rozek, T., Prins, W. Macromolecules 5 (1972) 100–102.

[132] James, H.M., Guth, E.G. J. Chem. Phys. 15 (1947) 669–683.

[133] Haeringer, A., Hild, G., Rempp, P., Benoit, H. Makromol. Chem. 169 (1973) 249–260.

[134] Rempp, P., Herz, J., Hild, G., Picot, C. Pure Appl. Chem. 43 (1975) 71–96.

[135] Benoit, H., Decker, D., Duplessix, R., Picot, C., Rempp, P., Cotton, J.P. et al., J. Polym. Sci.: Polym. Phys. Ed. 14 (1976) 2119–2128.

[136] Sarazin, D., Herz, J.E., Francois, J. Polymer 23 (1982) 1317–1321.

[137] Candau, S., Bastide, J., Delsanti, M. Adv. Polym. Sci. 44 (1982) 48–71; in K. Dušek (Ed.) Polymer Networks Springer-Verlag, Berlin, 1982, pp. 27–50.

[138] Flory, P.J. Principles of Polymer Chemistry, Cornell Univ. Press, Ithaca, New York, 1953.

[139] De Gennes, P.-G., Scaling Concepts in Polymer Physics, Mir, Moscow, 1982.

[140] Hayashi, H., Flory, P.J. Physica B120 (1983) 408–412.

[141] Herz, J.E., Rempp, P., Borchard, W. Adv. Polym. Sci. 26 (1978) 105–135.

[142] Tsitsilianis, C., Pierri, E., Dondos, A. J. Polym. Sci.: Polym. Lett. Ed. 21 (1983) 685–691.

[143] Haeringer, A., Hild, G., Rempp, P., Benoit, H. C. R. Acad. Sci. (Fr.) C27 (1973) 1711–1713.

[144] Sarazin, D., Francois, J. Polymer 19 (1978) 699–704.

[145] Bauer, D.R., Ullman, R. Macromolecules 13 (1980) 392–396.

[146] Nerger, D., Eisele, M., Kajiwara, K. Polym. Bull. 10 (1983) 182–186.

[147] Swislow, G., Sun, S.-T., Nishio, I., Tanaka, T. Phys. Rev. Lett. 44 (1980) 796–798.

[148] Nishio, I., Sun, S.-T., Swislow, G., Tanaka, T. Nature 281 (1979) 208–209.

[149] Vasserman, A.M., Aleksandrova, T.A., Kirsh, Yu.E. Vysokomol. Soedin. A22 (1980) 282–291.

[150] Benoit, H. J. Macromol. Sci.: Phys. B12 (1976) 27–40.

[151] Daoud, M., Cotton, J.P., Farnoux, B., Jannink, G., Sarma, G., Benoit, H. et al., Macromolecules 8 (1975) 804–818.

[152] Cotton, J.-P., Farnoux, B., Jannik, G., Mous, J., Picot, C. C. R. Acad. Sci. (Fr.), Ser. C, 275 (1972) 175–178.

[153] Ballard, D.G.H., Wignall, G.D., Schelten, J. Eur. Polym. J. 9 (1973) 965–969.

[154] Schelten, J., Wignall, G.D., Ballard, D.G. Polymer 15 (1974) 682–685.

[155] Schelten, J., Ballard, D.G.H., Wignall, G.D., Longman, G., Schmatz, W. Polymer 17 (1976) 751–757.

[156] Hayashi, H., Hamada, F., Nakajima, A. Makromol. Chem. 178 (1977) 827–842.

[157] Shaulov, A.Yu., Yan'kova, M.A., Vasserman, A.M., Shapiro, A.B., Enikolopyan, N. S. Dokl. Akad. Nauk SSSR 225 (1975) 364–367.

[158] Lindenmeyer, P.H. J. Macromol. Sci. B8 (1973) 361–366.

[159] Tager, A.A., Dreval, V.E. Uspekhi Khimii 36 (1967) 888–910.

[160] Kuzub, L.I., Irzhak, V.I., Bogdanova, L.M., Enikolopyan, N.S. Vysokomol. Soedin. B16 (1974) 431–433.

[161] Grebentchikov, Yu.B., Irzhak, V.I., Kuzub, L.I., Kush, P.P., Enikolopyan, N.S. Dokl. Akad. Nauk SSSR 210 (1973) 1124–1126.

[162] De Vos, D.E., Bellemas, A. Macromolecules 8 (1975) 651–655.

[163] Andrianova, G.P., Narozhnaya, E.L. Vysokomol. Soedin. A17 (1975) 923–928.

[164] Andrianova, G.P., Krasnikova, N.P. Vysokomol. Soedin B14 (1972) 4.

[165] Privalko, V.P., in: Yu.S. Lipatov (Ed.) Structural Peculiarities of Polymers, Naukova Dumka, Kiev, 1978, pp. 3–32.

[166] Ziabicky, A. Colloid Polym. Sci. 252 (1974) 49–53.

[167] Belkebir-Mrani, A., Herz, J.E., Rempp, P. Makromol. Chem. 178 (1977) 485–504.

[168] Rempp, P., Herz, J.E. Angew. Makromol. Chem. 76–77 (1979) 373–391.

[169] Wang, X., Xu, Z., Wan, Y., Huang, T., Pispas, S., Mays, J.W., et al., Macromolecules 30 (1997) 7202–7205.

[170] Tanaka, T., Donkai, N., Inagaki, H. Macromolecules 13 (1980) 1021–1023.

[171] Kayillo, S., Gray, M.J., Shalliker, R.A., Dennis, G.R. J. Chromatogr. A1073 (2005) 83–86.

[172] Hong, P.P., Boerio, F.J., Smith, S.D. Macromolecules 26 (1993) 1460–1464.

[173] Bates, F.S., Fetters, L.J., Wignall, G.D. Macromolecules 21 (1988) 1086–1094.

[174] Bates, F.S., Wignall, G.D. Phys. Rev. Lett. 57 (1986) 1429–1432.

[175] Hild, G., Froelich, D., Rempp, P., Benoit, H. Makromol. Chem. 151 (1972) 59–81.

[176] Bastide, J., Candau, S., Leibler, L. Macromolecules 14 (1981) 719–726.

[177] Weiss, P., Herz, J.E., Rempp, P., Gallot, Z., Benoit, H. Makromol. Chem. 145, (1971) 105–121.

[178] Rietsch, F., Daveloose, D., Froelich, D. Polymer 17 (1976) 859–863.

[179] Helfferich, F., Ion Exchange, McGraw-Hill, New York, 1962.

[180] Tremillon, B., Les Séparations par les Résines Echangeuses d'Ions, Gauthier-Villars, Paris, 1965.

[181] Soldatov, V.S., Bychkova, V.A., Ion Exchange Equilibrium in Multicomponent Systems, Minsk, 1988.

[182] Samuelson, O. Ion Exchange Separations in Analytical Chemistry, Almquist & Wiksell, Stockholm, 1965.

[183] Ion Exchange and Solvent Extraction, A Series of Advances, Ed. Marinsky J.A. and Marcus, Y., Marcel Dekker, New York.

[184] Dorfner, K. Ion Exchange, Walter de Gruyter (Ed.), Berlin & New York, 1991.

[185] Saldadze, G.K., Varentsov, V.K., Lavrent'ev, Yu.G., Pevnitskaya, M.V., Pospelova, L.K. Izvestia AN SSSR, Ser. Khim. Nauk 9 (1974) 133–137.

[186] Varentsov, V.K. Dokl. Akad. Nauk SSSR 186 (1969) 330–332.

[187] Wiley, R.H., Jun, J.-I., Tae-Oan, A. J. Macromol. Sci. A2 (1968) 407–409.

[188] Burenin, A.A., Selemenhev, V.F., Sharuda, B.A., in: Theory and Practice of Sorption Processes. Issue 15, Voronesh, Russia 1982, pp. 6–13.

[189] Saldadze, K.M., Belfer, S.I., Kotov, A.V., Kudryavtsev, E.A. Plastmassy N2 (1973) 70–71.

[190] Belov, P.S., Belfer, S.I., Ivanova, I.I., Saldadze, K.M. Zh. Priklad. Khim. 46 (1973) 2031–2034.

[191] Sabzali, J., Michaud, C.F., in: J.A. Greig (Ed.) Ion Exchange at the Millennium, Imperial College Press, SCI, UK, 2000 pp. 269–278.

[192] Barrett, J.H., Howell, T.J., Lein, G.M. Patent USA 4,419,245 (1983).

[193] Harris, W.I. Patent USA 4,564,644 (1986).

[194] Harris, W.I. Patent USA 5,231,115 (1993).

[195] Scheffler, A., in: J.A. Greig, (Ed.), Ion Exchange at the Millennium, Imperial College Press, SCI, UK, 2000 pp. 77–84.

[196] Timm, E.E., Leng, D.E. Patent USA 4,623,706 (1986).

[197] Timm, E.E. Eur. Pat. Appl. 0051210 (1982).

[198] Kressman, T.R.E. J. Phys. Chem. 56 (1952) 118–123.

[199] Belfer, S.I., Saldadze, K.M., Gintsberg, E.G., Kovarskaya, B.M. Zh. Fiz. Khim. 44 (1970) 1104–1105.

[200] Brutskus, T.K., Saldadze, K.M., Uvarova, E.A., Fedtsova, M.A., Belfer, S.I. Zh. Fiz. Khim. 47 (1973), 1528–1530.

[201] Samsonov, G.V., Moskvichev, B.V., Yurchenko, V.S., Genedi, A.Sh., Chokina, B.Sh., Ion Exchange and Ion Exchange Resins, Nauka, Leningrad, 1970, pp. 142–146.

[202] Genedi, A.Sh., Moskvichov, B.V., Samsonov, G.V. Zh. Prikl. Khim. 43 (1970) 1171–1174.

[203] Genedi, A.Sh., Samsonov, G.V. Colloid Zh. 31 (1969) 674–678.

[204] Musabekov, K.V., Dinaburg, V.A., Samsonov, G.V. Zh. Prikl. Khim. 42 (1969) 82–87.

[205] Moskvichov, B.V., Yurchenko, V.S., Genedi, A.Sh., Chokina, B.Sh., Samsonov, G.V. Synthesis, Structure and Properties of Polymers, Nauka, Leningrad, 1970, pp. 263–266.

[206] Klikh, S.F., El'kin, G.E., Samsonov, G.V. Kolloid. Zh. 37 (1975) 1167–1171.

[207] Bakaeva, R.M., Samsonov, G.V., Selective Ion Exchange Sorption of Antibiotics, Chem.-Pharm. Inst., Leningrad, N25, 1968, pp. 63–68.

[208] Nemtsova, N.N., Pasechnik, V.A., Samsonov, G.V. Zh. Phys. Khim. 47 (1979) 2398–2400.

[209] Hale, D.K., Packham, D.I., Pepper, K.W. J. Chem. Soc. (1953) 844–851.

[210] Millar, J.R., Smith, D.G., Marr, W.E., Kressman, T.R.E. J. Chem. Soc. (1963), 2779–2784.

[211] Trostyanskaya, E.B., Tevlina, A.S., Naumova, F.A. Vysokomol. Soedin. A5 (1963) 1240–1244.

[212] Genedi, A.Sh. Samsonov, G.V. Selective Ion Exchange Sorption of Antibiotics, Chem.-Pharm. Inst., Leningrad, N25,1968, pp. 164–170.

[213] Gregor, H.P., Hoeschele, G.K., Potenza, J., Tsuk, A.G., Feinland, R., Shida, M. J. Am. Chem. Soc. 87 (1965) 5525–5534.

[214] Tsuk, A.G., Gregor, H.P. J. Am. Chem. Soc. 87 (1965) 5534–5538.

[215] Tsuk, A.G., Gregor, H.P. J. Am. Chem. Soc. 87 (1965) 5538–5542.

[216] Gregor, H.P., Teyssie, P., Hoeschele, G.K., Feinland, R., Shida, M., Tsuk, A. J. Am. Chem. Soc. Polym. Prepr. 5 (1964) 873–877.

[217] Haklits, I., Szanto, J. Plaste Kautsch. 18 (1971) 175–177.

[218] Makarova, S.B., Pakhomova, E.M., Babina, O.V., Egorov, E.V. Plastmassy N8 (1969) 15.

[219] Shostenko, Yu.V., Simon, I.S., Pletneva, T.A., Gubina, T.N., Zavadovskaya, A.S., Lustgarten, E.I. Zh. Phys. Khim. 50 (1976) 123–126.

[220] Hild, G., Haeringer, A., Rempp, P. C. R. Acad. Sc. (Fr.) C280 (1975) 1405–1407.

[221] Popov, G., Schwachula, G., Gehrke, K. Plaste Kautsch. 27 (1980) 307–309.

[222] Popov, G., Schwachula, G., Gehrke, K. Plaste Kautsch. 27 (1980) 65–68.

[223] Popov, G., Schwachula, G., Gehrke, K. Plaste Kautsch. 27 (1980) 367–371.

[224] Popov, G., Schwachula, G., Gehrke, K. Plaste Kautsch. 28 (1981) 66–69.

[225] Millar, J.R. J. Chem. Soc. 263 (1960) 1311–1317.

[226] Shibayama, K., Suzuki, Y. Rubber Chem. and Techn. 40 (1967) 467–483.

[227] Thiele, J.L., Cohen, R.E. J. Am. Chem. Soc. Polym. Prepr. Div. Polym. Chem. 19 (1978) 133–136.

[228] Siegfried, D.L., Thomas, D.A., Sperling, L.H. Macromolecules 12 (1979) 586–589.

[229] Sperling, L.H., Interpenetrating Polymer Networks and Related Materials, Moscow, Mir, 1984.

[230] Siegfried, D.L., Manson, J.A., Sperling, L.H. J. Polym. Sci.: Polym. Phys. Ed. 16 (1978) 583–597.

[231] Millar, J.R., Smith, D.G., Marr, W.E. J. Chem. Soc. (1962) 1789–1794.
[232] Kunin, R., Meitzner, E.F., Bortnick, N. J. Am. Chem. Soc. 84 (1962) 305–306.
[233] Seidl, J., Malinsky, J., Dušek K., Heitz, W. Adv. Polym. Sci. 5 (1967) 113–213.
[234] Sakodynskii, K.I., Panina, L.I., in: K.V. Chmutov (Ed.), Polymeric Sorbents for Molecular Chromatography, Nauka, Moscow, 1977, pp. 5–26.
[235] Guyot, A. in: Synthesis and Separations Using Functional Polymers, D.C. Sherrington, P. Hodge (Eds.), John Wiley & Sons, 1988, pp. 1–42.
[236] Tager, A.A., Tsilipotkina, M.V. Uspekhi Khimii 47 (1978) 152–175.
[237] Okay, O. Progr. Polym. Sci. 25 (2000) 711–779.
[238] IUPAC Recommendations for the Characterization of Porous Solids, Pure & Appl. Chem., 66 (1994) 1739–1758.
[239] Tsilipotkina, M.V., in: G.L. Slonimskii (Ed.), Modern Physical Methods of Investigation of Polymers, Moscow, Khimia, 1982, pp.198–209.
[240] Gregg, S.J., Sing, K.W., Adsorption. Specific Surface Area. Porosity, Moscow, Nauka, 1970, p. 407.
[241] Belyakova, L.D. Uspekhi Khimii 60 (1991) 374–397.
[242] Brunauer, S., Emmett, P.H., Teller, E. J. Am. Chem. Soc. 60 (1938) 309–319.
[243] Barrett, E.P., Joyner, L.G., Halenda, P.P. J. Am. Chem. Soc. 73 (1951) 373–380.
[244] Brunauer, S., Deming, L.S., Deming, W.S., Teller, E. J. Am. Chem. Soc. 62 (1940) 1723–1732.
[245] Stadnik, A.M., Goncharov, A.I., Stadnik, A.S., Zavadovskaya, A.S., Tkach, N.D. Plastmassy 5 (1981) 36–37.
[246] Giles, C.H., Nakhwa, S.N. J. Appl. Chem. 12 (1962) 266–273.
[247] Tager, A.A., Askadskii, A.A., Tsilipotkina, M.V. Vysokomol. Soedin. A17 (1975) 1346–1352.
[248] Polyanskii, N.G., Gorbunov, G.V., Polyanskaya, N.L., Methods of Investigation of Ionits, Moscow, Khimiya, 1976, p. 207.
[249] Uvarova, E.A., Saldadze, K.M., Panina, L.I., Brutskus, T.K. Plastmassy N5 (1972) 72.
[250] Brutskus, T.K., Saldadze, K.M., Uvarova, E.A., Fedtsova, M.A. Zavodskaya Laboratoria 44 (1978) 985–986.
[251] Karnaukhov, A.P. Adsorption. Texture of Dispersed and Porous Materials, Novosibirsk, Nauka, 1999, 469p.
[252] Gagarin, A.A., Ferapontov, N.B. Sorption and Chromatographic Processes (Russia), 4 (2004) 541–549.
[253] Vol'fkovich, Yu. M., Bagotskii, V.S., Sosenkin, V.E., Shkol'nikov, E.I. Elektrokhimiya 16 (1980) 1620–1652.
[254] Shkolnikov, E.I., Volkov, V.V. Dokl. Akad. Nauk. 378 (2001) 507–510.
[255] Detcheverry, F., Kierlik, D., Rosenberg, M.L., Tarus, G. Adsorption 11 (2005) 115–119.
[256] Ravikovitch, P.I., Neimark, A.V. Adsorption 11 (2005) 265–270.
[257] Lubda, D., Lindner, W., Quaglia, M., Hohenesche, C., Unger, K.K. J. Chromatogr. A1083 (2005) 14–22.
[258] Nechaeva, O.V., Tsilipotkina, M.V., Tager, A.A.,. Netimenko, T.P., Vysokomol. Soedin. A, 17 (1975) 2347–250.
[259] Kun, K.A., Kunin, R. J. Polym. Sci.: B2 Polymer Lett. 2 (1964) 587–592.
[260] Gorbunov, A.A., Solovyova, L.Ya., Pasechnik, V.A. J. Chromatogr. 448 (1988) 307–332.
[261] Schloegl, R., Schurig, H.L. Z. f. Electrochemie 65 (1961) 863–870.
[262] Brun, M., Ouinson, J.F., Le Parlouer, P., Spitz, R., Bartholin, M. J. Calorimetr. et analyse therm. Barcelona, Vol. XI, S.1., s.a., 3.10/1–3.10/10.
[263] Ogino, K., Sato, H. J. Polym. Sci. Part B: Polym. Phys. 33 (1995) 445–451.
[264] Ogino, K., Sato, H. J. Appl. Polym. Sci. 58 (1995) 1015–1020.

[265] Dušek, K. J. Polym. Sci. B, Polym. Lett. 3 (1965) 209–212.
[266] Dušek K. in: A.J. Chompff, S. Newman, (Eds.), Polymer Networks. Structure and Mechanical Properties, Plenum Press, New York, 1971, pp.245–260.
[267] Seidl, J., Dušek, K., Coll. Czech. Chem. Commun. 31 (1966) 2695–2700.
[268] Dušek, K., Sedláček, B., Coll. Czech. Chem. Commun. 34 (1969) 136–157.
[269] Wong, Q.C., Svec, F., Fréchet, J.M.J. J. Polym. Sci.: Part A Polym. Chem. 32 (1994) 2577–2588.
[270] Seidl, J., Malinsky, J., Dušek, K., Plastmassy N12 (1963) 7–11.
[271] Saldadze, K.M., Brutskus, T.K., Uvariva, E.A., Fedtsova, M.A. Plastmassy N2 (1971) 29–30.
[272] Tager, A.A., Tsilipotkina, M.V., Makovskaya, E.B., Pashkov, A.B., Lyustgarten, E.I., Pechenkina, M.A. Vysokomol. Soedin. A10 (1968) 1065–1073.
[273] Lyustgarten, E.I., Brutskus, T.K., Pashkov, A.B., Artyushin, G.A., Grigor'ev, V.A., Plastmassy N12 (1980) 8–9.
[274] Tager, A.A., Tsilipotkina, M.V., Pashkov, A.B., Makovskaya, E.B., Lyustgarten, E.I., Itkina, M.I. Plastmassy N3 (1966) 23–27.
[275] Brutskus, T.K., Saldadze, K.M., Galitskaya, N.B., Fedtsova, M.A., Stebeneva, I.G. Production and Processing of Plastics and Synthetic Resins, Moscow, NIIPM, N7, 1973, p. 35.
[276] Galitskaya, N.B., Brutskus, T.K., Stebeneva, I.G., Fedtsova, M.A., Zak, V.E., Pashkov, A.B. Plastmassy N6 (1974) 8–10.
[277] Brutskus, T.K., Saldadze, K.M., Uvarova, E.A., Lyustgarten, E.I. Kolloid. Zh. 34 (1972) 509–513.
[278] Uvarova, E.A., Brutskus, T.K., Saldadze, K.M., Lyustgarten, E.I. Plastmassy N8 (1982) 54–55.
[279] Wojaczynska, M., Kolarz, B.N. Angew. Makromol. Chem. 86 (1980) 65–82.
[280] Svetlov, A.K., Tsvetkov, Yu.S., Tavobilov, M.F., Petrova, N.A. Zh. Prikl. Khimii 47 (1974) 2376–2378.
[281] Svetlov, A.K., Demenkova, T.D., Kryuchkov, V.V., Tsvetkov, Yu.S., Chemistry and Chemical Technology, Kemerovo, N26 (1971) 214–218.
[282] Svetlov, A.K., Rakhovskaya, S.M., Khomutov, L.I., Borimskaya, V.S., Demenkova, T.N. Kolloid. Zh. 33 (1971) 264–267.
[283] Oza, K.S., Joshi, K.M. Indian J. Technol. 21 (1983) 81–85.
[284] Wolf, F., Schaaf, R. Plaste Kautschuk 17 (1970) 323–326.
[285] Brutskus, T.K., Uvarova, E.A., Zavadovskaya, A.S., Saldadze, G.K. Plastmassy N8 (1989) 68–70.
[286] Góźdź, A.S., Kolarz, B.N. Makromol. Chem. 180 (1979) 2473–2481.
[287] Wieczorek, P., Ilavsky, M., Kolarz, B.N., Dušek, K. J. Appl. Polym. Sci. 27 (1982) 277–288.
[288] Baldrian, J., Kolarz, B.N., Galina, H. Coll. Czech. Chem. Commun. 46 (1981) 1675–1681.
[289] Galina, H., Kolarz, B.N., Wieczorek, P.P., Wojczynska, M. Brit. Polym. J. 17 (1985) 215–218.
[290] Yu, Z., Shi, Z. Geochimica N3C (1988) 207–212.
[291] Hilgen, H., De Jong, G.J., Sederel, W.L. J. Appl. Polym. Sci. 19 (1975) 2647–2654.
[292] Brutskus, T.K., Saldadze, K.M., Uvarova, E.A., Lyustgarten, E.I. Kolloid. Zh. 35 (1973) 445–450.
[293] Brutskus, T.K., Saldadze, K.M., Fedtsova, M.A., Uvarova, E.A., Lyustgarten, E.I., Itkina, M.I. et al., Vysokomol. Soedin. A17 (1975) 1247–1251.
[294] Brutskus, T.K., Saldadze, K.M., Lyustgarten, E.I., Uvarova, E.A., Semenova, T.S., Gorshkova, G.N. Vysokomolek. Soedin. A23 (1981) 1853–1857.

[295] Howard, G.J., Midgley, C.A. J. Appl. Polym. Sci. 26 (1981) 3845–3870.
[296] Jacobelli, H., Bartholin, M., Guyot, A. J. Appl. Polym. Sci. 23 (1979) 927–939.
[297] Smith, D.H. J. Colloid Interface Sci. 108 (1985) 471–483.
[298] Brutskus, T.K., Uvarova, E.A., Saldadze, K.M., Lyustgarten, E.I. Zh. Prikl. Khimii 55 (1982) 1687–1690.
[299] Tsilipotkina, M.V., Tager, A.A., Makovskaya, E.B., Pashkov, A.B., Lyustgarten, E.I., Palekhova, T.N. et al., Plastmassy, N5 (1967) 15–18.
[300] Brutskus, T.K., Galitskaya, N.B., Fedtsova, M.A., Stebeneva, I.G. Vysokomol. Soedin. A17 (1975) 54–57.
[301] Wojaczynska, M., Kolarz, B.N. J. Chromatogr. 196 (1980) 75–83.
[302] Hradil, J. Angew. Makromol. Chem. 66 (1978) 51–66.
[303] Mikes, J.A. Proc. Conf. Ion Exchange in the Process Industries, SCI, London, 1969, pp. 16–21.
[304] Richter, H.-P., Kődderilzsch, H., Schwachula, G., Häupke, K., Plaste Kautsch. 28 (1981) 133–135.
[305] Pelzbauer, Z., Forst, V. Coll. Czech. Chem. Commun. 31 (1966) 2338–2343.
[306] Svetlov, A.K., Lazarenko, V.D., Tavobilov, M.F., Yarigina, G.P. Vysokomol. Soedin. B17 (1975) 293–297.
[307] Krška, F., Pelzbauer, Z. Coll. Czech. Chem. Commun. 32 (1967) 4175–4178.
[308] Kalló, A., Faserforsch. Textiltechn. 25 (1974) 490–492.
[309] Gusev, I., Huang, X., Horvath, C. J. Chromatogr. A855 (1999) 273–290.
[310] Jacobelli, H., Bartholin, M., Guyot, A. Angew. Makromol. Chem. 80 (1979) 31–51.
[311] Kun, K.A., Kunin, R. J. Polym. Sci. A1 (6) (1968) 2689–2701.
[312] Häupke K., Pientka, V. J. Chromatogr. 102 (1974) 117–121.
[313] Chung, D.-Y., Bartholin, M., Guyot, A. Angew. Makromol. Chem. 103 (1982) 109–123.
[314] Guyot, A., Bartholin, M. Progr. Polym. Sci. 8 (1982) 277–332.
[315] Brutskus, T.K., Saldadze, K.M., Uvarova, E.A., Lyustgarten, E.I. Kolloid. Zh. 34 (1972) 672–676.
[316] Brutskus, T.K., Saldadze, K.M., Fedtsova, M.A., Galitskaya, N.B., Stebeneva, I.G. Kolloid. Zh. 36 (1974) 643–648.
[317] Papkov, S.P., Jelly State of Polymers, Moscow, Khimiya 1974, 255p.
[318] Millar, J.R., Smith, D.G., Marr, W.E., Kressman, T.R.E. J. Chem. Soc. (1962) 218–225.
[319] Brun, M., Quinson, J.-F., Spitz, R., Bartholin, M. Macromol. Chem. 183 (1982) 1523–1531.
[320] Khirsanova, I.F., Pokrovskaya, A.I., Soldatov, V.S., Artamonov, V.A. Izvestia AN BSSR Ser. khim. nauk 1 (1974) 19–23.
[321] Fang, F.T., Golownia, R.F. J. Am. Chem. Soc. Polym. Prepr. 8 (1967) 374–379.
[322] Góźdź, A.S., Kolarz, B.N., Cichosz, M., Chem. Stosowana 24 (1980) 399–408.
[323] Galina, H., Kolarz, B.N. Polym. Bull. 2 (1980) 235–239.
[324] Wieczorek, P.P., Kolarz, B.N., Galina, H. Angew. Makromol. Chem. 126 (1984) 39–50.
[325] Poinescu, I.C., Rotaru, E.J.L., Carpov, A., Dimitriu, R. Pat. Romania 70544 (1980).
[326] Coutinho, F.M.B., Cid, R.C.A. Eur. Polym. J. 26 (1990) 915–918.
[327] Smith, J.R.L., Tameesh, A.H.H., Waddington, D.J. J. Chromatogr. 148 (1978) 353–363.
[328] Lloyd, W.G., Alfrey, T. J. Polym. Sci. 62 (1962) 301–316.
[329] Jun, Y., Xun-hua, W., Yanshan, Y. React. Funct. Polym. 43 (2000) 227–232.
[330] Davankov, V.A., Tsyurupa, M.P., in: S.M. Aharoni (Eds.), Synthesis, Characterization and Theory of Polymeric Networks and Gels,, Plenum Press, New York, 1992, 179–200.

[331] Huxham, I.M., Rowatt, B., Sherrington, D.C. Makromol. Chem. 192 (1991) 1695–1703.
[332] Sederel, W.L., De Jong, G.J. J. Appl. Polym. Sci. 17 (1973) 2835–2846.
[333] Kolarz, B.N., Wieczorek, P.P., Wojaczynska, M. Angew. Makromol. Chem. 96 (1981) 193–200.
[334] Witte, G., Starnik, J. Angew. Makromol. Chem. 84 (1980) 7–36.
[335] Millar, J.R. J. Polym. Sci.: Polym. Symp. 68 (1980) 167–177.
[336] Wood, C.D., Cooper, A.I. Macromolecules 34 (2001) 5–8.
[337] Dušek, K., Seidl, J., Malinsky, J. Coll. Czech. Chem. Commun. 32 (1967) 2766–2778.
[338] Seidl, J., Malinsky, J. Chem. Průmysl 13 (1963) 100–104.
[339] Exner, J., Bohdanecky, M. Coll. Czech. Chem. Commun. 31 (1966) 3985–3989.
[340] Dragan, S., Nichifor, M., Petrariu, I. Makromol. Chem. 180 (1979) 2085–2093.
[341] Svetlov, A.K., Nikitina, T.M., Pushkareva, I.A., Tavobilov, M.F., Zakharova, L.I. Zh. Prikl. Khimii 48 (1975) 2037–2041.
[342] Borovsky, R., Krais, S., Malinsky, J., Seidl, J. Chem. Průmysl 13 (1963) p. 446.
[343] Kolarz, B. Chem. Stosowana 18 (1974) 575–583.
[344] Azzola, F.K., Schmidt, E. Angew. Makromol. Chem. 10 (1970) 203–207.
[345] Krska, F., Stamberg, J., Pelzbauer, Z. Angew. Makromol. Chem. 3 (1968) 149–159.
[346] Luca, C., Neagu, V., Simionescu, B.C., Rabia, I., Zerouk, J., Bencheikh, Z. React. Funct. Polym. 36 (1998) 79–90.
[347] Kolarz, B. Chem. Stosowana 18 (1974) 543–555.
[348] Kolarz, B. Chem. Stosowana 19 (1975) 71–79.
[349] Poinescu, I.C., Vlad, C.-D., Eur. Polym. J. 33 (1997) 1515–1521.
[350] Macintyre, F.S., Sherrington, D.C. Macromolecules 37 (2004) 7628–7636.
[351] Kunin, R., Meitzner, E.F., Oline, J.A., Fisher, S.A., Frich, N. Ind. Eng. Chem. Prod. Res. Develop. 1 (1962) 140–144.
[352] Sharma, M.M. React. Funct. Polym. 26 (1995) 13–23.
[353] Saldadze, K.M., Brutskus, T.K., Fedtsova, M.A., Uvarova, E.A., Lyustgarten, E.I., Semenova, T.S. Zh. Fiz. Khimii 44 (1970) 2815–2819.
[354] Brutskus, T.K., Saldadze, K.M., Uvarova, E.A., Lyustgarten, E.I. Zh. Fiz. Khimii 47 (1973) 353–357.
[355] Saldadze, K.M., Brutskus, T.K., Fedtsova, M.A., Mekvabishvili, T.V. Zh. Fiz. Khimii 43 (1969) 2679.
[356] Soldatov, V.S., Martsinkevich, R.V., Pokrovskaya, A.I. Zh. Fiz. Khimii 49 (1975) 2366–2369.
[357] Millar, J.R., Smith, D.G., Kressman, T.R.E. J. Chem. Soc. (1965), 304–310.
[358] Tsyurupa, M.P., Davankov, V.A., Yamskov, I.A., Budanov, M.V. Patent USSR 804671 (1980).
[359] Shi, Y., Dong, X.-Y., Sun, Y. Chromatographia 55 (2002) 405–410.
[360] Zhang, M., Sun, Y. J. Chromatogr. A922 (2001) 77–86.
[361] Yamskov, I.A., Budanov, M.V., Davankov, V.A. Bioorganich. Khim. 5 (1979) 757–767.
[362] Yamskov, I.A., Budanov, M.V., Davankov, V.A., Nys, P.S., Savitskaya, E.M. Bioorganich. Khim. 5 (1979) 604–610.
[363] Yamskov, I.A., Budanov, M.V., Davankov, V.A. Biokhimiya 46 (1981) 1603–1608.
[364] Shi, Y., Sun, Y. Chromatographia. 57 (2003) 29–35.
[365] Menger, F.M., Tsuno, T., Hammond, G.S. J. Am. Chem. Soc. 112 (1990) 1263–1264.
[366] Zhu, X.X., Banana, K., Yen, R. Macromolecules 30 (1997) 3031–3035.
[367] Zhu, X.X., Banana, K., Liu, H.Y., Krause, M., Yang, M. Macromolecules 32 (1999) 277–281.

[368] Menger, F.M., Tsuno, T. J. Am. Chem. Soc. 112 (1990) 6723–6724.
[369] Guyot, A., Hodge, P., Sherrington, D.C., Widdecke, H. React. Polym. 16 (1991/92) 233–250.
[370] Menner, A., Powell, R., Bismarck, A. Macromolecules 39 (2006) 2034–2035.
[371] Hainey, P., Huxham, I.M., Rowatt, B., Sherrington, D.C. Macromolecules 24 (1991) 117–121.
[372] Small, P.W., Sherrington, D.C. J. Chem. Soc. Chem. Commun. 21 (1989) 1589–1591.
[373] Cameron, N.R., Sherrington, D.C. Macromolecules 30 (1997) 5860–5869.
[374] Afeyan, N.B., Gordon, N.F., Maszaroff, I., Varady, L., Fulton, S.P., Yang, Y.B. et al. J. Chromatogr. 519 (1990) 1–29
[375] Fulton, S.P., Afeyan, N.B., Gordon, N.F. J. Chromatogr. 547 (1991) 452–456.
[376] Wren, S.A.C., Tchelitcheff, P. J. Chromatogr. A1119 (2006) 140–146.
[377] MacNair, J.E., Patel, K.D., Jorgenson, J.W. Anal. Chem. 71 (1999) 700–708.
[378] Lippert, J.A., Xin, B., Wu, N., Lee, M.L. J. Microcolumn Sep. 11 (1997) 631–643.
[379] Patel, K.D., Jerkovich, A.D., Link, J.C., Jorgenson, J.W. Anal. Chem. 76 (2004) 5768–5777.
[380] Wu, N., Liu, J., Lee, M.L. J Chromatogr. A1131 (2006) 142–150.
[381] Hjerten, S., Liao, J.-L., Zhang, R. J. Chromatogr. 473 (1989) 273–275.
[382] Svec, F., Frechet, J.M.J. Anal. Chem. 64 (1992) 820–822.
[383] Svec, F., Frechet, J.M.J. Macromolecules 28 (1995) 7580–7582.
[384] Svec, F., Frechet, J.M.J. Chem. Mater. 7 (1995) 707–715.
[385] Peters, E.C., Petro, M., Svec, F., Frechet, J.M.J. Anal. Chem. 70 (1998) 2288–2295.
[386] Grafnetter, I., Coufal, P., Tesarova, E., Suchankova, J., Bosakova, Z., Sevcik, J. J. Chromatogr. A1049 (2004) 43–49.
[387] Svetlov, A.K., Lazarchenko, V.D., Tavobilov, M.F., Yarichina, G.P. Vysokomol. Soedin. B17 (1975) 293–297.
[388] Wang, Q.C., Svec, F., Frechet, J.M.J. Anal. Chem. 65 (1993) 2243–2248.
[389] Viklund, C., Svec, F., Frechet, J.M.J. Chem. Mater. 8 (1996) 744–750.
[390] Santora, B.P., Gagne, M.R., Moloy, K.G., Radu, N.S. Macromolecules 34 (2001) 658–661.
[391] Peters, E.C., Svec, F., Frechet, J.M.J. Chem. Mater. 9 (1997) 1898–1902.
[392] Peters, E.C., Svec, F., Frechet, J.M.J. Viklund, C., Irgum, K. Macromolecules 32 (1999) 6377–6379.
[393] Meyer, U., Svec, F., Frechet, J.M.J., Hawker, C.J., Irgum, K. Macromolecules 33 (2000) 7769–7775.
[394] Viklund, C., Nordström, A., Irgum, K., Svec, F., Frechet, J.M.J. Macromolecules 34 (2001) 4361–4369.
[395] Hubband, K.L., Finch, J.A., Darling, G.D. React. Funct. Polym. 36 (1998) 1–16.
[396] Peters, E.C., Petro, M., Svec, F., Frechet, J.M.J., Anal. Chem. 69 (1997) 3646–3649.
[397] Zou, H., Huang, X., Ye, M., Luo, Q. J. Chromatogr. A954 (2002) 5–32.
[398] Xiong, B., Zhang, L., Zhang, Y., Zou, H., Wang, J. J. High Resolut. Chromatogr. 23 (2000) 67–72.
[399] Vidic, J., Podgornik, A., Strancar, A. J. Chromatogr. A1065 (2005) 51–58.
[400] Wang, Q.C., Svec, F., Frechet, J.M.J. J. Chromatogr. A669 (1994) 230–235.
[401] Petro, M., Svec, F., Gitsov, I., Frechet, J.M.J. Anal. Chem. 68 (1996) 315–321.
[402] Glökner G, Gradient HPLC of Copolymers and Chromatographic Cross-Fractionation, Springer, Berlin, 1991, pp. 45–74.
[403] Gusev, I., Huang, X., Horvath, C. J. Chromatogr. A835 (1999) 273–290.
[404] Wang, Q.C., Svec, F., Frechet, J.M.J. Anal. Chem. 67 (1995) 670–674.
[405] Xie, S., Allington, R.W., Svec, F., Frechet, J.M.J. J. Chromatogr. A865 (1999) 169–174.

[406] Zhang, S., Huang, X., Zhang, J., Horvath, C. J. Chromatogr. A887 (2000) 465–477.
[407] Sykora, D., Svec, F., Frechet, J.M.J. J. Chromatogr. A852 (1999) 297–304.
[408] Svec, F., Frechet, J.M.J. J. Chromatogr. A702 (1995) 89–95.
[409] Strancar, A., Koselj, P., Schwinn, H., Josic, D. Anal. Chem. 68 (1996) 3483–3488.
[410] Josic, D., Schwinn, H., Strancar, A., Podgornik, A., Barut, M., Lim, Y.-P. et al., J. Chromatogr. A803 (1998) 61–71.
[411] Uzun, L., Say, R., Denizli, A. React. Funct. Polym. 64 (2005) 93–102.
[412] Sun, X., Chai, Z. J. Chromatogr. A943 (2002) 209–218.
[413] Matsui, J., Kato, T., Takeuchi, T., Suzuki, M., Yokoyama, K., Tamiya, E. et al., Anal. Chem. 65 (1993) 2223–2224.
[414] Sellergren, B. Anal. Chem. 66 (1994) 1578–1582.
[415] Xie, S., Svec, F., Frechet, J.M.J. Chem. Mater. 10 (1998) 4072–4078.
[416] Svec, F., Peters, E.C., Sykora, D., Yu, C., Frechet, J.M.J. J. High Resolut. Chromatogr. 23 (2000) 3–18.
[417] Tennikova, T.B., Freitag, R. J. High Resolut. Chromatogr. 23 (2000) 27–38.
[418] Maruska, A., Kornysova, O. J. Biochem. Biophys. Methods 59 (2004) 1–48.
[419] Tennikova, T.B., Reusch, J. J. Chromatogr. A1065 (2005) 13–17.
[420] Belenkii, B.G. Russian J. Bioorg. Chem. 32 (2006) 323–332.
[421] Sinner, F., Buchmeiser, M.R. Macromolecules 33 (2000) 5777–5786.
[422] Zalusky, A.S., Olayo-Valles, R., Taylor, C.J., Hillmyer, M.A. J. Am. Chem. Soc. 123 (2001) 1519–1520.
[423] Advances in Monoliths, Ed. A.M. Siouffi, J. Chromatogr. A1109 (2006).
[424] Crittenden, B., Patton, A., Jouin, C., Perera, S., Tennison, S., Angel, J. et al., Adsorption 11 (2005) 537–541.
[425] Korolev, A.A., Shiryaeva, V.E., Popova, T.A., Kozin, A.V., Dyakov, I.A., Kurganov, A.A. Vysokomol. Soedin. A48 (2006) 1373–1382.
[426] Kanatyeva, A., Kurganov, A., Viktorova, E., Korolev, A. Russ. Chem. Rev. 77 (2008) 373–379.
[427] Kurganov, A., Korolev, A., Viktorova, E., Kanatieva, A. Russ. J. Phys. Chem. A83 (2009), 303–307.
[428] Al-Bokari, M., Charrak, D., Guiochon, G. J. Chromatogr. A975 (2002) 275–284.
[429] Anderson, R.E., Ind. Eng. Chem., Prod. Res. Develop. 3 (1964) 85–89.
[430] Brutskus, T.K., Saldadze, K.M., Fedtsova, M.A., Kaminsky, M.Ya., Semenova, T.S. Zh. Prikl. Khimii 47 (1974) 840–843.
[431] Wheaton, R.M., Hatch, M.J. in: J.A. Marinsky (Ed.), Ion Exchange. A Series of Advances, Marcel Dekker, New York, 2 (1969), pp. 191–233.
[432] Wolff, J.J. Tribune du Cebedeau 22 (1969) 387–396.
[433] Kolarz, B., Stespien-Firkowicz, A., Marciniak, A., Prace Naukove Institutu Technologii Organicznej i Tworzyw Sztucznych Politechniki Wroclawskiej, N13, Konferencje N2 (1973) 105–118.
[434] Noda, I., Kagava, I. J. Chem. Soc. Japan Ind. Chem. Sect. 66 (1963) 854–857.
[435] Asami, R., Siolata, T., Tokura, N. Bull. Chem. Res. Inst. Non-Aqueous Sol. Tohoku Univ. 10 (1961) 99–103.
[436] Hauptmann, R., Schwachula, G. Z. Chem. 8 (1968) 227–228.
[437] Schwachula, G, Hauptmann, R., Kain, J., J. Polym. Sci., Polym. Symp. N47 (1974) 103–109.
[438] Laskorin, B.N., Fedorova, L.A., Stupin, N.P. Dokl. AN USSR 204 (1972) 1411–1414.
[439] Trushin, B.N., Tyurikova, V.K. Vysokomol. Soedin. B16 (1974) 823–826.
[440] Trushin, B.N., Vel'dyaskina, V.L. Collection of Sientific Works of Kuzbass Polytechnic Institute, Chemistry and Chemical Technology Kemerovo (36) 1971, pp. 139–140.

[441] Trushin, B.N., Tyurikova, V.K., Kolmagorov, E.N., Galaktionov, Yu.I. Collection of Scientific Works of Kuzbass Polytechnic Instituite Kemerovo, (69) (1974), pp. 189–192.

[442] Belfer, S., Glozman, R., Deshe, A., Warshawsky, A. J. Appl. Polym. Sci. 25 (1980) 2241–2263.

[443] Belfer, S., Warshawsky, A. IUPAC Makro Mainz: 26th Int. Symp. Macromolecules, 1979, Prepr. Short Commun. Vol. 1, Mainz, s.a., 718–721.

[444] Belfer, S. IUPAC Macro, Florence, 1980, Int. Symp. Macromolecules, 1980, Prepr. Vol. 4, Pisa, s.a., 230–233.

[445] Belfer, S., Glozman, R. J. Appl. Polym. Sci. 24 (1979) 2147–2157.

[446] Grebenyuk, V.D. Zh. Phiz. Khimii 44 (1970) 3149–3152.

[447] Veverka, P., Jeřabek K. React. Funct. Polym. 41 (1999) 21–25.

PART Two

Hypercrosslinked
Polystyrene Networks

CHAPTER *6*

Preparation of Macronet Isoporous and Hypercrosslinked Polystyrene Networks

1. BASIC PRINCIPLES OF FORMATION OF HYPERCROSSLINKED POLYSTYRENE NETWORKS

Based on several decades of an intensive use of polymeric adsorbing materials in chromatography, ion exchange, and large-scale adsorption technologies, the most important requirements for an ideal column packing material can be formulated. It should combine a high adsorption capacity toward the target compounds with the high permeability of its structure to these adsorbates, thus providing both high productivity and high rate of the sorption process. Second, when packed in a column, the material should preserve its volume on a rather constant level when using percolating solutions with different concentrations, compositions, or pH values. Otherwise, strong expansion–contraction ("breathing") of the sorbent bed would unavoidably result in channeling of the packing or complete plugging of the column. Also, substantial swelling and shrinking ("deswelling") effects during the sorption–regeneration cycles lead to an early breakdown of the sorbent beads. Finally, the sorbent particles should exhibit high mechanical and abrasive strength to endure the column bed weight and the mobile phase pressure drop. All these properties of the sorbent depend on the structure of its matrix, whereas the equally important selectivity of the sorption process entirely depends on the chemistry of its sorption sites.

Evidently, gel-type sorbents, mostly conventional ion-exchange resins, do not satisfy these requirements. Their low resistance to

Comprehensive Analytical Chemistry, Volume 56
ISSN 0166-526X, DOI 10.1016/S0166-526X(10)56006-X

osmotic shock relates to their strong swelling. Indeed, a sudden change of the outer liquid results in a rapid swelling (or deswelling) of the outer layers of the bead, whereas the interior of the bead needs much more time for equilibration with the environment. This local change in the network expansion causes internal stresses that cannot be tolerated by the matrix. Even the smoothly conducted swelling–shrinking cycles, accompanying the many times repeated sorption–regeneration cycles, reduce considerably the lifetime of the material. The extremely nonuniform distribution of divinylbenzene (DVB) crosslinks throughout the bead volume of the gel-type sorbent causes different swelling of the slightly and the highly crosslinked network domains. One cannot eliminate completely this inherent drawback of styrene–DVB copolymers because it is conditioned upon the different reactivity of the comonomers. Reduction of the extent of network swelling by increasing its degree of crosslinking enhances the durability of the sorbent, but simultaneously compromises the network permeability and the rate of mass transfer.

Gel-type isoporous (see Chapter 5) anion-exchange resins suffer from these drawbacks to a lesser degree since the variations of their volume in different media are less pronounced. However, it is difficult to control at a pre-described level the extent of the post-crosslinking reaction of the initial copolymer. Even the smallest deviations in the swelling degree of the copolymer, the temperature, quality, or quantity of the catalyst, or any other synthesis conditions may result in a substantial change of the network topology and, hence, the properties of the resulting materials. In other words, it is difficult to manufacture isoporous ion-exchange resins with exactly reproducible properties.

In many respects macroporous resins are free of the swelling-related problems and meet the above-mentioned requirements to an ideal adsorbent, including fast mass transfer and permeability to large organic molecules. Still, the sorbate-accessible functional groups mostly locate on the surface of the pores in the macroporous matrix, so that the adsorption capacity of macroporous resins often proves to be insufficient for industrial applications.

A fundamentally new approach was suggested by Davankov and Tsyurupa as far back as 1969 to prepare uniformly crosslinked polystyrene networks that allow the combination of a high permeability with very small volume changes in different liquid media [1].

The basic principle of their suggestion was formulated as intensive crosslinking of polymeric chains, in particular, chains of polystyrene, in solution or in swollen state by rigid bridges. This approach led to the creation of a new type of polymeric materials with special structure and peculiar properties that were not characteristic of any known types of networks. Therefore, a new term was needed to characterize these new materials; we have called these new polymers "hypercrosslinked polystyrene."

The following fundamental principles constitute the concept of hypercrosslinked polystyrene network formation.

First of all, it was assumed that the crosslinking of polystyrene chains in a solution or in highly swollen styrene–DVB gels has to result in a statistically homogeneous, that is, random distribution of crosslinks throughout the entire volume of the resulting networks. This character of the links' distribution is provided primarily by the homogeneous distribution of all components of the crosslinking reaction throughout the volume of the initial system containing long polymer chains and the crosslinking agent. Indeed, the macromolecular coils of linear polystyrene with molecular weights of 100,000–700,000 Da begin to overlap one another already in a rather diluted solution in any good solvent, for example, toluene [2] (Table 6.1).

Table 6.1 Critical overlap concentration, C^*, for polystyrene coils in solution measured by energy transfer between carbazole- and anthracene-labeled polystyrene chains

Molecular weight of PS (kDa)	Solvent	C^* (g/L)	$[\eta]$ (dL/g)
100	Tol	56 (38–74)	0.395
100	EA	76 (57–95)	0.286
670	Tol	18 (13–25)	1.653
670	EA	35 (20–50)	0.840
670/100*	Tol	17 (12–22)	0.952
670/100*	EA	33 (25.5–37)	0.568

Tol = toluene.
EA = ethyl acetate.
Reprinted from [2] with kind permission of the American Chemical Society.
*Mixture of two homopolymers.

Based on the data presented in Table 6.1, we may assume that the polystyrene coils with a molecular weight of ~300,000 Da and broad molecular weight distribution begin to overlap one another at a concentration of ~2% (w/v) if dissolved in ethylene dichloride (EDC), a thermodynamically good solvent (the intrinsic viscosity of the polymer in EDC is 0.85 dL/g). In a 10% solution, the degree of coil interpenetrating should be considerable. If the polystyrene coils interpenetrate freely, as predicted by the classical theories of polymer solutions, the distribution of the polymer segments throughout the 10% solution should be regarded as homogeneous. At least, concentrated solutions of polystyrene in good solvents were found to contain no large clusters of chains [3]. If the interpenetration of macromolecules is limited to their outer regions, the polymeric coils do not lose completely their individuality. In this case the distribution of segments in the solution could be less homogeneous. Still, whatever the structure of a 10% solution is, it is most important that all polystyrene coils incorporate, in addition to many fragments of neighbor coils, large amounts (an about 10-fold excess) of the solvent. Therefore, a low molecular weight crosslinking agent and a catalyst have to be distributed homogeneously throughout the entire polymer solution prior to start of the reaction. The same must be true for swollen styrene–DVB copolymers. On heating the mixture the crosslinking reaction must start simultaneously in many points of the system. First intermolecular crosslinking bridges introduced in the very beginning of the reaction substantially reduce the mobility of the polymeric coils, thus preventing them from separation or contraction. Notably, the short-range order in the dislocation of neighboring phenyl rings in a dissolved polystyrene chain is such that the formation of several subsequent links between two chains is practically ruled out (see below). The crosslinking agents thus simultaneously participate in both the intra- and the intermolecular reactions, and no thermodynamic or kinetic factors interfere that could disturb the statistical character of bridge distribution in the final system.

The second fundamentally important principle is that the polymeric chains maintain their highly solvated state throughout the reaction. It implies that the reaction solvent used presents a thermodynamically good medium with respect to both the initial chains and the final network. To enforce this condition, on introducing crosslinking bridges one should not substantially change the chemical nature of the material. Preferably, the crosslinking bridges should have a similar composition and polarity as the starting polymer. Two significant consequences follow from this requirement. First, no thermodynamic reasons should arise for any microphase separation in the course of network formation. Second, the crosslinking bridges will fix the favorable, undistorted conformations of the macromolecular chains, thus minimizing the probability of the appearance of any inner stresses in the final swollen network.

Finally, the third principle of hypercrosslinked polystyrene network formation is that the resulting network is conformationally rigid. This requires both the very intensive crosslinking of the rather flexible polystyrene chains and conformational rigidity of the formed links. Numerous rigid bridges (spacers) should hold the polystyrene chains at a substantial distance from one another in the open network-type molecular construction, which, in its turn, should diminish significantly the changes in network volume when replacing one solvent for another.

Naturally, the crosslinking reaction should be strictly controllable, which is easily achievable if the chosen crosslinking agent reacts with polystyrene actively and completely.

2. CROSSLINKING AGENTS AND CHEMISTRY OF POST-CROSSLINKING

Many bifunctional substances can react with polystyrene with the formation of bridges between two phenyl rings belonging to different chains or different segments of the same chain. Still, according to the above-formulated principles, bis-chloromethyl derivatives of aromatic hydrocarbons were found to be the preferable crosslinking agents. They are highly reactive under the conditions of Friedel–Crafts reaction and can form bridges of sufficient rigidity. Thus, 1,4-bis-(chloromethyl)-- diphenyl (CMDP) [4], p-xylylene dichloride (XDC) [5–10], 1,4-bis-(p-chloromethylphenyl)-butane (DPB) [11], or 1,3,5-tris-(chloromethyl)-mesitylene (CMM) [11] reacts with two polystyrene chains (three chains for CMM) in the presence of Friedel– Crafts catalysts forming crosslinking bridges that are two phenyl rings longer than the crosslinking molecule itself:

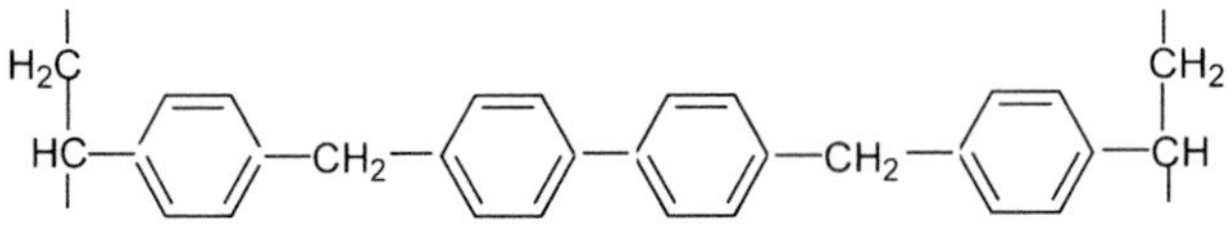

The most rigid networks form when the polystyrene chains are crosslinked with tris-(chloromethyl)-mesitylene (CMM) [12], since one molecule of this compound connects in space with not two but three polystyrene chains at one junction point. Contrary to this, four subsequent methylene groups in the CMB molecule should strongly contribute to the conformational mobility of cross-bridges and the network as a whole.

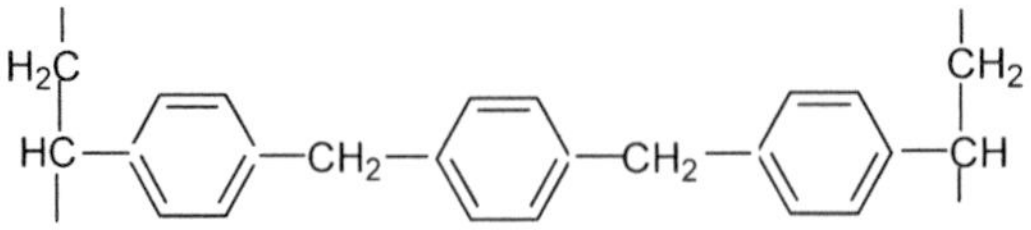

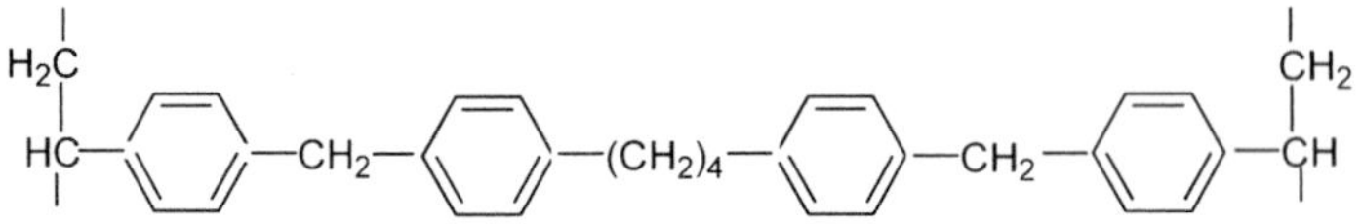

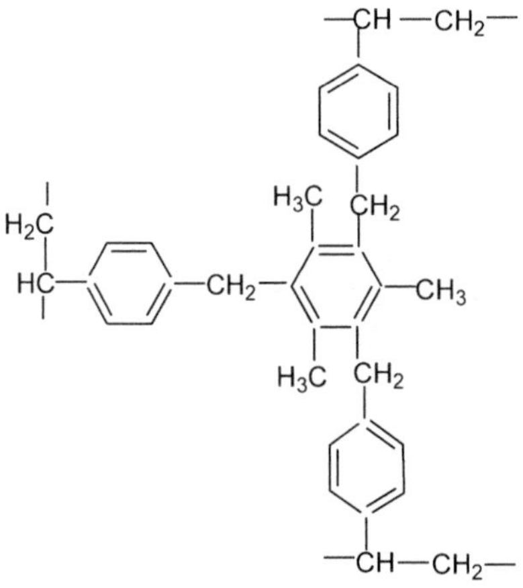

Monochlorodimethyl ether, CH_3OCH_2Cl (MCDE) [1, 13, 14], though a rather toxic compound, is also considered to be a suitable bifunctional crosslinking reagent. Its reaction with polystyrene proceeds quantitatively through the intermediate stage of chloromethylation; the introduced chloromethyl groups can react further with another phenyl ring forming an equivalent number of diphenylmethane-type rigid bridges:

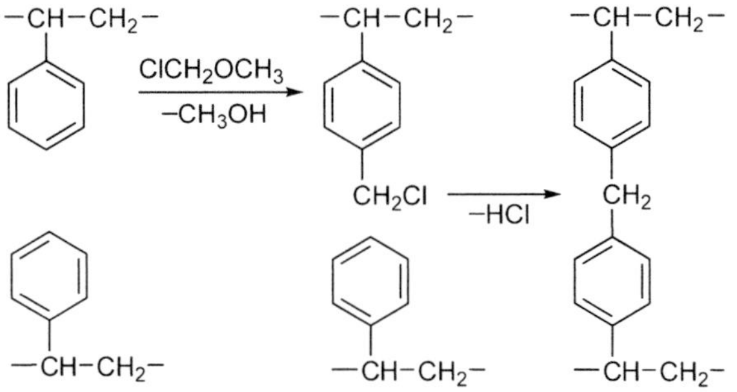

It is quite evident that the chloromethyl groups can be introduced by any chloromethylation method known for polystyrene, prior to the actual post-crosslinking process. For example, in excess MCDE in the presence of stannic tetrachloride, almost all phenyl rings of the polymer are substituted with chloromethyl functions under mild conditions [15–17]. When 2.1 weight parts of H_2SO_4 (83%) per 1 weight part of polystyrene were taken as the catalyst for this reaction, approximately 70% of the phenyl rings participate in the reaction within 6 h at 50°C [18]. Chloromethylation can be comprehensive (23% of chlorine introduced) when a mixture of 3.46 mol of MCDE, 2 mol of H_2SO_4, and 0.05 mol of ferric trichloride per 1 mol of styrene-repeating units is used at 57°C for 6 h. Compounds such as $Fe_2(SO_4)_3$ or Fe_2O_3 were also found to successfully catalyze chloromethylation with MCDE [19]. The mixtures of methylal, chlorosulfonic acid, and sulfuryl dichloride [19], methylal and chlorosulfonic acid [20, 21], and trioxane, methylene dichloride, and hydrogen chloride [22] are also assumed to be suitable for chloromethylation. Warshawsky et al. [23] found a safe method of halogenomethylation that is not accompanied by the formation of any traces of the extremely toxic [24] bis-chloromethyl ether. According to this method, halogenomethylation proceeds via long-chain halogenomethylalkyl ethers $ROCH_2X$ produced by the reaction of long-chain alcohols with para-formaldehyde in the presence of a catalyst and gaseous hydrogen chloride or hydrogen bromide. The transfer of $-CH_2X$ residues from the mixed ether to the aromatic rings releases the starting alcohol ROH that can then be separated and reused. The long-chain alcohols do not cause deactivation of Lewis acids used as catalysts. Monochloromethyloctyl ether was found to introduce 20% chlorine into the gel-type styrene–2% DVB copolymer under mild conditions (28°C, 2 h). Bacak et al. [25] suggested the use of excess methylene dichloride and aluminum trichloride for the introduction of chloro-methyl groups. However, the extent of substitution of the aromatic rings amounts only to 1 mmol of chlorine (3.5%) per gram of styrene–2% DVB copolymer. This procedure, certainly, is not suitable to obtain hypercrosslinked networks but appears to be useful for preparing supports for solid-phase peptide syntheses where polymers with a low degree of chloromethylation are needed. Safe chloromethylation of aromatic com-pounds can also be performed by methoxy acetyl chloride in the presence of AlCl$_3$ in nitromethane [26]. The reaction also takes place with SnCl$_4$ in methylene dichloride:

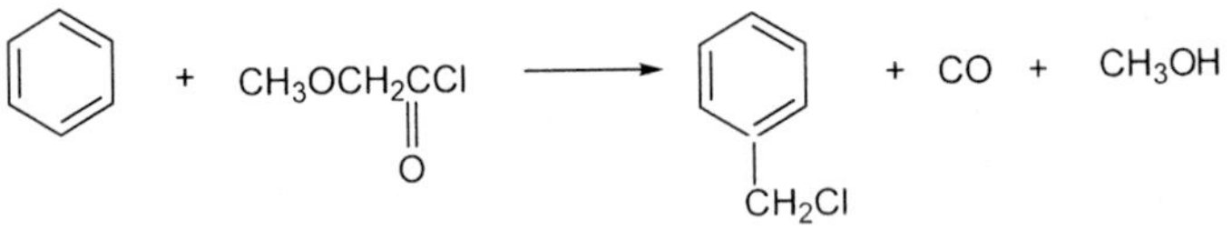

After introducing chloromethyl groups into initial styrene–DVB copolymers according to one of the above chloromethylation proce-dures, heating of the beads, highly swollen in EDC, in the presence of Friedel–Crafts catalysts, AlCl$_3$, FeCl$_3$, ZnCl$_2$, SnCl$_4$, or TiCl$_4$, results in the post-crosslinking of the copolymer and formation of hypercrosslinked material [27–47]. In this case the degree of additional bridging is difficult to control and often remains unknown in the final materials, but this technically simple approach is most suitable for the industrial production of hypercrosslinked sorbents.

Trying to completely avoid the technically unpleasant process of chloromethylation, Negre et al. [48, 49] prepared a linear styrene copolymer with p-vinylbenzyl chloride and then subjected the product to self-crosslinking. Alternatively to the earlier-mentioned crosslinking of linear polystyrene with MCDE, this procedure results in local inhomo-geneity of crosslinks' distribution, because of the uneven distribution of the two comonomers along the initial chain (the monomer reactivity ratios of vinylbenzyl chloride and styrene are 1.41 and 0.71, respectively). Nevertheless, vinylbenzyl chloride became a popular comonomer for styrene and DVB in the preparation of beaded hypercrosslinked products [50–52].

Dimethoxymethane (DMM), also called methylal, acts similar to MCDE as a bifunctional compound and alkylates two phenyl rings yielding networks crosslinked with diphenylmethane-type rigid bridges [53, 54]:

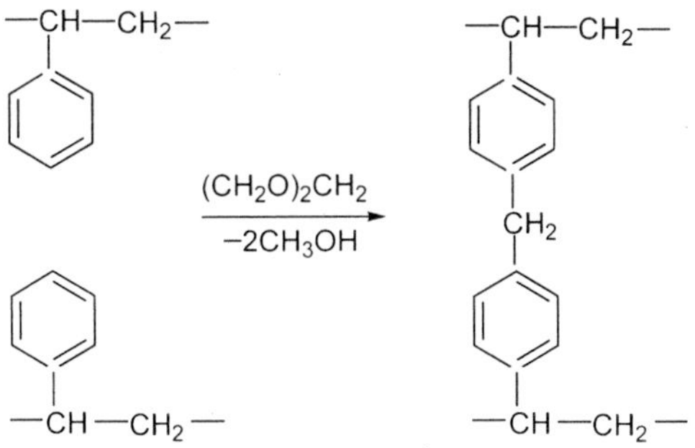

Similarly to XDC, its analogue, 1,4-dimethyl-2,5-dichloromethylbenzene as well as linear oligomers obtained by its self-condensation [55–58] react with relatively low molecular weight polystyrene dissolved in acetic acid, methylethyl ketone, or butyl acetate in the presence of H_2SO_4 and $HClO_4$ as catalysts. The reaction in a diluted solution results in the formation of soluble branched macromolecules with enlarged glass transition temperature [59]. An effective gelation was observed in more concentrated solutions in EDC on using $SbCl_5$ as the catalyst [60]. To prepare the final product in the form of spherical beads, the solution of all components was suspended in polydiethylsiloxane oil saturated with EDC. Naturally, the crosslinking of polystyrene with these reactive oligomers proceeds easily in the presence of a more active catalyst, stannic tetrachloride [61].

Ando et al. [62, 63] suggested using unreacted double bonds in swollen styrene–DVB copolymers, which also react with aromatic rings via Friedel–Crafts reaction forming rigid crosslinks:

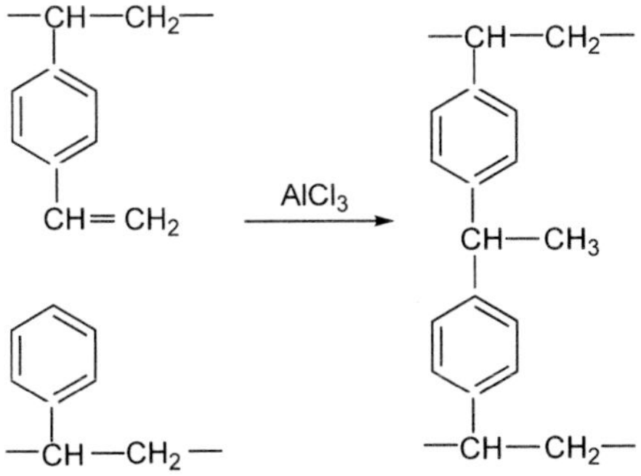

Still, the reactive ability of residual double bonds is relatively low and requires stronger catalysts, such as $AlCl_3$ or $FeCl_3$, to make the reaction run with acceptable rate. However, aluminum trichloride gives rise to side reactions of dealkylation and transalkylation:

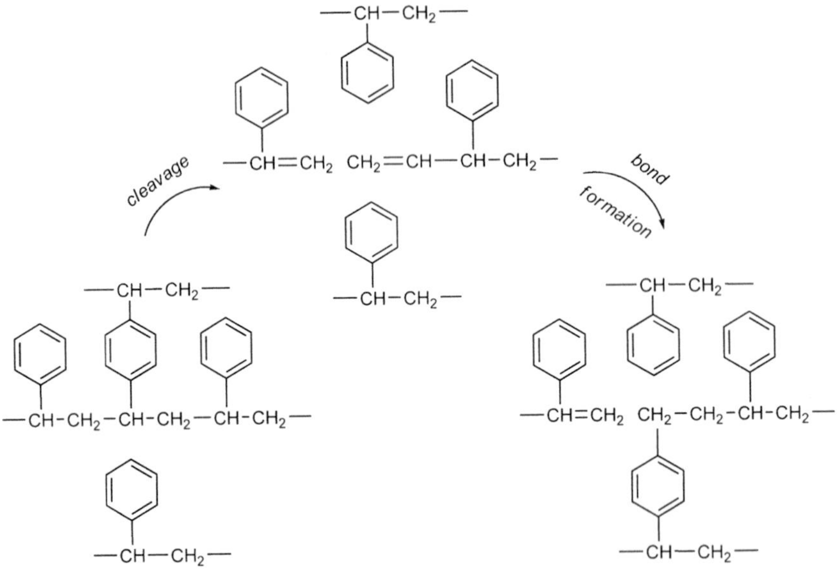

For this reason, a styrene–4% DVB copolymer (containing very few unreacted double bonds) either converts into a slightly crosslinked gel or completely decomposes to soluble oligomers. A gel-type styrene copolymer with 16% DVB withstands the treatment with $AlCl_3$ and even yields low-porosity material with a surface area of $80\,m^2/g$. It has been possible to obtain a highly porous resin ($1000\,m^2/g$) on the basis of macroporous poly(55% DVB) only. Such highly crosslinked macroporous copolymers, however, poorly swell in EDC and so the permeability of the resulting resins is rather low for organic compounds. To increase the swelling of the polymeric phase of macroporous materials in EDC, their synthesis was suggested to be conducted in a solvating media, toluene, or its mixtures with a precipitant, such as n-hexane [64]. Many years later, Hao et al. [65] made use of this earlier idea of involving pendent vinyl groups (PVGs) into post-crosslinking in order to enhance the specific surface area of macroporous poly(DVB) monosize beads. The same effect may also be achieved by entering the unrelated double bonds of poly(DVB) in a radical-initiated post-crosslinking reaction in the presence of di-*tert*-butyl peroxide [66].

Nyhus et al. [67, 68] and Gawdzik et al. [69] assumed that in the presence of $AlCl_3$ additional bridging through the pending double bonds is caused by the "cationic polymerization" of the latter occurring with a simultaneous addition of hydrogen chloride:

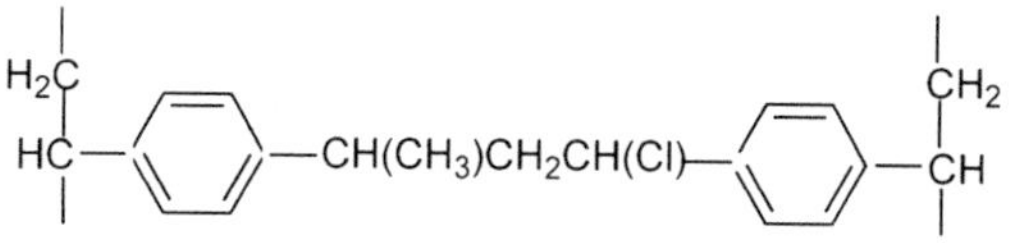

$$H_2C \quad\quad\quad\quad\quad\quad\quad\quad\quad CH_2$$
$$HC-\langle\ \rangle-CH(CH_3)CH_2CH(Cl)-\langle\ \rangle-CH$$

With the setting aside of the latter reaction, we can state that the idea of post-polymerization of PVGs in styrene–DVB copolymers represents a rather widespread misunderstanding. Some vinyl groups remain unreacted during the free radical crosslinking polymerization for the only reason that they were unable to find a suitable partner to attach. Taking into account the random dislocation of PVGs in the polymeric phase, it is impossible to imagine the formation of new long aliphatic chains with their participation. At the same time, if the reaction would happen between just two PVGs located close to each other as shown above, it would result in the formation of flexible bridges, which, as we will see later, do not provide the network with all the peculiar properties of hypercrosslinked polystyrene. If still these properties emerge in the materials containing a high proportion of DVB units, this implies only that a major part of the PVGs will alkylate the phenyl rings rather than add to the neighbor PVGs.

All the above-mentioned crosslinking agents form bridges, the chemical nature of which is very similar to that of polystyrene itself. Therefore they do not form ballast in the total network structure: both the initial polystyrene and the bridges formed can enter into the same reactions. For example, it is likely to expect that ion-exchange resins having a high degree of functionalization and high exchange capacity can be prepared even with a densely crosslinked matrix. Being of an aromatic hydrocarbon nature, hypercrosslinked polymers obtained with all the above-mentioned crosslinking agents should be strongly hydrophobic and possess high hydrolytic stability.

Halogen anhydrides of dicarboxylic acids such as terephthalic or diphenyldicarbonic acid [1], as well as mesitylene tricarbonic acid [29], also react with polystyrene in the presence of AlCl$_3$, with the only difference being that the crosslinking bridges will incorporate carbonyl groups:

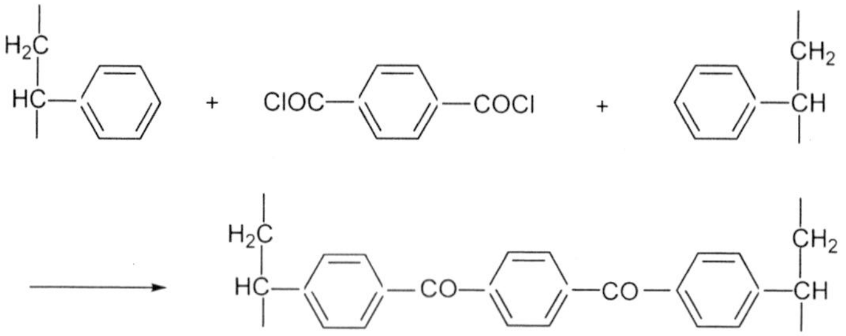

No detailed information is available about networks crosslinked by carbonyl-containing bridges. It is believed that large amounts of carbonyl groups would coordinate and partially inactivate the Friedel–Crafts catalyst and also slightly passivate the adjacent phenyl rings in the electrophilic substitution reaction. In this case it should be more difficult to form true hypercrosslinked structures with these crosslinking reagents.

Many other bi- and polyfunctional compounds, such as chloroform, carbon tetrachloride [70, 71], sulfuryl chloride, thionyl chloride, paraformaldehyde [30], and cyanuric chloride [29], have been suggested as potential crosslinking agents, but, except that some of them are rather inexpensive, there are no convincing data about their advantages over the earlier suggested chloromethylated crosslinking agents.

Linear polystyrene of different molecular weights or gel-type copolymers of styrene with 0.3–4.0% DVB (or any other crosslinking agent needed only to obtain the initial copolymer in the form of beads) can be used as the suitable starting polymer for the post-crosslinking reaction. Naturally, macroporous styrene–DVB copolymers may serve in this capacity as well. It is only necessary to remember that, in compliance with the basic principles of hypercrosslinked network preparation, both the initial gel-type and the macroporous copolymers should exhibit a sufficiently high swelling capacity in the reaction medium. Slightly crosslinked gel-type copolymers easily meet this requirement, while macroporous materials containing relatively high DVB proportions may prove unsuitable for obtaining useful hypercrosslinked polymers. An interesting means of decreasing the threshold DVB content in the preparation of macroporous polymers had been described [40], consisting of conducting the suspension copolymerization under pressure at elevated temperatures of about

$100-130°C$. Thus, styrene copolymers with 1.5% DVB obtained in the presence of 35–50 wt% of *iso*-octane were reported to display cellular texture, with the pore size ranging from 100 to 2000 Å. Chloromethylation and subsequent post-crosslinking of such copolymers should result in highly permeable hypercrosslinked sorbents.

3. NEW TERMS FOR POLYMERIC NETWORKS

Notably, all the above-described post-crosslinking procedures yield networks with long-chain bridges between the initial polystyrene chains. Indeed, the bridge incorporates two phenyl rings and a residue from the crosslinking agent and exceeds the dimension of a DVB molecule that delivers only one phenylene residue to the junction point. Networks with long crosslinking bridges were labeled as "macronet." In the 1960s through the 1970s, such macronet ion-exchange resins were intensively studied in the former USSR by the research groups of Samsonov [72–74], Ergozhin [75–86], and others [87–91]. Macronet ion exchanger were typically prepared by the copolymerization of acrylic or methacrylic acids with ethylene glycol or even diethylene glycol dimethacrylate, or the corresponding amides, or bis-acryloyl derivatives of various aromatic bisphenols or bis-amines. The macronet resins were primary aimed at the isolation of antibiotics from the cultural liquids of their microbiological production and, indeed, they exhibited an enhanced permeability and sorption capacity.

At the same time, crosslinking of polystyrene chains in solution or in swollen state is expected to lead to a homogeneous distribution of links, that is, to materials referred to as "isoporous" (see Chapter 5).

For these reasons, the products of the post-crosslinking reaction were initially termed "macronet isoporous" networks, because they combine all the distinguishing features of the two types of networks. We still believe that this name represents an adequate definition of the products of post-crosslinking of polystyrene chains up to a degree of crosslinking of about 25%. Beyond this value expanded networks acquire substantial rigidity and start to exhibit peculiar properties, first of all the ability to swell in any liquid media and even water, and show microporosity in dry state. However, neither the term "macronet" nor "isoporous" or any other known terms used for three-dimensional networks reflect these properties. This is the reason why we have introduced the new term "hypercrosslinked" for the open networks with degree of crosslinkings of over 40%.

4. SYNTHESIS OF MACRONET ISOPOROUS AND HYPERCROSSLINKED POLYSTYRENE NETWORKS

4.1 Choice of solvents and catalysts

Alkylation of aromatic compounds, including polystyrene, by halogen alkyls, α-halogen-activated ethers, or acetals represents the reaction of electrophilic substitution on the aromatic ring, which occurs in the presence of Lewis acids (Friedel–Crafts reaction):

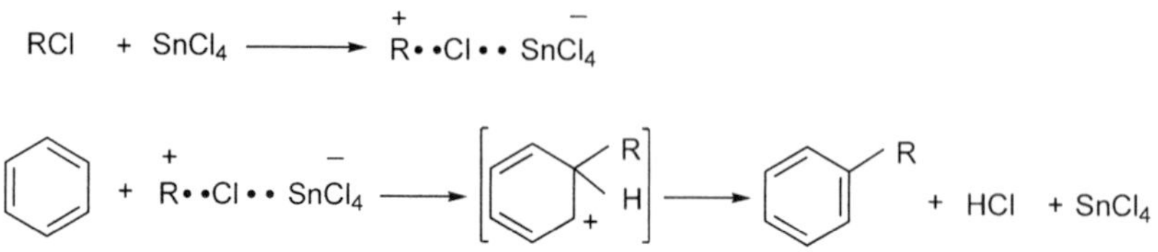

Friedel–Crafts reaction can proceed in a rather limited number of inert solvents, such as aliphatic hydrocarbons, benzene, nitrobenzene, CS_2, and chlorinated or fluorinated hydrocarbons. Among these solvents, however, only chlorinated hydrocarbons, particularly EDC, nitrobenzene, as well as cyclohexane at an elevated temperature, are suited for the synthesis of hypercrosslinked polystyrene, since they are thermodynamically good solvents for both the initial polystyrene and the final products of crosslinking.

The choice of a catalyst is dictated by two main requirements: it should dissolve in the selected solvent and provide a sufficiently high rate of chemical reaction between polystyrene and a crosslinking agent. The activity of anhydrous Friedel–Crafts catalysts is well known to decrease in the following sequence:

$$AlBr_3 \approx AlCl_3 > FeCl_3 > SnCl_4 > TiCl_4 > ZnCl_2 \gg BF_3 > HCl > H_2SO_4 > P_2O_5$$

$AlCl_3$ and $AlBr_3$ are the most active catalysts but they cause degradation of polystyrene chains. $FeCl_3$ represents a rather strong oxidative agent, and so there always exists an anxiety for introducing some functional groups into the final products. Both $AlCl_3$ and $FeCl_3$ possess low solubility in the solvents suitable for the reaction. Still, $FeCl_3$ and $AlCl_3$ are often used to catalyze the post-crosslinking of preliminary chloromethylated styrene–DVB copolymers. The salts, though scarcely soluble in EDC, readily migrate into the swollen copolymer beads due to a fairly fast formation of a reaction complex with an aromatic species [92, 93]. A substantial shortcoming of both $AlCl_3$ and $FeCl_3$ is that they also catalyze the reaction of the solvent, EDC, with polystyrene [94]. $SnCl_4$ and $TiCl_4$ may be considered as most suitable catalysts for the post-crosslinking reaction, at least for laboratory-scale studies. Finally, hydrochloric and sulfuric acids, though catalyzing the crosslinking of polystyrene with the most reactive 1,4-dichloromethyl-2,5-dimethylbenzene [95], are too weak to provide a sufficiently high rate of the reaction.

It must be emphasized here that all Friedel–Crafts catalysts are strongly hygroscopic and decompose in improperly dried solvents, although indications can be found in the literature that small traces of water or alcohols act as activating co-catalysts. In any case attention must be paid to the purity of the solvent selected for the post-crosslinking reaction.

In laboratory experiments we opted, for convenience reasons, for the use of stannic tetrachloride, which dissolves in EDC, cyclohexane, and nitrobenzene, possesses sufficiently high activity at elevated temperatures, and at the same time initiates a rather smooth beginning of the reaction.

4.2 Synthesis conditions of macronet isoporous and hypercrosslinked polystyrene networks

The main protocol for crosslinking linear polystyrene in our studies was as follows: polystyrene, an industrial product of ~300,000 Da molecular weight and wide molecular weight distribution, and the required amounts of a crosslinking agent were dissolved in EDC. Before the addition of stannic tetrachloride, the solution was cooled down to about −20°C in order to have enough time for homogeneously distributing the catalyst in the viscous polymeric solution, prior to gelation of the mixture. As a rule, the polystyrene concentration in the final mixture of all components was 12.5% (w/v) (1 g in 8 mL). On raising the temperature, the solution quickly transformed into a transparent, intensively colored gel. On heating, the volume of the gel progressively decreases and a certain amount of free solvent separates (macrosyneresis occurs). Naturally, the crosslinking of linear polystyrene results in obtaining a relatively rigid block of gel, which has to be disintegrated into irregularly shaped particles.

Besides linear polystyrene, spherical beads of slightly crosslinked copolymers of styrene with 0.3–4% DVB, swollen to a maximum in EDC, have also been used as an initial polymer for the subsequent intensive post-crosslinking with MCDE. Spherical particles are preferable for many practical applications of the final products.

By conducting the reaction at 60–80°C in 5–10 h and in the presence of 0.1–1 mol of the catalyst per 1 mol of reactive groups (e.g., chloromethyl groups) of the crosslinking agents, it is possible to achieve the complete conversion of the latter [4]. The final polymers contain no pendent chloromethyl groups, and no free crosslinking agents are left in the reaction media. When volatile crosslinking agents such as MCDE and methylal are used, it seems to be unlikely that their loss is noticeable. Indeed, they should be involved almost immediately in the formation of reactive charge transfer complexes with the catalyst, before the temperature of the reaction mixture attains or exceeds their boiling points, 59.5 and 42 °C, respectively. The crosslinking of both linear polystyrene and beaded styrene–DVB copolymers with MCDE was conducted under identical conditions as reported in Table 6.2. Importantly, EDC was shown not to react with polystyrene under these optimal conditions.

Table 6.2   Conditions of synthesis of hypercrosslinked networks based on linear polystyrene

Crosslinking agent	X (%)	[SnCl$_4$]^a	$T_1{}^b$ (°C)	$T_2{}^c$ (°C)	τ^d (h)	Cl^e (%)	UCA^f (%)
CMDP	10–100	0.2	20	60	4	No	No
				80	+2		
MCDE	10–25	1.0	20	50	3	No	No
MCDE	40–100	1.0	−20	80	10	<0.5	1–2
XDC	10–100	2.0	20	80	10	No	No
DMM	10–25	1.0	20	80	10	–	No
DMM	40–100	1.0	−20	80	10	–	0–3
CMB	10–25	2.0	20	80	10	No	No
CMB	40–100	2.0	−20	80	10	0.2	No
CMM	5–20	3.0	20	80	15	No	No
CMM	25–100	3.0	−20	80	15	<1	No

^{a}[SnCl$_4$] = amount of catalyst, mole per mole of crosslinking agent.
^{b} T_1 = temperature of the added catalyst.
^{c} T_2 = temperature of the synthesis.
^{d} t = reaction time.
^{e}Cl = unreacted chlorine content in the final product, as determined by elemental microanalysis.
^{f}UCA = amount of unreacted crosslinking agent in the separated solvent, percent of the initial amount, determined by thin layer chromatography (CMDP, XDC, CMM, CMB) or gas chromatography (MCDE, DMM).
CMDP = 1,4-bis-chloromethyldiphenyl; CMM = tris-(chloromethyl)-mesitylene; CMB = 1,4-bis-(p-chloromethylphenyl)-butane; DMM = dimethoxymethane (methylal); MCDE = monochlorodimethyl ether; XDC = p-xylylenedichloride.
Reprinted from [96] with kind permission of Elsevier.

Complete conversion of the crosslinking agent under anhydrous conditions (Table 6.2) permits the calculation of the degree of crosslinking, based on the molar ratio of the two reacting components. We define the degree of crosslinking, X, as the ratio of crosslinking bridge numbers to the sum of the bridges and unsubstituted phenyl rings, expressed in percent:

$$X(\%) = \frac{Z}{Z + (1 - fZ)} \times 100 \qquad [6.1]$$

Here, Z is the number of moles of the crosslinking agent taken for 1 mol of styrene-repeating units and f is the functionality of the crosslinking agent. Thus, if 0.5 mol of a bifunctional reagent binds to 1 mol of polystyrene, the formal degree of crosslinking of the network is 100%. It implies that, theoretically, all phenyl groups of the initial polystyrene chains are involved in the bridge formation. However, a perfect pair-wise arrangement of all the chain-bound phenyl groups for a subsequent interaction with the bifunctional molecules of the crosslinking agent is impossible for steric and statistic reasons. Indeed, in real networks, with 100% of crosslinking, a certain portion of unsubstituted phenyl rings remains. Still, the complete consumption of the crosslinking reagent in the above reaction is a well-documented fact. Therefore, we have to assume that some phenyl rings of the initial polystyrene become involved in the reaction with two molecules of the crosslinking agent, thus producing trisubstituted aromatic rings. Obviously, along with the major disubstituted rings in the final structure, trisubstituted and monosubstituted rings must also be present, the latter two structures in equivalent amounts. Multiple substitutions of the initial phenyl rings of polystyrene are potentially possible under the conditions of the Friedel–Crafts reaction as new alkyl substituents in the aromatic ring do not reduce its reactivity.

The number of remaining phenyl rings not involved in crosslinking bridge formation can be strongly reduced by using more than 0.5 mol of the crosslinking agent, for example, 0.75 or even 1 mol per mole of styrene repeat units. With increasing the degree of crosslinking beyond the formal 100%, the structure of crosslinking bridges becomes more and more complicated, and the portion of the trisubstituted benzene rings constantly increases until the network gets predominantly composed of trisubstituted bridges, when 1 mol of the bifunctional crosslinking agent is used. It is important to mention that even in the case when MCDE is taken in the above high proportions, it reacts almost comprehensively; the obtained products contain not more than 1% of unreacted chlorine. It is quite evident in such cases that Eq. [6.1] cannot be used any more to calculate the crosslinking density of the networks that exceeds 100%. For the sake of simplicity of presentation, however, let us agree to use formal values of the conventional degree of crosslinking of 150 and 200% when, respectively, 0.75 and 1 mol of a bifunctional crosslinking agent react with 1 mol of styrene-repeating units. Naturally, it is also clear that the above consideration does not distinguish between intra- and intermolecular links. For the hypercrosslinked networks, the very differentiation between these two types of crosslinking loses completely its initial sense.

4.3 FTIR spectra of hypercrosslinked polystyrenes

Figure 6.1 demonstrates the Fourier transform infrared (FTIR) spectra of linear polystyrene, linear poly(p-methylstyrene), and several typical spectra of hypercrosslinked polystyrenes based on both a linear precursor and styrene–0.5% DVB copolymer. Usually the type of substitution on the benzene rings is revealed in the absorbance range below $900 \, cm^{-1}$. In this range the spectra of all the polymers in question exhibit three main bands. Absorbance at $811–814 \, cm^{-1}$ is characteristic of the out-of-plane deformational vibration of two adjacent hydrogen atoms in 1,4-disubstituted benzene ring (p-substitution). This band is very characteristic of the spectrum of poly(p-methylstyrene) at $816 \, cm^{-1}$. Its presence in the spectra of all hypercrosslinked polystyrenes confirms the formation of 1,4-disubstituted benzene rings, that is, the preferred p-substitution in the post-crosslinking process. The intensity of this band increases in parallel with the degree of crosslinking. A small shift of the band at $1492 \, cm^{-1}$ in the spectrum of polystyrene toward $1511 \, cm^{-1}$ in the spectra of the hypercrosslinked networks also corroborates the formation of p-disubstituted benzene rings.

In the spectrum of linear polystyrene two strong bands, 756 and $700 \, cm^{-1}$, unambiguously characterize the out-of-plane conformational vibrations of monosubstituted benzene rings. They are absent in the spectrum of poly(p-methylstyrene), but can be found in the spectra of many hypercrosslinked networks, thus revealing residual monosubstituted benzene rings, not involved in crosslinking. The intensity of these bands substantially decreases with increasing crosslinking density. Both bands disappear almost completely in the spectrum of the polymer with the formal degree of crosslinking of 200% prepared by using 1:1 stoichiometric ratio of polymer-repeating units to the crosslinking agent, MCDE. At the same time in the styrene–DVB copolymer crosslinked to 100%

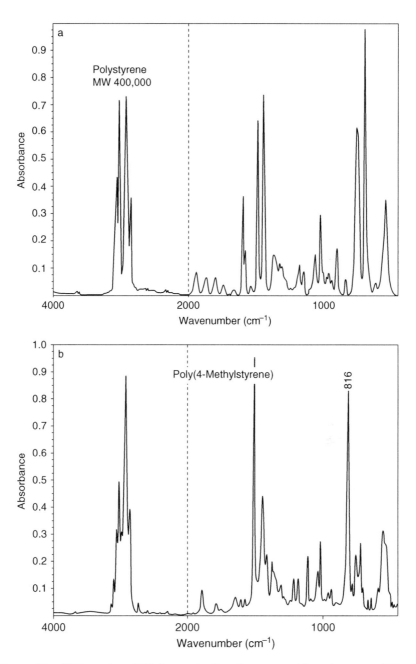

Figure 6.1 FTIR spectra of (a) linear atactic polystyrene with molecular weight of 400 kDa, (b) poly(p-methylstyrene), and hypercrosslinked polystyrenes prepared by crosslinking styrene–0.5% DVB copolymer with (c) 0.3, (d) 0.5, (e) 1.0, and (f) 1.5 mol of monochlorodimethyl ether per styrene repeating unit.

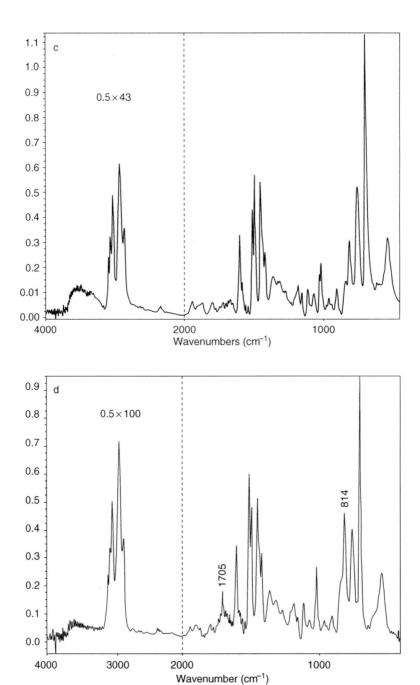

Figure 6.1 *(Continued)*

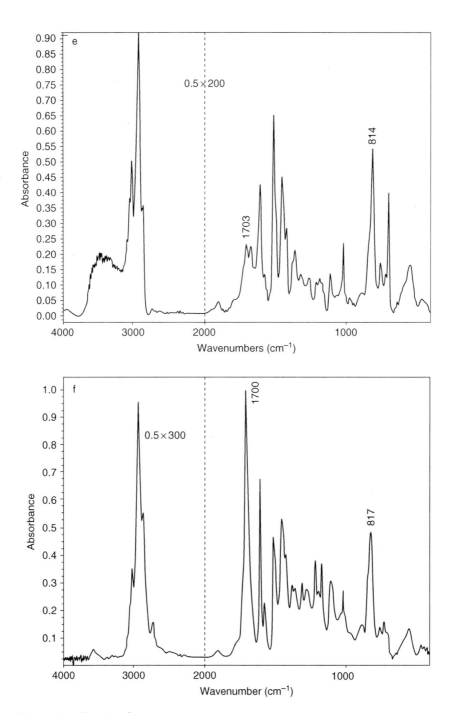

Figure 6.1   (*Continued*)

with 0.5 mol of MCDE, the residual unreacted phenyl rings can be easily seen by the bands at 758 and 699 cm^{-1}.

Regarding the 1,2,4–trisubstituted phenyl rings that have to be present in equivalent amounts to that of residual phenyl units, they have unfortunately no characteristic strong bands in the FTIR spectra. Sometimes a relatively small peak at 1020 cm^{-1} and a shoulder at 850 cm^{-1} are ascribed to the 1,2,4–substitution, but these peaks are also present in the spectra of disubstituted benzenes.

The above-postulated types of substitution on the benzene rings are also characteristic of the hypercrosslinked sorbent MN-200 currently produced by Purolite International Ltd. (Llantrisant, Wales, UK) on an industrial scale. In its spectrum the band at 760 cm^{-1} is barely visible and the absorbance at 700 cm^{-1} is also small. Most probably the crosslinking density of this polymer is high and it contains only very few free phenyl groups.

Unfortunately, the quantitative determination of unreacted phenyls and trisubstituted units in solid polymers by FTIR represents a rather difficult task. Surprisingly, solid–state ^{13}C NMR method, which is believed to give more reliable information, revealed unexpectedly large amounts of unreacted phenyl groups, up to 40%, in a network crosslinked with MCDE to an extent of 100% [97]. We consider this value to be rather overestimated, but there are no other data to compare with.

4.4 Some chemical groups in the structure of hypercrosslinked polystyrene

In accordance with the data of elemental microanalysis, MN-200 resin contains ~2.5% oxygen when calculated as O% = 100 - (C% + H% + Cl%). Interestingly, this quantity is markedly smaller than that reported elsewhere [98, 99]. We have to remember, however, that the elemental analysis of polymers, especially highly crosslinked polymers, is never as exact as that of low molecular weight compounds. Moreover, the determination of oxygen in the sorbent would require a thorough elimination of adsorbed water and air.

Some part of the chemically bonded oxygen may belong to hydroxymethyl groups $-CH_2OH$, which can easily form through the hydrolysis of residual chloromethyl groups. The band at 3581 cm^{-1} in the FTIR spectrum of MN-200 may be ascribed to such isolated HO groups, not involved in the formation of hydrogen bonds with any other functional groups.

In the FTIR spectrum of the industrial MN-200 resin there is a strong band at 1706 cm^{-1}. Usually, it testifies to the presence of carbonyl-containing functional groups. Grassie et al. [100] have found this band in the infrared (IR) spectrum of the product of self-condensation of α,α'-dimethoxy xylene via Friedel–Crafts reaction and explained its appearance by the side reaction of oxidation. Law et al. [99] and Streat et al. [98] also hold the idea of oxidation as the origin for this band in the industrial MN-200 resin; they attributed it to ketone functions. A similar band appears at 1701 cm^{-1} in the spectrum of the network obtained by crosslinking linear polystyrene with 0.5 mol of MCDE or methylal. This band becomes even more intensive at 1705 cm^{-1} (Fig. 6.1(f)) in products of a more dense post-crosslinking of linear polystyrene or styrene–DVB copolymers with 1.5 mol of MCDE. Indeed, it is possible to imagine the oxidation of pendent chloromethyl groups to aldehyde, as well as the oxidation of $-CH_2-$ bridges to ketone groups. However, no absorption can be found in the range of 1700–1705 cm^{-1} in the IR spectra of products obtained under identical conditions, but with less than 0.5 mol amounts of MCDE. Thus, assigning the absorbance at 1701–1705 cm^{-1} to carbonyl functions cannot be accepted without certain doubt. The more so as information exists [101] that vibration overtones of substituted benzene rings may show up in the wide range of 1600–2000 cm^{-1} and they have been sometimes used for the identification of the type of substitution.

The above doubts are strongly supported by the results of more detailed investigation of the origin of carbonyl groups in the products of post-crosslinking styrene–0.5% DVB copolymer with 1.5, 2.0, and 2.5 mol MCDE to nominal degree of crosslinking of 300, 400, and 500%, respectively. In FTIR spectra of these polymers the band at 1701–1703 cm^{-1} is exceptionally intensive and broad (Fig. 6.1), though distorted by the presence of two shoulders at 1675 and 1715 cm^{-1}. Contrary to this, the solid state ^{13}C NMR method reveals no carbonyl groups in these materials, at all. Estimation of C=O group concentration by the addition of sodium bisulfite resulted in a value of about 0.3 mmol/g. On the other hand, the standard reaction with hydroxylamine [102] showed values as high as 1.0–1.3 mmol/g. However, hydroxylamine was found to be largely adsorbed by the polymer. Elimination of possible carbonyl groups by reaction of the polymers with ethyl orthoformate or hydroxylamine or oxidation with nitric acid did not change the form and intensity of the band at 1701 cm^{-1} in the products. Even in the spectrum of a three-dimensional polymer prepared by self-condensation of XDC through Friedel–Crafts reaction in an inert atmosphere, followed by treatment under conditions excluding any contacts with water, one can observe the same band under consideration. Note that there was no source of oxygen for the formation of carbonyl groups in this system.

The band between 1700 and 1705 cm^{-1} also presents itself in polystyrene networks post-crosslinked with methylal, even in the case when the reaction is carried out in an inert atmosphere and with a nonoxidizing catalyst, $SnCl_4$. Interestingly, in products crosslinked at 20 °C, the concentration of carbonyls (titrated with hydroxylamine) was found to be higher (~0.25 meq/g) than in products obtained at 80°C (~0.1 meq/g). In the latter case up to 1.5% of all phenyl groups may carry aldehyde fragments.

Thus, all the above-discussed findings suggest the absorbance at 1701 cm^{-1} to mainly reflect the out-of-plane deformational vibration of poly-substituted benzene rings, most likely having distorted conformations caused by shrinkage of rigid hypercrosslinked networks on drying. This band may well overlap with or even mask the carbonyl absorption band. Therefore, the problem on the presence of carbonyls in several hypercrosslinked polystyrene products still remains to be investigated in more detail.

When studying two industrial hypercrosslinked resin samples, MN-200 and MN-270, by ^{13}C magic angle spinning (MAS) NMR technique, Law et al. [99] uncovered a marked portion of dihydroanthracene fragments that may have resulted from the following possible reaction sequence [103]:

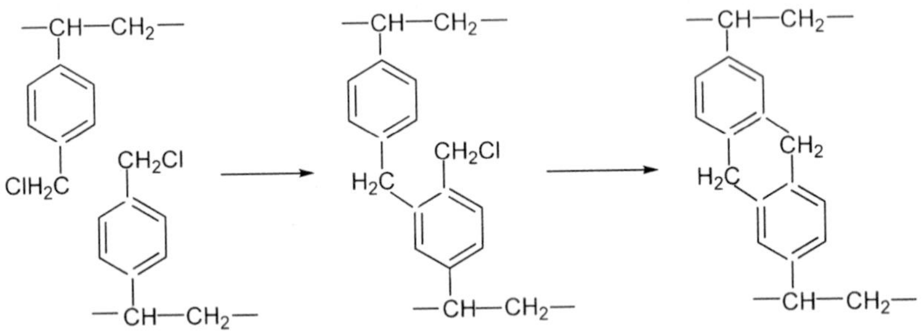

Here, two phenyl rings become connected through two methylene bridges, thus increasing the number of trisubstituted benzene rings without adequately increasing the actual degree of crosslinking of the network. It is appropriate to add here that these condensed structures must be prone to dehydration to form anthracene-type bridges. It is also possible that fragments of these or similar polyaromatic hydrocarbons cause the beige color of many hypercrosslinked products.

Finally, one more side reaction appears to proceed during post-crosslinking. According to elemental microanalysis, MN-200 contains ~1.3% of chlorine atoms, but the band at 1260 cm^{-1}, characteristic of an aromatic (PhCH$_2$Cl) chloromethyl group [104], does not show up in the FTIR spectrum. On the other hand, this band is revealed in the spectrum of the styrene–0.6% DVB copolymer intensively crosslinked with 1 mol of MCDE, which contains a comparable amount (1%) of pending chlorine. It is quite possible that industrial MN-200 contains chlorine of a different kind. A part of rather inert chlorine in the form of chloroethyl groups may have originated by the involvement of EDC into the Friedel–Crafts reaction. One may also assume that the initial copolymer of that product incorporates more DVB; thus the unreacted pendent double bonds may have interacted with MCDE during the post-crosslinking reaction, leading to the formation of methyl chloropropyl ether fragments [105]:

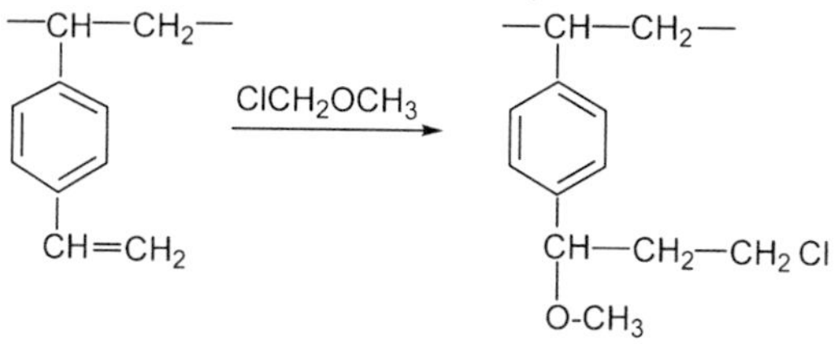

MN-200 may have also gained additional oxygen in this way.

Bis-chloromethylated aromatic compounds are well known to enter into self-condensation reaction in the presence of Lewis acids, yielding macromolecular products [106–116]. Opinion has been expressed [117] that the use of 1,4-dichloromethyl-2,5-dimethylbenzene as a crosslinking agent may first result in the formation of oligomeric molecules, before the oligomers react with polystyrene. Enhanced swelling of the final networks may thus be attributed to the long-chain structure of the formed bridges. Collette et al. [118] reported the formation of dimeric bridges during the crosslinking of polystyrene with 9,10-bis-chloromethylanthracene, although Krakovyak et al. [119] did not observe any oligomerization of this reagent under the same conditions. Trying to shed light on the possibility of formation of oligomeric bridges, Hausler et al. [120] subjected the networks crosslinked with XDC to pyrolysis. However, no specific substances could be found by the gas-chromatographic analysis of the volatile pyrolysis products, which would indicate the self-condensation of this crosslinking agent. Although the partial self-condensation of the cross-linking agent cannot be ruled out, we never observed any qualitative difference in the properties of hypercrosslinked polystyrenes crosslinked with reagents capable of self-condensation under conditions of

Friedel–Crafts reaction (XDC, CMDP, CMB) and with reagents that are not prone to form oligomers (MCDE, CMM, DMM).

4.5 Synthesis of hypercrosslinked networks in the presence of aqueous solutions of Friedel–Crafts catalysts

It has long been known that water traces serve as a co-catalyst in Friedel–Crafts reactions [121–123]. Therefore, it was not so surprising that $FeCl_3 \cdot 6H_2O$ also provoked the reaction of MCDE with polystyrene dissolved in EDC [124]. The resulting products demonstrate the main features of the macronet isoporous and even hypercrosslinked polystyrene, namely, a twofold increase in the volume on swelling in ethanol or hexane and true porosity with an apparent specific surface area ranging from 50 to 170 m^2/g. This implies that the Friedel–Crafts reaction can run at a sufficiently high rate even in the presence of marked amounts of water. This finding provided a paradoxical way for the preparation of beaded hypercrosslinked materials from linear polystyrene using aqueous solutions of Lewis acids. The latter play simultaneously the role of a suspension medium and a source of the catalyst. By suspending a solution of linear polystyrene and a crosslinking agent in EDC or nitrobenzene (which are immiscible with water) in the solution of a catalyst in concentrated HCl, crosslinked beaded products can easily be obtained [125]. The post-crosslinking of preformed styrene–DVB copolymers may also be carried out under the same conditions without questions.

$FeCl_3$, $SnCl_4$, and $ZnCl_2$ more or less retain their catalytic activity while dissolved in concentrated HCl, while $AlCl_3$ will completely hydrolyze in the acid. $ZnCl_2$ dissolved in HCl possesses a very low catalytic activity if EDC is used as the solvent for polystyrene, but it initiates the formation of acceptable products in nitrobenzene. For example, a 70% aqueous solution of $ZnCl_2$ acidified with hydrochloric acid to pH 0.1 will catalyze the crosslinking of polystyrene with 1,4-dichloromethyl-2,5-dimethylbenzene in nitrobenzene [117]. In contrast, $FeCl_3$ exhibits excessive solubility in nitrobenzene and will precipitate polystyrene from its solution in this solvent.

The best activity has been obtained with a solution of 70 g $ZnCl_2$ and 3.5 g $FeCl_3$ in 50 mL of concentrated (36%) HCl (molar ratio of $ZnCl_2$: $FeCl_3$: H_2O : HCl is 1.0:0.04:4.10:1.14). Quite interesting is the fact that the addition of 3.5 g $ZnCl_2$ to 70 g $FeCl_3$ does not enhance the catalytic activity of $FeCl_3$ in concentrated HCl.

The reaction of polystyrene with MCDE or methylal in EDC dispersed in excess of the above catalytic mixture at 80°C yields beaded

Table 6.3 Conditions of synthesis and properties of hypercrosslinked networks obtained by suspending a solution of linear polystyrene (**I**) in EDC (12.5 w/v) or swollen styrene–0.7% DVB copolymer (**II**) in the solution of 70 g $ZnCl_2$, 3.5 g $FeCl_3$ in 50 mL concentrated HCl at 80°C

Type of polymer	Type of crosslinking agent	Amount of crosslinking agent (mol/mol PS)	V_{in}/V_{org}	Time (h)	Toluene regain (g/g)	Surface area (m^2/g)
I	DMM	0.5	3	25	1.57	550
I	DMM	1.0	3	10	1.05	1010
I	DMM	1.5	3	10	1.60	850
II	MCDE	0.5	4	10	0.83	370
II	MCDE	0.6	4	7	0.84	1200
II	MCDE	0.7	4	7	0.89	870
II	MCDE	0.8	4	7	0.90	1000
II	MCDE	1.0	4	7	0.90	1000
II	MCDE	1.0	2	7	0.65	1100
II	MCDE	1.0	1	7	0.60	870
II	MCDE	1.0	0.5	7	0.55	810

V_{in}/V_{org} = volume ratio of inorganic to organic phases.

porous materials, the apparent specific surface area of which is as high as $1000-1200 \, m^2/g$, provided, however, that large amounts of the crosslinking agents, 0.6 mol or more per styrene-repeating unit, are used in the reaction (Table 6.3).

The crosslinking of linear polystyrene as well as styrene–DVB copolymers dispersed in excess aqueous solutions of Lewis acids is accompanied by a certain loss, due to a noticeable solubility of both EDC and the crosslinking agents MCDE or DMM, in the catalytic mixture. The unspecified change in the concentration of the main reaction partners, particularly at an elevated temperature, makes it more difficult to correctly evaluate the relationships between the conditions of synthesis and the properties of the resulting hypercrosslinked products. Still, the use of very inexpensive aqueous catalytic solutions for the preparation of beaded hypercrosslinked materials from industrially available linear polystyrene may be of practical significance.

CHAPTER 7

Properties of Hypercrosslinked Polystyrene

1. FACTORS DETERMINING THE SWELLING BEHAVIOR OF HYPERCROSSLINKED POLYSTYRENE NETWORKS

With regard to a linear polymer and its three-dimensional networks, all solvents can be divided into three groups. Thermodynamically good solvents are those the interaction of which with the polymeric chains is stronger than the interactions between the segments of the polymeric chains. Solvation of polymeric chains with such good solvent results in the rupture of the interactions between the chains and dissolution of the linear polymer. The same process of chain solvation leads to an increase in volume, that is, swelling, of the crosslinked polymeric network. In contrast, solvents which cannot overcome the interactions between the chains fail to dissolve the linear polymer or cause swelling of its networks. Such "non-solvents" precipitate the linear polymer from its solutions in a good solvent and cause shrinkage of a swollen network. The third group of solvents, the so-called Θ-solvents, is characterized by equality, at a specific temperature, of all polymer–polymer, polymer–solvent, and solvent–solvent interactions (see Chapter 1, Section 2.3). Cyclohexane at 34.5°C represents a typical Θ-solvent for polystyrene. Toluene and ethylene dichloride (EDC) are very good solvents [126, 127], while methanol and hexane are representatives of typical precipitating media for polystyrene.

The swelling of network polymers in thermodynamically good solvents is of particular interest for both the practical application of such materials and the theoretical consideration regarding their structure. First of all, the extent of swelling or, more precisely, the total solvent uptake represents a measure of the network's permeability, a parameter which is highly important when the polymer is used as a matrix for an ion-exchange resin or as a sorbent operating in a liquid media. Also, as far back as in the middle of the last century, it was suggested to use the extent of swelling of networks in good solvents to characterize the structure of networks, that is, the number of elastically active chains (the degree of crosslinking), the presence or absence of dangling chains, intramolecular loops, entanglements, etc. In Chapter 2, Section 2.2,

Comprehensive Analytical Chemistry, Volume 56
ISSN 0166-526X, DOI 10.1016/S0166-526X(10)56007-1

we have already discussed the equation suggested by Dušek and Prins, describing the swelling behavior of network polymers. They stated that the swelling of a network is determined by four parameters: the degree of crosslinking, the thermodynamic quality of the good solvent, the degree of dilution of monomers with inert solvents during network formation, and the functionality of junction points. These notions are still generally accepted as solely correct by the scientific community studying the swelling behavior of networks. According to these concepts, the networks' equilibrium swelling characterizes the capability of chains between neighbor junction points to change their conformation. From this viewpoint, the hypercrosslinked networks with degree of crosslinkings as high as 100–200% should be rigid as a crystal and should not respond to the presence of any solvent. Indeed, nearly each phenyl ring in such networks participates once or even twice in bridge formation and there are no polymer segments between junctions. These networks would not have to swell, at all. However, the hypercrosslinked polystyrene networks refute this "classic" theory, by demonstrating unusually high mobility. This manifests itself in both the high swelling capacity and remarkable mechanical flexibility of the hypercrosslinked polymers. It becomes obvious that, in addition to the above-specified four important factors governing the swelling of polymeric networks in liquid media, other crucial structural parameters also exist that have been overlooked or ignored by the classic approach to conventional weakly crosslinked networks. Let us consider these factors.

1.1 The influence of dilution of the initial system

Figure 7.1 presents the swelling capacity in toluene of networks obtained by crosslinking linear polystyrene with monochlorodimethyl ether (MCDE) or xylylene dichloride (XDC), as a function of the concentration of the initial polystyrene solution in EDC [128]. At any given degree of crosslinking, usually calculated from the molar ratio of the crosslinking agent to the initial polystyrene according to Eq. [6.1], the swelling dramatically increases with diluting the system taken for the post-crosslinking. A similar dependence is also valid for the products of crosslinking styrene–divinylbenzene (DVB) copolymers swollen with EDC to maximum (Fig. 7.2): the higher is the DVB content and, accordingly, the smaller the swelling of the starting copolymer, the smaller is the amount of EDC that remained incorporated within the final network, and this is also reflected by the proportional decrease of the toluene uptake capacity of the dry material. (Interestingly, in both systems, substitution of the conventional MCDE by substantially different crosslinking agents such as XDC or CMDP does not distort the overall picture of the swelling dependence on the concentration of the initial polymer.)

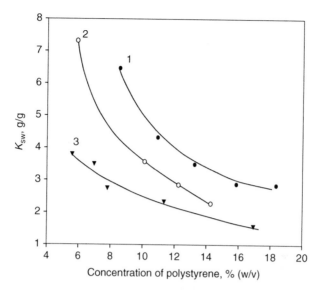

Figure 7.1   Equilibrium swelling in toluene vs concentration of initial polystyrene solution for the networks crosslinked with (1, 2) *p*-xylylene dichloride and (3) monochlorodimethyl ether to crosslinking degrees: (1) 11, (2) 100, and (3) 43%.

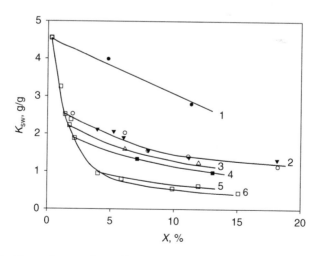

Figure 7.2   Dependence of equilibrium swelling in toluene on the summarized crosslinking degree for the networks prepared by crosslinking gel-type styrene–DVB copolymers with (1, 2, 4, 5) 1.4-bis(chloromethyl)diphenyl or (3) monochlorodimethyl ether; (6) styrene–DVB copolymers. *(After [128]).*

The enlarged swelling capacity of networks prepared from diluted polymer solutions is well known. The scientific community uncritically accepted the simple explanation that in a diluted system, many intramolecular crosslinks or loops form, which are thought to be ineffective in reducing the swelling capacity of the network. However, one can accept this explanation only for very diluted starting systems, with polymer concentrations of 1–2%, where the transition from an interpenetrating coil system to individual macromolecular coils takes place and the probability of loop formation strongly increases. Even there, the intramolecular crosslinks can only render denser segment packing in polymeric coils. They must also reduce the effective chain length between junctions, thus restricting the conformational mobility of the network. In our experiments shown in Figs. 7.1 and 7.2, the polymer concentration amounted up to 18% for linear polystyrene and between 28 and 47% for the swollen copolymers of styrene with 1–4% DVB. At such a high concentration the modern theories require a complete interpenetration of macromolecular coils. If this is the case, the probability of meeting two styrene units belonging to the same chain to form an intramolecular loop should be regarded as negligible. More importantly, for hypercrosslinked networks with the degree of crosslinking as high as 43–100% (plots 2 and 3 in Fig. 7.1), where no loops could remain unbound to the total network, there is no reason at all to distinguish between intramolecular and intermolecular crosslinks.

An alternative approach to the explanation of the role of the dilution factor in the system under post-crosslinking is the consideration of the network as a collection of mutually condensed and interpenetrating meshes. Each mesh comprises crosslinking bridges and sections of polymeric chains between them. For highly crosslinked networks, it does not matter whether these sections belong to the same initial chain or to different chains. Very important, however, is that each mesh can embrace chains belonging to the neighboring meshes. This understandable interpenetration of meshes in the randomly formed network is the factor that strongly depends on the concentration of polymeric chains in the initial solution during the closure of meshes. Surprisingly, up to now, this topological factor, namely, the unavoidable interpenetration of network meshes, has not been recognized and considered as the major factor determining the swelling ability of networks.

This decisive topological factor largely accounts for the fact that, at equivalent degree of crosslinkings, the swelling abilities of the products of post-crosslinking linear polystyrene in solution or swollen styrene–DVB copolymer are much higher than that of conventional gel-type styrene–DVB copolymers. The latter products are prepared in the absence of any solvent and, therefore, each mesh of their network must be densely filled with alien polymeric chains and meshes. The topologically interpenetrating meshes can never disentangle on swelling and must severely restrict any expansion of the network. Contrary to this, in the loose hypercrosslinked network that is formed in the presence of an about 10-fold excess of a solvent, the degree of interpenetration of meshes is considerably reduced, by about a factor of 10, and this must facilitate large volume changes of the network on swelling.

1.2 The role of the initial copolymer network

It seems to be very interesting to compare the behavior of two slightly different series of copolymers prepared by additional crosslinking with MCDE under identical conditions. One is made from the copolymer of styrene with 1% DVB, swollen with EDC to the maximum, and another is based on the copolymer of styrene with 0.3% DVB swollen partially with the same solvent in such a manner that both polymers absorbed exactly identical amounts of EDC before the post-crosslinking reactions. For the two series, one could expect the swelling ability of the corresponding final hypercrosslinked products to be identical, due to the identity of the concentration of polymeric chains (0.485 mg/mL) in the two starting systems, the identity of final degree of crosslinkings, and the type of the crosslinking agent. Surprisingly, the products obtained on the basis of the partially swollen copolymer having 0.3% DVB absorb toluene to a markedly higher extent (Fig. 7.3) [129]. It looks as if the negligible difference of 0.7% DVB between the two series of products still remained visible after introducing numerous additional crosslinks.

Figure 7.3 also presents swelling data for two more series of products. One was obtained by post-crosslinking of the styrene–0.3% DVB copolymer swollen to its maximum and the other by crosslinking soluble linear polystyrene dissolved in EDC to approximately the same concentration (in fact, a slightly higher concentration, 0.125 against 0.111 mg/mL, respectively). According to the decisive role of the dilution factor of the system during the crosslinking reaction, discussed in Section 1.1, these two series of products swell substantially stronger than the previous two series crosslinked at a higher concentration of 0.485 mg/mL. But, again, the swelling of networks on the basis of the fully swollen styrene–0.3% DVB copolymer appears to be considerably lower than that of networks based on the linear polystyrene (although the latter materials were obtained in more concentrated solution). The 0.3% DVB of the initial copolymer continues to strongly reduce the swelling of the products additionally crosslinked up to 100%.

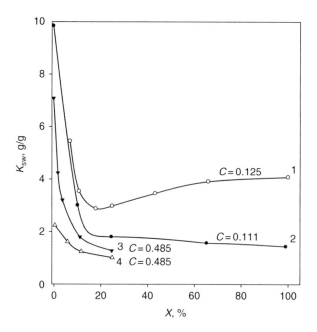

Figure 7.3 Dependence of equilibrium swelling in toluene on the crosslinking degree of networks prepared by crosslinking with monochlorodimethyl ether of (1) linear polystyrene at the concentration of $C_o = 0.125\,mg/mL$; (2) styrene–0.3% DVB copolymer swollen to a maximum, $C_o = 0.111\,mg/mL$; (3) styrene–0.3% DVB copolymer partially swollen to $C_o = 0.485\,mg/mL$; (4) styrene–1% DVB copolymer swollen to a maximum, $C_o = 0.485\,mg/mL$. *(After [129]).*

The unproportionally high "crosslinking efficiency" of the initial DVB links compared to the $-C_6H_4-CH_2-C_6H_4-$ bridges formed later can be understood in terms of the stress experienced by the initial styrene–DVB network on swelling and retained in the structure of the final hypercrosslinked network.

Conventional copolymers of styrene with 0.3 or 1% DVB were prepared in the absence of any solvent, and their structure is unstressed when in dry state. On swelling these products with EDC, their network expands until the growing stresses of the elastically active chains stop further swelling at the maximal equilibrium level. It is in this maximally pre-strained state that the network receives numerous additional crosslinks on reacting with the bifunctional MCDE. The pre-strained structural fragments prove to be immobilized in the final hypercrosslinked framework that actually receives no additional strains during post-crosslinking in thermodynamically good media. The internal stress of the initial copolymer network has no ways to disappear or relax. It will always manifest itself as a serious limiting factor for the solvent uptake of the final hypercrosslinked product. Only materials prepared from dissolved linear polystyrene will not incorporate any pre-strained units. These materials will always display the highest swelling ability.

1.3 Influence of the uniformity of crosslink distribution

Insofar as the initial reaction mixture during the crosslinking of linear polystyrene with monochlorodimethyl ether represents a homogeneous solution, the distribution of crosslinking bridges in the final network could be expected to be more or less uniform. On the other hand, if instead of MCDE partially chloromethylated polystyrene chains are taken for the reaction with unsubstituted polystyrene chains, a certain deviation must be expected from the statistical distribution of crosslinks [129]. The chloro-methylated polystyrene coils must give rise to the formation of more densely crosslinked micro domains in the final network. The more chlorine atoms are present in the chains of the polymeric crosslinking agent, the smaller quantities of the latter are needed to attain a desired net crosslinking degree and the less homogeneous will be the structure of the final gel.

Figure 7.4 illustrates the plot of swelling with toluene versus the degree of crosslinking of the products thus obtained. The toluene uptake of 3.5 g/g relates to the swelling capacity of the network prepared by the self-crosslinking of the chloromethylated polystyrene containing 3% chlorine, the degree of

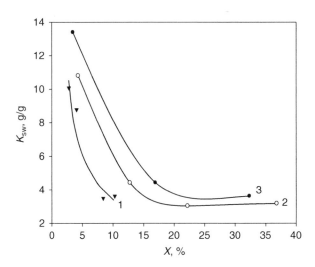

Figure 7.4 Dependence of equilibrium swelling in toluene on the crosslinking degree of networks prepared by crosslinking linear polystyrene with partially chloromethylated polystyrene containing (1) 3; (2) 11.5 and (3) 14.8% chlorine. *(After [129])*.

crosslinking of this homogeneously crosslinked network being 10%. Products with the same nominal 10% or higher degree of crosslinking can be also obtained by involving chloromethylated polystyrene having more reactive chlorine, for example, 11.5 or 14.8% Cl, as the crosslinking agent and mixing it with an appropriate amount of unsubstituted polystyrene. Naturally, from such mixtures less homogeneous networks will be formed. As can be seen from Fig. 7.4, at any given value of network crosslinking, the inhomogeneous distribution of crosslinks manifests itself in the higher swelling of the networks.

1.4 The role of inner stresses of the hypercrosslinked network and the structure of crosslinking bridges

Linear polymers are well known to dissolve only in solvents exhibiting strong affinity to the polymers, and traditional three-dimensional polymers can also swell with only such kind of media. Contrary to this rule hypercrosslinked polystyrene networks swell both with thermodynamically good solvents, such as toluene or methylene dichloride, and with typical precipitating media for linear polystyrene, such as methanol or n-hexane (Fig. 7.5, Tables 7.1 and 7.2) [129–131].

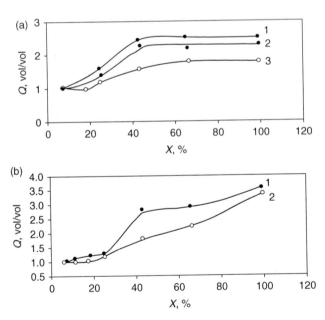

Figure 7.5 Dependence of volume swelling in (1) n-hexane; (2) methanol and (3) water on the crosslinking degree of networks prepared by crosslinking linear polystyrene with (a) p-xylylene dichloride and (b) methylal.

Table 7.1 Swelling (mL/g) of networks prepared by crosslinking linear polystyrene (**I**) and styrene–1% DVB copolymer (**II**) with different crosslinking agents (measured by weights of dry and swollen networks)

Solvent	Degree of crosslinking, (%)					
	5	11	25	43	66	100
I, Monochlorodimethyl ether						
Toluene	6.23	3.97	3.10	3.86	4.25	4.37
n-Hexane	0.06	0.06	0.48	3.48	4.30	4.12
Methanol	0.04	0.06	0.25	3.08	4.10	3.84
Water	–	0.17	0.29	0.72	1.49	2.27
I, Methylal						
Toluene	6.42	3.76	2.81	3.14	3.51	3.49
n-Hexane	0.07	0.07	1.00	1.64	3.05	3.23
Methanol	0.04	0.04	0.63	1.34	2.92	2.90
Water	–	–	0.34	0.62	0.96	1.23
I, 1,4-Bis(chloromethyl)diphenyl						
Toluene	5.51	3.90	2.98	3.17	3.38	3.60
n-Hexane	–	–	0.12	1.00	2.48	2.68
Methanol	–	–	0.20	1.09	2.17	2.91
Water	–	–	0.14	0.27	0.77	1.23
I, 1,4-Bis(*p*-chloromethylphenyl)butane						
Toluene	1.23	4.51	3.67	3.29	3.71	3.91
n-Hexane	0.07	0.07	0.17	0.26	0.32	0.18
Methanol	3.91	0.09	0.10	0.15	0.19	0.09
Water	–	–	–	–	–	–
I, 1,3,5-Tris(chloromethyl)mesitylene						
Toluene	3.86	2.83	2.91	3.08	2.88	2.91
n-Hexane	0.08	0.20	2.21	3.10	3.10	3.10
Methanol	0.06	0.06	1.81	3.06	3.06	3.05
Water	–	–	0.37	0.81	0.87	0.94
II, Monochlorodimethyl ether						
Toluene	1.98	1.39	1.00	1.05	1.12	1.08
n-Hexane	–	–	0.69	0.86	0.95	1.02
Methanol	–	–	0.45	0.89	1.04	1.04
Water	–	–	0.03	0.13	0.20	0.30

Conventional gel-type styrene copolymers taken in the form swollen with toluene rapidly shrink when placed in methanol. Hypercrosslinked gels swollen with toluene do not exhibit any noticeable change in their volume when this solvent is replaced by any other organic solvent, including any precipitating media. It is only the complete removal of the solvent by drying which results in the shrinkage of the

Table 7.2 Swelling (mL/g) of polymers based on linear polystyrene crosslinked with
p-xylylene dichloride (measured by the weight of the dry and swollen product)

Solvent	Degree of crosslinking, (%)				
	11	25	43	66	100
Water	0.05	0.10	0.30	0.80	0.84
Methanol	0.15	0.32	1.13	1.71	1.67
Ethanol	0.10	0.29	1.15	1.86	1.77
Propanol	0.11	0.26	1.43	1.99	1.78
Isopropanol	0.07	0.15	1.23	1.83	1.76
Butanol	0.08	0.22	1.52	2.0	1.79
Isobutanol	0.08	0.11	1.25	1.88	1.72
Heptane	0.16	0.58	1.70	2.0	1.81
Toluene	3.69	2.46	2.19	2.19	1.92
Benzene	4.00	2.65	2.23	2.21	1.96
Dichloroethane	3.33	2.32	2.22	2.19	1.92
Ethanol → water → dichloroethane	–	–	2.18	2.19	1.80
Dichloroethane → heptane	–	–	1.72	1.99	1.81
Ethanol → water → heptane	–	–	1.69	1.99	1.78

Source: After [134].

swollen polymer. This shrinkage is totally reversible, implying that the
dry polymer rapidly swells with all organic liquids. The equilibrium
nature of the swelling is further corroborated by the fact that the
swelling values are independent of the sequence of solvents that contact
the polymer. Indeed, polymers being either in direct contact with
n-heptane or after replacing EDC with the solvent take up the same
amount of n-heptane (Table 7.2).

Table 7.2 also demonstrates that the volume of organic solvents taken up
by the networks derived from linear polystyrene and p-xylylene dichloride
(XDC) is nearly independent of the solvent nature, amounting to
1.84 ± 0.15 and 1.99 ± 0.22 mL/g for networks with degree of crosslinking of 100 and 66%, respectively. When considering separately the groups of
solvating and non-solvating liquids, one can notice a distinct stability of the
swelling values within each group. For the group of solvating solvents they
amount to 1.94 ± 0.02, 2.20 ± 0.01, and 2.21 ± 0.02 mL/g for the
networks crosslinked to 100, 66, and 43%, respectively. On the other
hand, for non-solvating media (except for water), the swelling equals
1.76 ± 0.03 and 1.90 ± 0.09 mL/g at the degree of crosslinking of 100
and 66%, respectively. We have to stress here that all these values of solvent
uptake exceed many times the total pore volume of the dry materials, so that

swelling really means a significant volume increase of the polymer (contrary to the solvent uptake by macroporous materials). These results suggest that the swelling of hypercrosslinked polystyrenes is strongly facilitated by certain peculiarities of their structure, that is, by certain driving force that prompts the hypercrosslinked network to take up any liquid and increase in volume. This force acts jointly with the solvation of the polymeric chains. Notably, the difference in the solvation power of good solvents and non-solvents plays a minor role, contributing a very small increment of about 0.2–0.3 mL/g to the equilibrium state of the polymer/solvent system.

Among the non-solvating liquids, water represents a special medium with respect to the hypercrosslinked polystyrene. It reflects the fact that the hydrophobic surface of the aromatic polymer contains no hydrophilic functions and cannot be directly wetted by water. Being placed on the surface of water, dry particles of hypercrosslinked polymers float for many days and take up no more than 0.01 g/g of water. For this reason, the values of swelling in water have always been obtained after washing with water the materials pre-swollen with a water-miscible organic solvent. Again, the precursor does not affect the volume of water finally retained. Therefore, we consider the amount of retained water as the equilibrium swelling capacity with the only difference that because of the lack of wetting, the equilibrium cannot be reached from the side of the dry polymer and water. It is appropriate to mention here that water can easily replace methanol or acetone in the internal space of the initially swollen network, but cannot push out EDC, hexane, or other water-immiscible liquids. On the other hand, the latter solvents completely replace water from the water-swollen hypercrosslinked networks, leading to ejection of water and an increase in the sample volume until equilibrium is reached for that solvent.

Critical consideration of the data presented in Table 7.1 and Figs. 7.5 and 7.6 reveals no suggestive correlation between the swelling of networks and structural parameters of the crosslinking agent such as its length or the number of aromatic phenylene rings. Quite evident is, however, that swelling in typical precipitating media requires a certain threshold rigidity of the network as a whole, that is, a sufficient degree of crosslinking and a definite conformational rigidity of the crosslinking bridges. This threshold rigidity is already attained at the degree of crosslinking of about 25% if the trifunctional reagent tris-(chloromethyl)-mesitylene is applied, because it connects three polystyrene chains at every junction point. With bifunctional reagents a more intensive crosslinking is required, about 40% when methylene ($-CH_2-$) or p-xylylene ($-CH_2C_6H_4CH_2-$) bridges are

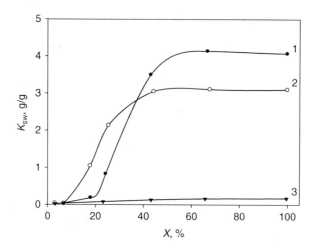

Figure 7.6 Dependence of equilibrium swelling in *n*-hexane on the crosslinking degree of networks prepared by crosslinking linear polystyrene with (1) monochlorodimethyl ether; (2) tris-(chloromethyl)-mesitylene and (3) 1,4-bis(*p*-chloromethylphenyl)butane. *(Reprinted from [132] with permission of Elsevier Publishing Company.)*

introduced between phenyl rings of two polystyrene chains. On the other hand, 1,4-bis(*p*-chloromethylphenyl)butane containing a flexible tetra-methylene fragment in the molecule does not yield a hypercrosslinked polymer that could swell to a noticeable extent with methanol or hexane (Fig. 7.6), even at a very intensive crosslinking.

Similar to the pattern of toluene uptake, the swelling of hypercrosslinked polystyrenes in the precipitating media is strongly affected by the topological factor of the interpenetration of network meshes discussed above, which manifests itself in the increasing swelling capacity with the dilution of the starting polystyrene solution (Fig. 7.7). The non-relaxing inner stresses of the fully swollen styrene–0.3% DVB copolymer that are captured by the post-crosslinks reduce the regain of hexane or methanol, as compared with the swelling of networks based on the stress-free coils of linear polystyrene (Fig. 7.8).

The outstanding ability of hypercrosslinked polystyrene networks to increase several times their volume on contacting the dry polymer with "non-solvents" is a logical consequence of the main principle of network formation. The polymers have been obtained through the fixation of favorable unstressed chain conformations of the initially highly solvated polystyrene coils, by introducing numerous rigid crosslinks between them.

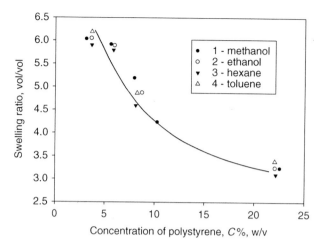

Figure 7.7 Dependence of swelling ratio in (1) methanol; (2) ethanol; (3) *n*-hexane and (4) toluene on the concentration of starting solution for the networks prepared by crosslinking linear polystyrene with monochlorodimethyl ether to 66% crosslinking degree. *(Reprinted from [133] with permission of Elsevier Publishing Company.)*

The resulting swollen solvated networks are thus free of any serious inner stresses. The network must "remember" and seek after this relaxed state. Contrary to this, the shrinkage of the network on the removal of the solvent from the hypercrosslinked gel must cause strong stresses in the final dry material. On drying, two opposite trends have to be taken into consideration: (i) the tendency to shrink and achieve a dense packing of polystyrene chains by a cooperative rearrangement of conformations of constituent network meshes, and (ii) the resistance of rigid meshes and the highly crosslinked network as a whole to any deviations from the relaxed equilibrium state. Stresses accumulate in the desolvated network in the form of distorted valence angles and bond lengths and they eventually equilibrate the attraction forces between the non-solvated network fragments. The finally balanced or equilibrated state of the dry material corresponds to an extremely loose chain packing in the open–network type "construction". The dispersion interactions between chains are dramatically weakened in the hypercrosslinked network. Any organic liquid, even those weakly solvating polystyrene units, can easily break in the strained hypercrosslinked network the impaired interactions between the chains, and cause expansion of the network that thus returns to its favored largely unstrained state.

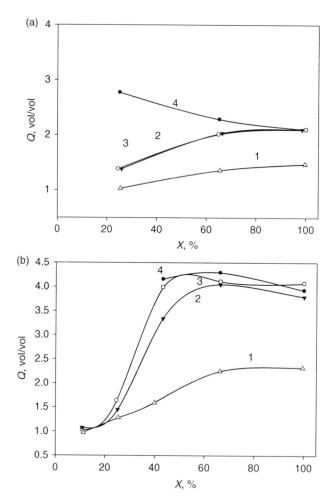

Figure 7.8 Dependence of volume swelling in (1) water; (2) methanol; (3) *n*-hexane and (4) toluene on the crosslinking degree of networks prepared by crosslinking (a) styrene–0.3% DVB copolymer and (b) linear polystyrene with monochlorodimethyl ether. *(After [128]).*

Theoretically a fully relaxed network state should correspond to the swelling in a Θ-solvent. Relative to this state, thermodynamically good solvents would cause a slight expansion of the network, while the non-solvents would cause a slight contraction. The above-mentioned difference of about 0.2–0.3 mL/g in the swelling ability will thus reflect the difference in the solvation energies of polystyrene network by these extreme types of liquids.

The smaller values of the equilibrium swelling of hypercrosslinked hydrophobic networks in water, as compared to organic solvents, arise from the disadvantageousness of replacing the strong interactions between water molecules in bulk liquid by extremely weak water–polystyrene dispersive interactions, if water would be placed into the numerous nano-sized compartments within the polystyrene network. Still, according to Fig. 7.8, even water does solvate the hypercrosslinked polystyrene network to a certain extent, causing an up to twofold expansion of the latter compared to its dry state. Naturally the internal strains of the dry network partially relax on this expansion.

It is of interest to cite here a Chinese-authored publication [135]. By conducting copolymerization of DVB with methyl acrylate in the presence of toluene, a thermodynamically good diluent, the Chinese authors prepared a porous (actually, hypercrosslinked) hydrophobic material incorporating some slightly polar ester groups. The copolymers showed a surprising ability of being directly wetted with water, and swelling with water to nearly the same extent as with toluene. This striking effect is caused by a combined action of strong inner stresses in the dry rigid network and an enhanced affinity of the more polar methyl acrylate units to water. However, the threshold polarity and rigidity of the network requires the use of at least 46% of methyl acrylate and at least 50% DVB. Meeting the latter condition is only possible if almost pure p-DVB or $tech$-DVB is used.

The notion of reduced energy of the polymer–polymer interactions in the low-density hypercrosslinked polystyrene networks is strongly supported by the results of calorimetric measurements of the integral heats of solvation of these materials with various solvents [136]. The integral heat evolved on contacting the dry polymer with the solvent presents the difference between the energy of new polymer–solvent, E_{ps}, interactions, on the one hand, and, on the other hand, the polymer–polymer, E_{pp}, and solvent–solvent, E_{ss}, dispersion interactions that are lost in the process of swelling:

$$\Delta H = E_{ps} - E_{pp} - E_{ss} \qquad [7.1]$$

In all the systems examined and listed in Table 7.3, the interaction between the polymers and solvents is an exothermic process.

Numerous polymer–polymer contacts are realized in the dry gel-type styrene–2% DVB copolymer (type IV). Only a thermodynamically good solvent, such as toluene, can overcome these strong interactions between the polymeric chains, resulting in the generation of a small quantity of heat, 25 J/g. This value is similar to the heat effect of dissolution of linear polystyrene in a good solvent [137]. Increasing the DVB content to 20%

Table 7.3 Integral heats of interaction of polystyrene networks with good and bad solvents

Network type	X, (%)	Integral enthalpy, (J/g)		Solvent uptake, (mL/g)		Pore volume, (cm³/g)
		Toluene	Methanol	Toluene	Methanol	
I	5	33.5	0	6.29	0.04	0
I	11	36.0	3.4	3.43	0.06	0
I	25	75.4	27.2	3.25	0.25	0.21
I	43	100.6	66.6	3.71	3.08	0.36
I	66	119.8	101.4	4.15	4.10	0.44
I	100	127.0	104.3	3.81	3.84	0.51
II	100	131.6	96.4	0.96	1.02	0.32
III	30	43.6	18.4	1.44	1.16	1.15
IV	2	25.6	0	2.23	0.05	0
IV	20	39.4	0	0.46	0.08	0

I = linear polystyrene crosslinked with MCDE, X = 100%.
II = styrene–0.7% DVB copolymer crosslinked with MCDE, X = 100%.
III = macroporous copolymer of styrene with 30% DVB.
IV = gel-type copolymers of styrene with 2 and 20% DVB.
Reprinted from [136] with kind permission of Wiley & Sons Inc.

leads to a decrease in the density of chain packing. For this reason the heat effect of the interaction between this highly crosslinked gel-type network and toluene is slightly higher (whereas the equilibrium swelling extent is lower).

In macroporous styrene–30% DVB copolymer (type III), the density of chain packing is further disturbed because of the higher DVB content. Moreover, this material is porous which implies that a part of the polymer–polymer-air contacts is replaced by weaker polymer– air contacts. This is the reason why the interaction of the macroporous copolymer with toluene is accompanied with the generation of a larger heat quantity, 44 J/g. Hypercrosslinked networks of type I are prepared in the presence of 10-fold excess of a good solvent and contain a large number of rigid bridges. The latter prevent the polystyrene chains from approaching one another, thus reducing both the number and strength of the polymer–polymer interactions (and minimizing the value of E_{pp} in Eq. [7.1] by up to 100 J/g). As a result, rigid hypercrosslinked networks exhibit an extremely large enthalpic effect on contacting toluene, up to 130 J/g, compared to $\Delta H \sim 33$ J/g for a slightly crosslinked gel-type copolymer. We can also say that the unrealized energy of missing polymer–polymer interactions in the hypercrosslinked network is accumulated in the form of the internal stress of the network. The stress energy will inevitably appear on swelling, acting as the additional driving force for swelling and contributing to the total enthalpy of the process.

It is important to emphasize that the heat effects of solvating hypercrosslinked polystyrene with methanol are also very high, 100 J/g, only 30 J/g lower than the heat effects of the network–toluene interaction. Two conclusions follow from these data. According to the first conclusion the major fraction of the polystyrene chains in the loose hypercrosslinked networks is accessible to methanol or becomes accessible on swelling. In contrast, only those chains of the gel and macroporous copolymers which are located on the surface of the beads and on the surface of the macropores can interact with methanol. As a result, the heat effect on wetting with methanol is negligible or very small, 0 or 18 J/g for type IV and type III polymers, correspondingly. The latter value is in good agreement with the heat of interaction between many organic solvents and the macroporous polystyrene adsorbents XAD-1 and XAD-2 [138].

Another important conclusion is that methanol solvates the accessible polystyrene chains almost as vigorously as toluene. The above difference in the energy of dispersive interactions with polystyrene is actually small when compared to the absolute value of the energy, although this difference markedly differentiates between the groups of thermodynamically good solvents and non-solvents. The very definitions of "solvating" and "non-solvating" media, while based on the evident fact that only the former is being able to dissolve a linear polymer (or cause swelling of a network), actually result from the small difference between three large energy values stated in Eq. [7.1].

These conclusions are further confirmed by the very large free energy of interactions of the hypercrosslinked polystyrenes with vapors of both good and bad solvents, as calculated from their adsorption isotherms [139] (Table 7.4). With the inner surface area of the polymers increasing, the free energy of adsorption also increases, particularly for hypercrosslinked polymers, while the difference between the corresponding values for good solvents and non-solvents may remain nearly constant. Obviously the increase in the energy of interaction of liquids or vapors with porous materials reflects the deficiency in the energy of polymer–polymer interactions in the networks of these materials.

Many different methods used to characterize hypercrosslinked polystyrene networks testify that the affinity of various media to accessible polystyrene fragments decreases in the following order:

aromatic hydrocarbons, chlorinated hydrocarbons > aliphatic hydrocarbons, ketones, esters and ethers, alcohols > perfluorocarbons >> water > liquid N_2 and Ar (at $-196°C$) >> CO_2, N_2, Ar (at $20°C$).

Table 7.4 Free energy of adsorption of organic solvent vapors onto polystyrenes at the molar fractions of 0.8 and 0.2 for the polymer and adsorbate, respectively

Type of polymer	S, (m²/g)	G, (J/mol)				
		Benzene	Methanol	Cyclohexane	n-Hexane	n-Heptane
PS	7.6	578	59	100	–	–
PS*	50.6	838	335	209	–	–
III	38.1	922	293	587	–	–
I	950	2135	–	1886	–	–
II	1200	–	2260	–	2721	2260

PS = linear polystyrene.
PS* = polystyrene precipitated from benzene into methanol.
III = macroporous styrene–20% DVB copolymer.
I = linear polystyrene crosslinked with XDC, X = 100%.
II = styrene–0.7% DVB copolymer crosslinked with MCDE, X = 100%.
Source: After [139].

Our long-term experience of dealing with hypercrosslinked polystyrenes indicates that there is no medium that would not cause swelling of hypercrosslinked polystyrene networks to a larger or smaller extent. Naturally, some sieving effects can be noticed. While even oligomers of epoxy resins cause swelling and soak into the polymer beads, when exposed to vapors of n-perfluorooctane having large molecules, only 0.6 cm³/g of that compound was found to be absorbed by the hypercrosslinked polymer [140], in contrast to 1.2 cm³/g of smaller protonated hydrocarbons [141]. It is only natural that a certain portion of the smallest network meshes may remain inaccessible to large sorbate molecules.

In conclusion, it would be well to mention again that, as was already discussed in Chapter 3, Section 1.4, Alfrey and Lloyd [142], Lloyd and Alfrey [143] seem to have been the first to face the phenomenon of swelling in precipitating media of poly(DVB) obtained with toluene as the porogen, although they only measured the weight of the dry and swollen polymer sample and did not notice the increase in the polymer volume. Millar et al. [144] also reported the large weight-swelling values of styrene–DVB copolymers, modified with toluene, in precipitants that proved to be far above the pore volume of the initial dry material. The data indicated a noticeable increase in the network volume. The authors have explained this unusual fact by the ability of non-solvents to solvate those polystyrene chains that are connected to highly crosslinked microgel domains of the overall macroporous structure. Although this explanation obviously contradicts the fact that the same non-solvents fail to solvate linear polystyrene or chains connecting microgel domains in gel-type copolymers, Rabelo et al. [145] still followed this erroneous point of view. Negre et al. [48] prepared hypercrosslinked networks by self-crosslinking linear styrene–vinylbenzyl chloride copolymers dissolved in EDC in the presence of stannic tetrachloride as the catalyst of the Friedel–Crafts reaction. They also took notice of the three- to fivefold volume increase with typical precipitants as a striking feature of the hyper-crosslinked networks. Again, they tried to explain this feature using a rather heterogeneous model of the networks and relaxation of stressed loose poly-styrene chains connecting the highly crosslinked and unswollen nodules of the copolymer. The reason for these loose chains being stressed was not specified, however. Only when studying the swelling behavior of macro-porous styrene–DVB-type copolymers prepared in the presence of solvating porogens did a number of researchers [135, 146–150] not only point out the ability of the copolymers to swell in precipitants for linear polystyrene but also accept and support our perception of inner stresses as an inherent property of all rigid networks prepared with solvating porogens. It is this stressed state of the dry hypercrosslinked polystyrene network that plays a role of the driving force for the swelling of the latter in all non-solvents.

Experimental visualization of emerging inner stresses, both on shrinking and expanding a hypercrosslinked network, will be presented in Section 4.2.

1.5 The role of the reaction rate of polystyrene with crosslinking agents

Swelling ability of conventional gel-type styrene–DVB copolymers (as well as other polymerization and polycondensation networks) is well known to sharply drop with increasing amounts of crosslinking bridges. Thus the copolymer with 20% DVB does not take any more any noticeable amount of toluene. Swelling of hypercrosslinked polymers prepared from linear polystyrene displays a fundamentally different character of the dependence between swelling and degree of crosslinking [131]. The swelling capacity in toluene presented in Fig. 7.9 was measured by weights of dry and swollen polymer samples and then calculated per gram of the starting linear polystyrene (rather than per gram of final products, as usual). In this way, the varying weight contributions from the introduced crosslinking bridges to the weight of the final polymer are eliminated. As can be seen, the toluene uptake dramatically drops with the degree of crosslinking increas-ing from 5 to about 20%. A further increase in the crosslinking density

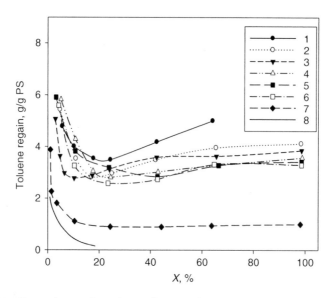

Figure 7.9 Dependence of weight swelling in toluene calculated per gram of starting polystyrene on the crosslinking degree of networks prepared by crosslinking (1-6) linear polystyrene and (7) styrene–1% DVB copolymer with (1) 1,4-bis-chloromethyldiphenyl, (2, 7) monochlorodimethyl ether, (3) tris-(chloromethyl)-mesitylene, (4) *p*-xylylene dichloride; (5) 1,4-bis(*p*-chloromethylphenyl)butane and (6) dimethylformal; (8) gel-type styrene–DVB copolymers.

surprisingly results in a striking increase in the swelling ability of hyper-crosslinked networks: polymers with 100% degree of crosslinking absorb 4–5 vol of toluene. The same swelling is characteristic of conventional copolymers containing only 0.5–1% DVB, a small amount. When the measured values are plotted directly in grams of toluene per 1 g of the final polymer, as shown in Fig. 7.10, the dependency will level out; thus an increase in the degree of crosslinking from about 60 to 200% (when 1 mol of the cross-linking agent reacts with 1 mol of styrene repeating units) does not change anymore dramatically the properties of the final polymers.

If beaded styrene–DVB copolymers (in maximum swollen state) are taken for the further post-crosslinking, this unusual increase of the swelling ability in toluene at higher degree of crosslinkings can be noted only in the case of starting styrene copolymers containing 0.17–0.6% DVB (Fig. 7.11). With an increased DVB content of 1% or higher, the toluene uptake of hypercrosslinked networks is nearly independent of the post-degree of crosslinking in the whole range of 40–200%. The swelling ability of the

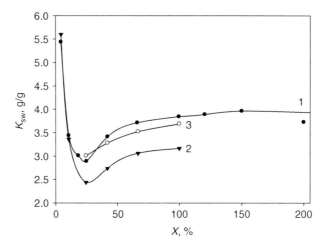

Figure 7.10 Dependence of weight swelling in toluene on the crosslinking degree of networks prepared by crosslinking linear polystyrene with (1) monochlorodimethyl ether and (2, 3) methylal in the presence of (1, 2) 1 and (3) 2 mol SnCl$_4$ per 1 mol of styrene repeating units. *(After [53]).*

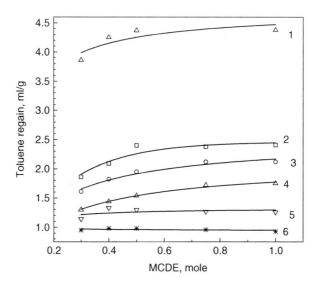

Figure 7.11 Toluene regain for the hypercrosslinked networks prepared by crosslinking with MCDE of (1) linear polystyrene of about 300 kDa molecular weight and (2–6) styrene copolymers with (2) 0.17, (3) 0.3, (4) 0.6, (5) 1.40, and (6) 2.7% DVB. *(Reprinted from [151] with permission of Wiley & Son, Inc.)*

final hypercrosslinked materials will be determined only by the initial copolymer network, that is, its DVB content and the extent of swelling during the post-crosslinking (as discussed in Section 1.4).

Contrary to the above strongly expressed role of the DVB crosslinks in the initial copolymer, the structure and even the functionality of the cross-linking agents used for the conversion of initial polymers into a rigid hypercrosslinked network show no correlation with the swelling capacity of the latter. Fig. 7.9 demonstrates that hypercrosslinked networks cross-linked with trifunctional tris-(chloromethyl)-mesitylene absorb more toluene as compared to networks crosslinked with bifunctional methylal, although all "classic" theories predict the opposite relationship. The swel-lings of networks crosslinked with MCDE or methylal differ significantly, although both crosslinking agents produce bridges of identical structure. In other words, another factor rather than the crosslinking density or func-tionality of the crosslinking agent plays a dominating role.

Unexpectedly a clear correlation has been found with the reactive ability of crosslinking agents in the Friedel–Crafts reaction [152]: the higher the rate of the chemical reaction with linear polystyrene, the higher the swelling capacity of the final polymer. This follows from the comparison of the rate of HCl generation, $d(HCl)/dt$, in the reaction of linear polystyrene with 0.1 mol of different crosslinking agents, and of the weight swelling K_{sw} in toluene of the obtained networks:

	CMDP	>>	MCDE	>	CMM	~	XDC	~	CMB
$d(HCl)/dt$	6.60		2.78		2.02		1.90		1.87
K_{sw}, (mL/g)	>5.0		4.0		3.5		3.8		3.4

This finding can be explained by considering that the conversion of a polymer solution into a single block of highly crosslinked homogeneous gel may be accompanied by macrosyneresis. Indeed, a certain portion of the initial solvent was always observed to depart from the gel. However, the process of macrosyneresis must be relatively slow, because the expulsion of the excess solvent (EDC) through simple diffusion of the solvent molecules from the center of the gel block to its periphery requires much time. Meanwhile, the network very soon becomes rigid and the macrosyneresis of EDC stops at a rather early stage of the crosslinking, depending on the reactivity of the crosslinking agent: the higher its reactivity, the more solvent remains captured in the gel, which is equivalent to an enhanced swelling ability of the final dry product.

This hypothesis also explains the abnormal dependence of the swelling of hypercrosslinked gels on their degree of crosslinking, as illustrated in Figs. 7.9 and 7.10. One only has to consider that enhanced amounts of crosslinking agents require larger concentrations of Friedel–Crafts catalyst to be introduced into the reaction mixture, which immediately accelerates the conversion of the polystyrene solution into a rigid gel and increases the amount of the solvent entrapped in the product. This phenomenon must be of general nature. Indeed, it was reported recently [153] that hydrogels obtained by crosslinking sodium alginate macromolecules ($M = 170,000$ Da) in semidiluted aqueous solutions with adipic dihydrazide, L-lysine methyl ester, or polyethylene glycols fitted on both ends with amine functions exhibit a similar extremal dependence of swelling on the crosslinking density, a minimum of swelling corresponding to 10–30 molar percent of the crosslinking agent. A similar nonlinear swelling behavior was also described for rigid chitin-based gels [154], which were obtained by crosslinking chitosan with nitrilotriacetic acid in the presence of water-soluble carbodiimide in a homogeneous aqueous solution and were then fully N-acetylated.

In many other instances the reaction rate also causes a similar outcome. For example, by crosslinking linear polystyrene with 0.5 mol of MCDE in the presence of 1.0 mol $SnCl_4$, a network having a swelling capacity as high as 3.8 g/g was obtained. In this reaction the conversion of the crosslinking agent was close to 100%, so that the final polymer contained only traces of pending chloromethyl groups. On the other hand, in the presence of 0.1 mol of the catalyst, the same reaction produced a product with only 1.8 g/g swelling capacity, in spite of the fact that in this case the conversion of the crosslinking agent was lower and the final product contained 3.7% of unreacted chlorine atoms.

A similar situation is observed for the reaction of polystyrene with MCDE in cyclohexane (Table 7.5) [155]. We can see again that the lower is the catalyst concentration in the reaction medium, the lower is the ability of the resulting networks to absorb both good and poor solvents, in spite of the fact that the conversion of the initially introduced chloromethyl groups is also lower. The paradox here consists in that smaller swelling corresponds to smaller crosslinking density, rather than vice versa.

Presumably the reactive ability of methylal (dimethoxymethane) is lower than that of MCDE. This is why methylal produces in the presence of 1 mol of stannic tetrachloride hypercrosslinked networks with smaller

Table 7.5 Influence of the amount of $SnCl_4$ on the equilibrium swelling capacity of networks prepared by crosslinking linear polystyrene with MCDE to 100% in cyclohexane (measured by weight of dry and swollen polymer)

$SnCl_4^*$	Swelling, (mL/g)				Cl^{**} %
	Toluene	Methanol	n-Hexane	Water	
3	5.40	–	–	2.93	1.30
2	4.40	4.04	4.19	2.46	1.78
1	3.03	2.46	2.82	1.60	2.65
0.5	2.94	1.54	2.34	1.38	2.76
0.25***	9.0	0.15	0.04	0.17	4.86

* $SnCl_4$ = amount of the catalyst, mol per mol MCDE.
** Cl = content of unreacted chlorine atoms, wt%.
*** incomplete conversion.
The concentration of linear polystyrene is 12.5% w/v.
Source: After [155].

swelling [53]. When reacting polystyrene with methylal in the presence of 2 mol of $SnCl_4$, the toluene uptake of the networks increases and becomes nearly equal to that of a network crosslinked with MCDE (Fig. 7.10). Methanol released in the reaction with methylal can also play a definite role, because the catalytic activity of the 1:2 complex of $SnCl_4$ with methanol is known to be lower than that of the 1:1 complex [114].

Finally, this concept of the important role of the reaction rate is also applicable to hypercrosslinked beaded materials prepared from styrene–DVB copolymers. Here, because of the very small size of gel beads, the ratio of the removal rate of the solvent from the gel to the crosslinking reaction rate is much more favorable. This explains the much lower amounts of the solvent finally entrapped in the beads and the reduced swelling ability of the beaded materials.

It should be useful to mention here that some antagonism exists between our data and information that appeared later in the literature. Thus, Popov et al. [156] consider methylal generally unsuitable for the Friedel–Crafts crosslinking reaction, contradicting all of our experience. By introducing XDC portion by portion, a series of hypercrosslinked materials were prepared earlier by the same group [9], which did not exhibit a minimum on the plot of toluene regain versus the degree of crosslinking. Due to the above-discussed dependence of toluene uptake on the rate of the chemical reaction, it becomes clear that under such conditions the conversion of the polystyrene solution into a gel was slow and the macrosyneresis had much more time to proceed. Negre et al. [48] examined the self-crosslinking of a styrene–vinylbenzyl chloride copolymer by using only 0.25 mol of stannic tetrachloride per mole of the $-CH_2Cl$ groups. Obviously, due to the slow reaction rate, the authors did not find any minimum on the plot of toluene regain versus crosslinking density. In the post-crosslinking experiments of Jeřábek with chloromethylated styrene–DVB copolymers [157], the low reaction rate in the presence of only 0.2 mol $SnCl_4$ per one chloromethyl group was the possible reason for obtaining hypercrosslinked materials with a relatively large amount of pending chloromethyl groups. Thus, none of the above reports can compromise our findings and their interpretation.

1.6 The effect of the reaction medium

As already mentioned in Chapter 6, Section 4, appropriate solvents for conducting Friedel–Crafts crosslinking reaction of polystyrene are EDC, tetrachloroethane, nitrobenzene, and cyclohexane (at an elevated temperature). While the products prepared in the first two solvents are identical in all their swelling behavior, the character of swelling of materials obtained in nitrobenzene is different.

Nitrobenzene accelerates the reaction of XDC with polystyrene. In this solvent the rate of hydrogen chloride generation, $d(HCl)/dt$, increases to 4.26 (arbitrary units) as compared with 1.90 for the reaction rate in EDC under identical conditions. The minimum on the curve of swelling versus degree of crosslinking becomes now well marked (even in the case of calculating the toluene uptake per gram of resulting networks). This result agrees well with the finding that the fast formation of a rigid hypercrosslinked network produces polymers with enlarged amounts of entrapped nitrobenzene and enhanced swelling capacity. The bridging of polystyrene by means of MCDE seems to be also accelerated: $d(HCl)/dt$ increases from 2.78 in EDC to 4.5 when the reaction is carried out in nitrobenzene. However, MCDE was found to be unstable in nitrobenzene: the mixture of MCDE and $SnCl_4$ in nitrobenzene at 80–90°C evolves HCl without adding any polystyrene to the mixture. This unknown side reaction may be responsible for obtaining products with an unusual swelling ability in both good solvents (Fig. 7.12) and precipitating media (Fig. 7.13).

Cyclohexane at 34.5°C is well known to be a Θ-solvent for polystyrene. At this fairly low temperature the crosslinking reaction proceeds too slowly; therefore, the synthesis of hypercrosslinked polystyrenes was performed at 60°C in this solvent. Although the affinity of cyclohexane to polystyrene increases with temperature, in terms of thermodynamic affinity the solvent still remains not as good as EDC at 80°C. Alternatively to the crosslinking processes in EDC and nitrobenzene, the formation of hypercrosslinked

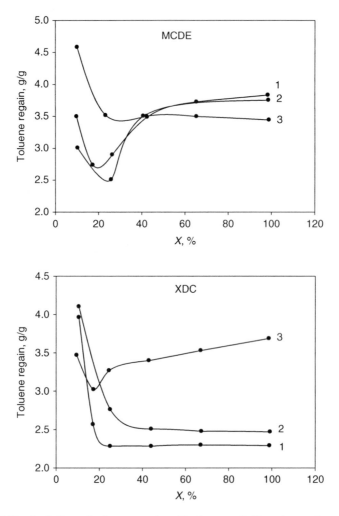

Figure 7.12 Evolution of toluene regain with the crosslinking degree for networks prepared by crosslinking linear polystyrene with monochlorodimethyl ether and *p*-xylylene dichloride in (1) ethylene dichloride, (2) ethane tetrachloride, and (3) nitrobenzene.

polystyrene in cyclohexane is accompanied by microphase separation (the reasons for this will be discussed in details in Section 7). It should be only noted here that the products prepared in cyclohexane exhibit the same character of the swelling versus degree of crosslinking relationship as those obtained in EDC (Fig. 7.14) [158].

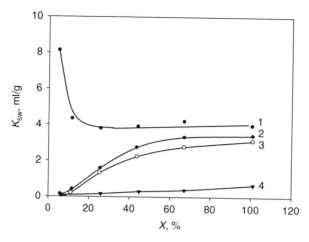

Figure 7.13   Dependence of swelling in (1) toluene, (2) *n*-hexane, (3) methanol, and (4) water on the crosslinking degree of networks prepared by crosslinking linear polystyrene with monochlorodimethyl ether in nitrobenzene.

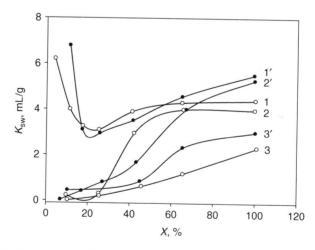

Figure 7.14   Dependence of swelling in (1, 1′) toluene, (2, 2′) methanol, and (3, 3′) water on the crosslinking degree of networks prepared by crosslinking linear polystyrene with monochlorodimethyl ether in (1–3) ethylene dichloride and (1′–3′) cyclohexane. *(After [158]).*

1.7 The influence of polystyrene molecular weight

The molecular weight of linear polystyrene taken for the subsequent bridging was also found to influence the swelling ability of both macronet isoporous and hypercrosslinked networks. As indicated by Figs. 7.15 and 7.16, the higher the molecular weight of the starting polystyrene, the lower the swelling capacity of the resulting networks, regardless of the solvent of

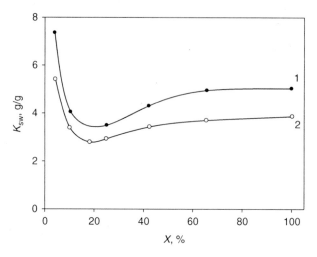

Figure 7.15 Toluene regain of the networks prepared by crosslinking with monochlorodimethyl ether in ethylene dichloride of polystyrene with molecular weight of (1) 8800 and (2) 300,000 Da.

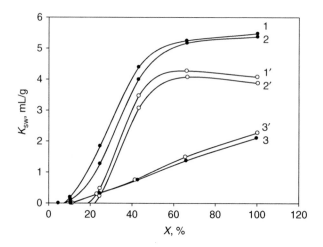

Figure 7.16 Swelling in (1, 1′) n-hexane, (2, 2′) methanol, and (3, 3′) water of the networks prepared by crosslinking with monochlorodimethyl ether in ethylene dichloride of polystyrene with molecular weight of (1–3) 8800 and (1′–3′) 300,000 Da.

the reaction. It is difficult to find a rational explanation for this finding. We can only speculate that, obviously, the microstructure of the initial solutions at the concentration of about 10% used during the post-crosslinking process depends on the size of the polystyrene coils. If all polymeric coils would attain a voluminous, unperturbed, or even expanded size and, hence, completely interpenetrate each other to give a rater homogeneous system, as the classic theories of polymer solutions would require, then the chain length would not play any noticeable role.

2. THE KINETICS OF SWELLING OF HYPERCROSSLINKED POLYSTYRENE

The fundamental difference in the structure of conventional gel-type, macroporous, and hypercrosslinked networks is clearly indicated not just by the fact that the first two swell with thermodynamically good solvents whereas the latter responds equally to both good solvents and non-solvents. The three types of networks also differ fundamentally in both the extent of swelling and swelling kinetics. These parameters have been recently examined by automatically registering the diameter of a single polymeric bead of about 0.6 mm after contacting the bead with the solvent. This technique permits precise measurements of very fast bead deformations during the swelling process. Simultaneously, the movement of the solvent front toward the bead center was observed by a microscope [159].

Gel-type styrene–DVB copolymers swell rather slowly, with the deformation versus time plot being almost linear (Fig 7.17). Increasing the DVB content decreases both the extent and rate of swelling. This character of the swelling deformation corresponds to the slow diffusion of the solvent from the bead surface toward its center. A good solvent initially causes swelling of the peripheral layers of the polymer, where the rather long polymeric segments between junction points immediately start to expand. It will take 1 h or even more for the bead to fully expand. Interestingly, a macronet isoporous polymer containing 0.2% DVB and 25% of subsequently introduced methylene links between the phenyl rings behaves similarly to typical gel-type copolymers and swells at about the same rate as a styrene–2.7% DVB copolymer, although to a substantially larger end volume.

Due to their highly developed porous structure, macroporous polymers become completely wetted with toluene within a few seconds. Simultaneously, the bead slightly expands. Afterward, a slower expansion takes place, where, probably, the highly crosslinked polymeric phase

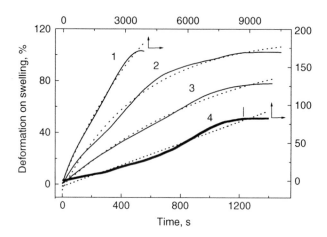

Figure 7.17 Kinetics of swelling in toluene of gel-type copolymers: (1) the network prepared by crosslinking styrene–0.2% DVB copolymer with monochlorodimethyl ether to 25% and conventional styrene copolymers with (2) 1.4, (3) 2.7, and (4) 5.3% DVB, as measured on individual beads (full lines) and averaged over 10 measurements (dotted lines). *(Reprinted from [159] with permission of Akademizdattsentr "Nauka" RAN.)*

(that comprises the walls of pores) continues to swell. Thus, the overall plot of the swelling kinetics has an exponential shape (Fig. 7.18). The swelling equilibrium is attained within 10–60 s, but the extent of the bead expansion remains relatively insignificant, from 10 to 30%.

The volume of a hypercrosslinked bead expands by at least 150% within about 1 min, even for the most rigid networks with $X = 200\%$ in which each phenyl ring participates twice in bridging neighboring chains. Interestingly, the rate of swelling increases with the degree of crosslinking, while the extent of volume expansion drops only insignificantly (Fig. 7.19). Contrary to macroporous polymers, hypercrosslinked networks take a slow start within the first 5–15 s needed for the solvent front to reach the bead center. After that, the rate of volume swelling rapidly increases, so that the overall kinetic plot attains an S-shape. We explain this self-accelerating process by the extreme rigidity of the hypercrosslinked network. It does not contain polymer segments that could move separately from near and distant neighbors. The surprisingly high increase in the volume of such a rigid open framework can only take place by a cooperative simultaneous rearrangement of the whole ensemble of network meshes. This requires the entire structure to be wetted with the solvent before the conformations of the meshes start changing. The driving force for the rearrangement is the

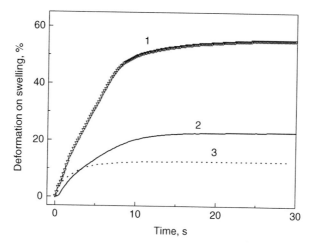

Figure 7.18   Kinetics of swelling in toluene of macroporous copolymers: (1) styrene–15% DVB copolymer prepared in the presence of 60 wt% *n*-hexane, (2) polydivinylbenzene prepared in the presence of 60 wt% toluene, (3) polydivinylbenzene prepared in the presence of 60 wt% *n*-hexane. *(After [159])*.

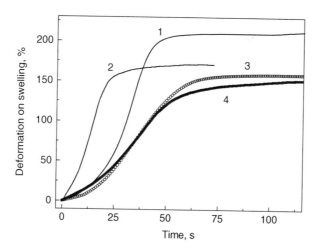

Figure 7.19   Kinetics of swelling in (1, 2) toluene and (3, 4) ethanol of hypercrosslinked networks prepared by crosslinking styrene–0.2% DVB copolymer with monochlorodimethyl ether to (1, 3) 100 and (2, 4) 200%. *(After [159])*.

inner stress of the network. Expansion of the network releases the stress, and this stress relaxation enhances the diffusion of the solvent, the latter being simply sucked in by the expanding bead.

Another distinguishing feature of hypercrosslinked polymers is that they swell with the non-solvent, ethanol, to almost the same extent, but at a somewhat slower rate. Probably toluene, by readily solvating polystyrene chains, acts as a plasticizer, whereas ethanol cannot facilitate conformational rearrangement to the same extent.

3. SOME REMARKS CONCERNING THE SWELLING ABILITY OF THREE-DIMENSIONAL POLYMERS

The keystone for theoretical considerations of swelling of three-dimensional polymers in good solvents is the declaration that the network swelling is governed above all by its degree of crosslinking. Accordingly, in all mathematical formulas, the most important value of M_c, which is the size of the chain confined between two neighboring junction points, appears and is directly connected with the swelling ability. The theoretical value, M_c^{th}, can be easily calculated from the molar composition of the copolymer mixture as

$$M_c^{th} = \frac{2M_m}{f} \left(\frac{1}{N} - 1 \right) \qquad [7.2]$$

where f is the functionality of the crosslinking agent ($f = 4$ for DVB), N is the molar fraction of that component, and M_m is the molecular weight of the main monomer, for example, styrene.

It logically follows from the above general notion that a network is defective if its swelling differs from the theoretical predictions. Three types of network topological defects are generally under consideration:

(i) Chain loops produced by the connection of two chain ends in one junction point or loops formed by two chains of equal or different lengths binding two junction points. It is assumed that such a ring system does not contribute exhaustively to network crosslinking.

(ii) Appearance of dangling chains, which is caused by chain functionalities remaining unreacted for steric or other reasons. Dangling chains are believed to reduce the number of elastically active chains in the network and, hence, favor network swelling.

(iii) Permanent chain entanglements acting as additional links, thus increasing the apparent density of crosslinking.

Undoubtedly, all these types of defects are more or less peculiar to any real network polymer. However, one may question to what extent can the discrepancy between the theoretical predictions of swelling and the behavior of a real network be explained by these defects? An even more important question is how correct is the current theory and whether it adequately assigns priorities to the most important factors that determine the swelling ability of polymer networks.

When assuming the dominant role of the degree of crosslinking, the experimental value M_c^{ex} of the molecular weight of the chain between neighboring junction points is calculated from the experimentally measured swelling capacity of the network. One may use the Flory–Rehner equation [160], or the approaches of James and Guth [161], Hermans [162], and Dušek and Prins [163], or similar considerations (see, for example, [164–167]), or Flory's latest theory according to which the extent of "non-affine" deformation of a chain between adjacent junctions on swelling depends on the looseness with which crosslinks are embedded in the network [168].

In practice, the experimentally derived M_c^{ex} values only occasionally coincide with the theoretical expectations. Fig. 7.20 illustrates a typical example of discrepancy between the values of M_c^{th} and M_c^{ex} for styrene–DVB copolymers [169]. For the networks with small amounts of DVB crosslinks M_c^{ex} usually exceeds M_c^{th}, and a large number of intramolecular

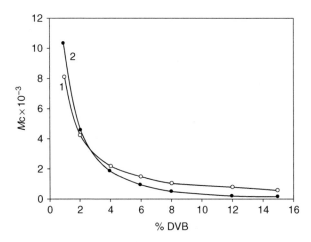

Figure 7.20   Dependence of (1) M_c^{th} and (2) M_c^{ex} on DVB content for the conventional styrene–DVB copolymers. *(After [169])*.

loops or dangling chains are said to be the most probable defects responsible for the enlarged network swelling. On the contrary, at a relatively high degree of crosslinking, the theoretical distance between neighboring junctions proves to be larger than that calculated from the experimentally measured swelling. Then, the enhanced contribution of chain entanglements to the effective network crosslinking is postulated.

Certain doubts exist concerning the validity of these explanations. Indeed, the formation of elastically ineffective intramolecular loops, that is, cyclization, occurs by the reaction of a macroradical with pendent DVB vinyl groups of the same chain, and so the probability of cyclization would have to decrease with decreasing DVB content, rather than vice versa. It is true that at a very low crosslinking stage, styrene–DVB copolymers contain many dangling chains [170]. But, as we have seen in Chapter 1, Section 2.2, even a large amount of dangling chains will not increase the network swelling. Thus the explanation of the enlarged swelling of slightly crosslinked networks by these two defects seems to be quite problematic.

The notion that chain entanglements are defects especially characteristic of the stronger crosslinked networks does not seem to be motivated either. If a network was obtained without any porogen, every mesh of the network would be filled completely with other network parts, which must restrict the mobility of both slightly and highly crosslinked polymers. There are no arguments for speculating whether slightly or highly crosslinked networks are more affected by entanglements. Entanglements should be regarded as an important inherent feature of such type of networks, rather than just as a defect.

The traditional statement that swelling is mainly determined by the degree of crosslinking which, in its turn, determines the number of elastically active chains of a network is further questioned by the work of Sarin et al. [171]. When employing conventional styrene–1% DVB copolymer for solid-phase peptide synthesis, they paid attention to an unusual fact. The swelling capacity of the starting copolymer with dimethylformamide amounted to 3.3 mL/g, whereas after binding 80% (by weight) of polypeptides with a molecular weight of 6000 Da, it increased to a very high value, 28 mL/g. The authors have taken an interest in this phenomenon and calculated the maximum volume of the copolymer under the condition that its polymeric chains between junction points would exhibit a completely stretched conformation. The result proved to be fairly surprising: the volume amounted to 196 mL/g.

Naturally, from the thermodynamic point of view, polymeric chains cannot exhibit a completely stretched conformation on swelling. Rather, the above given result unambiguously shows that chains in the starting swollen network are stretched to an extremely small extent. If additional driving force arises for swelling, such as the strong solvation of polypeptide fragments with dimethylformamide, the network will swell until the free energy of mixing of the network and peptide with the solvent becomes again equilibrated with the free energy of elastic network deformation. Now, on stretching, additional numbers of chains and entanglements will be involved in deformation, together with those which were elastically active in the starting copolymer. In other words, the increase in the number of elastically active chains does not lead automatically to a decrease of swelling. Similarly, the swelling of conventional styrene–DVB copolymers in toluene, especially of those with low crosslinking, is always lower than the swelling of their sulfonates in water, because the solvation of ionic functional groups with water is stronger than the van der Waals interactions of toluene with polystyrene chains.

To our point, none of the existing network theories describes adequately the swelling behavior of real polymers and none of the existing approaches can be applied to describe the swelling properties of hypercrosslinked networks. Naturally, these theories have been basically developed for networks composed of Gaussian chains, that is, long chains between punctual junctions. On the other hand, in hypercrosslinked polystyrene networks, every monomer unit participates once or even twice in the formation of crosslinks. Actually the networks entirely consist of crosslinking bridges with no chains between them. Nevertheless, with respect to swelling (as well as to elasticity in swollen state), these networks may behave very similar to slightly crosslinked copolymers. It would be senseless to involve concepts of network defects or low efficiency of post-crosslinking reactions for explaining the abnormally high swelling of hypercrosslinked networks. Obviously, the theoretical treatment of networks misses or underestimates certain most important factors that influence the swelling ability even stronger than the degree of crosslinking.

Summarizing our experience with hypercrosslinked polystyrene networks, as well as with conventional gel-type and macroporous copolymers, we believe that the topology of the network was undeservedly neglected by all theoretical treatments. By this we particularly mean that a network must be considered as an ensemble of interconnected and interpenetrating meshes or cycles, rather than as a combination of chains and junctions.

Then, the extent of the interpenetration of cycles that cannot be disentangled becomes the most important factor determining the equilibrium swelling ability in a thermodynamically good solvent rather than just the crosslinking density. The latter merely determines the size of the chain segments between junctions and, more indirectly, the size of network meshes. Paradoxically, it follows from this concept that the degree of interpenetration of the cycles must be higher for large cycles, since they embrace more alien chains and other structural units of the network. On the other hand, at the degree of crosslinking of 200%, very small network meshes could be almost free of foreign segments if the network was formed in the highly swollen state. It is mainly for the lack of entanglements that hypercrosslinked polystyrene networks having a crosslinking density of 100 or even 200% swell to the same extent as conventional gel copolymers with only 1% DVB prepared without a solvating diluent do.

At a given crosslinking density, the extent of interpenetration of network meshes is determined by the dilution of the initial comonomer mixture or initial polymer solution with the thermodynamically good porogen. Here such a factor as partial macrosyneresis sometimes interferes: it may depend on the rate of conversion of the initial solution into the final gel. The thermodynamic quality of the porogen may also become visible, if a tendency to micro- or macro-phase separation becomes evident on progressive consumption of the comonomers. Factors such as the inclusion of preliminary strained fragments into the final network or the uneven distribution of crosslinks may affect the equilibrium swelling, the latter factor, however, being less important.

Swelling in non-solvents, which is not characteristic at all for classical gel-type and macroporous networks but only for hypercrosslinked structures, is entirely determined by the inner stresses in the dry material. The intensity of stresses and the material expansion in non-solvents now depends on the network rigidity. Tendency to the relaxation of stress will always try to bring about an increase in the polymer volume toward the volume of the gel formed at the end of network formation. Networks with flexible crosslinking bridges or flexible chain segments cannot store stresses and will not swell in non-solvents, even in the case of an extremely high degree of crosslinking and low extent of interpenetration of the network meshes. This factor of inner stresses was never recognized before hypercrosslinked networks were synthesized and examined.

The large number and the mutual interplay of factors determining the swelling ability of real networks dramatically complicate any mathematical description of the dependence of swelling on all the above-discussed factors. Even if the role of each factor will be exactly revealed and expressed mathematically, one would still need an exact knowledge of the synthesis protocol of a polymer, in order to characterize its network topology and structural parameters and predict its swelling behavior. However, it will never be possible to unambiguously characterize the structure of a network while knowing only one of its properties, the extent of swelling with any solvent.

Still, a fundamentally new approach is needed to describe the swelling behavior of polymeric networks, which would also include hypercrosslinked networks. This must be based on the presentation of a network as a system of meshes, the size, structure, and conformational and thermodynamic state of which, and, most importantly, their topological relations with neighbor meshes would be regarded as the most important parameters. It can be mentioned here that Erukhimovich has already suggested [172] to consider a network as a hierarchy of topologically complicated cycles, but this approach has had a pure mathematical, abstract character and has resisted experimental verification. We do not suggest our own mathematical description of the swelling behavior of networks; nevertheless, while we fully recognize the importance of theory, we are guided by the thought of Flory formulated in 1979: "It may be sufficient ... to comprehend the physical principles governing the behavior and properties of real networks in lieu of more precise theory" [173].

4. SWELLING AND DEFORMATION OF HYPERCROSSLINKED NETWORKS

4.1 Physical background of photoelasticity phenomenon

Deformation of a polymer sample on compression, extension, or bending is well known to be accompanied by unavoidable emergence of significant inner stresses because of changes – compared to equilibrium state – in the mutual arrangement of atoms and groups, as well as conformations of polymeric chains. At that, some initially symmetric elements of the material internal structure distort and become asymmetric. These elements acquire the ability to rotate the plane of polarization of transmitted polarized light. In other words, a local optical activity emerges in the material. Traditional explanation of this phenomenon is that under the influence of stresses the initially isotropic polymer becomes optically anisotropic. In such material a light beam propagates along the lines of stresses and across them with slightly different velocities. Thus the material acquires the property of birefringence. One could say that a plane-polarized beam, when passing through anisotropic medium, falls into two beams, usual and unusual, which are polarized in mutually perpendicular planes. After passing the anisotropic domain, these beams appear to be shifted in phase, which finally results in a certain local turn of the polarization plane of the beam. In any case, the initially plane-polarized beam (it is also called linear-polarized beam) experiences partial depolarization; the higher is the extent of depolarization, the stronger are the inner stresses present in the deformed sample.

This phenomenon is the basis for the photoelasticity method of studying inner stresses in transparent materials.

In an optical polarizing microscope the polarizer film or crystal transmits only a plane-polarized beam. The optical analyzer quenches this beam if the analyzer is turned under right angle against the polarizer. When one places an optically transparent isotropic material between the polarizer and analyzer crossed under the right angle, the material remains invisible on dark background. However, if the studied material has optically active fragments, that is, zones exhibiting the property of birefringence, the plane of polarization of the transmitted beam is disordered and the analyzer fails to quench the beam completely. Dark and light interference stripes emerge on the picture of a stressed plate. Using the distance between the stripes, it is even possible to calculate the local stresses appearing on the sample deformation. The photoelasticity method has usually been applied to investigate the distribution of local stresses in deformed models of building units and machinery parts when direct measurements of stresses are impossible in actual constructions.

An ordered packing of macromolecules may also cause an optical anisotropy and birefringence, which are characteristic, for instance, of polymer spherulites. Because of the radial anisotropy of a spherulite and the convergence of beams in the spherical structure, the interference picture represents the so-called Maltese cross, the center of which is located in the center of spherulite. No calculations are performed using such a picture but the photoelasticity method is very efficient in revealing qualitatively the presence of any spherulites, or a mesomorphic or ordered sate of polymeric chains.

In a polarizing optical microscope with crossed polarizer and analyzer, a transparent spherical isotropic sample looks like a bright object on the dark background since the reflection and refraction on spherical surfaces partially depolarize the plane-polarized transmitted light. It is of great importance that the emergence of an interference picture in the form of a cross for such spherical objects unambiguously testifies either to radial order of polymeric chains (as in spherulites) or to radial tension of their internal structure. Indeed, if one observes in the polarizing microscope with crossed polarizer and analyzer a water-swollen (i.e., strongly expanded and stretched) bead of styrene–DVB sulfonate, it is possible to see a clearly-cut Maltese cross. We met the mention of this phenomenon in some publications of the 1960s and 1970s.

4.2 Visualization of inner stresses in networks on swelling

We took advantage of the photoelasticity technique for studying in detail conditions under which inner stresses emerge or, vice versa, fully disappear in polymeric networks, depending on their swelling state. Three types of polystyrene networks have been chosen for comparison, namely a spherical conventional gel-type styrene–0.5% DVB copolymer, the hypercrosslinked network obtained by crosslinking this copolymer with MCDE to $X = 100\%$, as well as the styrene–5% DVB copolymer prepared in the presence of 30% (by volume) of toluene. Table 7.6 presents their characteristics.

As can be seen, the gel-type polymer is non-porous and can swell only in such thermodynamically good solvents as toluene. The properties of the expanded network prepared in the presence of toluene are similar, although some incipient swelling in n-hexane can be noted. The hypercrosslinked polystyrene is a highly porous material capable to swell both in good solvent, toluene, and in precipitants, methanol and hexane. The difference between swelling values in the above solvents is small, in comparison with the swelling value as such, but, nevertheless, a general tendency can be

Table 7.6   Porosity and swelling of polystyrene-type networks

Type of network	S, (m^2/g)	W_o, (cm^3/g)	Volume on swelling			
			Toluene	Hexane	Methanol	Water
Gel-type 0.5%-DVB*	0	0	7.3	1.0	1.0	1.0
Expanded 5%-DVB*	0	0	1.73	1.09	1.03	1.0
Hypercrosslinked	1300	0.21	1.91	1.72	1.65	1.25

*Non-porous products.
Source: After [174].

observed for all polystyrene-type networks, namely that the volume of material increases in the order water < methanol < *n*-hexane < toluene.

Figure 7.21 shows micrographs taken in polarized light for dry beads of the initial styrene–0.5% DVB copolymer (Fig. 7.21a) and hypercrosslinked polymer on its base (Fig. 7.21b). Both the polymers are transparent, though the hypercrosslinked one is a highly porous material with a significant free volume and apparent specific surface area as large as $1300\,m^2/g$. If now one crosses the polaroids, the dry bead of the isotropic gel-type polymer remains transparent (Fig. 7.21c), while the hypercrosslinked polymer reveals its inner stains and optical anisotropy in the form of a clear Maltese cross (Fig. 7.21d).

The single-phase uniformly crosslinked open network can be stressed in two opposite ways, on shrinkage and extension. As was discussed earlier, stresses of shrinkage appear in the dry hypercrosslinked material as a result of confrontation between the network rigidity and the attraction forces

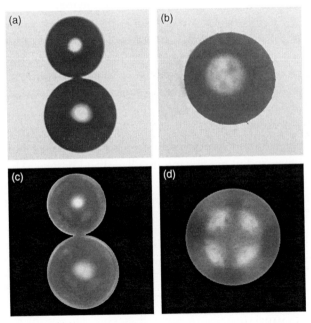

Figure 7.21   Micrographs made in polarized light with parallel (a, b) and crossed (c, d) polaroids of dry beads of (a, c) gel-type styrene–0.5% DVB copolymer and (b, d) a hypercrosslinked network based on this copolymer. *(Reprinted from [174] with permission of Akademizdattsentr "Nauka" RAN.)*

between the non-solvated network fragments. This type of strong stress is responsible for the emergence of Maltese cross in the dry bead, and it reveals itself in many ways providing very specific properties to hypercrosslinked polymers, including their strong swelling in all non-solvating media.

However, a strong solvation of the network by thermodynamically good solvents, in particular, toluene, must lead to its expansion and emergence of a stress having an opposite sign. Toluene is known to cause an increase in volume of linear polystyrene coils in comparison with their unperturbed dimensions in Θ-state. Accordingly, an ultimate swelling in toluene must exert the stress of extension on any polystyrene-type network. Indeed, in the toluene-swollen beads of hypercrosslinked polystyrene, a well-formed Maltese cross has always been seen (Fig. 7.22) at crossed polaroids, though looking somewhat different from the cross characteristic of the dry bead. The same interference pictures are characteristic of all polystyrene networks swollen in good solvents, such as toluene, benzene, chloroform, dioxane, and even cyclohexane.

Accordingly, it is legitimate to assume that in the course of swelling with toluene in a dry bead of a hypercrosslinked polymer the initial stress of shrinkage should be replaced by a final stress of extension. Indeed, as a droplet of toluene falls on the dry bead and the latter starts to swell, one can observe in the microscope the cross disappearing for a while at the moment the solvent reaches the bead center, and, then emerging again, in the form shown in Fig. 7.22. Here, it is important to emphasize that the transition of

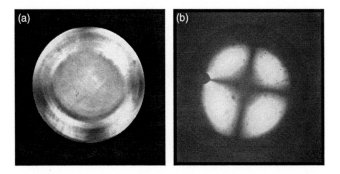

Figure 7.22 Micrographs made with crossed polaroids of swollen-in-toluene (a) gel-type styrene–0.5% DVB copolymer and (b) hypercrosslinked polystyrene on its basis. *(Reprinted from [174] with permission of Akademizdattsentr "Nauka" RAN.)*

the hypercrosslinked network from one anisotropic state to another must proceed through an intermediate isotropic state, where the network is totally free from any stress.

The type of solvent in contact with the polymer determines the sign and intensity of inner stresses of the hypercrosslinked network. Water exhibits the smallest affinity to the aromatic polymer. The ultimate uptake of water by the above hypercrosslinked polystyrene exceeds the total pore volume of the dry polymer, 0.62 and 0.21 cm^3/g, respectively. Swelling with water brings about certain expansion of the network, but it is insufficient for relaxation of inner stresses and the network remains compressed. A water-swollen bead exhibits a clear Maltese cross which is much alike the cross of a dry polymer (Fig. 7.23). It is particularly remarkable that methanol causes a more significant increase of the volume of the bead (1.65 times) and converts the initially anisotropic bead into fully isotropic swollen material that is fully transparent in crossed polaroids. Hexane does not dissolve linear polystyrene, either, but its affinity to the hypercrosslinked aromatic network seems to be higher (swelling ratio 1.72). Due to this reason, we see the first sign of stretching of the network in the bead swollen with hexane (Fig. 7.23). Of course, the Maltese cross is most expressed in beads swollen with toluene and expanded by a factor of 1.91.

Thus, by using the photoelasticity effect, we managed to ascertain for the first time that the hypercrosslinked polystyrene network can experience inner stresses of shrinkage on drying, stresses of extension on swelling in good solvents, and also can acquire a fully relaxed state, totally free from any stresses, when wetted with methanol. Dušek and Prins [163] have postulated a peculiar "reference" state when polymeric chains do not exert any forces on network junction points (see also Chapter 1, Section 2.2), but this state of

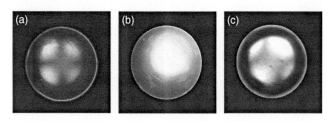

Figure 7.23   Micrographs made with crossed polaroids of beads of hypercrosslinked polystyrene swollen in (a) water, (b) methanol, and (c) *n*-hexane. *(Reprinted from [174] with permission of Akademizdattsentr "Nauka" RAN.)*

networks has never been realized experimentally. By analogy with the unperturbed state of linear polymer chains in a Θ-solvent, we may claim now that the network of hypercrosslinked polystyrene acquires the Θ-state in methanol.

Methanol does not belong to the family of solvents for linear polystyrene. Nevertheless, the energy of interaction between methanol and the hypercrosslinked aromatic polymer proves to be sufficient for increasing the network volume to exactly the state where all compression stresses fully relax, but no excessive network stretching takes place. In all probability, methanol is the Θ-solvent for the hypercrosslinked polystyrene (at room temperature) to the same extent as cyclohexane is the Θ-solvent for linear polystyrene (at 34.5°C). However, cyclohexane proves to be a good solvating liquid for hypercrosslinked polystyrene, which may be accounted by the (small) difference between the chemical composition of that network and linear polystyrene. By the way, cyclohexane gives rise to swelling of the gel-type styrene–DVB copolymer, too, thus transforming it into a stretched form from the non-swollen state, the latter, most likely, being the unstrained reference state for copolymers prepared in the absence of any diluent. As to the "expanded" styrene–5% DVB network obtained in the presence of 30% of toluene, it also swells in cyclohexane into an even more expanded and stressed state. It is interesting that n-hexane, too, causes certain swelling of this polymer, as well as of the hypercrosslinked network. While gel-type copolymers remain indifferent to both methanol and n-hexane, the expanded network exhibits affinity to that last solvent (Fig. 7.24). (The expanded copolymer after synthesis needs relaxation at 100°C in order to attain the non–porous unstressed structure.)

Thus, hypercrosslinked polystyrene networks gradually expand while their void volume becomes occupied with the following media:

$$\text{air} \ll \text{water} \ll \text{methanol} \ll n-\text{hexane} < \text{cyclohexane} < \text{toluene}$$

Methanol happens to be the Θ-solvent for the examined hypercrosslinked sample, where its network relaxes completely. Solvents to the right from methanol cause additional swelling and stretching of the network. When filled with water and, especially, air, the network contracts significantly and acquires strong inner stresses.

It is not yet possible to estimate quantitatively the stresses developed in the network on its deformation. However, some idea can be inferred from the fact that the volume of the bead, taken for unity in its relaxed methanol-filled state, expands in toluene by a factor of 1.13, but contracts on drying to

Figure 7.24 Micrographs made with crossed polaroids of expanded styrene–5% DVB copolymer swollen in *n*-hexane. *(Reprinted from [174] with permission of Akademizdattsentr "Nauka" RAN.)*

0.59 of its volume. It is also appropriate to recall here that the heat of swelling of the dry hypercrosslinked polystyrene with $X = 100\%$ in methanol is about 100 kJ/g and in toluene only by 30 kJ/g more (Table 7.3). Thus we can imagine that the van der Waals attraction forces between non-solvated polystyrene fragments in the dry hypercrosslinked network cause very significant contraction strains of the network and provide it with high potential energy that facilitates all processes related to swelling and adsorption, that is, expansion.

5. POROSITY OF HYPERCROSSLINKED POLYSTYRENE

5.1 Apparent density of hypercrosslinked polystyrenes

Similar to the gel-type styrene copolymers, the hypercrosslinked networks based on both linear polystyrene and styrene–DVB copolymers are transparent materials, being in dry or swollen state. Nevertheless, alternatively to the conventional copolymers, the hypercrosslinked networks exhibit very low apparent density (ρ_{app}) when isolated from the reaction media and dried. The apparent density of the polymers (**I**) based on linear polystyrene (Table 7.7) decreases with increasing the degree of crosslinking;

Table 7.7 Apparent density (ρ_{app}, g/cm³) and total pore volume (W_o, cm³/g) of hypercrosslinked networks based on linear polystyrene with a molecular weight of 3×10^5 Da and crosslinked with p,p-bis-chloromethyl-diphenyl, p-xylylene dichloride, or monochlorodimethyl ether

X, (%)	CMDP			XDC			MCDE			
	ρ_{app}	W_o^p	W_o^{Hg}	ρ_{app}	W_o^p	W_o^{Hg}	ρ_{app}	W_p^p	W_p^{Hg}	$W_o^{N_2}$
I, 25	0.86	0.27	0.18	0.84	0.30	0.28	0.91	0.21	0.18	0.09
I, 43	0.85	0.28	0.20	0.77	0.41	0.42	0.80	0.36	0.25	0.46
I, 66	0.81	0.34	0.27	0.74	0.46	0.43	0.75	0.44	0.39	0.54
I, 80	0.79	0.37	0.31	0.73	0.48	0.44	0.71	0.51	0.41	0.64
I, 100	0.79	0.37	0.27	–	–	–	0.71	0.51	0.48	0.68
II, 100	–	–	–	–	–	–	0.75	0.43	–	0.65

X is the degree of crosslinking.
W_p^p (cm³/g) is the total pore volume calculated from the "true" (1.12 cm³/g) and apparent densities.
W_o^{Hg} (cm³/g) is the total pore volume determined by mercury intrusion.
$W_o^{N_2}$ (cm³/g) is the total pore volume measured by nitrogen adsorption at $p/p_s \sim 1$.
Reprinted from [175] with kind permission of Elsevier.

for networks with 100% rigid bridges it drops to approximately 0.7 g/cm³, compared to 1.05 g/cm³ for atactic polystyrene. The same low density is characteristic of the hypercrosslinked polymers (**II**) prepared on the base of styrene copolymers with very small amounts of DVB (Fig. 7.25). With increasing the initial DVB content over 1% and decreasing the number of

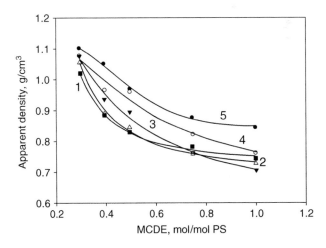

Figure 7.25 Apparent density of the networks prepared by post-crosslinking with MCDE of styrene copolymers with (1) 0.17, (2) 0.3, (3) 0.6, (4) 1.4, and (5) 2.7% DVB. *(Reprinted from [151] with permission of Wiley & Sons, Inc.)*

additional diphenylmethane bridges below 43%, the gradually increasing apparent density of post-crosslinked materials reaches the normal density of the initial copolymers. The values of apparent density of about $0.7\,g/cm^3$ are known to be characteristic of many opaque macroporous styrene–DVB copolymers, where the low density value is associated with the true porosity. Thus, the transparent hypercrosslinked polystyrene also has to be some kind of porous material [175].

5.2 Apparent inner surface area of hypercrosslinked polystyrenes

Hypercrosslinked polymers derived from either linear polystyrene or gel-type styrene–DVB copolymers take up large amounts of inert gases at low temperatures. Figure 7.26 illustrates a typical sorption isotherm for nitrogen at 77 K onto the hypercrosslinked material prepared by crosslinking styrene–0.3% DVB copolymer with 0.5 mol of MCDE per 1 mol of styrene

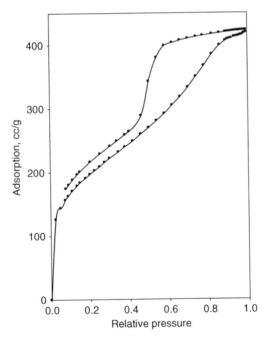

Figure 7.26 Sorption isotherm of nitrogen at 77 K on the hypercrosslinked polymer prepared by post-crosslinking styrene-0.3% DVB copolymer with monochlorodimethyl ether to 100%. Measured by Micromeritics Instrument Corporation.

repeating units (corresponding to 100% degree of crosslinking). In accordance with the classification of Brunauer, Deming, Deming, and Teller (BDDT), this isotherm may be attributed to type IV which is characteristic of mesoporous sorbents and for which the method of Brunauer, Emmet, and Teller (BET) is deemed to provide a real surface area value. The formal sign of applicability of this method is the straightening of the initial section of the isotherm plotted in coordinates of the BET equation (Eq. [3.1]). This is valid for the isotherm in the usual range of relative pressures, p/p_s from 0 to at least 0.2. The B-point on the sorption isotherm of Fig. 7.26 fits this straight-line dependence, which is another sign of the applicability of the BET method (although sorption at the B-point, 197 cm^3/g, does not fully correspond to the calculated monolayer capacity of 165 cm^3/g). Moreover, the C constant of the BET equation was found to have a reasonable value of 95 for this sample, and 100–140 for other hypercrosslinked sorbents [176]. Thus, sorption of nitrogen at low temperature on hypercrosslinked polystyrenes formally meets the requirements of the BET method, justifying the calculation of the specific surface area of these adsorbing materials, and Table 7.8 summarizes some of the data [175].

Table 7.8   Apparent specific inner surface area (m^2/g) of hypercrosslinked networks (measured by the single-point BET method)

Cross-agent[a]	Initial polymer[b]	Degree of crosslinking, (%)						
		11	25	43	66	100	150[c]	200[c]
MCDE	I	0	240	650	1000	1000	–	–
XDC	I	0	0–100	500	850	1000	–	–
DMM	I	0	250	600	750	900	–	–
CMM	I	10	700	1000	1200	1200	–	–
CMDP	I	0	0–100	650	800	1000	–	–
CMB	I	0	0	0	0	0	–	–
MCDE	II, 0.17% DVB	0	–	150	490	770	1150	1200
MCDE	II, 0.3% DVB	0	130	–	1050	1350	1250	1400
MCDE	II, 0.6% DVB	0	0	0	200	900	1500	1800
MCDE	II, 1.4% DVB	0	0	0	60	450	1200	1600
MCDE	II, 2.7% DVB	0	0	0	0	480	850	1140

[a]XDC = p-xylylene dichloride; DMM = dimethoxymethane; CMM = tris(chloromethyl)mesitylene; CMDP = p,p-bis-chloromethyl-diphenyl; CMB = bis-(p,p-chloromethylphenyl)butane-1,4.
[b] Initial polymer, **I** = linear polystyrene of 3 × 10^5 Da, **II** = styrene–DVB copolymers.
[c]More than one cross-bridge to each phenyl group of polystyrene.
Reprinted from [175] with kind permission of Elsevier.

Permanent porosity is only characteristic of rigid highly crosslinked networks. Thus, the most rigid networks obtained by crosslinking linear polystyrene with the trifunctional tris-(chloromethyl)-mesitylene start to exhibit high porosity already at a moderate degree of crosslinking of 25%. The other networks incorporating less conformationally stiff bridges should be crosslinked to at least 40% to display a measurable surface area. With the number of rigid bridges increasing, the apparent specific surface area, S_{app}, increases and reaches very high values of $1000–1500\,m^2/g$, sometimes even $1800\,m^2/g$ (for the networks with a formal degree of crosslinking of 100–200%). Importantly, when polystyrene is crosslinked with really flexible bridges formed by bis-chloromethyl derivatives of diphenylbutane, the networks exhibit no porosity at all at any degree of crosslinking.

Compared to solutions of linear polystyrene, swollen styrene–DVB copolymers need a more intensive post-crosslinking to acquire typical features of hypercrosslinked materials. The strongly swelling in the reaction medium styrene–0.17% DVB copolymer will produce a low-density material having a specific surface area of $150\,m^2/g$ only at a post-degree of crosslinking as high as $X = 43\%$. The smaller is the swelling of the initial DVB-containing copolymers in EDC (the higher the concentration of polystyrene chains to be post-crosslinked), the larger is the amount of rigid bridges that have to be introduced between the polystyrene chains, in order to obtain networks with permanent porosity in dry state (Table 7.8). A gel-type copolymer with 2.7% DVB takes up a relatively small amount of EDC, and, therefore, polystyrene chains in this swollen gel locate rather close to one another. On removing the solvent they will easily achieve dense packing if the number of newly introduced crosslinks is insufficient. Still, being crosslinked to 100 or 200%, this copolymer too will result in materials with an exceptionally high inner surface. However, at any given degree of crosslinking, the surface area is always lower for networks based on the copolymers, especially those having increased DVB contents.

In contrast to the main tendencies known for truly macroporous copolymers, the dilution of initial linear polystyrene solutions was observed to have no dramatic influence on the apparent inner surface area of hypercrosslinked polymers. The surface area rather shows a tendency to decrease with dilution, contrary to the pore volume (Table 7.9).

The replacement of one good solvent, EDC, for another good solvent, nitrobenzene or ethylene tetrachloride, used to synthesize hypercrosslinked polymers, exerts no influence upon the apparent inner surface area of the resulting networks. However, the synthesis of hypercrosslinked

Table 7.9 The influence of the concentration of polystyrene solution on the values of the apparent specific surface area and total pore volume of the resulting hypercrosslinked networks

Crosslinking agent	X, (%)	C, (vol.%)	S_{app}, (m^2/g)	W_o, (cm^3/g)
XDC	43	6.9	475	0.83*
XDC	43	11.0	530	0.38*
XDC	43	15.2	550	0.19*
XDC	100	5.3	675	–
XDC	100	11.0	1000	0.62**
MCDE	66	3.7	–	0.54**
MCDE	66	4.6	–	0.50**
MCDE	66	6.0	–	0.48**
MCDE	66	10.2	–	0.42**
MCDE	66	17.4	–	0.36**

The total pore volume was measured by bulk weight (*) or calculated from the true and apparent densities (**).
Reprinted from [175] with kind permission of Elsevier.

materials in a poorer solvent, cyclohexane, results in products with significantly reduced surface area, the latter again being the smaller, the higher is the dilution of the starting polystyrene solutions (Table 7.10) [155].

The porous structure of hypercrosslinked polymers is rather stable and tolerates heating at 150°C for at least 1 h, that is, well above the glass transition point of polystyrene. After heating, the surface area does not change markedly (Table 7.11).

Similarly, treatment of polymers with various solvents followed by drying at elevated temperatures (Table 7.12) will also not affect noticeably the porous structure of the hypercrosslinked polymers. On the other

Table 7.10 The influence of dilution on the values of the apparent specific inner surface area (S_{app}, m^2/g) and total pore volume (W_o, cm^3/g) of networks prepared by crosslinking linear polystyrene with MCDE up to 100% in cyclohexane in the presence of 2 mol of SnCl$_4$ per 1 mol of MCDE

C, (vol.%)	S, (m^2/g)	W_o, (cm^3/g)	Cl, (%)
14.0	592	0.3	1.50
11.0	483	0.5	1.78
9.0	438	0.9	1.85
7.7	350	1.2	2.20

W_o was measured by bulk weight; Cl is the concentration of unreacted chlorine atoms in the final network, %.
Source: After [155].

Table 7.11 Heat resistance of hypercrosslinked polymers prepared in ethylene dichloride [133]

Network type, crosslinking agent	X, (%)	Specific surface area after heating, (m²/g)		
		80°C, 10 h	100°C, 1 h	150°C, 1 h
I, XDC	43	530	510	133
I, XDC	66	820	830	700
I, XDC	100	1000	1000	1050
I, MCDE	25	240	220	67
I, MCDE	43	640	650	460
I, MCDE	66	1000	1000	800
I, MCDE	100	1000	920	900
II, 1% DVB, MCDE	100	1000	980	1000

I and II = the networks based on linear polystyrene and styrene–1% DVB copolymer, respectively.

Table 7.12 The effect of pretreatment with solvents and the temperature of drying on the apparent specific inner surface area

Solvent	T_d, °C	Surface area, m²/g	
		I, XDC	II, MCDE
Toluene	125	700	940
n-Hexane	125	840	980
Water	125	950	1000
Water	80	720	1000

Hypercrosslinked networks were prepared by crosslinking linear polystyrene (**I**) with p-xylylene dichloride and styrene–0.7% DVB copolymer (**II**) with monochlorodimethyl ether; $X = 100\%$.
Source: After [133].

hand, the structure of many macroporous styrene–DVB copolymers is known to depend much stronger on the pretreatment protocol. Obviously, the passage from a swollen to dry state of the hypercrosslinked polymers, although connected with very large volume changes, should be regarded as totally reversible transition between two equilibrium states (swollen and dry) of the network structure.

The apparent specific surface area of hypercrosslinked polystyrenes is much higher than that of traditional macroporous styrene–DVB copolymers. However, when discussing this subject, we have to pay attention to the fact that the adsorption and desorption branches of the isotherm given in Fig. 7.26 do not coincide in the whole range of relative pressures, including the range of very small pressures. The sorption hysteresis at low p/p_s is known to be accounted for the swelling of the sorbent in sorbate vapors [177], rather than to adsorption–desorption inequivalence. Indeed, hypercrosslinked polystyrenes are mobile networks and are capable of markedly increasing their volume on taking up organic and inorganic sorbates [141, 176]. It will be shown below that these materials swell even with liquid nitrogen. Unfortunately, there are no experimental data indicating when the swelling starts, either immediately after sorption of the very first portions of nitrogen, that is, in the region of BET calculations, or only at high relative pressures when capillary condensation of nitrogen starts. It is evident, however, that the swelling has to cause some uncertainty in the surface area determination. We should also keep in mind that hypercrosslinked networks are single-phase materials, and so there is no real interface in the sorbent bead for nitrogen to adsorb. "Pores" have no walls; they simply represent voids between loosely packed polystyrene chains. For this reason we refer to the calculated specific surface area as "apparent inner specific surface area," rather than true surface area, and interpret the very large calculated values of S_{app} as a reflection of high sorption capacity of the hypercrosslinked sorbents to nitrogen.

With respect to the values of the apparent specific surface area, we have to note here that modern instrumentation permitting the measurement of nitrogen adsorption isotherms is usually provided with software permitting surface area calculation according to several mathematical models, including the BJH (corresponding to the names of Barrett, Joyner, Halenda) method, Langmuir adsorption isotherms, and the BET theory; these methods may provide for the same sample of a porous material, values that may differ by as much as 30%.

5.3 Pore volume of hypercrosslinked polymers

The total pore volume (W_o) of porous materials may be determined by several methods. It is believed that most precise values result from measuring the true, ρ_{tr}, and apparent, ρ_{app}, densities of the material, the specific pore volume being equal to $W_o^p = 1/\rho_{app} - 1/\rho_{tr}$. Data obtained in this manner for hypercrosslinked polystyrenes are given in tables 7.9 and 7.10. The apparent density presented there was measured with a mercury pycnometer, while the true density was arbitrarily set to equal the density of the isotactic polystyrene crystal, $1.12\,g/cm^3$ [178]. By chance, this value coincided with the value of ρ_{tr} measured in nitrogen for the product of post-crosslinking of styrene–0.3% DVB with MCDE to 100% (Table 7.7).

But, the whole notion of the "true" density and its measurement for porous materials in nitrogen (or, better, in helium) requires critical reconsideration. Indeed, density of atactic polystyrene is known to be $1.05\,g/cm^3$, as polymeric chains in that material are packed less densely than in isotactic polystyrene. In fact, the coefficient of chain packing in the atactic material is estimated to be as low as 0.674 [179]. This means that the free volume in the bulk polystyrene phase may amount to about $0.326\,mL/g$. This free volume, however, is not accessible to helium or nitrogen and is not considered as pore volume. The situation changes with the introduction of rigid crosslinks between the chains, which, being inserted into the swollen state of the network, rather function as spacers that gradually make the network loose and more permeable. The gradually increasing free volume of the material finally becomes accessible to gas molecules, which, paradoxically, will then manifest itself in *higher* measured values of the "true" density of the loose network. Thus, for a hypercrosslinked polystyrene, a true density value as high as $1.37\,g/cm^3$ was reported in the literature [180]. This situation resulted in a wrong impression that crosslinks enhance the density of the polymeric phase. In reality, rather the opposite is true: crosslinking (in the presence of a solvent) reduces the density of a network, but makes its free volume accessible to measurement by gas permeation. It would probably be better to define the true density of a polymer based on the weight of its repeating unit and the calculated van der Waals volume [181] of the latter.

The apparent density is also badly defined. It actually uses volumes of a polymer sample measured either by microscopy or by mercury pycnometry under 1 atm pressure. With pores large enough, these methods can provide different results. From this consideration it follows that the "most reliable" method of measuring the pore volume W_o^p, of a porous material, by comparing its apparent and true densities, bears serious uncertainty.

According to Table 7.7 the estimated pore volume, W_o^p, of hypercrosslinked polymers prepared from linear polystyrene is rather small. It increases with the degree of crosslinking, but does not exceed $0.5\,cm^3/g$ for rigid networks crosslinked with MCDE to 100%. For somewhat more flexible networks crosslinked with bis-chloromethylated diphenyl to the same extent, the value of the estimated pore volume is even smaller, $0.37\,cm^3/g$.

The method of mercury intrusion provides less exact, but still acceptable results for the pore volume: values of W_o^{Hg} were found to be slightly smaller than W_o^p.

Frequently the total pore volume of porous materials is also measured by adsorption of inert gases at a low temperature and relative pressures approaching unity. Correct results, however, are obtained only if the adsorption isotherm levels off in the vicinity of $p/p_s = 1$. Meeting this condition means that all pores of the sorbent are filled in with the liquefied sorbate, and the volume of the latter equals the sorbent's pore volume. Low-temperature sorption of nitrogen on hypercrosslinked polystyrene sorbents has always produced isotherms of the desired IV type (see Fig. 7.26). However, at any degree of crosslinking, values of $W_o^{N_2}$ prove to markedly exceed the porous volumes measured by mercury intrusion or calculated from the density of the materials, pointing to swelling of the hypercrosslinked network in liquid nitrogen. Only in the case of a degree of crosslinking as low as 25% will the method of densities give a larger value than that calculated from the sorption isotherm of nitrogen. At a low degree of crosslinking, polystyrene does not form expanded hypercrosslinked networks fully accessible to nitrogen.

As opposed to the apparent specific surface area, pore volume of hypercrosslinked polymers is strongly dependent on the conditions of their synthesis. Tables 7.9 and 7.10 demonstrate that the pore volume noticeably increases with increasing dilution of the initial polystyrene solution, both in EDC and cyclohexane.

5.4 Pore size and pore size distribution of hypercrosslinked polystyrenes

As documented above, the apparent inner surface area of the hypercrosslinked sorbents is exceptionally large, while the pore volume is relatively small. This suggests that the majority of the pores are small. If the pores were cylindrical in shape and equal in size, then for a typical polymer with an average specific surface area of $1000 \, \text{m}^2/\text{g}$ and pore volume of $0.5 \, \text{cm}^3/\text{g}$ their diameter calculated as $4W_o/S_{app}$ would be only 20 Å; this is the case with a sample crosslinked to 100% (tables 7.7 and 7.8). Since, on the one hand, the apparent surface area of hypercrosslinked polystyrenes is almost independent of the conditions of their preparation such as the molecular weight of the linear polystyrene precursor (in the range from 8.8×10^3 to $3 \times 10^6 \, \text{Da}$ examined) and the reaction media, and, on the other hand, the change in the pore volume with changing dilution of the starting solutions is not dramatic, the pore size of all hypercrosslinked products derived from linear polystyrene will remain small.

The apparent specific surface area of networks based on styrene–DVB copolymers is also very large. It increases up to $1800 \, \text{m}^2/\text{g}$ with the increase

Table 7.13 Comparison of apparent inner specific surface area (m^2/g) measured by low-temperature sorption of nitrogen and argon on hypercrosslinked polymers prepared by crosslinking a linear polystyrene having a molecular weight of 3×10^5 Da, in EDC solution with MCDE and p-XDC

X, (%)	MCDE		XDC	
	N_2	Ar	N_2	Ar
25	74	240	2	0
43	478	640	260	530
66	600	1000	440	820
100	740	1000	670	960

in the amount of a bifunctional crosslinking agent in the post–crosslinking reaction from 1.0 to 1.5 and 2.0 mol per 2 mol of phenyl groups in the initial polystyrene (with the respective formal degrees of crosslinking of 100, 150, and 200%). Extremely large values of S_{app} indicate the predominance of very narrow pores. In this connection, it is quite informative that nitrogen sorption provides smaller values of S_{app} than sorption of argon (Table 7.13). Judging by the values of the molar volumes of nitrogen and argon, 34.7 and 28.53 cm^3/mol, respectively, atoms of argon are smaller and, perhaps, penetrate into very narrow pores, inaccessible for the larger ellipsoid–like nitrogen molecules with dimensions of 4.1×3.0 Å (effective diameter 3.54 Å) [182]. (In addition, liquid argon shows a higher surface tension, 13.20 dyne/cm, as compared to 8.88 dyne/cm for nitrogen, which, however, should not distort the pore diameter calculations according to Eq. [3.4].)

From these data we have all reasons to expect that the size of the pores characteristic of all hypercrosslinked polystyrene networks will be in the range of few nanometers. Estimation of such small pore sizes presents special problems, particularly since the very term "pore" becomes a rather loose concept for an open framework structure. Let us now consider the results of pore size measurements of hypercrosslinked polystyrene by the available different experimental methods.

Low–temperature nitrogen adsorption. By taking capillary condensation as the reason for the hysteresis loops of the sorption isotherms (Fig. 7.26), it is possible to calculate the pore size distribution in accordance with the well-known Kelvin equation. This equation relates the relative pressure of vapor above a concave meniscus of liquid in a capillary to the curvature radius of the meniscus or, as a first approximation, to the radius of the capillary. In order to apply this equation to the calculation of pore size distribution in a polymeric sorbent, we have to assume that the porous structure of the material represents an ensemble of independent ("non-crossing") cylindrical capillaries with different diameters. For the single-phase open framework of a hypercrosslinked polymer, such model is far from reality, and the calculated pore size distribution function concerns only a model structure rather than the real polymer. We also should remember that the method totally ignores the chemistry of the surface and involves several other assumptions (many of them were discussed in Chapter 3, Section 1.1).

Figure 7.27 shows four plots of pore volume distribution as a function of the pore diameters: two plots were calculated from the desorption branch

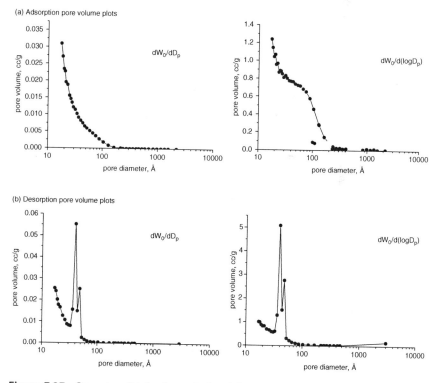

Figure 7.27   Pore size distribution calculated from (a) adsorption and (b) desorption branches of the sorption isotherm (see Fig. 7.26) of nitrogen on the hypercrosslinked polymer prepared by post-crosslinking styrene–0.3% DVB copolymer with MCDE to 100%. (Reprinted from [175] with permission of Elsevier Publishing Company.)

of the hysteresis loop of the nitrogen sorption isotherm given in Fig. 7.26, while the other two were obtained from the adsorption branch of the same isotherm. For each pair, the cumulative pore volume plots were differentiated either over the pore diameter or the logarithm of the pore diameter. All four ways of presenting pore size data are common, but it immediately becomes obvious from Fig. 7.27 that the four plots generate different impressions of the porous structure of the same polymer. Both the position of the maximum and the balance between smaller and larger pores depend on the way of calculation and plotting the data. Unfortunately, the comparison of results presented by different authors most often overlooks this fact.

Theoretically the use of the desorption branch of the isotherm is applicable for materials with pores predominantly comprised of independent capillaries, because this was the system used for the deduction of the Kelvin equation. If bottle-like pores are anticipated, where the liquid in the narrow opening prevents evaporation of the condensate from larger compartments, it is better to calculate pore size distribution using the adsorption branch of the hysteresis loop. Since the shape of pores is usually *a priori* unknown, the selection of the adsorption or desorption branch of the hysteresis loop largely remains arbitrary. Concerning the choice between dW_o/dD or $dW_o/dlog(D)$ plots, which by definition are different functions, one may note that the difference may be small for materials with narrow pore size distribution, but become more significant for a broad and unsymmetrical pore size distribution.

In the case of Fig. 7.27 the dW_o/dD plot obtained from the adsorption branch characterizes a polymer with wide pore size distribution, with pores ranging from 17 to about 200 Å. The same plot calculated from the desorption branch would indicate narrow pore size distribution, with pores having diameters from 17 to 60 Å with two pseudo-maxima located between 40 and 50 Å. A similar shape of pore distribution results from the $dW_o/dlog(D)$ presentation of the desorption branch, while the $dW_o/dlog(D)$ plot evaluated from the adsorption branch would indicate a wider pore size distribution, with pore diameters from 17 to 300 Å and a noticeable predominance of pores with diameters of 30–100 Å. Thus, the desorption branch of the isotherm points to a narrower pore size distribution compared to results of the analysis of the nitrogen adsorption isotherm of the same polymer sample.

Mercury intrusion. Mercury porosimetry represents another popular method for the characterization of porous structure. Its use is fully justified for rigid inorganic adsorbents, but may lead to artifacts with relatively soft polymeric porous materials. Thus, regardless of the type and amount of crosslinking bridges and the concentration of polystyrene chains in the initial solution for the post-crosslinking, all hypercrosslinked polystyrene samples seem to exhibit "pores" with diameters ranging from 100 to 2000 Å, with a maximum located around 200–300 Å. Undoubtedly, no such big pores exist in our single-phase transparent polymers and the mercury intrusion method obviously provides here incorrect information. Indeed, when applying pressure to a system consisting of mercury and immersed polymeric beads, there is no way to distinguish between the intrusion of mercury into the pores of the beads and simple compression of the porous material; we just cannot know what we measure. In all probability, the "pore" diameters calculated in this way correspond to the pressure causing compression of the expanded hypercrosslinked network that was found to be rather mobile and susceptible to mechanical forces (see Section 8).

Inversed size-exclusion chromatography. To characterize the porous structure of hypercrosslinked polystyrenes by chromatographic porosimetry, a resin, swollen in chloroform, was packed into a column (for details see [183]); diluted solutions of polystyrene standards of known molecular weight M and narrow molecular weight distribution were injected, and the retention volumes V_e of each standard were measured. From the calibration plot of $\log M$ versus V_e, the molecular weight (size) of the polystyrene coils excluded even from the largest pores of the resin phase may be easily determined. The entire pore-size distribution curve may be obtained by a graphical differentiation of K_d against D_c (D_c and K_d being the diameter of the coils and the coefficient of phase distribution of the coils, respectively), assuming $K_d = 1$ for toluene and acetone molecules and $K_d = 0$ for the excluded coils. The results obtained are presented in Table 7.14. Attention should be paid to the fact that, contrary to mercury intrusion and nitrogen adsorption techniques, inversed size exclusion chromatography examines the porous structure of the maximally swollen material.

In spite of the different types of crosslinking agents, the degree of bridging, and the concentration of polystyrene chains in the initial solution, hypercrosslinked polymers exhibit narrow pore-size distribution; the diameter of the majority of the pores is about 10 Å. The exclusion limit of the polymeric gels is rather small; polystyrene standards with coils of 50–60 Å in diameter in chloroform solution (molecular weight of 12,000–16,000 Da) are completely excluded from the polymer phase. It should be emphasized,

Table 7.14 Conditions of synthesis and exclusion limits of hypercrosslinked polystyrene networks

Initial polymer	Crosslinking agent	X, (%)	Solvent	$C_o{}^a$, (% w/vol)	$K_{sw}{}^b$, (mL/g)	Exclusion limit	
						$M_w \times 10^{-3}$	D, (Å)
I	MCDE	18	EDC	11	2.53	16	65
I	MCDE	43	EDC	11	2.99	12.5	57
I	MCDE	100	EDC	16.9	2.56	14.5	62
I	MCDE	100	EDC	11	3.55	10.5	51
I	MCDE	100	EDC	5.9	4.79	126	214
I	MCDE	43	CH	11	4.98	300	354
I	MCDE	66	CH	11	4.72	190	272
I	MCDE	100	CH	11	4.69	160	247
I	XDC	43	EDC	10	2.19	20	74
I	XDC	100	EDC	9.6	1.89	20	74
II	MCDE	25	EDC	–	1.72	4	29
II	MCDE	100	EDC	–	1.81	10	50

^{a} C_o = polystyrene concentration in the mixture of initial reagents.
^{b} K_{sw} = weight swelling in chloroform.
II = styrene–1% DVB copolymer as initial polymer.
Source: After [183].

however, that the $K_d(D_c)$ plot presents the distribution of K_d over dimensions of the polystyrene coils, rather than over the pore size, and the very important question of how pore size is related to the diameter of a polymer coil that can enter the pore still remains to be answered. If the diameter of the pores has to exceed the diameter of the coils by a factor of 2–2.5, as was suggested in [184–186], then the dominating pores of our swollen hypercrosslinked polystyrene would have a diameter of about 20–25 Å, with the largest pores being smaller than 100–150 Å.

Gorbunov et al. [187] suggested a scheme of processing the inversed size-exclusion chromatography data that are based on presenting K_d as a function of the ratio of coil gyration radius to pore radius. They considered the pore's surface-to-volume ratio as a universal and model-independent pore characteristic, and derived mathematical functions aiming at the calculation of the pore size in the maxima of pore volume distribution, D_v, and pore surface distribution, D_s, as well as the polydispersity parameter, U, characterizing the width of distribution. We used our experimental data to perform calculations according to this approach; the obtained results are given in Table 7.15. The hypercrosslinked networks

Table 7.15 Chromatographic porosimetry data according to Gorbunov model

Network type	X, (%)	C₀, (vol.%)	Porosimetry data		
			D_{vol}, (Å)	D_{surf}, (Å)	U
I, MCDE	18	11.0	52	46	1.13
I, MCDE	43	11.0	46	46	1.0
I, MCDE	100	16.9	45	36	1.26
I, MCDE	100	11.0	48	34	1.43
I, DMM	43	11.0	46	44	1.05
I, DMM	100	11.0	53	43	1.25
I, XDC	43	10.0	56	41	1.38
I, XDC	100	9.6	53	31	1.72
I, CMB	66	11.0	54	43	1.27
II, MCDE	25	–	27	27	1.03
II, MCDE	100	–	48	36	1.36

Reprinted from [175] with kind permission of Elsevier.

now seem to have the maximum of pore size distribution at 40–50 Å, with no pores larger than 120 Å. These values seem to us to be overestimated.

Graphical differentiation of the $K_d(D_c)$ function demonstrated that unlike the hypercrosslinked networks obtained in EDC, the networks prepared in cyclohexane display bimodal pore size distribution. In addition to the pores of the same size as those in the polymers obtained in EDC, these polymers have larger pores with diameters of up to 350 and 250 Å for the samples with degrees of crosslinking of 40 and 100%, respectively. The formation of such large pores appears to be a sign of phase separation during network preparation. Indeed, these polymers do not look transparent any more.

Both the graphical differentiation of the K_d versus D_s function and the Gorbunov's model testify to narrow pore size distribution in hypercrosslinked polystyrene prepared in a concentrated solution in a good solvent (Fig. 7.28 and Table 7.15). It implies that crosslinks are homogeneously distributed throughout the whole volume of these networks. When in swollen state, these polymers have micropores to small mesopores, and as the polymers absorb large amounts of the mobile phase (chloroform), the number of these pores must be really large. Obviously, the solvation involves the whole polymer network rather than the individual domains of its structure; in the latter case, the strong swelling of the polymers would require presence of large pores or channels. We should also conclude that porous hypercrosslinked networks possess a single-phase structure, except those obtained in cyclohexane and, maybe, in diluted EDC solutions.

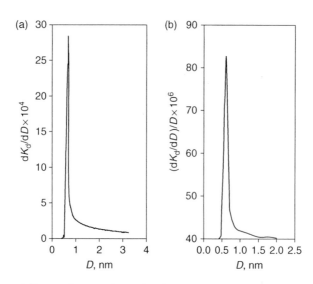

Figure 7.28   Differential pore volume distribution over the diameters of polystyrene-standard coils, D, for the network prepared by crosslinking linear polystyrene with p-xylylene dichloride to 43% crosslinking degree; (a) volume distribution and (b) numerical distribution. *(After [183]).*

Note that the homogeneous structure of swollen polymers crosslinked with flexible bis-(p,p-chloromethylphenyl)butane-1,4 bridges (which are non-porous when in dry state) is very similar to that of swollen typical nanoporous hypercrosslinked polymers.

In contrast to these results, Jeřabek and Setinek [188], Jeřabek [189], employing Ogston's model [190] for chromatographic porosimetry, arrived at the conclusion of an extremely heterogeneous distribution of crosslinks in the network prepared by crosslinking linear polystyrene with XDC to 66%, as well as in two products of its partial and complete sulfonation. Strangely, the distributions were found to be distinctly different for these tree variants of the same matrix. However, we believe that this conclusion results from an incorrect mathematical treatment of scattered experimental points, rather than from real changes in the pore size distribution of the hypercrosslinked polymer matrix on the functionalization of the latter.

The major problem in inversed size-exclusion chromatography is the arbitrary judgment that a pore must be much larger, in order to allow a polymeric coil to enter it. On the other hand, large and flexible coils of polystyrene have been reported to penetrate into narrower pores, due to the change of their spherical conformation and reptation-type movements

[191] into small channels. All these uncertainties substantially reduce the value of inversed size-exclusion chromatography for the determination of sizes of pores. One also has to acknowledge that this technique does not cover possible bottle-type pores of the material. Larger voids remain inaccessible to polymeric test molecules if the latter fail to pass to the voids through narrower channels. This fact alone makes the pattern of the pore size distribution obtained by chromatographic techniques substantially different from those estimated from nitrogen adsorption or mercury intrusion.

Annihilation of positronium. Positron annihilation lifetime (PAL) spectroscopy represents still another method for testing elementary free volumes in polymeric materials. Positronium is a bound atomic system consisting of an electron (e^-) and positron (e^+). It tends to be localized in domains with lowered electron density and annihilates when encountering the electrons of surrounding molecules that belong, for example, to the pore walls. The lifetimes of positronium and the corresponding intensities of annihilation depend on the size and concentration of elementary free volumes, respectively, and, therefore, PAL spectroscopy may be used to examine the size distribution of very small voids in the polymeric material. As the study of Shantarovich et al. [192] showed, in the dry hypercrosslinked polystyrene incorporating 66% of diphenylmethane-type links in addition to the initial 0.3% DVB, in parallel with small holes of 2–3 Å in radius (that are inherent to the initial copolymer), larger holes of 5 and 14 Å in radius appear. The concentration of the three types of holes is estimated as 7.3, 4.5, and $2.3–3.0 \times 10^{19}$ cm^{-3}, respectively. With the degree of crosslinking rising to 100%, the largest lifetime and corresponding intensity remain practically unchanged, signifying that the size of pores in the networks with 66 and 100% degree of crosslinking may be of the same order of magnitude.

Miscellaneous techniques. Recently the rate of evaporation of water or other liquids from a wetted porous material under diffusion-controlled conditions was shown to depend on the size of pores of the material [193]. This fact was interpreted in terms of the dependence of the equilibrium vapor pressure above the sample on the radius of the meniscus of the liquid in the pores. This technique is sensitive and relatively simple. However, estimation of the pore radius from the rate of evaporation is again based on the use of the Kelvin equation in a wide range of relative pressures and, therefore, inherits all the uncertainties of the technique and incorporates some additional assumptions. This technique reveals two maxima in the pore size distribution of the water-wetted hypercrosslinked material prepared from a styrene–0.3%

DVB copolymer with $X = 100\%$. An approximate evaluation locates the maxima at diameters of 40 and 200 Å. No other technique indicates an enhanced proportion of pores of 200 Å in that material.

Water-swollen hypercrosslinked polystyrene networks were also tested by a relatively new structural method, pulse broad band ^{1}H nuclear magnetic resonance (NMR) of water protons. This method registers the appearance and growth of water proton mobility on slow heating of a preliminary deep-frozen polymer–water system from very low to room temperature. In accordance to theoretical considerations [194], the depression of the temperature of freezing (thawing) of water standing in a thin capillary is reciprocally proportional to the diameter of the latter. This has made possible the evaluation of pore diameters by registering the intensity of proton signals in a thawing system. The method proved to be very sensitive to the smallest differences in the structure of very similar polymers. As an example, Fig. 7.29 demonstrates the change in relative integral intensities of ^{1}H NMR water signals for two laboratory and two industrial samples of hypercrosslinked polystyrenes, all of them having the same number of chemically identical bridges.

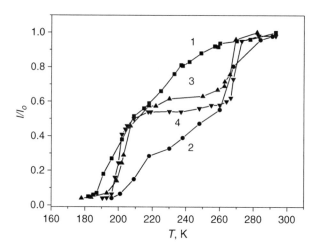

Figure 7.29 Relative integral intensity (normalized to the intensity at 298 K) of NMR ^{1}H H$_2$O signals on heating pre-frozen water-swollen hypercrosslinked polystyrenes from 178 to 298 K; (1) styrene–1.4% DVB copolymer crosslinked with MCDE, $X = 200\%$; (2) styrene–0.17% DVB copolymer crosslinked with MCDE, $X = 200\%$; (3, 4) MN-270 samples taken from two different batches.

Finally, it should be recalled here that hypercrosslinked materials display high permeability and quickly arrive at adsorption equilibrium. Thus, a hypercrosslinked sulfonated cation exchanger prepared on the base of linear polystyrene will readily sorb tetrabutyl ammonium ions having a hydrated radius of 4.94 Å [195]. For 0.5 mm particles, no more than 1.5 min are required to substitute half of the functional groups with these voluminous ions [196]. The fast diffusion suggests that the majority of pores of the material are larger than 10 Å. On the other hand, in the novel technique of frontal size-exclusion chromatography of mineral electrolytes [197], large ions of Al^{3+} and Fe^{2+} were found to elute from the column almost with the interstitial volume. The radii of these hydrated ions was estimated [195] as 4.75 and 4.28 Å, respectively. Therefore, the majority of pores of the hypercrosslinked polystyrene column packing material (based on a styrene–DVB copolymer) should not exceed 15 Å.

In conclusion, it should be pointed out that none of the physico-chemical techniques discussed above permits the direct measurement of the elements of the polymeric materials' porous structure; we measure the properties of the systems where the polymers interact with certain test substances (nitrogen, mercury, water, polystyrene standards, ions, etc.), and not the dimensions of the pores or other supramolecular elements of the material. Therefore, the evaluation of the surface area and diameters of pores available to the molecules of these substances must be considered as indirect methods of examining the porous structure. Because of this, all calculations are based on assuming certain models of the structure of the material and accepting certain assumptions as to the mechanism of interaction between the material and test molecules. Only transmittance, scanning, and, in particular, atomic force microscopy can be considered as direct methods of measuring dimensions and distances. However, up to now the last technique has not been applied to microporous hypercrosslinked polymers.

Hypercrosslinked polystyrene is the first representative of a new class of polymeric networks and materials, displaying a special type of porosity. The indirect techniques mentioned above, being applied to probe the porous structure of the sorbents, produced rather contradicting results. Neither of the methods proved sufficiently sensitive or precise to reveal any dependence of the pore size on the degree of crosslinking (in the range of 50–200% examined), the type of the crosslinking agent, and the concentration of starting polystyrene chains under post-crosslinking.

Nevertheless, some generalized conclusions may be drawn at this stage. Hypercrosslinked polystyrene sorbents represent basically microporous single-phase materials, the apparent specific surface area of which is exceptionally high (up to $1000-1500 \, m^2/g$), while the pore volume is relatively small (up to $0.5 \, cm^3/g$). All methods testify to narrow pore size distribution, implying a rather uniform distribution of crosslinks throughout the network volume. The size of the pores is small, probably around $20 \, \mathring{A}$. Dry hypercrosslinked polystyrenes contain no noticeable portion of pores larger than $30 \, \mathring{A}$ (provided that they have not been intentionally introduced by special means). The size of pores slightly increases in swollen networks. Since the size of the pores is just a few nanometers, it is not incorrect to call the new materials as nanoporous. Nanoporous polymers have not been known before, only macroporous and mesoporous. Though the traditional classification recommended by IUPAC sets a border between micropores and mesopores at about $20 \, \mathring{A}$ (2 nm), thus leaving no room for nanopores, the term "nanoporous structures" is being frequently used in the modern scientific literature.

6. MORPHOLOGY OF HYPERCROSSLINKED POLYSTYRENES

Porous polymeric adsorbing resins, particularly styrene–DVB copolymers, have a complex texture which is well known to result from the microphase separation of an inert diluent from the polymer phase during the polymerization of the initially homogeneous solution composed of the comonomers and the diluent. Two types of diluents and, correspondingly, two reasons have been recognized for the phase separation.

One type of the porogens is non-solvents for the growing polymer chains; usually this is an aliphatic hydrocarbon or alcohol. These liquids barely solvate the styrene–DVB polymeric species forming one by one; therefore, the latter shrink with the gradual disappearance of the monomer and finally precipitate from the reaction solution in the form of spherical micro-nodules. This microphase separation often proceeds before the point of gelation of the initial solution. It is also facilitated by the fact that the initially formed intramolecularly crosslinked copolymer nodules preferentially incorporate the more reactive DVB molecules and acquire an enhanced degree of crosslinking. Subsequent aggregation and, consequently, agglomeration of the primary non-porous nodules, accompanied by the segregation of the diluent between them, lead to the formation of a typical cauliflower texture of the final macroporous product.

Thermodynamically good solvents such as toluene represent the second type of diluents. In this case the growing chains remain solvated throughout the whole time of polymerization. However, their swelling ability may prove insufficient to accommodate all the diluent present in the system. At certain threshold DVB proportions and toluene contents, microphase separation occurs, because of the segregation between the swollen microgels and excess solvent. The final heterogeneous material forms again through aggregation and agglomeration of the primary swollen species and, therefore, consists of macropores (filled with the diluent) and swollen polymeric phase that has an "expanded network" structure. As a result, the overall texture of the macroporous copolymer prepared in a good solvent is similar to the cauliflower texture obtained in a precipitant, with the difference that the primary polymeric nodules in the former material must be microporous, while in the latter they are non-porous.

   Hypercrosslinked networks are prepared from preformed, long polystyrene chains in solution (or in a highly solvated gel of styrene–DVB beads) by introducing numerous rigid bridges between the chains. The initial polymeric chains extend though the whole initial solution (or gel) and remain strongly solvated over the whole period of network formation. The crosslinking bridges emerge statistically throughout the entire volume of the system, thus converting it into a homogeneous rigid network swollen with the solvent. One must expect that these materials are microporous with a rather narrow pore-size distribution. Examining the hypercrosslinked materials by nitrogen adsorption measurement, inversed size-exclusion chromatography, and ^{1}H NMR technique supports this suggestion, as was discussed in the previous section. Additional information can be obtained by electron microscopy and small-angle X-ray scattering technique.

6.1 Investigation of polymer texture by electron microscopy

Electron microscopy represents the only direct method that permits to see "with our own eyes" the interior of a sample with a resolution of a few nanometers. To characterize the texture of hypercrosslinked polystyrenes both transmission and scanning electron microscopy have been applied. In the former case, ultrathin sections with a thickness of about 600 Å were cut from a sample fixed in epoxy resin, and then directly examined in transmission mode. Alternatively, two-step replicas have been prepared from the cleavage face. To prepare the replica the surface was first coated with a collodion film applied from amyl acetate solution, and then with a carbon–platinum film. Finally, the collodion support was dissolved and the free carbon–platinum replica examined under a transmission microscope.

In the case of scanning electron microscopy, a fresh cleavage face was coated with gold and examined.

Electron microscopy has revealed obvious differences in the internal structure of hypercrosslinked networks derived from dissolved linear polystyrene and those derived from swollen styrene–DVB copolymers [198]. Both the highly porous hypercrosslinked materials and the non-porous slightly crosslinked (macronet isoporous) polymers obtained from linear polystyrene display a characteristic supramolecular internal structure composed of well-defined spherical species (Fig. 7.30). Their size is nearly independent of the degree of crosslinking or the type of cross-linking agent. One can recognize these spherical species having a diameter of no less than 200 Å on the two-step replicas (Fig. 7.30a and 7.30b) as well as in the ultrathin sections. However, on the scanning electron microscopy micrographs of sample cleavage surface, many

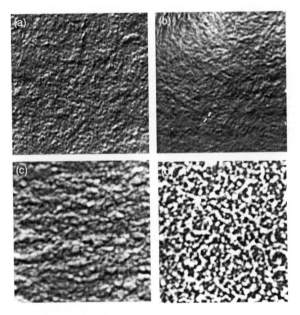

Figure 7.30 Electron macrographs of the networks prepared by post-crosslinking linear polystyrene ($M = 300,000$ Da) with (a, b) p-xylylene dichloride: (a) $X = 100\%$, (b) $X = 43\%$; and (c, d) monochlorodimethyl ether: (c) $X = 100\%$, (d) $X = 5\%$; (a, b) two-step replicas, transmission electron microscopy, 46,600×; (c, d) scanning electron microscopy, (c) 40,000×, (d) 100,000×. *(After [198])*.

smaller spherical species with approximately $100\,\text{Å}$ diameter (Fig. 7.30c) can also be recognized.

A distinctly different structure is characteristic of hypercrosslinked polymers on the basis of styrene–2% DVB copolymers. Crosslinking of this fully homogeneous copolymer with MCDE to 100% induces no signs of microphase separation or formation of any elements of a supramolecular structure. The interior of the beads appears totally homogeneous both before and after the intensive crosslinking (Fig. 7.31a and 7.31b), although the final material becomes highly porous, with the apparent inner specific surface area being as high as $1000\,\text{m}^2/\text{g}$. Similarly, the heterogeneous structure of another initial material, a popcorn copolymer, remains unchanged after the additional intensive bridging (Fig. 7.31c and 7.31d). Obviously, while introducing microporosity, post-crosslinking causes no change in the overall structure of the initial copolymer, and the

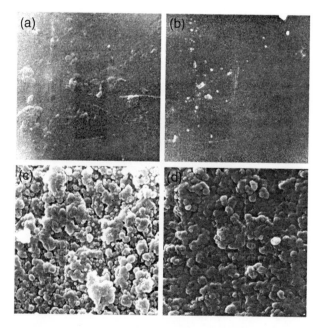

Figure 7.31 Electron micrographs of (a) gel-type styrene–2% DVB copolymer and (b) the product of its post-crosslinking with monochlorodimethyl ether to 100%; (c) popcorn-containing gel-type styrene–1% DVB copolymer; and (d) the same polymer crosslinked with monochlorodimethyl ether to 100%; scanning electron microscopy, (a, b) 10,000×, (c, d) 100,000×. *(After [198]).*

morphology of the final network is determined only by the morphology of the starting polymer–solvent system. Therefore, the globular relief on the cleavage surfaces of polymers based on linear polystyrene can be assumed to reflect the microstructure of the initial concentrated polystyrene solutions. The structure of the solution could become fixed by the randomly introduced crosslinks and thus made accessible to observation by microscopy or other techniques. Since the size of the spherical species is small, they likely correspond to individual coils of dissolved polystyrene macromolecules. This assumption is strongly supported by the fact that with the molecular weight of the starting polystyrenes increasing from 3×10^5 to 3×10^6 Da, the diameter of the smallest visible spheres increases from 100 to 450 Å (Fig. 7.32).

According to simple calculations, the diameter of a dry densely packed polystyrene species with a molecular weight of 300000 Da and density of $1.05 \, \text{g/cm}^3$ corresponds to 98 Å. Intensive crosslinking of such species in solution introduces $0.5 \, \text{cm}^3/\text{g}$ of voids in the dried material, and, therefore, the diameter of the dry porous coils should increase to 113 Å. As a rule, the dimensions of spherical species observed on the cleavage exceed this

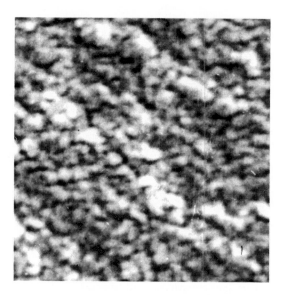

Figure 7.32 Texture of the network prepared by post-crosslinking linear polystyrene of 3,000,000 Da molecular weight with monochlorodimethyl ether to 100% in 6.7% ethylene dichloride solution; scanning electron microscopy, 40,000×. *(After [198])*.

calculated value, suggesting that association of several coils could be possible. However, when comparing the calculated and observed dimensions, one should take into account several circumstances. First, an insufficient contrast of micrographs complicates the measurement of smallest species. The initial polymer was highly polydisperse. The smallest macromolecules can be easily overseen on the micrographs, while the larger species show up much better, thus shifting the average size of the observed species toward larger values. Second, the methods used for sample preparation undoubtedly distort the real structure, since the hypercrosslinked polymers quickly swell during the application of the collodion solution and even during fixation in the epoxy resin. These effects lead to overestimated sizes of the primary components of the material's structure.

Taking into account the above remarks we are inclined to believe that what we see on the cleavage surface of samples based on linear polystyrene is the image corresponding to individual polystyrene coils in the initial polymer solution. In all probability, the polymeric coils in the concentrated solution maintain their identity to a certain degree and overlap only partially with each other. The disintegration of the final hypercrosslinked material then predominantly proceeds via rupture of weaker zones between the individual coils. Naturally, we have to add here that this assumption contradicts with the generally accepted classical theory of polymer solutions, according to which the coils maintain in solution their unperturbed or even expanded dimensions and, therefore, are totally overlapped with many neighbor coils.

Still, our conclusion on the restricted overlapping of individual polystyrene coils in the hypercrosslinked network and in the initial polymer solution was further corroborated by a direct experiment, namely by the preparation of a hypercrosslinked material deliberately constructed from individual polystyrene macromolecules. The initial linear polystyrene was first crosslinked intramolecularly in a very dilute solution and then the individual intramolecularly crosslinked coils, nanosponges, were additionally crosslinked intermolecularly in a concentrated solution, thus giving a bulk hypercrosslinked material. Both the nanosponges and the final polymeric material were subjected to electron microscopic investigation.

Soluble intramolecularly hypercrosslinked polystyrene (nanosponges) represent a fundamentally new macromolecular object that will be described in detail in Chapter 8. For the above two-step modeling experiments, nanosponges were prepared from the industrial linear polystyrene having a molecular weight of 300 000 Da and a wide molecular weight distribution, that is, from the same starting material as used for the direct one-step synthesis of hypercrosslinked networks. According to the first protocol, a dilute polymer solution (0.5–1.0%) in EDC was reacted with MCDE (0.4–0.5 mol MCDE per styrene repeating unit) at 80°C for 50 h in the presence of stannic tetrachloride (1 mol per mole of MCDE). The crosslinking agent was found to be completely consumed in the reaction with polystyrene. However, the product contained 0.8–1.2% of pending chlorine atoms. The pendent residual chloromethyl groups were assumed to mostly reside on the surface of nanosponges, and they were eliminated by adding toluene to the reaction mixture and heating it for an additional 20 h. The product was precipitated with methanol as a thin powder containing no unreacted chlorine any more. When in dry state, the product exhibited a specific surface area of $1000 \, m^2/g$ (1 g of polystyrene spheres with the diameter of 113 Å would display an outer surface area of only $510 \, m^2$). The polymer swelled with methanol by a factor of about 3.

The average dimensions of the intramolecularly crosslinked coils were expected to amount to 113 Å when in dry state and 162 Å when swollen by a factor of 3. Transmission electron microscopy visualized individual microspherical particles of about 180 ± 30 Å, whereas small-angle X-ray scattering by the nanosponge powder gave an average diameter of 120 Å. When examined in chloroform solution by size-exclusion chromatography using styrogels as the separating medium, the diameter of the species was found to be about 150 Å. Thus, all three methods showed reasonable correlation between the measured and calculated dimensions of the nanosponges, confirming their individuality.

Finally, consolidation of the individual nanosponges into a block of hypercrosslinked material was achieved with MCDE in a 3–9% solution in EDC according to a standard protocol. Data on swelling and apparent surface area of the polymers obtained are given in Table 7.16. Neither the concentration of the nanosponge solutions nor the degree of additional crosslinking affects the swelling ability of the bulk three-dimensional material. It absorbs on average 5.5 mL/g of any organic solvent, both good and bad for the linear precursor. This value significantly exceeds the swelling of hypercrosslinked networks, approximately 4 mL/g, obtained through crosslinking the same linear polystyrene in a 12.5% w/v solution with MCDE to 40–100%; however, it is not a surprising result. The nanosponges were crosslinked at a much lower concentration. Besides, while being rigid species, they cannot be consolidated into a voidless homogeneous material in the second step of intermolecular crosslinking.

Table 7.16 Properties of networks based on the intramolecularly hypercrosslinked polystyrene

X, during the second step, (%)	C_o, (vol.%)	Swelling, (mL/g)		S, (m²/g)
		Toluene	n-Hexane	
25	2.8	5.6	5.5	750
25	3.9	5.8	5.0	900
25	4.8	5.9	5.5	1000
43	4.8	5.4	4.9	800
66	4.8	5.9	5.0	1500

The morphology of the product prepared by combining nanosponges in an entire network with 0.1 mol MCDE in a 9% EDC solution proved to be much like the morphology of the usual hypercrosslinked polymer prepared in a one-step procedure from a 12.5% solution. In both cases the diameters of the visible spherical species amount to 250–350 Å (Fig. 7.33). Due to the above-discussed insufficient resolution of the technique and swelling during the sample preparation, the identity of spherical species comprising the morphologies of the two products – although larger than

Figure 7.33 Texture of three-dimensional material based on hypercrosslinked nanosponges; additional crosslinking degree with monochlorodimethyl ether is 11%; ultrathin section, transmission electron microscopy, 42,000×. *(After [198]).*

expected – strongly supports our doubts on the classical description of concentrated polymer solutions and speaks for the compressed conformation of macromolecular coils in real solutions.

Thus, electron microscopy did not reveal any visible pores in porous hypercrosslinked materials obtained from linear polystyrene or styrene copolymers in a good solvent. Consequently, the size of these pores is smaller than the resolution of electron microscopy. Also, this technique did not reveal any supramolecular constructions, the appearance of which could be attributed to microphase separation. Hence, hypercrosslinked polymers prepared in thermodynamically good solvents, such as EDC, have to be considered as single-phase materials. (As will be shown in Section 7, microphase separation can occur during hypercrosslinking in a poorer solvent, cyclohexane.)

6.2 Investigation of hypercrosslinked polystyrenes by small-angle X-ray scattering

The method of small-angle X-ray scattering (SAXS) is known to be very sensitive to any type of structural heterogeneities in both dry and swollen polymers, and it was thought to give useful information on the structure of hypercrosslinked polystyrenes [199]. Figure 7.34 compares the

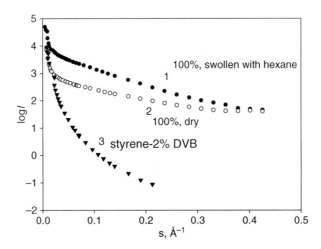

Figure 7.34   Plots of small-angle X-ray scattering (logarithm of intensity vs modulus of scattering vector $s = 4\pi \sin \varphi/\lambda$), normalized on the thickness of a sample and attenuation for (1, 2) the network obtained by post-crosslinking styrene–2% DVB copolymer with monochlorodimethyl ether to 100% and for (3) initial copolymer; (1) the sample swollen in n-hexane and (2, 3) dry samples.

intensities of X-ray scattering from three polymer samples: dry beads of a gel-type styrene–2% DVB copolymer (plot 3), dry hypercrosslinked network with $X = 100\%$ (plot 2) based on this copolymer, and the same hypercrosslinked polymer swollen in n-hexane (plot 1). SAXS from the gel-type copolymer points to the presence of rather large scattering spherical domains with a diameter of 500–600 Å. Since the volume portion of this heterogeneity in the beads is negligible, only 0.005–0.01%, it most likely represents traces of the popcorn-type polymer. Thus the overall structure of the initial styrene–2% DVB copolymer can be considered to be completely homogeneous, which is also in full agreement with the electron micrograph presented earlier (see Fig. 7.31).

The introduction of 100% rigid brigs in the homogeneous styrene–2% DVB copolymer results in obtaining a highly porous material, which nevertheless maintains a homogeneous structure, according to electron microscopy. The scattering ability of this hypercrosslinked polymer increased, but the character of the SAXS curves for both the dry and swollen material remained typical for dispersed systems (in terms of difference in the electron densities). A mathematical treatment of the plots points to the appearance of 3% (by volume) of scattering heterogeneities in the dry hypercrosslinked sample, while their volume increases to 10% in the swollen polymer. These values are unrealistically small, compared to the real volume of voids introduced into the copolymer by the post-crosslinking and by subsequent swelling of the product. More important is that the shape of X-rays scattering testifies to an extremely narrow pore size distribution.

In spite of the stronger X-ray scattering ability the hypercrosslinked polymers do not fit into the category of classical heterogeneous materials that, by definition, must comprise of pores and polymeric phase. This statement finds a strong support in Fig. 7.35, representing the X-ray diffraction under large angles for the starting and the post-crosslinked samples. The dry non-porous styrene–2% DVB copolymer exhibits two amorphous halo corresponding to $d_1 = 4.35$ Å and $d_2 = 8.64$ Å. Both are characteristic of atactic polystyrene and its copolymers. Contrary to this only one broad halo with $d = 4.85$ Å appears for the X-ray diffraction on both the dry and swollen hypercrosslinked network. This indicates the disappearance of the typical short-range ordering in the disposition of the phenyl rings. In other words, the polystyrene phase as such disappeared in the material after the post-crosslinking, and it has been converted entirely

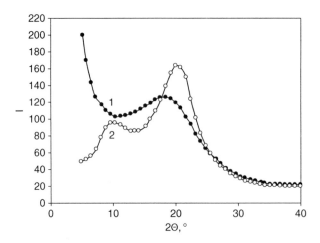

Figure 7.35   Diffraction patterns under large angles for (1) the dry hypercrosslinked network obtained by post-crosslinking styrene–2% DVB copolymer with monochlorodimethyl ether to 100% and (2) for the conventional copolymer precursor.

into a fundamentally different hypercrosslinked phase material. Thus, the SAXS method corroborates once more the assumption that the high porosity of hypercrosslinked networks does not result from microphase separation, which would disintegrate rather than completely abolish the polystyrene phase. We have all reasons to assume that hypercrosslinked polystyrene is basically a single-phase, isomorphous material, and it also remains isomorphous in the swollen state. The 3–10% heterogeneities do not constitute a different phase, but merely result from some other types of density fluctuations.

7. BIPOROUS HYPERCROSSLINKED POLYSTYRENE NETWORKS

The basic tendencies of hypercrosslinked network formation in media having high thermodynamic affinity to polystyrene (e.g., EDC or nitrobenzene) are also obeyed when the post-crosslinking of polystyrene chains proceeds in a poorer solvent, such as cyclohexane [152]. Thus, the data presented in tables 7.10 and 7.17 illustrate that the properties of networks crosslinked with MCDE to 100% strongly depend on both the rate of chemical reaction of the ether with polystyrene and the extent of dilution of the latter with cyclohexane. Indeed, at a constant linear polystyrene

Table 7.17 Properties of hypercrosslinked polystyrene networks prepared in cyclohexane and EDC + n-octane (1:1.6 v/v) mixture (crosslinking with MCDE to 100% degree of crosslinking at 60°C within 11 h) [152]

Initial polymer	C_{in}, (vol. %)	[SnCl$_4$], (mol/mol PS)	Swelling, (mL/g)				S_{app}, (m^2/ g)	W_o, (cm^3/ g)	Cl, (%)
			Toluene	Methanol	Heptane	Water			
I	11	0.25	9.0	0.15	0.04	0.17	0	–	4.86
I	11	0.5	2.94	1.54	2.34	1.38	247	0.1	2.76
I	11	1	3.03	2.46	2.82	1.60	338	0.3	2.65
I	11	2	4.40	4.04	4.19	2.46	483	0.5	1.78
I*	11	2	5.42	4.97	5.37	3.36	642	1.5	0.75
I	14	2	3.10	2.52	2.87	1.71	592	0.3	1.50
I	9	2	5.16	4.61	4.22	3.11	438	0.9	1.85
I	7.7	2	5.60	5.27	4.92	3.50	350	1.2	2.20
II*	–	2	0.98	0.87	0.87	0.67	300	0	–
II	–	2	0.89	0.79	0.74	0.52	66	0	–

I = linear polystyrene.
II = styrene–0.7% DVB copolymer.
I*, II* = networks were prepared by crosslinking in the mixture of EDC with n-octane.
W_o was measured by bulk weight.
Swelling was measured by weight of dry and swollen samples.

concentration in the mixture of MCDE, SnCl$_4$, and cyclohexane, the swelling of resulting networks in good and bad solvents increases in parallel with the amount of the catalyst introduced into the starting reaction solution. Note, when a very large amount of SnCl$_4$, 2 mol per 1 mol of MCDE, is involved, the hypercrosslinked network takes up almost equal amounts of toluene, methanol, and heptane, while networks formed in the presence of relatively small quantities of the catalyst swell in methanol or heptane to a substantially smaller extent than in toluene. This implies that larger amounts of the catalyst accelerate the crosslinking process, ensuring a complete conversion of the crosslinking agent (reducing the concentration of unreacted chlorine in the product), and guaranteeing the formation of a more rigid hypercrosslinked network with an enlarged apparent specific surface area (Table 7.17). The total pore volume in the product also increases with the SnCl$_4$ concentration, reaching a value typical for the products synthesized in good solvents.

Similarly to the crosslinking in EDC, increasing dilution of polystyrene chains with cyclohexane reduces the extent of mutual network mesh entanglements and, therefore, significantly enhances the swelling capacity of the resulting products. However, judging from the rising content of pendent chloromethyl groups in the final products, the real degree of cross-linking drops with the dilution, thus leading to diminishing specific surface area values. On the other hand, with the dilution of the initial polystyrene solution the total pore volume of dry materials substantially increases, reaching values characteristic of real macroporous polymers (Table 7.17).

Still, the use of cyclohexane, a poor solvent for polystyrene, during the post-crosslinking process leaves a mark on the properties of the final materials. Contrary to transparent products of post-crosslinking in good solvents, those obtained in cyclohexane are completely opaque. The intensive light scattering immediately suggests the availability of large pores in the polymer, comparable in size with the wavelength of visible light. This assumption also follows from the fact that crosslinking in cyclohexane results in products with significantly smaller values of the apparent specific surface area and, conversely, much larger pore volume, compared to networks obtained in EDC.

The presence of larger pores is further corroborated by two methods: inversed size-exclusion chromatography and electron microscopy. The chromatographic porosimetry reveals a bimodal pore size distribution in materials prepared in cyclohexane from linear polystyrene (Fig. 7.36). In addition to narrow pores of the same size which are characteristic of networks obtained in EDC, these biporous polymers have additionally large pores with diameters up to 250–350 Å (in the swollen state) [183]. On the micrographs of the polymer one can easily see relatively large spherical fragments (Fig. 7.37) having a diameter not less than 1000 Å. They are much larger than those spherical species that form the internal structure of products derived from EDC solution of polystyrene of the same molecular weight. Undoubtedly, such large fragments forming the network texture as well as large pores between them can only result from microphase separation during the post-crosslinking of polystyrene coils in cyclohexane.

The reasons for microphase separation during the formation of macro-porous polystyrene networks through crosslinking copolymerization of monomers in both good solvents and in polymer-precipitating media are well understood and they have been discussed in detail in Chapter 3 Sections 1.2–1.4. Formation of hypercrosslinked networks in non-solvating media is principally impossible. Good solvents give no cause for phase separation in the post-crosslinking reaction. Obviously, cyclohexane, as a Θ-solvent for polystyrene, allows the crosslinking reaction to start in a true polymer solution and finish it in the situation where the solvent is rejected from the

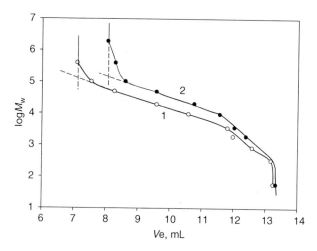

Figure 7.36   SEC calibration curves for the networks prepared by crosslinking linear polystyrene with monochlorodimethyl ether to 100% crosslinking degree in (1) ethylene dichloride at volume concentration of 5.9% and in (2) cyclohexane at volume concentration of 11%.

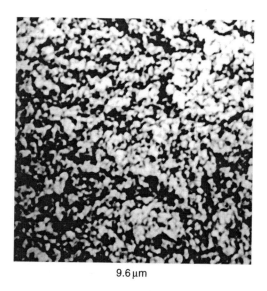

9.6 μm

Figure 7.37   Texture of the network prepared by post-crosslinking linear polystyrene of 300,000 Da molecular weight with MCDE to 100% in cyclohexane; scanning electron microscopy. *(After [152]).*

crosslinked network. It has been observed previously [200, 201] that the affinity of a poor solvent to the polymer decreases with the introduction of bridges between the polymer chains. This has been reported as an increase of χ, the Flory–Huggins parameter, with crosslinking, reflecting the gradual conversion of cyclohexane from a solvent to a precipitating media. If this transformation of a Θ-solvent is a general phenomenon, microphase separation also has to occur in any other solvent possessing a similar poor thermodynamic affinity to polystyrene.

A 1:1.6 (v/v) mixture of EDC and n-octane is such a Θ-solvent for polystyrene at 40°C that it is pretty close to the Θ-temperature for polystyrene solution in cyclohexane. When reacting linear polystyrene with MCDE in this mixed solvent at 60°C (as was done in cyclohexane) [152], the microphase separation was also observed to occur that resulted in obtaining opaque polymers. Notably, the polymers swell in various organic solvents to a higher extent than those prepared in cyclohexane (Table 7.17). Even more important is the fact that the total pore volume in dry polymers proved to be 3 times larger than that in polymers obtained in pure EDC or cyclohexane. This implies the formation of rather large pores in the polymer and the substantially reduced tendency of the network to contract on removing the solvents from the crosslinked gel. The tendency to microphase separation in the mixed solvent may be additionally enhanced by the weak capability of n-octane to solvate the intermediate reaction complex of polystyrene–ether-catalyst, because of the insolubility of stannic tetrachloride in n-octane. During phase separation n-octane appears to be rejected predominantly, whereas the hypercrosslinked polystyrene phase remains enriched with EDC. Interestingly, the rate of the reaction of MCDE with polystyrene was found to be slower in both the mixed system and cyclohexane, compared to the rate in EDC, as marginally demonstrated by the time of gelation: 30 s at −20°C in pure EDC, 5 min at 60°C in the mixed solvent, and 15 min at 60°C in cyclohexane.

Surprisingly, the structure of the hypercrosslinked networks fully changes if a slightly crosslinked beaded copolymer is used as the starting material instead of a dissolved linear polystyrene (Table 7.17). Although the beads obtained both in cyclohexane and in the mixed solvent are capable of swelling in good and bad solvents nearly to the same extent, their porosity proved to be too small to be measured by the rather rough method of bulk weight. This suggests that no microphase separation occurred in the initially swollen copolymer beads. The only reason which could explain such a sharp difference in the structure of the hypercrosslinked networks based on the beaded copolymer and on the dissolved linear polystyrene is the very large difference between the volumes of the two starting systems subjected to post-crosslinking. Indeed, the volume of the polystyrene solution, as a rule not less than 40 mL, exceeds the volume of the initial copolymer bead of 0.1–0.3 mm in diameter by 6–8 orders of magnitude. As crosslinks are introduced, the affinity of the poor solvent to the network is expected to decrease. There are only two opportunities for the separating solvent to depart from the polymer phase: either to be separated into the outer space in the form of macrosyneresis, or to form a microphase inside the swollen gel. In the case of the small beads, the first opportunity appears to be easily realized. Therefore, the networks formed within the beads exhibit a rather dense packing of polystyrene chains and an absence of permanent porosity. When linear polystyrene is crosslinked in bulk solution, the diffusion path of cyclohexane from the center of the large gel block proves to be long; the solvent has not enough time to separate in the form of macrosyneresis and is forced to form the separate microphase within the gel, resulting in the formation of biporous networks.

Thus, the above observation suggests that the volume of the system to be crosslinked is one additional factor that can determine the structure and properties, including the swelling ability, of hypercrosslinked polystyrenes. However, it can manifest itself only at fundamentally different volumes of the initial systems. With this respect, it is worthwhile to mention here that a biporous network structure has also been observed to be formed in EDC when a very large hypercrosslinked block of 0.5 L in volume was prepared from a more diluted polystyrene solution with a concentration of 5.9%. Obviously, the final network with 100% crosslinking had problems in accom-modating the 17-fold excess of the otherwise good solvent. It was also not possible to reject the excess from the big block, so that conditions emerged for an incipient microphase separation of the system. Figure 7.36 illustrates the bimodal pore size distribution in this polymer that looks similar to the biporous structure of the material prepared in cyclohexane.

One of remarkable distinguishing features of heterogeneous polymers, that is, biporous hypercrosslinked as well as macroporous networks, compared to homogeneous and gel-type networks, is that the former are shape resistant while the latter change their volume on swelling and drying. Thus, the industrial biporous material MN-200 can be packed into a column in dry state and then wetted by alcohol without gaining in its volume by more than 3%. Contrary to this, homogeneous Styrosorbs based on gel-type styrene–DVB beads have to be first swollen with any liquid and only then packed into a column. They retain their volume on changing the solvent, but not on drying. This property of macroporous sorbents and ion-exchange resins is well familiar to practicians, but has not found any scientific explanation. We believe that the polymeric phase comprising the walls of the macropores cannot be different from that of homogeneous gel-type network having the same degree of crosslinking, and, therefore, it must swell with thermodynamically good solvents. However, expansion of the polymeric phase in the macroporous material can proceed at the expense of the pore volume, with the overall size of the bead remaining nearly constant. Homogeneous materials do not have this possibility and swell with solvents with an increase in the bead size. Though this explanation seems to be logical and simple, it is difficult to corroborate it with any convincing experimental proof. Comparing the pore size distribution of a macroporous or biporous hypercrosslinked material in the dry and swollen states could shed light on that proposed mechanism of swelling of the hetero-geneous materials, but no precise methods of doing this exist thus far.

8. THERMOMECHANICAL PROPERTIES OF HYPERCROSSLINKED POLYSTYRENES

The classic theory basically explains the loss of the rubber elasticity state on progressive crosslinking by the chain length between two neighbor joints becoming shorter than the segment length that can provide certain mobi-lity. From this point of view, hypercrosslinked networks with ultimately short distances between joints should not swell in solvents and maintain the glassy state up to their degradation temperature. However, they show a remarkable ability to increase their volume by a factor of 3–5 on swelling, which must be regarded as a marked manifestation of the unusually high network mobility. This fact alone raises the question whether the hyper-crosslinked polystyrene materials could also pass at elevated temperatures to the rubber elasticity state from the theoretically prescribed glassy state.

The enhanced mobility of hypercrosslinked networks must also reveal itself under the action of mechanical forces in combination with the influence of temperature. Generally, the character of material deformation, the actual physical state of a polymer, and the possible phase transitions can be examined by two methods, thermomechanical and thermodilatometric, both giving an idea about the mobility of the polymer network.

8.1 Thermomechanical tests and the physical state of hypercrosslinked networks

The behavior of hypercrosslinked polymers under uniaxial compression and/or heating was studied using a technique that was specially developed for testing beaded polymer samples [202, 203]. One individual spherical bead of a crosslinked polymer having a diameter of 0.7 ± 0.03 mm was placed in a shallow cavity drilled in a quartz plate. The uniaxial compression of the bead was carried out with a quartz rod having a complementary cavity with a radius of 0.89 mm and connected to a sensor. The relative change in the bead diameter, ΔD_{o}, along the axis of compression was calculated as

$$\Delta D = \frac{D_{\mathrm{m}} - D}{D_{\mathrm{o}}} \times 100 \; (\%) \qquad [7.3]$$

where D, D_{o}, and D_{m} are the current bead size, its initial diameter, and its size at the moment of applying the initial pressure, respectively. D_{m} is smaller than D_{o} by about 4–7% due to the immediate elastic response of the bead to the applied weight. The thermomechanical plots presenting the bead deformation as a function of temperature were registered automatically using a special equipment model UIP-70 (Russia) at the heating rate of 5°C/min (if not specially stated).

In order to understand the peculiarities in the deformation of hypercrosslinked polystyrenes, let us first consider the deformation of conventional network styrene copolymers (Fig. 7.38). The copolymer incorporating 3% DVB exhibits two physical states, glassy and rubbery, with a narrow transition zone

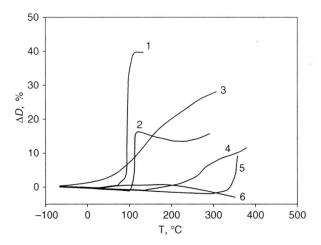

Figure 7.38 Thermomechanical curves for (1–3) styrene copolymers containing (1, 2) 3% DVB and (5) 34% DVB, (6) polydivinylbenzene, and (3, 4) hypercrosslinked network prepared by crosslinking styrene–0.57% DVB copolymer with monochlorodimethyl ether to 100%; Loading: (2, 4, 5) 10 g and (1, 3, 6) 400 g per bead. *(Reprinted from [202] with permission of Wiley & Sons, Inc.)*

centered around $100°C$ (glass transition temperature), provided that the pressure applied to the sample is small, that is, 10 g per bead. An increase in the load to 400 g per bead causes a $10°C$ shift in T_g toward lower temperatures, because of the development of forced high elasticity. It also results in an increase from 17 to 40% in the deformation when the copolymer reaches rubber-like elasticity. The highly crosslinked conventional copolymer with 34% DVB and poly-DVB retain their glassy state up to the temperature of chemical degradation (over $300°C$), while only exhibiting small thermal expansion up to this point.

The hypercrosslinked network with 100% degree of crosslinking shows a noticeable deformation already at $140°C$ under a load of 10 g. Under 400 g pressure (which still is 20 times smaller than the breakdown limit at $25°C$) the deformation starts at the temperature as low as $-50°C$. Note that the deformation of the slightly crosslinked styrene–3% DVB copolymer under this load starts in the temperature zone almost $130°C$ higher.

The maximum deformation ΔD_b of this 100% hypercrosslinked polymer achieves the value as high as 30% when the temperature approaches $300°C$. Although this deformation is characteristic of rubber-like elasticity (Fig. 7.38, plots 1 and 2), no typical plateau can be observed (plot 3). This statement is also valid for the whole set of plots obtained at varying loadings in the interval from 10 to 450 g per bead (Fig. 7.39). All plots have a flat

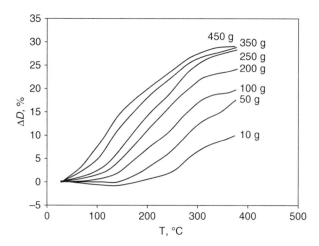

Figure 7.39 Effect of the loading on the form and position of the thermomechanical curves of the network obtained by crosslinking styrene–0.57% DVB copolymer with MCDE to 100%; heating rate, 5°C/min. *(Reprinted from [202] with permission of Wiley & Sons, Inc.)*

S-shaped form. The position of the inflection points (maximum of the deformation rate) strongly depends on the applied pressure, steadily shifting about 200°C toward lower temperature with increasing force.

Interestingly, the rate of heating, ranging from 2.5 to 20°C/min, does not change the shape or position of the thermomechanical plots; the non-systematic scattering of data remains within the limits of measurement reproducibility on different beads (Fig. 7.40). This implies that the deformation of a hypercrosslinked network rapidly approaches the equilibrium value determined by the applied temperature and stress. At the same time, the glass transition temperature for linear polymers is well known to linearly depend on the logarithm of the heating rate. In our experiments with the styrene–3% DVB copolymer, the systematic shift in T_g by 10°C toward higher temperatures was also observed with the increase in the heating rate.

Figure 7.41 illustrates another interesting characteristic of the thermomechanical plots of hypercrosslinked polystyrene: with increasing crosslinking density there is no shift in the inflection points toward higher temperatures. Rather an opposite tendency can be observed in that the noticeable deformation starts at the lowest temperatures for the polymer having the highest degree of crosslinking. In the series of plots presented in Fig. 7.41, only the decrease in the ultimate deformation at a given

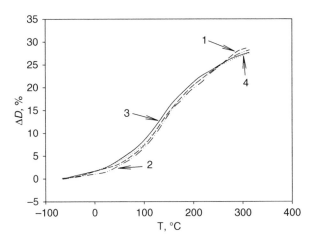

Figure 7.40 Effect of the heating rate on the position of thermomechanical curves of the network obtained by crosslinking styrene–0.57% DVB copolymer with MCDE to 100%: (1) 2.5, (2) 5, (3) 10, and (4) 20°C/min; loading 400 g per bead. *(Reprinted from [202] with permission of Wiley & Sons, Inc.)*

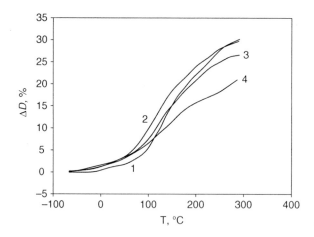

Figure 7.41   Effect of the degree of crosslinking on the position of thermomechanical curves; the networks were prepared by crosslinking styrene–0.57% DVB copolymer with (1) 0.3, (2) 0.4, (3) 0.5, and (4) 0.75 mol of MCDE; loading, 400 g per bead. *(Reprinted from [202] with permission of Wiley & Sons, Inc.)*

temperature, for example, 300°C, with increasing degree of crosslinking can find a trivial explanation.

As shown in Fig. 7.41, the deformation of all hypercrosslinked beads noticeably increases at a temperature above 100°C. However, this phenomenon is not related to trivial transition of polystyrene segments from glassy to rubber-like state. The very nature of deformation of hypercrosslinked materials differs fundamentally from the deformation of conventional networks under rubber-like elasticity.

This follows from Fig. 7.42, where the thermomechanical plots for beads of 100% crosslinked network were registered under three different protocols: under constant loadings of 400 and 50 g and under a periodically (every 3 min) varying loading between 400 and 50 g. The pattern presented implies that the total deformation of a bead consists of a constant (at any temperature) portion of elastic deformation and, with the temperature rising up to 250°C, an increasing portion of non-elastic deformation. The non-elastic residual deformations are responsible for a gradual change in the form of the initially spherical bead. When deformed with a flat rod, we can easily see under a microscope a circular print of the rod on the top of the bead, without any signs of destruction or fragmentation of the sample (provided that the temperature of the experiment does not exceed

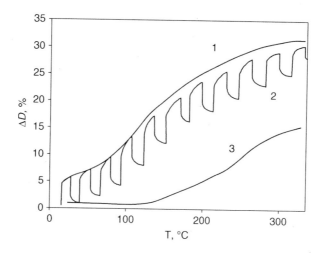

Figure 7.42 Thermomechanical curves of the sample obtained by crosslinking styrene–0.57% DVB copolymer with MCDE to 100%, registered under constant loading of (1) 400 and (3) 50 g and under a varying (every 3 min) loading between 400 g and 50 g. *(Reprinted from [202] with permission of Wiley & Sons, Inc.)*

250°C). This picture could resemble a plastic deformation of the bead of linear polystyrene. However, the residual deformation of the hypercrosslinked network cannot be regarded as plastic.

Figure 7.43 gives sufficient evidence for the reversibility of the residual deformation. Several hypercrosslinked beads were heated individually to 136°C under a pressure of 400 g and then were allowed to cool down to ambient temperature under that pressure. During the second cycle of heating under the same pressure, the deformation of these pretreated beads was observed to start only at 136°C (Fig. 7.43, plot 4). The sample obviously stored the information on the first cycle of deformation, possibly by achieving higher mechanical strength along the axis of compression. Such information storage would be unthinkable in a plastically deformed sample. Plots 2 and 3 in Fig. 7.43 demonstrate that both the non–elastic deformation and the "memory" effect in this precompressed network largely disappear if the deformations are allowed to relax. Plot 2 shows the behavior in the second cycle of deformation of beads relaxed at 164°C for 2 h without any pressure applied. Plot 3 relates to another set of beads with "frozen" deformations that was allowed to swell with acetone, followed by washing with water and drying at 80°C. As can be seen, the deformation behavior of compressed and then relaxed beads does not differ

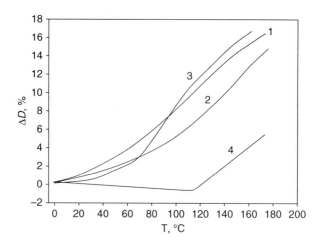

Figure 7.43 Effect of the pretreatment of the network obtained by crosslinking styrene–0.57% DVB copolymer with monochlorodimethyl ether to 100% on the position and form of thermomechanical curves: (1) control sample; (2) the sample heated up to 136°C under a loading of 400 g and relaxed at 164°C for 2 h without pressure; (3) the sample heated up to 136°C and then cooled under a loading of 400 g (12% residual deformations), then subjected to swelling and drying; (4) the sample heated up to 136°C and then cooled under a loading of 400 g (10% residual deformations). *(Reprinted from [202] with permission of Wiley & Sons, Inc.)*

substantially from the plots registered in the first cycle of deformation (Fig. 7.43, plot 1). Taking into account the normal scattering of the results registered for individual beads, one can state that residual "frozen" deformations in hypercrosslinked materials are completely reversible.

The reversibility of deformations is further confirmed by the results of another experiment. The compression of porous hypercrosslinked polymers based on linear polystyrene results in reducing their apparent specific surface area (Table 7.18). With increasing applied pressure, the samples having moderate degree of crosslinkings become less porous or even nonporous, irrespective of the type of crosslinking agent. The stronger crosslinked polymers with high apparent specific surface area of $1000\,m^2/g$ sustain a high pressure much better, although the decrease in the porosity is noticeable. This is not surprising because high outer pressure helps the mobile hypercrosslinked networks to realize their natural tendency to be packed densely. More important is the fact that the surface area is restored after swelling with acetone, washing with water and drying. The most rigid networks crosslinked with MCDE completely restore their porosity

Table 7.18 Change in the apparent inner specific surface area (m²/g) after applying pressure P (kg/cm²), and subsequent swelling, for the networks based on linear polystyrene

S_{in}	P 442 kg/cm²		P 2012 kg/cm²		P 7078 kg/cm²	
	S_1	S_2	S_1	S_2	S_1	S_2
Networks crosslinked with monochlorodimethyl ether						
160	20	180	10	160	0	40
600	180	630	70	680	70	560
1000	640	1000	630	1030	560	950
Networks crosslinked with p-xylylene dichloride						
150	30	120	23	90	0	90
430	130	450	90	410	50	340
750	480	680	380	615	250	640
830	775	890	750	780	670	750

S_{in}, S_1, and S_2 are the respective initial specific surface area, specific surface area after exerted pressure, and specific surface area restored by swelling with acetone followed by washing with water and drying.

(within the limits of measurement precision). In more flexible polymers crosslinked with XDC, the newly formed polymer–polymer contacts appear to be partially irreversible, so that the values of the restored surface area become somewhat smaller than the surface area of the initial sample.

The above-discussed results show that hypercrosslinked polystyrene exhibits unusually high mobility. Depending on the applied loading, the deformation of the polymers can start even at very low temperatures. Although the maximum deformation amounts up to 30%, which is characteristic of rubbery state rather than glassy state, the hypercrosslinked polymers are not elastic materials. The deformation behavior of hypercrosslinked polystyrene differs fundamentally from that of conventional styrene–DVB copolymers, the typical representatives of network polymers that exhibit classical transitions from glassy to rubbery state. In this connection the earlier reports observing glass transition temperature for a polymer having 100% degree of crosslinking by NMR [204] and spin rotation of a molecular probe [205] in the vicinity of 230 and 280°C, respectively, cannot be regarded fully correct. Most probably, the phenomena observed at these temperatures indicate unfreezing of some kind of local movements or an incipient chemical decomposition of the hypercrosslinked network.

8.2 Thermodilatometric analysis of hypercrosslinked polymers

Let us now consider information related to the mobility of the hypercrosslinked networks indicating volume changes of the polymers on heating [151]. The thermodilatometric experiments were carried out on single spherical polymer beads using UIP-70 (Russia), the same thermomechanical equipment as earlier. The flat quartz rod connected to a capacity sensor of the equipment touched the bead with a minimal load of 0.5 g. The thermal coefficient of volume expansion of the bead was calculated from the linear section of the dilatometric plot in the 200–300°C range at the heating rate of 5°C/min. Figure 7.44a–c shows the thermodilatometric plots for three groups of polymers prepared by post-crosslinking styrene copolymers containing 0.17, 0.6, and 2.7% DVB with MCDE. For better understanding the thermodilatometric plots are given as specific polymer

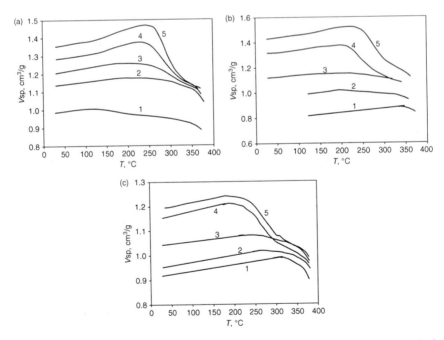

Figure 7.44 Dependence of specific volume on temperature for hypercrosslinked networks prepared by post-crosslinking copolymers of styrene with (a) 0.17, (b) 0.6, and (c) 2.7% DVB by means of (1) 0.3, (2) 0.4, (3) 0.5, (4) 0.75, and (5) 1.0 mol of MCDE per styrene repeating unit. *(Reprinted from [151] with permission of Wiley & Sons, Inc.)*

volume (V_{sp}) versus temperature. One can see that similarly to all glassy polymers, the investigated hypercrosslinked networks linearly expand on heating to 100°C. However, when the temperature exceeds this value, the thermal behavior of the above polymer groups starts to be different.

The hypercrosslinked polystyrenes obtained on the basis of the strongly swollen copolymer with 0.17% DVB represent the most porous products with substantially reduced density (Fig. 7.25). When the temperature exceeds 120°C, the V_{sp} value of the sample with $X = 43\%$ unexpectedly begins to decrease. The network keeps shrinking up to 325°C until V_{sp} reaches the value characteristic of conventional non-porous polystyrene. On the other hand the networks with 150 and 200% degree of crosslinking and naturally containing very large amounts of rigid bridges first demonstrate an unusually rapid increase in the specific volume, starting at about 110°C. This unprecedented expansion stops when the temperature approaches about 240°C. Above this temperature the polymer beads exhibit a dramatic, 17–25 vol.% shrinking, with the specific volumes decreasing from 1.4–1.5 to 1.2 cm^3/g at a temperature of about 300°C.

The polymer from the second group obtained by post-crosslinking styrene–0.6% DVB copolymer with 43% diphenylmethane-type bridges is a non-porous material with its V_{sp} value amounting to 0.9 cm^3/g. That is why this material shows only trivial slow thermal expansion up to 320°C, when an obvious destruction of the sample starts. In this group of hypercrosslinked networks, the reduced-density polymer having 66% degree of crosslinking demonstrates the shrinking, similar to the sample from the first group with $X = 43\%$, while the samples crosslinked to 150 and 200% show both the abnormal expansion on heating in approximately the same temperature range and the subsequent dramatic shrinking. Note that the shrinking of the samples from the first and the second groups, both having 200% degree of crosslinking, starts at slightly higher temperatures than for samples with 150% degree of crosslinking based on the copolymers with 0.17 and 0.6% DVB. In both groups, the thermodilatometric plots for the polymers crosslinked to 100% with 0.5 mol of the crosslinking agent (plot 3 in Fig. 7.44) are located between the shrinking and expanding extreme samples.

All networks of the third group (Fig. 7.44c) based on the copolymer with 2.7% DVB possess a relatively high density and, correspondingly, the lowest mobility. The samples with 43, 66, and 100% crosslinks experience the usual thermal expansion up to the temperature of chemical degradation ($\sim$300–320°C). The polymers with the highest degrees of post-crosslinking exhibit a very weak tendency to the enhanced expansion, and the adjacent shrinking process is also less pronounced.

Obviously the excessive expansion and the shrinking of hypercrosslinked polystyrenes have different origins. The samples with $X = 43$ and 66%, based on copolymers with 0.17 and 0.6% DVB, respectively, were obtained in the presence of significant amounts of solvent and were post-crosslinked with a relatively small number of rigid bridges. Both the reduced extent of chain entanglements and the moderate rigidity of the networks facilitate their mobility. These polymers have sufficient degree of freedom to realize their natural tendency toward a dense packing when they receive a certain portion of thermal energy and, therefore, gain additional possibilities for cooperative conformational rearrangements. As a result of this cooperative process, new polymer–polymer contacts appear and the specific volume of the polymers decreases (plot 1 in Fig. 7.44a and plot 2 in Fig. 7.44b).

The networks post-crosslinked to extreme extents by using 0.75 and 1.0 mol of MCDE are composed of disubstituted and trisubstituted rigid bridges. New chain contacts are unlikely to be realized here, even at very high temperatures, because the rigidity of the network is very high and the size of the network meshes is small. When prepared from copolymers with a low DVB content (0.17 and 0.6%), these materials have the largest free volume and V_{sp} values. Their network exhibits a permanent and very strong tendency to the minimization of the free volume, resulting in the strong distortion of conformations of the meshes toward minimization of the voids within each macrocyclic mesh. However, the heating enhances the vibration of the atoms and groups of the stress-distorted meshes of the network. With increasing amplitude of vibrations the conformations of the meshes change toward less distorted forms. In other words, with increasing thermal energy and smaller and less important role of the initial stress energy, the conformation of the macrocycles approaches an equilibrium conformation. A cooperative rearrangement of the whole ensemble of the mutually condensed meshes finally results in an unusually strong increase in the volume of the samples (plots 4 and 5 in Fig. 7.44a and 7.44b). Also a certain expansion of the hypercrosslinked network on heating reduces the inter-chain interactions and allows to partially eliminate the inner stresses.

In order to understand the dramatically strong shrinking, nearly collapse, of the networks at high temperatures, it is necessary to consider the results given in Fig. 7.45. When heated in air the product of a styrene–2.7% DVB copolymer post-crosslinked to 200% starts to shrink at a temperature slightly above 200°C. However, when heated in an inert atmosphere, the

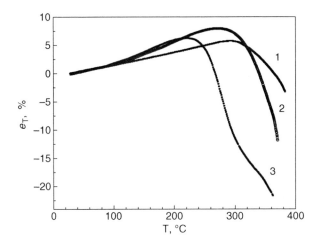

Figure 7.45 Thermodilatometric curves for (1) the gel-type styrene–34% DVB copolymer and (2, 3) the styrene–2.7% DVB copolymer crosslinked with MCDE to 200% and heated in (2) argon and (3) air; $e_T = (V - V_o)/V_o$, where V and V_o are the current and starting volumes of the beads, respectively. *(Reprinted from [151] with permission of Wiley & Sons, Inc.)*

polymer will shrink only above 300°C. Its behavior then does not differ much from the thermal degradation of the conventional non-porous copolymer of styrene with 34% DVB. This suggests that the highly permeable and stressed hypercrosslinked polymer networks are chemically attacked by oxygen when being exposed to air at temperatures above 200°C.

Indeed, when this (or a similar) sample was subjected to a preliminary heating at 280°C in air followed by recording the thermodilatometric plot, the network did not exhibit any unusual shrinking until the temperature reached 300°C (Fig. 7.46). No strong shrinking was either observed when this preheated sample was allowed to recover by swelling with acetone. On the other hand, if the sample was preheated at 280 or 325°C in an inert atmosphere, the thermodilatometric plots proved to completely coincide with the plot of the initial control sample. In the range of 220–320°C, all hypercrosslinked samples demonstrate a really impressive shrinking, amounting up to about 25% volume. Hence, the hypercrosslinked polymers irreversibly change their properties on heating in air above 200°C. Their oxidation is obviously facilitated by both the enhanced accessibility of the whole interior of the sample to oxygen molecules and by activation of certain bonds and groups due to intensive stresses of the network meshes.

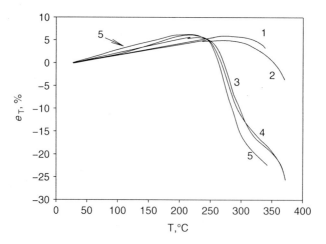

Figure 7.46   Effect of the pretreatment on the thermodilatometric behavior of the styrene–0.6% DVB copolymer crosslinked with MCDE to 100%: (1) the sample heated at 280°C for 30 min in air; (2) the sample heated at 280°C for 30 min in air, swollen in acetone, washed with water, and dried; (3) the sample heated at 325°C for 60 min in argon; (4) the control sample; (5) the sample heated at 280°C for 30 min in argon. $e_T = (V - V_o)/V_o$, where V and V_o are the current and starting volumes of the beads, respectively. *(Reprinted from [151] with permission of Wiley & Sons, Inc.)*

Further confirmation of network rearrangement and/or degradation in air, that is, in the presence of oxygen, can be found in Fig. 7.47, showing the change in the apparent specific surface area as a function of temperature and heating time. While heating in argon at 280 and even 325°C (within 1 h) does not affect the value of S_{app}, exposure to air results in a drastic decrease in the specific surface area, the porosity of the polymer completely disappearing in 30 min at 280°C.

Interestingly, the exposure of the polymers with 100% degree of crosslinking to air at 280°C is not accompanied by any weight loss, and no extractable low-molecular-weight products are formed by this treatment. (A very small decrease in mass may be caused by the removal of traces of adsorbed organic compounds. These polymers belong to highly porous materials with extreme adsorption ability.) Alternatively to the hypercrosslinked polystyrenes, a conventional styrene–34% DVB copolymer, when heated to 280°C, suffers from an intensive depolymerization, release of volatile compounds, and weight loss.

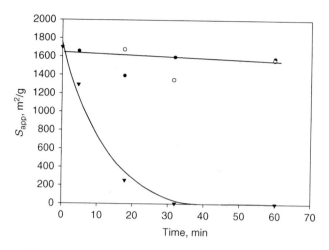

Figure 7.47 Effect of preheating on the apparent specific surface area of the sample prepared by crosslinking styrene–0.6% DVB copolymer with monochlorodimethyl ether to $X = 200\%$, and heated for different times to ($\bigcirc$) 280°C and ($\bullet$) 325°C in argon and at ($\blacktriangledown$) 280°C in air. *(Reprinted from [151] with permission of Wiley & Sons, Inc.)*

The chemical modification of hypercrosslinked polymers in air was found to be accompanied by the appearance of carbonyl-containing groups, which may reveal themselves at 1680, and 1726 cm^{-1} in the infrared (IR) spectrum of the network 0.6×100 exposed to air for 30 min at 280°C.

Thus, we can conclude that the shrinking of the hypercrosslinked polymers at temperatures in the range of 250–300°C is caused by oxidation, with the rupture of a certain portion of the stressed carbon–carbon bonds. Under the permanent tendency for reducing the free volume of the material, the scission of the most stressed bonds leads unavoidably to the formation of a more compact structure and to the decrease of the beads' volume and their inner specific surface area. The rupture of bonds can then be followed by their partial recombination or termination with oxygen. The latter process seems to proceed to a lower extent, since no noticeable weight change of the air-exposed samples was observed.

8.3 Thermal stability of hypercrosslinked polystyrene

As mentioned above, heating of hypercrosslinked polystyrene to 325°C for at least 1 h under protective argon atmosphere does not affect the porous structure and the apparent specific surface area of the material. Still, this temperature should be considered as the material's upper stability limit.

Following from Fig. 7.46, the polymer slowly starts to lose its weight at temperatures over 350°C. The polymer destruction proceeds most intensively in the range of 400–500°C, but nearly stops after 600°C. After further heating to 1000°C, a carbonaceous material is recovered, amounting to 45–55% of the initial weight of the polymer [206]. This final carbon yield is surprisingly high, compared to a complete depolymerization of a gel-type styrene–10% DVB copolymer and to a carbon yield of no more than 16% from poly-DVB. Obviously, in the latter material, both aliphatic chains are prone to radical depolymerization. On the other hand, in the hypercrosslinked structure, the radical-type destruction of the main polystyrene chains does not cause an immediate loss of the aromatic moieties. The latter remain connected to each other by residues of the crosslinking agents used during the post-crosslinking step of the synthesis. Indeed, volatile destruction products of hypercrosslinked materials contain no DVB and only very small amounts of styrene.

Interestingly, the presence of residual chlorine in the hypercrosslinked polymer enhances the yield of the final carbonizate (plot 1 in Fig. 7.48). It was found that the presence of carboxyl or sulfonic substituents in the aromatic rings of the polymer also enhances the yield of final carbons [207]. In fact it has been known [208] that sulfonated styrene–DVB copolymers can be used for the preparation of carbonaceous adsorbing materials. However, because of the elimination of heavy sulfonic substituents (over 40% of the material weight), the final yield of carbons is by no means higher than in the

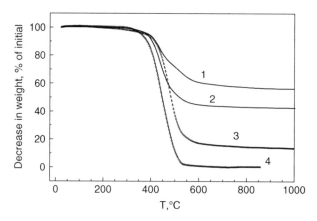

Figure 7.48 Thermogravimetric analysis in argon of hypercrosslinked polystyrene with $X = 200\%$ containing (1) 5.3% or (2) 1.3% residual chloride, (3) polydivinylbenzene–65%, and (4) styrene–10% DVB copolymer. *(Reprinted from [207] with permission of Akademizdattsentr "Nauka" RAN.)*

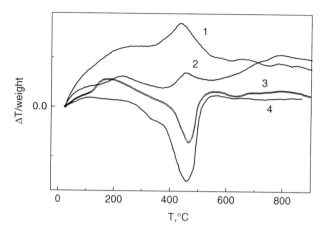

Figure 7.49 Differential thermal analysis of hypercrosslinked polystyrene with $X = 200\%$ containing (1) 5.3% or (2) 1.3% residual chloride, (3) polydivinylbenzene–65%, and (4) styrene–10% DVB copolymer. *(Reprinted from [207] with permission of Akademizdattsentr "Nauka" RAN.)*

above case of direct pyrolysis of hypercrosslinked polystyrene. In addition, the latter does not release harmful sulfur oxides during pyrolysis.

Figure 7.49 reveals another difference between conventional styrene copolymers and hypercrosslinked polystyrene during pyrolysis [207]. Differential thermal analysis of conventional styrene copolymers and poly-DVB characterizes the depolymerization of the material between 400 and 500°C as a strongly endothermic process. Contrary to this, the restructuring and destruction of hypercrosslinked polymers proves to be an exothermic process, although the material loses almost half of its weight in the same temperature range. Obviously thermal scission of strained meshes in the hypercrosslinked network and partial compaction of the material, which result in a loss of specific surface area and a decrease in the specific volume of the beads, release excess energy that was stored in the stressed network. This exothermic effect is sufficient to overcome the energy loss required for the elimination of the least stable, mostly aliphatic fragments of the polymer on converting it into a carbonizate.

This suggestion is directly corroborated by the emergence of very intensive signals of unpaired electrons in the electron paramagnetic (EPR) spectra of products pyrolyzed at 500–600°C [209]. Rupture of a stressed bond unavoidably results in a change of the conformation of the broken mesh, which places the formed radicals in positions unsuitable for their recombination. In this way many free radicals must build up in the pyrolyzed material. Indeed, even after 4 years of storage of materials pyrolyzed at 500–600°C, the concentration of unpaired electrons in them amounted to about 10^{19} spin/g.

Interestingly, the paramagnetic properties of the pyrolyzates have been found to strongly depend on the presence of oxygen. In the presence of oxygen, the EPR spectra are extremely broad and exhibit low intensity. When the oxygen is evacuated or is displaced by the sorption of certain small-molecule solvents such as methylene dichloride, the signal becomes sharp, with the intensity increased by several orders of magnitude (Fig. 7.50). This property of the system can be used for the preparation of an efficient oxygen EPR sensor in liquid media, including aqueous media [209].

In general, pyrolysis of certain hypercrosslinked polystyrene materials at a temperature of about 600°C may result in interesting carbonaceous adsorbing materials with a yield of up to 55–60% within 50–60 min. The products maintain the spherical form of the initial materials and their overall texture, and acquire an exceptional mechanical strength of up to 8 kg per bead. They are basically nanoporous and show size-dependent adsorbing properties [207, 210]. In particular, they efficiently separate mineral ions in accordance with the new frontal ion size exclusion process

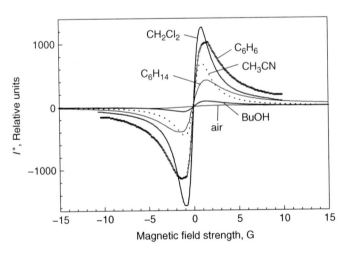

Figure 7.50 EPR spectra of a carbonizate obtained from a sulfonated hypercrosslinked polystyrene with $X = 200$ at 600°C for 60 min, taken in different media. $H_o = 4394\,G$. *(After [209]).*

(see Chapter 12). Their inner specific surface area of about 600–700 m²/g can be doubled by conducting the pyrolysis procedure in the presence of H_2O or CO_2 at higher temperatures. The semiconducting and paramagnetic properties of the materials may also present a special interest.

9. DESWELLING OF POROUS NETWORK POLYMERS

The experience accumulated on precisely measuring dimensions of a single small polymeric bead and registering its changes with temperature and with the applied stress proved to be useful to study the volume changes of the swollen bead when the solvent is evaporating from the polymer. The same method was previously applied to follow the fast kinetics of swelling of hypercrosslinked polymers (see Section 2). When examined on a single bead, the inverse process, deswelling (or volume shrinking), proved to be a complex process. Thus far, its details have been overseen since the generally accepted approach was to observe the behavior of a large sample of beaded material, where the effects of a volume increase in some individual beads are compensated by the shrinking of others, thus completely concealing the true behavior of the material.

The phenomenon of volume shrinking of various organic and inorganic materials during removal of water or other solvents by drying has been studied in detail. It is well known that when processing polymeric materials, the decrease in the volume of the products during the removal of residual solvents has to be accounted for. Therefore, it was quite unexpected to find out a noticeable (sometimes temporary) increase in the bead volume of various porous polymers, which, as a whole, steadily decreases on drying. The non-monotonic decrease in the size of swollen beads on drying was reliably and reproducibly detected as the anomalous shape of the recorded deswelling plots. This phenomenon is absent in the case of polymers which are non-porous in the dry state but can be observed to a greater or lesser extent in all samples with reduced apparent density, that is, in materials having a true porosity in the dry state. The abnormal deswelling profile was found to be characteristic not only of unfunctionalized hypercrosslinked polystyrenes, porous copolymers of styrene with DVB, and porous poly-DVBs, including the well-known commercially available sorbent Amberlite XAD-4, but also of ion exchange materials based on macroporous styrene copolymers and polyacrylate-type resins [159, 203].

The deswelling anomaly is particularly characteristic of the macroporous poly-DVB XAD-4, also containing a significant fraction of meso- and

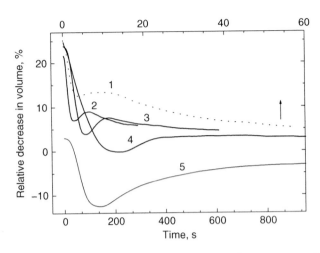

Figure 7.51 Deswelling of macroporous Amberlite XAD-4 on removal of various solvents: (1) diethyl ether, (2) ethanol, (3) toluene, (4) chlorobenzene, (5) water. *(After [159])*.

micropores. Figure 7.51 shows that the volume of the XAD-4 bead gradually decreases during the removal of various solvents; however, at a certain moment, it unexpectedly increases again, and will slowly decrease only after this period to a final value. Importantly, this phenomenon is independent of the drying rate and, therefore, has thermodynamic, rather than kinetic, nature.

One can imagine that at a certain point in time during evaporation, actually, almost immediately, the residual liquid in the capillaries and channels of the polymer matrix will form concave menisci at the outlets of the capillaries on the beads' surface. Because of the surface tension of the liquid film on the pore walls, the liquid within the pores finds itself under reduced pressure. Such a residual liquid pulls together the pore walls, leading to a marked decrease in the volume of the entire porous sample. When the liquid evaporates and air fills the pores, the excess contracting stress disappears and the bead expands again. In general, the phenomenon of capillary contraction is known: on the macro level it manifests itself, for example, in noticeable compression and strengthening of a powdered material after wetting it with a certain amount of a liquid. For example, moisture consolidates and makes sand suitable for the construction of complex bodies. However, the drying of sand or, conversely, excessive wetting leads to an instantaneous destruction of the bodies. A critical condition for the contracting action of a liquid is the presence of concave liquid menisci in the space between the particles of a powdered material or in the outlets of capillaries of a porous polymer material.

Undoubtedly, a specific pore structure, in combination with some elasticity of the network, favors the appearance of the anomaly in the deswelling plot of the XAD-4 sorbent to a greater extent than for other types of sorbents. Interestingly a minimum of the volume/time plot for the XAD-4 bead can be observed during the removal of any type of solvent (Fig. 7.51). Theoretically, the force of capillary contraction should increase with a decreasing pore radius, with an increase in the surface tension at the liquid–air interface, and with improvement of wetting of the pore's surface. It should be particularly noted that water, which has the maximum surface tension, causes the maximal decrease in the XAD-4 bead's volume, although water should not wet well this hydrophobic material.

The replacement of the liquid by air on further drying eliminates the contracting force, allowing the bead to markedly increase again its volume. After this sudden expansion, the bead continues to slowly shrink, which is definitely caused by the slow evaporation of the solvent from the polymeric phase of the pore walls. One can assume that the extent of the abnormal increase of the volume is determined by both the excess compression of the structure and the rate of relaxation of stresses arising from this excessive contraction within the spatial network. The degree of crosslinking and the parameters of the pore's space are undoubtedly decisive factors determining the fundamental differences in the relaxation rates of such stresses in macroporous and hypercrosslinked polymers. However, the plasticizing action of thermodynamically good solvents also plays a role in the acceleration of the relaxation processes. Probably the rapid relaxation of stresses in hypercrosslinked networks, in combination with the plasticizing action of toluene, causes the degeneration of the deswelling anomaly into a step in the plot of volume change for hypercrosslinked samples swollen by toluene (Fig. 7.52). The role of the plasticizing action of the solvent in these networks can be demonstrated by the fact that the elasticity modulus of the hypercrosslinked polymer 0.6×200 swollen in toluene is 200 MPa, but increases to 380 MPa after replacement of toluene by water. Anomalies that degenerate into steps in the deswelling plots are also characteristic of the macroporous poly-DVB plasticized by toluene and of the macroporous copolymer of styrene with DVB (Fig. 7.52).

On the other hand, water does not plasticize the polymer phase of the macroporous XAD-4 sample; therefore, the deformation of the rigid structure of the bead during drying from water has an elastic reversible character

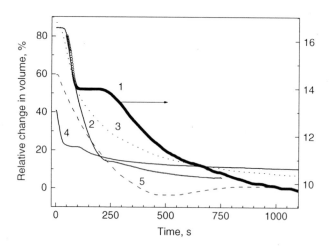

Figure 7.52 Deswelling of various polymers initially swollen with toluene. Macroporous polymers: (1) polydivinylbenzene, (4) styrene–15% DVB copolymer. Hypercrosslinked polymers: (2) styrene–0.6% DVB copolymer crosslinked to 200% with MCDE, (5) linear polystyrene crosslinked to 200% with XDC. (3) Gel-type nonporous styrene–2.7% DVB copolymer. *(After [159]).*

and the subsequent increase in the bead volume becomes particularly noticeable. Note that the bead volume at the strongly pronounced minimum is 12% smaller than the final equilibrium volume of the dry bead (Fig. 7.51, curve 5). An even stronger excess contraction, 18%, is also observed while the hydrophobic hypercrosslinked material prepared by crosslinking linear polystyrene to 200% with XDC is dried from water. The evaporation of residual water from the smallest pores and the return of this system from the excessively compacted state to the equilibrium state of the dry material occur relatively slowly, within 15–20 min.

After the introduction of polar sulfonic functions or quaternary ammonium groups into the polystyrene network, the material becomes hydrophilic; water solvates the polar groups of the polymer chains, thus plasticizing the polymeric phase. As in the case of removing toluene from nonpolar sorbents by drying, the removal of water from the hydrophilic matrices of ion-exchange resins is accompanied mainly by non-elastic, plastic, deformations of the polymer network stressed by capillary contraction. Under these conditions of plasticization of the polymer phase by water, the deswelling anomaly manifests itself primarily as an abrupt decrease in the bead volume, with a subsequent slow approach to an equilibrium value. Here, we should emphasize once again that for non-porous polymers the deswelling plot describes a gradual decrease in the bead volume without inflections, steps, and extremes (Fig. 7.52, curve 3). There is no capillary contraction here because of the absence of free pore space in which concave liquid menisci could form.

CHAPTER 8

Soluble Intramolecularly Hypercrosslinked Nanosponges

1. INTRAMOLECULAR CROSSLINKING OF POLYSTYRENE COILS

Among the numerous naturally occurring and synthetic macromolecular species, only four basic types can be distinguished when considering the overall shape of individual species in a dilute solution. These types are

- globular macromolecules, for example, proteins, largely spherical, dense species that contain no, or almost no, solvent molecules in their interior;

- rigid rod-like molecules, for example, helical polypeptides or RNA;

- flexible and voluminous coils which are characteristic of the majority of synthetic polymers and denatured proteins (the coils are formed by both linear or branched macromolecules); and

- highly branched (hyperbranched) macromolecules [211, 212, 213] and dendrimers [214] which are related topologically to branched polysaccharides (glycogen, amylopectin, and galactomannans) and which can be placed between the globules and coils with regard to their flexibility and capacity to incorporate the solvent.

These types of macromolecules are shown in Fig. 8.1. Many other variants of macromolecules have been synthesized; however, their overall shape represents combinations of the above four basic types, for example, a globule with one or several long-coiled chains grafted onto its surface. Similarly, all brush-type, worm-like, star-shaped molecules, dumbbells, etc., can be classified according to combinations of the above basic structures.

This chapter introduces a novel macromolecular species, the intramolecularly hypercrosslinked polystyrene (IHPS) or "nanosponge," which differs fundamentally from the above four basic types

Comprehensive Analytical Chemistry, Volume 56
ISSN 0166-526X, DOI 10.1016/S0166-526X(10)56008-3

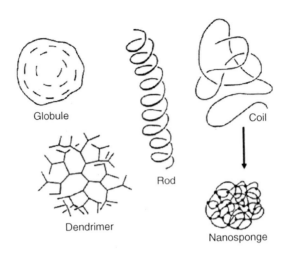

Figure 8.1   Schematic representation of four known types of macromolecular species and the nanosponge.

with respect to its topology as well as the behavior both in dry state and in solution.

When reacting a dissolved polymer with a bifunctional crosslinking reagent, there is only one way to significantly decrease the probability of the formation of intermolecular cross-bridges without affecting the probability of the formation of loops within the polymeric coil: diluting the initial polymer solution. For a polystyrene chain consisting of 2000 units (molecular weight of about 200,000 Da) and having 40 reactive sites, the probability of an intramolecular crosslinking at the total concentration of the polymer in a solution of 0.2 g/L was estimated to be 300 times higher than the probability of intermolecular binding [215]. When dealing with dilute solutions, it is thus possible to conduct the crosslinking reaction within individual macromolecular coils.

The synthetic path to the IHPS thus consists of introducing intramolecularly a large number of rigid crosslinks between the different repeat units of a single linear polymeric chain taken in the form of a coil in a dilute solution. In this way one can expect to arrive at a hypercrosslinked structure on a (macro)molecular level.

In general, intramolecularly crosslinked macromolecules, or microgels, can be prepared by aqueous microemulsion copolymerization of mono- and bifunctional comonomers using appropriate stabilizer systems [216–219]. This procedure results in nearly homogeneous spherical species, from about 20–200 nm in diameter and having molecular weights of several million daltons, and resembling the structure of networks of conventional gel-type styrene–divinylbenzene (DVB) copolymers [219].

Microgels with molecular weights over 10^5 Da, having a rather dense center and opening up continuously toward the outer regions, are produced by a crosslinking copolymerization procedure in a homogeneous solution in an organic medium below a certain critical concentration of the comonomers [220, 221]. Staudinger and Husemann [222] were probably the first to obtain in 1935 such colloidal solutions of poly-DVB microgels by a prolonged heating of DVB in a very dilute solution. In an important and detailed study [223], Downey et al. examined the conditions of microgel formation in the copolymerization of DVB with 4-methylstyrene at a low concentration, 4 vol.%, of the comonomers, in a mixture of methyl ethyl ketone, a thermodynamically good solvent, with heptane as the polymer-precipitating component. Soluble microgels have been only obtained at DVB content below 20%. At higher DVB concentrations, a continuous space-filling macrogel, microspheres of 2–4 μm in diameter, or just a coagulum have been obtained when reducing the methyl ethyl ketone content in the solvent mixture below 25 and 10%, respectively.

Another approach to crosslinked molecular species, introducing intramolecular crosslinks into dissolved macromolecules, has been examined by several groups. Allen et al. [215] reacted a partially chloromethylated polystyrene with butylamine and then introduced crosslinks by involving the secondary amino groups thus formed into a reaction with hexamethylene diisocyanate. A reasonable reduction of the radius of gyration of the coil (to about 0.5 for a degree of crosslinking of 2.5%) in solution has been registered [224, 225]. On the other hand, Antonietti et al. [226], by direct crosslinking of polystyrene coils with p-bis(chloromethyl)benzene under Friedel–Crafts conditions, observed a surprisingly small change in the dimension of the coiled species. Intramolecular crosslinking via the chloromethyl groups was also applied for the preparation of amine-functionalized polystyrene nanoparticles [227]. Mecerreyes et al. [228] have prepared linear copolymers containing acryloyl or methacryloyl functionalities, which were then involved in a self-crosslinking reaction in ultradiluted solutions in the presence of a radical initiator. Unimolecular species thus obtained exhibit a smaller hydrodynamic volume and a higher glass transition temperature than those of linear precursors.

An excellent review by Funke et al., "Microgels – Intramolecularly Crosslinked Macromolecules with a Globular Structure" [229], analyzed the synthesis conditions and properties of various microgels prepared according to the above-mentioned three approaches.

Our experiments fundamentally differ from those published in the literature in that the extent of crosslinking examined thus far was rather small, whereas our aim was to involve every phenyl group of the initial swollen polystyrene coil into an intramolecular crosslink. The open network–type products of such intensive crosslinking must become conformationally rigid. Due to the rigidity, the formed IHPSs differ from the flexible intramolecularly crosslinked microgels previously described as dramatically as the hypercrosslinked polystyrene networks differ from the conventional gel-type styrene–DVB copolymers.

As mentioned earlier (Chapter 6, Section 1), the chains of polystyrene with a weight-average molecular weight of about 300,000 Da and an intrinsic viscosity of 0.84 dL/g in dichloroethane become disentangled and barely touch each other when the concentration comes down to approximately 1.5 g/L. In more diluted solutions of the polystyrene, as well as in solutions of lower-molecular-weight polymers, there is a hope that the intramolecular reaction of a bifunctional reagent within a macromolecular coil would predominate over its reaction with two polystyrene coils. With this intention, polystyrene of 300 kDa has been subjected to intensive crosslinking in dilute dichloroethane solutions (0.05–0.5%) which, indeed, resulted in obtaining soluble materials, in contrast to single gel blocks obtained from semi-diluted or concentrated solutions.

There are two general approaches to intramolecularly hypercrosslinked polystyrene.

The first represents the direct interaction of polystyrene with, for example, 0.5 mol of monochlorodimethyl ether (MCDE) [230]. In this case, however, the reaction medium proves to be extremely diluted with respect to all reacting components, including the crosslinking reagent and the catalyst (stannic tetrachloride), and completing the process requires approximately 100 h. Also, small traces of impurities in the large amount of the solvent can deactivate the catalyst. These factors result in poor reproducibility of the reaction product, especially at very low concentrations of about 0.05% (Table 8.1).

The second approach [231] utilizes the fact that, under mild conditions, MCDE can safely introduce a desired amount of chloromethyl

Table 8.1 Properties of initial polystyrene, PS-300, and intramolecularly hypercrosslinked polystyrene, IHPS, prepared by treating PS-300 with MCDE [230]

Type	C (%)	$[\eta]$ (dL/g)		ν (cm^3/g)		S (m^2/g)	Swelling with methanol	
		Toluene	EDC	Toluene	EDC		(mL/g)	(v/v)
PS-300	–	0.95	0.84	0.840	0.918	0	–	–
IHPS-1	0.5	*	0.30	0.808	0.911	215	*	3.0
IHPS-2	0.5	0.57	0.54	*	*	115	*	*
IHPS-3	0.5	0.08	0.09	*	*	140	10.2	*
IHPS-4	0.5	*	*	*	*	1000	*	3.0
IHPS-5	0.25	0.20	0.10	0.856	0.907	260	6.4	*
IHPS-6	0.1	0.08	0.09	0.854	0.883	107	7.4	*
IHPS-7	0.1	0.09	0.08	*	–	45	9.2	*
IHPS-8	0.05	0.55	0.48	*	–	5	*	*

C = polymer concentration during the synthesis, %;
$[\eta]$ = intrinsic viscosity;
ν = specific partial volume.
EDC = ethylene dichloride
* Not measured.

groups into linear polystyrene chains, which then can interact with the phenyl rings of the same chain during the second stage of the reaction conducted at an elevated temperature or using a more efficient catalysis. Therefore, the preliminary partially chloromethylated polystyrene can serve more conveniently as the starting material. It already incorporates chloromethyl groups and can be treated in a diluted solution by an excess of the catalyst. The reaction is completed within less than 10 h at 80°C, since the reacting components already reside within the molecular coil and the catalyst tends to form complexes with polystyrene. It should be noted that the onset of the intramolecular crosslinking results in the rapid contraction of the initial coils, thus additionally favoring the intramolecular reaction as compared with the intermolecular reaction.

However, the intramolecularly hypercrosslinked soluble polymer obtained in this way always contained a certain amount of active chlorine, usually from 1.5 to about 6% (when a completely chloromethylated polystyrene was used). We assume that the unreacted chloromethyl groups predominantly locate on the surface of the IHPS species where they cannot find an appropriate intramolecular reaction partner. A similar assumption was also made by Kim and Webster [232] for the residual functional groups of hyperbranched polyphenylene macromolecules. The groups exposed to the surface can make the final product unstable in more concentrated solutions or during the storage of the dry material, because of possible intermolecular reactions with the participation of the residual functional groups. Fortunately, the pendent chloromethyl groups can be easily eliminated, if necessary, by end-capping with toluene on adding the latter to the reaction mixture after completing the intramolecular crosslinking. On the other hand, the surface-exposed functional groups open many ways to the surface modification of the IHPS species.

The intramolecularly hypercrosslinked and toluene-passivated polystyrene material can be precipitated from the synthesis solution with methanol in the form of a fine, slightly yellow powder. The latter remains soluble in dichloroethane, toluene, chloroform, and dioxane, that is, in the thermodynamically good solvents for the polystyrene precursor, and the solutions remain clear and stable for a long period.

The intrinsic viscosity $[\eta]$ of all IHPS samples in toluene and dichloroethane (Tables 8.1 and 8.2) was found to be much smaller than the value of both the starting unsubstituted polystyrene and chloromethylated polymer; the higher is the molecular weight of the starting polymer, the larger will be the difference. This fact unambiguously indicates conversion of the initial coils into significantly densified species.

Two distinguishing features are characteristic of the dry IHPS material:

(i)   It consists of highly porous species with the measured (apparent) specific surface area up to 1000 m^2/g (Tables 8.1 and 8.2); it should be noted that estimations of the outer surface of non-porous macromolecular globules (98 Å in diameter for PS-300 kDa) produce a noticeably smaller value of 650 m^2/g.

(ii)   The powder IHPS materials (even products with a relatively small inner surface area) increase in the volume by a factor of 3–4 when treated with methanol, the precipitating media for the linear polystyrene.

These two distinguishing properties unambiguously indicate the hypercrosslinked inner structure of the synthesized polystyrene species.

We introduce the term nanosponge for the IHPS species, since it refers to the macromolecular range of about 10–20 nm of the species' sizes; it is also associated with the low density and porosity of the material, its ability to readily incorporate large amounts of any liquid, as well as the rigidity of the species, both in dry and in swollen state.

Table 8.2 Properties of initial polystyrene (PS), chloromethylated polystyrene (CMPS), and intramolecularly hypercrosslinked polystyrenes (IHPSs), based on the latter

Type	Cl (%)	C (%)	$[\eta]$ (dL/g)		ν Toluene (mL/g)	S (m²/g)	Swelling with methanol	
			Toluene	EDC			(mL/g)	(v/v)
PS-300ᵃ	0	–	0.95	0.84	0.840	0	–	–
CMPS-300	18	–	0.66	*	0.834	0	–	–
IHPS-9	11	0.5	*	*	*	620	*	*
IHPS-10	8.9	0.5	0.35	0.22	*	1000	*	*
IHPS-11	18	0.2	0.13	0.18	*	960	*	*
IHPS-12	14.8	0.25	0.22	0.19	*	950	5.5	4.6
IHPS-13	18	0.1	0.10	*	*	750	*	*
PS-100ᵃ	0	–	0.48	*	0.893	0	–	–
CMPS-100	14	–	0.44	*	0.835	0	0	–
IHPS-14	14	0.1	0.10	*	0.847	670	6.2	2.75
PS-80ᵃ	0	–	0.38	*	0.916	0	0	–
CMPS-80	14.1	–	0.33	*	0.834	0	0	–
IHPS-15	14.1	0.2	0.09	*	0.846	450	4.2	2.75
PS-50ᵃ	0	–	0.30	*	0.917	*	0	*
CMPS-50	5	–	0.30	*	0.882	*	0	*
IHPS-16	5	0.1	0.08	–	0.893	*	4.2	*

C = chloromethylated polymer concentration during the crosslinking reaction, %; Cl = chlorine content of CMPS, %; $[\eta]$ = intrinsic viscosity; ν = specific partial volume.
Reprinted from [231] with permission of Elsevier.
* Not measured;
ᵃ The numbers indicate the molecular weight of starting polystyrene in kDa.

2. PROPERTIES OF POLYSTYRENE NANOSPONGES

As discussed, the intensive crosslinking of polystyrene coils in dilute solutions results in the formation of the soluble polymeric species, nanosponges, which can be expected to reproduce on the molecular level the major properties of the hypercrosslinked networks prepared from semi-diluted polystyrene solutions. Since the nanosponges present a novel type of macromolecular objects, their physico–chemical characterization could be of great theoretical interest.

From the data presented in Tables 8.1 and 8.2 it follows that the values of specific partial volumes, ν, of polystyrene in diluted solutions decrease on introducing chloromethyl groups into the polystyrene chains, which logically reflects the higher density (i.e., atomic weight) of chlorine-incorporating fragments. Less predictable is the fact that the specific partial volume of IHPS polymers in toluene, tetrahydrofuran, and dichloroethane is also smaller than that of the initial linear polystyrene precursor. The corresponding values for the IHPS and polystyrene solutions in tetrahydrofuran are 0.856 and 0.896 cm^3/g, respectively. This tendency cannot be accounted for the residual chlorine content of IHPS, since a similar difference was also observed for the toluene-end-capped products containing no residual chlorine. Obviously, 1 g of linear polymer displaces more solvent from a solution than does 1 g of an intramolecularly hypercrosslinked polymer. In our opinion, this tendency can be explained by the difference in the mobility of the structural units of the two compared species. While in the linear polystyrene the phenyl groups freely vibrate and rotate around the single bond to the carbon–carbon chain, thus requiring a definite additional space, all structural units in the hypercrosslinked network are highly restricted in any kind of movement. Here, the amplitude of all movements is small and the whole structure can be solvated more tightly by the solvent molecules, thus occupying less space in the solution.

Table 8.3 demonstrates the extremely high sedimentation constants, S_0, of IHPS molecules, compared to the parent linear polystyrene and chloromethylated polystyrene, all examined with an ultracentrifuge. The difference implies a significant contraction of polymeric coils in the course of their crosslinking. In this respect, the nanosponges resemble globular

Table 8.3 Molecular weights of parent polystyrene (PS), chloromethylated polystyrene (CMPS), and final intramolecularly hypercrosslinked polystyrenes (IHPSs)

Type	S_0 (Svedberg)	$\bar{M}_w$ kDa × 10^{-3}	
		Light scattering	Sedimentation
PS-300	5.8	*	340
IHPS-6	18.2	720	*
CMPS-300	10.1	*	*
IHPS-13	17.2	1200	*
PS-100	3.8	103	100
CMPS-100	6.5	175	100
IHPS-14	18.4	460	1000
PS-80	3.7	80	80
CMPS-80	6.5	*	92
IHPS-15	25.7	700	*
PS-50	2.95	50	50
CMPS-50	3.4	50	52
IHPS-16	6.5	80	120

S_0 = sedimentation constant.
Reprinted from [231] with permission of Elsevier.
* Not measured.

proteins; they are, however, differing from them fundamentally in that at least two-thirds of the IHPS volume remain occupied by a solvent.

The molecular weight of IHPS samples, when measured by both light-scattering and sedimentation-diffusion techniques, exceeds significantly those of the initial polymers. Surprisingly, the results of these two methods were found to poorly correlate with each other. Although the light-scattering Zimm's diagrams for IHPS proved to be distorted, they still should allow calculating the molecular weights from the scattering under larger angles [233]; it is this approach that was used for the IHPS samples. Usually the distorted Zimm's diagrams indicate the presence of a certain amount of microgels in the polymer solution. Nevertheless, by treating the data in accordance with the suggestions of Lange [234, 235], no microgel admixtures could be revealed: the light-scattering species proved to be rather uniform, with an average diameter of 13.6 nm for the IHPS-14 sample.

The abnormally high molecular weight of the IHPS samples could be easily understood in terms of a partial intermolecular crosslinking, which could not be ruled out completely even for a highly diluted starting polystyrene solution. If two or more macromolecules happen to combine during the crosslinking reaction, they should form hypercrosslinked particles of a more or less distorted shape, enlarged size, and multiple molecular weights. In order to facilitate detection of the species with multiple molecular weights, the subsequent experiments have been carried out [236] with a polystyrene standard having the molecular weight of 330 kDa and narrow molecular weight distribution of $M_w/M_n = 1.04$. The polymer was first subjected to partial chloromethylation (13.84% Cl) or complete chloromethylation (23.34% Cl) and then to intramolecular crosslinking at a concentration of 0.5 g/L in dichloroethane. The two final polymers (IHPS-17 and IHPS-18) were found to contain 2 and 6% residual chlorine, respectively. Notably, the hydrodynamic properties of the two batches of intramolecularly hypercrosslinked polystyrene were found to be very similar, in spite of the difference in the formal degree of crosslinking, 100 and 200%. Obviously, this difference is not really important, since the basic properties of bulk hypercrosslinked polystyrene networks do not change noticeably beyond the degree of crosslinking of about 50%. It should only be borne in mind that the introduction of different amounts of $-CH_2-$ bridges would cause a certain difference in the molecular weights of the final IHPS species, 350 and 370 kDa, respectively.

Intramolecular crosslinking causes a dramatic drop in the intrinsic viscosity of the solutions of polymer PS-330, from 0.98 down to 0.10

dL/g or even less. Moreover, the viscosity becomes nearly independent of the concentration (as well as of the molecular weight of nanosponges [231]), indicating that the polymeric species behave in solution as rigid spheres. Though very low, this intrinsic viscosity still exceeds the 0.0236 dL/g value for hard impermeable spheres calculated according to Einstein's equation and assuming a density of 1.1 g/cm^3. Obviously, the nanosponges differ from hard spheres in that they incorporate an appreciable amount of the solvent.

On crosslinking, the molecular weight of the polystyrene macromolecules of 330 kDa should rise to 370 kDa for sample IHPS-18, due to the introduction of one additional $-CH_2-$group per styrene repeat unit (at the formal degree of crosslinking of the nanosponge of 200% attained). It could even amount to 390 kDa if the residual 6% chlorine is also taken into consideration.

Comparing the initial linear polymer with the final IHPS-18 nanosponges by ultracentrifuge investigation showed that the diffusion coefficients in tetrahydrofuran increased by a factor of 2 and the sedimentation constants increased from 7.5 to 32.6 s, which is consistent with the expected strong compactization of the initial polymeric coil. However, when determined in accordance with the Archibald technique, the molecular weight of the nanosponges was found to be 460 kDa, again larger than the expected value of 390 kDa for IHPS-18.

Elastic light scattering on IHPS-18 resulted in a highly distorted Zimm's diagram (Fig. 8.2), which, in terms of Guinier [237], could approximately be interpreted as light scattering of a polymer with a molecular weight of 490 kDa containing an admixture of 0.3% of a polymer with $M = 85,000$ kDa. By contrast, the initial polystyrene gives a classical Zimm's diagram with a good match of the calculated and expected molecular weight of 330 kDa.

3. SELF-ASSEMBLING OF NANOSPONGES TO REGULAR CLUSTERS

In order to estimate the molecular dimensions of nanosponges, we applied dynamic light scattering and size-exclusion chromatography (SEC) to the solutions of the new species and low-angle X-ray scattering, electron microscopy, and scanning atomic force microscopy to the dry material [236, 238–240].

Simple calculations show that a fully collapsed crosslinked polystyrene molecule with a molecular weight of 370 kDa would occupy the volume of a sphere with a diameter of 10.4 nm if it could maintain the density of

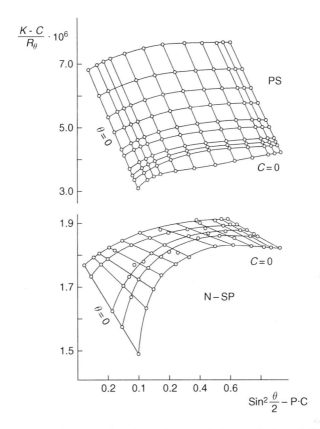

Figure 8.2 Zimm diagrams for the solution of PS-330 before (PS) and after intramolecular crosslinking (N–SP = nanosponge). Solvent: tetrahydrofuran. *(Reprinted from [236] with permission of the American Chemical Society.)*

conventional polystyrene, 1.04 g/cm^3. Since the nanosponge is expected to be a hypercrosslinked, "porous" low-density species of about 0.69 g/cm^3, it possibly would have a diameter of 11.9 nm when in dry state and 17.2 nm after having increased its volume by a factor of 3 on swelling with an organic solvent.

Indeed, when examined by low-angle X-ray scattering technique, the nanosponge powder scattered X-rays as a material with fluctuations in its density having a lattice spacing period of about 12 nm. SEC in tetrahydrofuran or chloroform, on both analytical and preparative scale [236], was carried out on a silica gel Diasorb Si 400 column having an acceptable performance (over 9000 theoretical plates for naphthalene in acetonitrile) and separation power (11 components could be identified in a test mixture of polystyrene standards with molecular weights from 0.68 to 7000 kDa). As follows from Fig. 8.3, the crude product of the intramolecular cross-linking of PS-330 contained two high molecular weight fractions, in addition to the main sharp peak. The elution volume of the main product corresponded to the elution volume of a linear polystyrene with a mole-cular weight of 70,800 Da. The mean gyration radius of a coil of such molecule can be calculated as equal to $2R_g = 18.7$ nm. This value coincides with the expected diameter of the swollen nanosponge particle of about 17.2 nm given above.

By preparative SEC of a concentrated solution of the IHPS-18, an attempt was made to obtain the major product, as well as the two by-products of higher molecular weight, in the purest possible form. Surprisingly, even after repeating many times the chromatography, each of the three fractions of interest proved to be contaminated with the neighboring fractions (Fig. 8.3). This unambiguously indicates a relatively fast association–dissociation process between the individual nanosponges of the main fraction #3, which leads to rather regular associates or clusters of

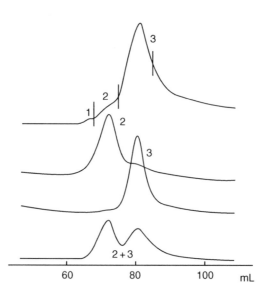

Figure 8.3 Preparative size-exclusion chromatography of the bulk intramolecular crosslinking product. Fractions 1, 2, and 3 correspond to microgels, clusters, and nanosponge, respectively. *(Reprinted from [239] with permission of Wiley & Sons, Inc.)*

fraction #2 and some kind of higher associates or microgels comprising the minor by-product in fraction #1. When taken separately, clusters in fraction #2 elute with the elution volume of a polystyrene having a molecular weight of about 158,500 Da, which corresponds to an average diameter of about 30 nm for the cluster.

Dynamic light scattering [236] of fractions #3 and #2 allowed estimation of the diameter of the nanosponges as 17 nm and that of the clusters as 34 nm. Indeed, transition electron micrographs reveal the spherical shape and a distinct size difference of the two types of dry species, approximately 15–20 nm and up to 40–50 nm, respectively [236, 238].

Formation of clusters explains the fact that all attempts of determining the molecular weight and size of the nanosponges resulted in values which were higher than expected. Most interestingly, one specific self-assembling associate, that is, a cluster of nanosponges, predominates in the high molecular weight fractions of the reaction mixture. It was suggested [238] that the most probable cluster of identical spherical species consists of 13 subunits (with 12 species surrounding the central one). In this case the associate should belong to the series of regular clusters incorporating $N = 1 + \sum(10n^2 + 2)$ identical spheres, where n is the number of shells around the central sphere ($n = 1$ in our case). Such shells can be recognized around any given ball in a volume which is densely packed with identical balls. It is natural that the discrete association character could only be expected with nanosponges of a very narrow size distribution, but not with the previously examined polydispersed nanosponge material.

Because of the reversibility of the association process, it proved impossible, at least at the beginning, to obtain pure fractions of the individual nanosponges and their clusters, even by repeating chromatography several times. However, it has been noted that during the two-year-long period of storage of the reaction mixture in ethylene dichloride, the cluster fraction rose to about one-fifth of the total polymer amount and the rate of its dissociation gradually decreased. This finding can only be explained by taking into consideration the fact that the initial polystyrene standard with $M = 330,000$ Da and $M_w/M_n = 1.04$ was a polymer of narrow molecular weight distribution, but not a perfect monodisperse sample. Therefore, the clusters formed through self-association were not perfect at the beginning, but they were given sufficient time to improve their structure and stability by exchanging their subunits with the equilibrium solution of individual nanosponges. This could be expected to take place by the clusters losing their least appropriate subunits and replacing them by the better fitting species of a more convenient size. Anyway it is now possible to chromatographically separate the nanosponge and cluster fractions and subject them to the sedimentation analysis.

Figure 8.3 represents a typical chromatogram in chloroform of the aged reaction mixture (injected in ethylene dichloride) and shows the cutting of the equilibrated mixture of the crosslinking products into three fractions. As the first fraction (less than 2% of the polymer) comprises larger conglomerates or associates and elutes outside the calibration range of the chromatographic column ($M > 7 \times 10^6$ Da), it was not further examined. The second fraction amounting up to 20% of the total polymer contains cluster species, whereas the third, major fraction contains nanosponges. A subsequent analysis of the separately collected fractions 2 and 3 visualizes the possibility of obtaining chloroform solutions of the clusters and nanosponges in a sufficiently pure state. After a quick concentrating step in vacuum to a concentration of about 2 g/L, these fractions were then subjected to a more thorough sedimentation analysis. In addition, a 1:1 mixture of the fractions represented in the Figure 8.3 was also prepared and analyzed.

By using the technique of an artificial boundary in the ultracentrifuge cell, diffusion coefficients D_0 for the clusters (fraction 2) and the nanosponges (fraction 3) were determined at a single concentration of 2 g/L (Table 8.4). It has been shown earlier that the hydrodynamic parameters of these species do not depend on the concentration. This is not the case with the initial linear polystyrene, where several measurements in the concentration range of 2.5–1.0 g/L were needed, in order to extrapolate to the limiting value of the diffusion constant D_0 at zero concentration.

In another set of experiments, sedimentation constants of all the species under consideration were determined. Because of the large difference in the flotation rates of the species in chloroform, the rotation speed of the ultracentrifuge had to be optimized. Sharp moving concentration zones could be observed at 40,000 rotations min^{-1} for the polystyrene coils, whereas only 16,000 min^{-1} for the clusters and 36,000 min^{-1} for the nanosponges were needed (Fig. 8.4). At an intermediate rotation rate of 25,000 min^{-1}, the mixture of fractions 2 and 3 gave two distinct concentration zones. Evaluation of the movement and broadening rates of these double peaks provided an additional possibility for the determination of the diffusion and sedimentation coefficients of the two components of the nanosponge–cluster mixture. As follows from Table 8.4, there is a good

Table 8.4 Molecular parameters of polymeric species determined through sedimentation analysis and theoretically expected corresponding values [238]

Polymer	$10^7 D_{sd}$ (cm²/s)	$10^{13} S_0$ (s⁻¹)	Found		Expected	
			M (Da)	d (nm)	M (Da)	d (nm)
Polystyrene PS-330	2.30	9.80	330×10^3	35.0	330×10^3	37.5
Nanosponge (fraction 3)	4.85	18.4	361×10^3	16.6	392×10^3	17.2
Cluster (fraction 2)	1.75	87.4	4.7×10^6	46.0	5.1×10^6	45
Mixture of fractions 2 and 3	4.77	18.95	373×10^3	17.0	392×10^3	17.2
Cluster	1.92	85.45	4.2×10^6	42.0	5.1×10^6	45

D_{sd} = self-diffusion coefficient; S_0 = sedimentation constant.

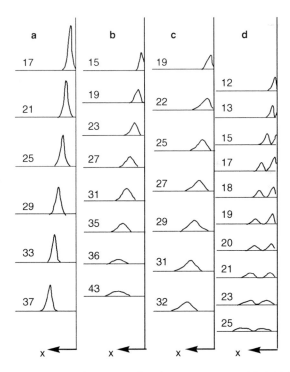

Figure 8.4 Sedimentation diagrams of polymeric species. (a) Initial polystyrene; (b) fraction 3 with nanosponges; (c) fraction 2 with clusters; (d) a 1:1 mixture of clusters and nanosponges. Solvent: chloroform, 25°C. Rotation rates (min⁻¹): (a) 40,000, (b) 36,000, (c) 16,000, (d) 25,000. *(Reprinted from [239] with permission of Wiley & Sons, Inc.)*

correspondence between the values obtained from analyzing the mixture and the individual species, which implies that there is no significant re-equilibration of the system during the centrifugation experiment.

Molecular weights of the polymers were calculated from the given diffusion and sedimentation coefficients according to Svedberg's formula [241]:

$$M_{SD} = \frac{S_0 RT}{D_0(1 - v\rho_0)} \qquad [8.1]$$

where $R = 8.314 \times 10^7$ erg/(mol grad) is the universal gas constant, T is the absolute temperature (in K), v is the specific partial volume of the polymer in the solution (in cm^3/g), and ρ_0 (= 1.473 g/cm^3) is the density of chloroform at 25°C. Finally, on the basis of the measured diffusion coefficients and with the assumption of the spherical shape of the species, the diameters of the latter could be calculated according to the following relationship [242]:

$$R = \frac{kT}{6D_0\pi\eta} \qquad [8.2]$$

where $\eta = 0.542$ cps is the viscosity of chloroform at 25°C.

The last two columns of Table 8.4 present the expected values of the molecular weights and sizes of the crosslinked species. In the case that the cluster consists of 13 subunits, its molecular weight should amount to 4.8 $\times$ 10^6 to 5.1 $\times$ 10^6 Da and the diameter to about 45 nm. The experimental values of Table 8.4 stay in good agreement with these expectations.

Atomic force microscopy [240] allows visualizing both the individual nanosponges from fraction 3 and their clusters from fraction 2 of the SEC separation (Fig. 8.5 left and right panels, respectively). The sizes of the species do not contradict the expected values of about 12 nm for the dry nanosponge and about 36 nm for the cluster of 13 subunits. Fig. 8.5 right presents the cluster species in a large number, besides the less-abandoned individual nanosponges. The figure is also remarkable in that the mica surface displaces a dislocation indicated by the form of a broad white line in the nanograph. Close to this dislocation spherical clusters can be observed that definitely exceed the size of the $N = 13$ species. Obviously, these clusters must contain two or three shells around the central nanosponge. With two shells, $n = 2$, and $N = 1 + 12 + 42 = 55$, the second-generation cluster must have a diameter of 60 nm and a molecular weight of 21,450,000 Da. The third generation of clusters should incorporate $N = 1 + 12 + 42 + 92 = 147$ subunits and rise

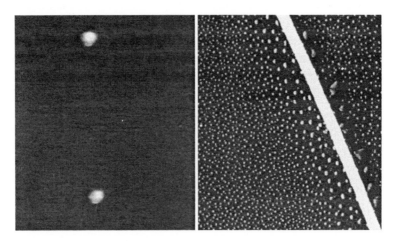

Figure 8.5   AFM micrographs of nanosponges and thier clusters. *(Reprinted from [240] with kind permission of Wiley & Sons, Inc.)*

to 80 nm in diameter. Fig. 8.5 right possibly demonstrates species of that size, which could comprise the minor not-examined fraction #1 of the crude product of the intramolecular crosslinking of polystyrene. On the mica surface they may have formed during the evaporation of the applied dilute chloroform solution of the $N = 13$ clusters, due to the enhanced concentration of polymeric species near the dislocation step.

It should be noted here that individual metal clusters of the $N = 1 + \sum(10n^2 + 2)$ series are known, for example, a Pd-1,10-phenanthroline cluster incorporating 561 Pd atoms which are arranged in a five-layered icosahedron [243]. The structural aspects of aggregation of globular proteins in solutions have long been an area of intensive investigation (see review [244] and references therein). More recently, the association behavior of synthetic hyper-branched macromolecules and dendrimers [245–247] as well as that of microgels and microemulsions [218, 248] have attracted attention. Self-assembling phenomena with nanoparticles [249] and formation of various micelles [250] and nanospheres [251] from diblock copolymers have been observed and examined. To the best of our knowledge, however, no examples of regular cluster formation through self-assembling of synthetic homopolymer species have been described prior to our experiments on polystyrene nanosponges.

At present, it is too early to talk of any practical application of the polystyrene nanosponges, considering the low productivity of their preparation in the low-concentration solutions. However, the presence of residual chloromethyl groups on the surface of nanosponges opens numerous ways for chemical transformations of the surface, which could provide species with biocompatibility and appropriate tissue specificity, while the porous interior of the molecular vehicle could provide a controlled delivery and release of efficient drugs.

CHAPTER *9*

Hypercrosslinked Polymers – A Novel Class of Polymeric Materials

1. DISTINGUISHING STRUCTURAL FEATURES OF HYPERCROSSLINKED POLYSTYRENE NETWORKS

Hypercrosslinked polystyrene networks owe their extraordinary structure and remarkable properties to the basic principle of their preparation; the latter consists of intensive bridging of strongly solvated polystyrene chains with conformationally rigid links. This approach differs fundamentally from the traditional free radical copolymerization of styrene and divinylbenzene (DVB), which is also often carried out in the presence of an inert diluent. In the latter process, after a certain induction period, a network forms through gradual formation of primary styrene–DVB microgel species, which then separate from the initially homogeneous comonomer solution and finally merge into a gel. Depending on the type of solvent used, two reasons have been recognized for the above micro-phase separation that takes place close to the gel point. The first is the thermodynamic incompatibility of growing polymeric chains and microgels with the solvent, because the latter is a precipitant for polystyrene. The second reason manifests itself on conducting copolymerization in the presence of an excessive amount of a thermodynamically good solvent. At a certain threshold DVB content, the swelling ability of the growing network may prove to be insufficient to accommodate the entire amount of the good solvent present in the system. In this case, fully swollen styrene–DVB microgels separate from the solution through aggregation and agglomeration, while the excess solvent remains between the subunits of the emerging supramolecular construction. In both good solvents and precipitating media, the micro-phase separation results in the formation of a cauliflower-like texture of the final macroporous product that is generally composed of two segregated phases: the polymeric phase and the phase of the solvent.

Comprehensive Analytical Chemistry, Volume 56
ISSN 0166-526X, DOI 10.1016/S0166-526X(10)56009-5

The formation process of a hypercrosslinked polystyrene network is quite different. It starts with strongly solvated polystyrene chains, rather than monomers. It is very important to emphasize here that two basic factors prevent the starting homogeneous polymer solution (or gel) from micro-phase separation during the post-crosslinking process. First, long polymeric chains extend though the initial solution (or gel) and remain strongly solvated over the entire period of network formation. Second, the crosslinking bridges emerge throughout the volume of the reacting system with a statistically uniform distribution. The latter point implies that the probability of formation of intermolecular and intramolecular crosslinks depends only on the conformation of macromolecules and the probability of preexisting mutual contacts between different polymeric chains. The process of bridging does not cause any redistribution of the chains in the initial system. This statement requires further elucidation.

Since the post-crosslinking reaction is carried out in a thermodynamically good solvent, all polystyrene chains are strongly solvated and retain their extended relaxed conformation. The bifunctional (or polyfunctional) crosslinking agent first reacts with one of its active sites with a phenyl ring of a polymeric chain, thus providing the chain the ability to form a bridge on contacting an unsubstituted phenyl ring that belongs to a different chain or to a distant portion of the same chain. The crosslinking reaction starts simultaneously at many preexisting between-chain contacts and rapidly converts the initial solution into a gel. As this takes place, the favorable, strongly expanded conformations of the solvated polystyrene chains are fixed and preserved. It is generally believed that the chains exist in solution in the form of strongly swollen coils and that each chain runs through many neighbor coils. If this is so, a polystyrene chain of about 300,000 Da in a semi-concentrated solution of 0.1 g/mL must pierce at least 10 neighboring macromolecules. This makes the formation of intermolecular initial crosslinks more probable than the formation of intramolecular loops.

Another important question is whether or not the initial intermolecular crosslinking favors the formation of neighboring links between the same chains. Two neighboring links between same chains would bring the second chain in a more or less parallel position with the first one, thus favoring the formation of ladder-type structures. However, molecular modeling clearly shows that inserting two bridging methylene fragments between two neighboring pairs of phenyl groups on two chains is improbable, since it would imply a closure of a strongly stressed macrocycle. Indeed, according to the evaluation of the microstructure of an atactic polystyrene chain by X-ray scattering method [252], two neighboring phenyl rings in a zigzag chain are twisted by about 120°. Therefore, the phenyl ring next to the initial bridge would more likely interact with phenyls of a third chain, thus giving rise to formation of network meshes composed of several (more than two) chain segments and bridges. A logical consequence of this conformational consideration is that the initial bridge does not stimulate formation of neighboring bridges or domains with enlarged crosslinking density. Rather, uniformly (statistically) crosslinked open-work-type three-dimensional (3D) net must result from the post-crosslinking of solvated polystyrene chains. Obviously this is what happens in reality. All methods employed to characterize the porous structure of hypercrosslinked network testify to its single-phase nature and extremely narrow size distribution of network meshes.

When considering the distinguishing features of hypercrosslinked polystyrene, we are guided by the fact that all its peculiar properties become apparent at the degree of crosslinkings not less than 40%. At this threshold, 6 of every 10 phenyl rings are already involved in formation of bridges. At a degree of crosslinking of 100–200%, practically all phenyls of the starting polystyrene chains become connected with each other and then participate twice in the formation of interchain links. Undoubtedly, at such a high crosslinking density, traditional notions such as the chain length between neighboring junction points, number of elastically active chains, and the ratio between intermolecular and intramolecular crosslinks become completely meaningless. The only fruitful approach to understanding the properties of hypercrosslinked polystyrene networks is to consider their structure as a cycle-composed system.

A single-phase hypercrosslinked network is an ensemble of mutually condensed and interpenetrating non-planar cycles (meshes) composed of crosslinking bridges and chain segments between them (Fig. 9.1). Both the degree of crosslinking and the length of crosslinking bridges determine the size of the meshes (their contour length). In the studied systems the smallest unstressed mesh is composed of three pairs of neighboring phenyl rings belonging to three polystyrene chains (or distant chain segments) connected by three methylene groups. Certainly, any other arrangement of crosslinks would result in the formation of larger unstressed meshes, particularly when longer crosslinking agents such as p-xylylene dichloride

a   b

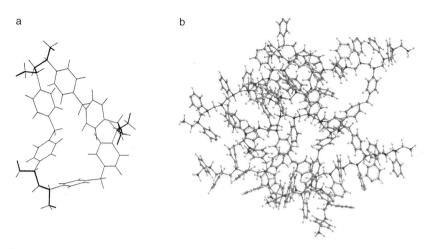

Figure 9.1 Computer simulation of (a) the smallest possible unstressed mesh *(after [174])* and (b) fragment of hypercrosslinked network *(from [253])*.

(XDC) or p,p'-bis(chloromethyl)-diphenyl are employed. It is, however, very important that even the smallest unstressed mesh incorporates many carbon–carbon single bonds and may, in principle, change its conformation.

Such a smallest non-planar network mesh is condensed with six adjacent meshes. Since in the hypercrosslinked network none of the meshes incorporates long flexible segments, any change in the conformation of one mesh immediately involves changes in the conformations of its neighbors. Conformational transitions in such a network are basically of a cooperative nature. Only a cooperative conformational rearrangement of all constituent network meshes can make the large alterations of the network volume possible, which take place on drying and swelling of the hypercrosslinked polymer.

However, the cooperative rearrangement is only possible, if all or the overwhelming majority of the meshes have enough freedom of movement. In the first place this freedom depends on the availability of free space inside the mesh, that is, on the number of alien chains (belonging to other meshes) that are embraced by the mesh. The extent of such interpenetration of meshes is unambiguously determined by the concentration of polystyrene chains in the initial solution subjected to crosslinking. The higher the polymer concentration in the unit volume of the solution, the more will alien mesh fragments be entrapped in the closing mesh, thus strongly restricting the conformational mobility of the latter. For this reason the swelling ability of the resulting network is largely independent of the crosslinking density, but rapidly decreases with increasing concentration of the starting polystyrene solution subjected to post-crosslinking. In the ultimate case, when a network forms with no diluent present in the system, as in the case of conventional styrene–DVB copolymerization, each network mesh is fully stuffed with alien polystyrene chains. These entrapped entanglements can never be disentangled. Even at a moderate degree of crosslinking of only 15–20%, such networks, although composed of really large meshes, are practically deprived of any conformational freedom and are unable to swell (increase their volume) on contacting solvents. In contrast to conventional copolymers, hypercrosslinked polystyrene is usually prepared in the presence of about 10-fold excess of a solvent. The drastically reduced extent of mesh entanglements provides the network with an unprecedented mobility and other unusual properties.

Thus, the size of network meshes and the extent of mesh entanglements are the basic topological factors specifying the mobility (including swelling ability) of polymeric networks. These factors – and not the chain length between adjacent junction points – must serve primarily for a theoretical description of network swelling, at least in the case of reasonably and densely crosslinked networks. If the crucial role of mesh entanglements would have been understood earlier, it would not have been necessary to speculate so much with such characteristics as elastically ineffective chains, loops, dangling chains, intramolecular links, etc., in order to explain the deviations in the swelling behavior of real networks from theoretical expectations. It is totally obvious now that no combination of these topological defects could ever elucidate the network mobility, particularly the swelling of hypercrosslinked polystyrene.

2. UNUSUAL STRUCTURE–PROPERTY RELATIONS FOR HYPERCROSSLINKED POLYSTYRENE

Being formed in the strongly solvating media and having avoided rejection of the diluent through micro-phase separation, the final hypercrosslinked polystyrene presents a single-phase rigid gel swollen with the initial solvent. The amount of the retained solvent determines the equilibrium swelling ability of the network in this solvent. However, it does not mean that the entire amount of the initial diluent remains included in the final product: a part of the solvent may escape entrapping through macro-phase separation, that is, syneresis.

A few words need to be said about the reason for the macro-phase separation. On the onset of post-crosslinking, the entire polymer solution is transformed into a gel. Still, full conversion of the bifunctional crosslinking agent requires time in the two-step Friedel–Crafts reaction. After attaching to one polymeric chain, most bifunctional reagent molecules have to spend some time until, through fluctuations, an aromatic ring of another chain approaches the reagent's pending second functional group to a distance suitable for reaction. The bridging immediately reduces the fluctuation amplitude of both the chains by fixing the relatively short interchain distance. The chain coils are thus forced stepwise to occupy less space by rejecting excessive solvent from the gradually shrinking gel. At a certain point of time, the network acquires sufficient rigidity that prevents further contraction. The residual functional groups are then consumed by reacting with the nearest phenyl groups or with just emerging crosslinking bridges, without causing further noticeable network shrinkage.

The macrosyneresis effect reduces the swelling ability of the final material in the solvent. In fact, an evident correlation between the volume of the separated solvent and the swelling capacity of the networks has always been observed in the experiments (Fig. 9.2). The extent of the macrosyneresis depends on several factors.

Many examples (see Chapter 7, Section 1.5) revealed that the solvent regain of the hypercrosslinked networks correlates with the rate of the crosslinking reaction of the polystyrene chains present in solution. In solvents facilitating the Friedel–Crafts reaction, with more reactive crosslinking agents, and at higher concentrations of the catalyst, polymers are obtained which demonstrate an enhanced swelling ability at any given degree of crosslinking. Most probably, the fast conversion of the system into a rigid unshrinkable network gives little time to the solvent to migrate from the gel into the outer space. Indeed, macrosyneresis is a relatively slow process as the solvent molecules have to diffuse from the center of a large gel block to its periphery. Obviously, in samples crosslinked to the same extent but at different reaction rates, the polystyrene chains are fixed with different conformations and the networks differ in their topology.

When higher degrees of crosslinking are desired, higher concentrations of the crosslinking agent and, correspondingly, larger amounts of the catalyst must be involved. The rate of the crosslinking reaction also increases, so that the fixation of initial chain conformations into a rigid

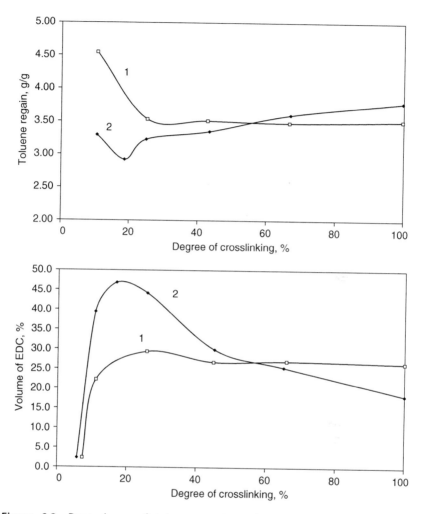

Figure 9.2 Dependence of toluene regain and volume of rejected EDC on crosslinking degree of the networks prepared by crosslinking linear polystyrene with (1) p-xylylene dichloride and (2) monochlorodimethyl ether.

network proceeds quickly and macrosyneresis occurs to a lesser extent. The highly unusual fact that the swelling of hypercrosslinked polystyrene increases parallel to the degree of crosslinking would be difficult to explain by any other reason than the higher synthesis rate and reduced extent of macrosyneresis.

It is also obvious that the macrosyneresis of the reaction media proceeds much easier from small beads of styrene–DVB copolymers than from a large-volume gel resulting from a solution of linear polystyrene; their volumes usually differ by several orders of magnitude. This size factor strongly contributes to the reduced swelling capacity of the hypercrosslinked networks derived from beaded styrene–DVB copolymers (along with the inherent inner stress introduced by the DVB units in the initial fully swollen beaded copolymer).

The hypercrosslinked polystyrene networks are rigid. They tend to retain the volume in which the crosslinking process has been completed, because any contraction of the rigid network would distort the basically unstressed mesh conformations and lead to the appearance and growth of inner stresses. Replacement of ethylene dichloride, the usual synthesis medium occupying the interior of the final network, by toluene, methanol, or any other organic solvent cannot noticeably affect the volume of the rigid network. Accordingly, the dry hypercrosslinked networks take up almost equal amounts of any organic solvent, regardless whether these are precipitants or solvating liquids.

Only a powerful effect like the total removal of the solvent from the swollen gel can change the network's volume. It is this impact that forces the network to shrink through a cooperative conformational rearrangement of constituent network meshes. However, the decrease in network volume is unavoidably accompanied by the appearance and rapid growth of strong inner stresses as a result of two opposite tendencies: the tendency to shrink and achieve a dense packing of the desolvated polystyrene chains, and the tendency to maintain the open-work structure of the rigid, highly crosslinked network. The stresses of the shrinking network accumulate in the form of distorted valence angles and bond lengths and they eventually equilibrate the attraction forces between the non-solvated network fragments. In any case, we consider both the systems, the initial strongly solvated gel and the final dry polymer, to be in equilibrium states. The high conformational mobility of the hypercrosslinked network allows the system network/media to reach the corresponding equilibrium state in a reasonably short time. Of course, the minimum of the energy of the equilibrated system does not imply minimization of inner stresses in the network (see Chapter 7, Section 4.2). It should be emphasized that one has to consider the equilibrium state of the whole system network/medium, rather than the state of the polymer alone.

The strong inner stresses accumulated in the dry network cannot relax with time on their own account. The relaxation of stresses is possible only through expansion of the network volume and elimination of distorted conformations of network meshes. Being an additional driving force for swelling, the relaxation of inner stresses accounts for the network swelling in badly solvating media and strongly contributes to the fast kinetics of network expansion on swelling.

Water, being characterized by very weak affinity to hydrophobic polystyrene, causes only a partial increase in the network volume and a partial elimination of the stresses existing in the dry material. The residual network stresses in aqueous media facilitate the absorption of all organic compounds, whether nonpolar or polar, with the sorption always accompanied by an additional volume increase of the polymer. As will be shown later, this is a supplementary driving force for absorption and a reason for the high loading capacity of hypercrosslinked adsorbing materials.

Shrinkage of rigid hypercrosslinked networks during solvent removal stops long before the moment when the polymeric chains achieve a dense packing that is normally characteristic of the initial polystyrene and its gel-type copolymers. The free volume remaining in the dry low-density single-phase hypercrosslinked material is so large, up to $0.5\,cm^3/g$, that it may be thought of as true porosity. However, this type of porosity is fundamentally new. Up to now, the porosity of the milky white two-phase conventional macroporous copolymers was well known to result from the microphase separation during their synthesis. Alternatively, the single-phase hypercrosslinked materials are transparent. They do not scatter visible light and scatter X-rays like a homogeneous polystyrene solution, rather than macroporous polystyrene copolymers.

The highly porous hypercrosslinked networks absorb very large amounts of inert gases at low temperature. It is important to understand, however, that the specific internal surface of $1500–2000\,m^2/g$ calculated from the adsorption data is not a real surface of pore walls. In the single-phase hypercrosslinked open-type network the "pores" are nothing else than spaces between loosely packed chains, while the high specific surface area values simply reflect the high capability of the polymer to absorb gases. That is why we use the term of "apparent inner surface area."

The dimensions of the pores are pretty small, and their precise determination proved to be a real challenge. Nevertheless, relying on data obtained using a number of methods, we arrived at the conclusion that, on average, the diameter of the pores is about 2–3 nm. This size is adjacent to real micropores (below 2 nm, in accordance with the IUPAC recommendations), and, using the modern terminology, we consider the hypercrosslinked polystyrene as the first nanoporous polymeric material.

In addition to the extremely large volume changes on swelling or drying, the high mobility of the low-density open-type hypercrosslinked networks is also the reason for large deformations of the polymer when exposed to an external mechanical force. This property is not trivial either, since it is not characteristic of conventional networks with a crosslinking density over 15–20%, which maintain their rigid glassy state up to very high temperatures. However, the very slow increase in the extent of deformation with temperature (at a constant loading force) for the hyper-crosslinked polystyrene has little in common with the normal sharp transition from a glassy state to a rubber elasticity state, which conventional gel-type copolymers with 2–3% DVB demonstrate at the glass transition temperature of about 100°C. Certainly, there are known polymeric systems characterized by a broad temperature range of transition from a glassy to a rubber elasticity state. This is the case with heterogeneous two-phase systems, such as interpenetrating networks composed of partially compatible components, semi-interpenetrating networks, or polymeric blends. In these cases phase transitions represent the superposition of several transition zones corresponding to all segregated constituent components [200, 254, 255]. This situation is not applicable to hypercrosslinked networks which are homogeneous single-phase polymers and, accordingly, would have to exhibit a much sharper phase transition.

Additionally, the thermomechanical deformation plots of hypercros-slinked polystyrene can be shifted along the temperature axis in the range of about 200°C, depending on the applied pressure. Although some forced high-elasticity deformations can be observed for a butadiene resin or polymethyl methacrylate [256] 70 and 90°C below their respective glass transition temperatures (until their brittle points are reached), still an analogous shift of the forced high-elasticity temperature by 200°C, downward to −70°C, seems to be unrealistic.

Finally, the inverse dependence of the deformation ability on the degree of crosslinking is another particular feature of the hypercrosslinked material that cannot be explained in terms of glass–rubber transitions. Thus, a classical glassy state is not characteristic of the hypercrosslinked polystyrene in the entire temperature range from −70 to +350°C, scanned in a conventional thermomechanical test. Again, we consider the mobility of network meshes, rather than the mobility of chain segments between the joints, to be responsible for the mobility of the entire dry hypercrosslinked network. Obviously, the cooperative conformational rearrangement of its meshes can already take place even at the very low temperature of $-70°C$.

Hypercrosslinked polystyrene cannot be considered to be an elastomer either. Although two of the most important features of rubber elasticity, namely large values of deformation and reversibility of the deformation, are characteristic of the hypercrosslinked polystyrene, the non-elastic part of their deformation is still definitely too high. The reversal of this non-elastic part of deformation requires long heating of the material or treating the sample with a solvent. Such a hindered relaxation is not characteristic of typical rubbers.

As was shown above, depending on the nature of the solvent, several solvent–hypercrosslinked polystyrene systems can be regarded as standing at a thermodynamic equilibrium. Similarly, for a dry hypercrosslinked network, several "quasi-equilibrated" states can be considered which, however, depend on the prehistory of the sample. Thus, the total pore volume and the inner specific surface area may depend on the type of the last solvent evaporated from the sample. Applying an outer pressure to the dry material may also reduce or totally eliminate its porosity. In the dry hypercrosslinked material the inner stresses are definitely equilibrated with the unsaturated attraction forces of its network fragments. However, such an "equilibration" can be achieved in many different sets of conformations of the network meshes. A uniaxial compression of the beaded sample distorts the initial structure of the open-network construction and results in replacing the initial strong interactions in the existing contact points of polymeric fragments by a new set of contacts and interactions. The new shape of the polymer bead with non-relaxed residual deformations may also be characterized by a new local minimum of the total network energy. Therefore, even a prolonged heating of the mechanically deformed sample should not necessarily result in a complete shape relaxation to the initially spherical bead. Only a cycle of swelling and drying of the precompressed sample will fully restore its initial conformational structure and spherical outer shape, and this conclusion corresponds to the experimental results.

In general, hypercrosslinked polystyrene is the first representative of polymeric materials the physical state of which cannot be defined in terms of customary perceptions of network behavior. Indeed, no conventional network polymer exists that would reduce its volume on heating or, conversely, increase it far beyond a usual thermal expansion. Several examples of such unusual behavior of hypercrosslinked polystyrene samples have been described in Chapter 7, Section 8.

We explained all of the aforementioned properties of hypercrosslinked polystyrene by its open-work molecular architecture and we see an undisputable corroboration of this explanation in the properties of nanosponges. The latter product has been obtained by applying the basic principle of hypercrosslinked network synthesis to individual polystyrene macromolecules in a highly diluted solution in ethylene dichloride. Intensive crosslinking with rigid bridges yields intramolecularly hypercrosslinked individual coils, or nanosponges. These species differ fundamentally from known macromolecular objects such as flexible and voluminous polymeric coils, dense globular macromolecules, hyperbranched molecules or dendrimers, and rod-like macromolecules. Similarly to bulk hypercrosslinked polystyrene, the dry nanosponge polymer exhibits low density and large apparent specific surface area. It increases its volume by a factor of about 3 in all organic solvents that fail to dissolve linear polystyrene. In good solvents, such as toluene, chloroform, and ethylene dichloride, the nanosponges will dissolve. Here, we use the term of "dissolution" for the process of formation of a thermodynamically stable, homogeneous liquid phase containing two molecularly dispersed components: the solvent and the crosslinked macromolecules. The unusual properties of such solutions are the extremely low viscosity and high rates of diffusion and sedimentation of the dissolved nanosponges. Furthermore, the monodisperse spherical nanosponges are prone to self-association and reversible formation of regular clusters, composed of 13 subunits (1 in the middle and 12 around it). Thus, the nanosponge represents a new macromolecular object that demonstrated for the first time the formation of regular clusters.

The post-crosslinking of the expanded individual polystyrene coils results in the strong contraction of these coils, as, on the molecular size scale, no diffusion problems can be expected to exist for the removal of excess solvent from the coil. For a polystyrene of 3×10^5 Da molecular weight, the diameter of the individual coils was calculated to be 393 Å in a highly diluted ethylene dichloride solution. The resulting intramolecularly hypercrosslinked nanosponge was found to have the diameter of about 160 Å (in solution). Hence, the diameter of the coil decreased by 2.5 times, while its volume decreased by a factor of about 15. In these data and in the similarity of properties of bulk hypercrosslinked polystyrene and the nanosponges we see strong evidence to the suggestion that the intramolecular crosslinks are as efficient as the intermolecular bridges with respect to all network properties. This idea contradicts with the still diffused opinion that intramolecular crosslinks contribute insignificantly to the network swelling ability.

Preparation of a solid block material from polystyrene nanosponges casts doubt on another classical notion about the structure of semi-concentrated polymeric solutions. That is, the overall morphology of the hypercrosslinked bulk material was found to be composed of microspherical species, both for the product of the crosslinking of linear polystyrene in semi-concentrated solutions and the product of consolidation of the nanosponges. This points to the idea that polystyrene macromolecules largely maintain their individuality in the initial semi-concentrated solution, rather than attain an unperturbed or even stronger swollen coil conformation, thus penetrating many (in the order of 10) neighbor coils.

Accomplishing the consideration of the structure and properties of hypercrosslinked polystyrene we can merely say once more that it really presents an amazing and extraordinary material. Probably, its distinguishing features revealed thus far are only the above-water part of the iceberg we refer to as hypercrosslinked polystyrene.

3. OTHER TYPES OF HYPERCROSSLINKED NETWORKS

If it is the rigidity of the open-framework structure that provides hypercrosslinked polystyrene with its unique set of specific properties, then several ways must exist to obtaining polymeric networks having a similar structure and, therefore, exhibiting similar properties. Indeed, various synthetic procedures can be suggested for obtaining a rigid, highly solvated polymeric network. When starting from (dissolved or strongly solvated) relatively flexible polymeric chains, such as the chains of polystyrene, introduction of a really large number of rigid bridges between the chains is required for attaining the hypercrosslinked network state. A smaller degree of crosslinking would be sufficient when starting from more rigid polymer chains. In addition to this straightforward post-crosslinking protocol, a conventional crosslinking polymerization or polycondensation of polyfunctional monomers carried out in a thermodynamically good (with respect to the final network) solvent must also produce desirable materials. The only condition is that the final network constitutes an expanded rigid molecular construction. Several examples discussed below confirm these predictions [132].

3.1 Macroporous hypercrosslinked styrene–divinylbenzene copolymers and related networks

Free radical copolymerization of styrene with 37 wt% DVB was carried out in the presence of 1, 4, and 8 vol. of ethylene dichloride per 1 vol. of the comonomer mixture. DVB forms rigid and short crosslinks, since in the network one DVB molecule belongs simultaneously to two polystyrene chains that cross each other in this junction point at a right angle [257]. When the polymerization of styrene with a large proportion of DVB proceeds in a good solvent taken in an amount exceeding the maximum swelling of the resulting copolymer, the excess solvent will separate in the form of micro-droplets in the interior of the growing network (micro-phase separation). After removing the solvent from the gel, a system of mutually connected pores remains. The size of these macropores may be relatively large. However, the rigid polymeric phase of the pore walls must possess additional, inherent microporosity of a hypercrosslinked structure, as it was formed in an ultimately solvated state. Indeed, similar to the real hypercross-linked polystyrene, these styrene–DVB copolymers swell in n-hexane or methanol [133]. The volume swelling increases with the dilution of the comonomers with toluene and reaches the value of about 3, which is very close to the volume swelling of other hypercrosslinked samples obtained from linear polystyrene at the same dilution (1 mL of the comonomers or 1 g of linear polystyrene in 8 mL ethylene dichloride (EDC)). The same property is characteristic of poly-DVB prepared in a diluted cyclohexane solution. This polymer is capable of increasing its volume by a factor of about 2, and its inner specific surface area amounts to $650 \, \text{m}^2/\text{g}$. Similar properties have been reported in the literature [133] for some poly-DVB and related networks. Undoubtedly, these styrene–DVB copolymers prepared in the presence of significant amounts of a thermodynamically good solvent have to be referred to as hypercrosslinked materials. The well-known Amberlite XAD-4 resin also belongs to the family of hypercrosslinked polymers: it has a specific surface area of approximately $1000 \, \text{m}^2/\text{g}$ and was found to take up $1.34 \, \text{mL/g}$ of toluene and $1.33 \, \text{mL/g}$ of methanol, which are noticeably higher values than the total specific pore volume of the dry polymer, $1.0 \, \text{cm}^3/\text{g}$. Obviously, this material changes its volume on wetting and then maintains it on a constant level in any liquid medium, which is characteristic of all hypercrosslinked polymers.

3.2 Hypercrosslinked polysulfone

As compared to polystyrene, polysulfone presents a more flexible polymer, because of oxygen links in its main chain. Computer simulation of the hydrodynamic behavior of an isolated polysulfone molecule results in the Kuhn segment of 16.5 Å, in contrast to 20.5 Å for polystyrene. Nevertheless, it is possible to obtain a hypercrosslinked polymer on the basis of polysulfone through introducing many rigid bridges between its chains dissolved in ethylene dichloride. Warshawsky et al. [23] developed a safe method for halomethylation of aromatic compounds, permitting the introduction of one bromomethyl group into each phenylene ring of the polysulfone. By a subsequent self-condensation of this modified polymer in solution, in the presence of a Friedel–Crafts catalyst, networks having typical properties of hypercrosslinked structures were obtained [132]:

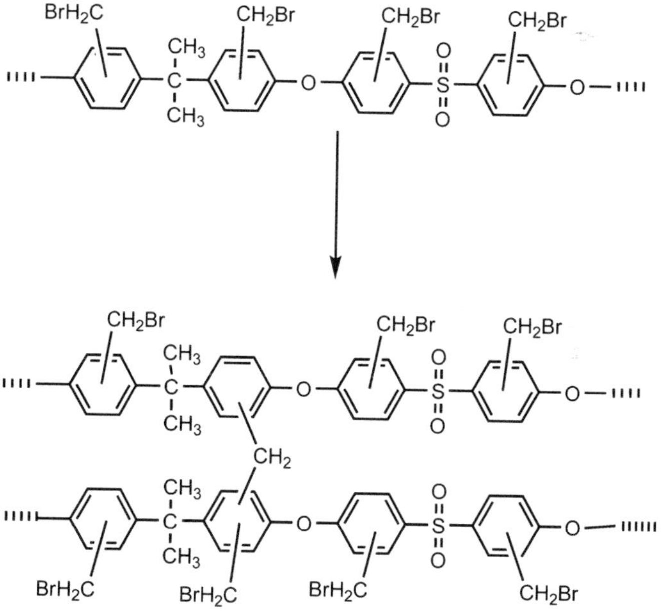

Their apparent inner surface area and swelling in non-solvents are not high (Table 9.1) but, still, they are noticeable and definitely exceed the experimental errors of measurements. Two reasons may be responsible for these low values: (a) the higher conformational flexibility of the polysulfone chains and network meshes facilitates achievement of a denser packing; (b) each sulfone group deactivates two neighboring benzene

Table 9.1   Properties of hypercrosslinked polysulfone. Synthesis conditions: 80°C, 12 h, ethylene dichloride, SnCl$_4$

Run	Charge (g)	EDC (mL)	SnCl$_4$ (mL)	S (m^2/g)	Weight swelling (mL/g)			
					Toluene	CH$_3$OH	C$_7$H$_{16}$	H$_2$O
1	1	3	0.3	32	1.03	0.48	0.43	0.34
2	1	4	0.3	46	1.28	0.59	0.56	0.51
3	1	8	0.3	42	0.86	0.49	0.40	0.43
5	1	4	0.6	20	1.36	0.67	0.61	0.62
4	1	8	0.6	9.5	1.04	0.48	0.43	0.44
6[a]	10	40	6	72	2.72	1.69	2.55	1.65

[a] Residual bromine content, 5.7%.
Reprinted from [132] with permission of Elsevier.

rings so that only two rings of the bis-phenol unit can participate in the Friedel–Crafts reaction. Because of this, many bromomethyl groups fail to find an appropriate phenylene partner for the reaction. The reduced conversion of the functional groups results in the formation of a less crosslinked and less rigid network. Unfortunately, no reliable experimental techniques exist to measure the actual degree of crosslinking of a polymeric network.

3.3 Hypercrosslinked polyarylates

Two polyarylates, I and II, composed of residues of isophthalic acids and two different bis-phenols, were taken for the preparation of networks [132]:

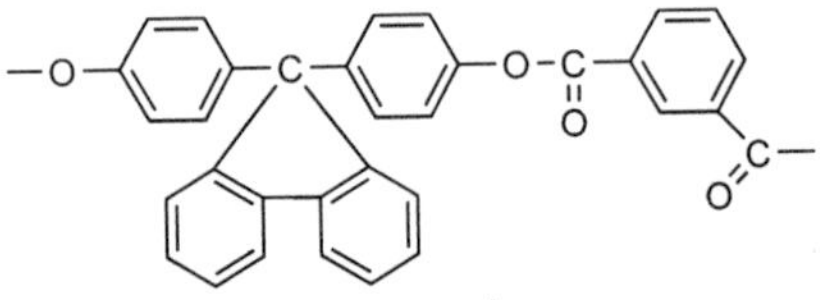

I

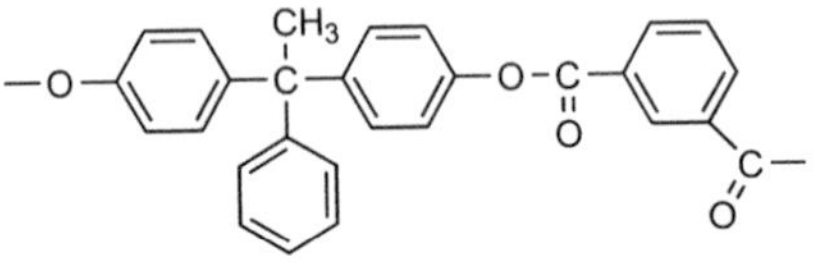

II

The macromolecule of polyarylate **I** has a voluminous side substituent in the bis-phenol unit and, therefore, it is more rigid than the chain of polyarylate **II**. In accordance with this, polyarylate **I** produces networks of higher porosity than polyarylate **II** crosslinked at the same concentration of the initial EDC solution (Table 9.2). Interestingly, a markedly higher dilution (0.055 g/mL) is required to obtain a porous polyarylate, as compared with that used for preparing hypercrosslinked polystyrene.

Table 9.2 Properties of hypercrosslinked polyarylates crosslinked with monochlorodimethyl ether (1 mol per repeating unit). Synthesis conditions: 80°C, 10 h, 1,2-ethylene dichloride, SnCl₄ (1 mol per mole of the ether)

Type of polyarylate	C	S	Weight swelling (mL/g)	
	(g/mL)	(m²/g)	n-Heptane	EDC
I	0.125	0	1.77	3.42
I	0.055	380	1.75	4.22
II	0.125	0	0.58	2.94
II	0.055	108	0.72	6.12

C = concentration of polyarylate in EDC.
Reprinted from [132] with permission of Elsevier.

(Usually crosslinking of linear polystyrene even in a more concentrated solution, 0.125 g/mL, in EDC results in the formation of highly porous materials with the apparent specific surface area of 1000 m²/g, while further dilution changes the porosity only insignificantly.) Porous crosslinked polyarylates swell in n-heptane, a precipitating medium for the polymer precursor, and, therefore, they also belong to the family of hypercrosslinked polymers.

3.4 Hypercrosslinked polyxylylene

This material was obtained [132] by self-condensation of p-XDC in ethylene dichloride solution in the presence of stannic tetrachloride:

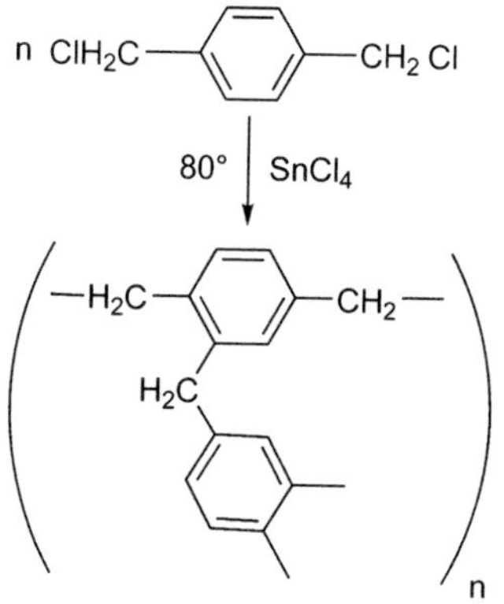

The reaction results in a typical hypercrosslinked polymer; its apparent specific surface area reaches a very high value, 1000 m²/g, and

Table 9.3   Properties of three-dimensional products prepared by self-condensation of p-xylylene dichloride (80°C, 20 h, 1,2-ethylene dichloride, SnCl₄)

Run	C(g/mL)	S (m²/g)	Volume swelling (mL/g)		
			Toluene	Heptane	Methanol
1	0.25	1000	1.4	1.5	1.4
2	0.125	900	1.7	1.4	1.7
3**	0.083	1000	1.4	1.4	1.3

C = concentration of p-xylylene dichloride in EDC.
**Residual chlorine content 3.4%
Reprinted from [132] with kind permission of Elsevier.

the networks swell in toluene, methanol, or heptane with the volume increasing by a factor of 1.4–1.7 (Table 9.3).

In these experiments we did not find any dependence of the swelling of polymers on the concentration of starting solutions of the monomer. Obviously, preparation of a hypercrosslinked material from a solution of long polymer chains differs from that of converting a solution of a low-molecular-weight compound into a polymer in that only in the first system the whole volume of the initial solution quickly converts into a bulk gel. Rejection of excess solvent from the gel is then restricted by the slow kinetics of solvent migration, so that the major part of the solvent remains entrapped in the final homogeneous gel block. On the contrary, formation of the polymer phase from monomer units takes place by stepwise addition of the latter to the initial microgel seeds, so that only those solvent molecules get included into the polymer gel which immediately participate in the solvation of the added monomer units, whereas the most part of the solvent easily departs via a macrosyneresis process. When microgel species coalesce, a certain solvent volume remains between them, thus causing macroporosity of the final material. This is the case with both the polymerization of DVB in toluene and polycondensation of XDC in EDC. Therefore, there is no correlation between the swelling ability of hypercrosslinked materials prepared from monomer units and the concentration of the initial solution of the latter.

It is appropriate to emphasize here that many similar condensation reactions can be suggested. Thus, we easily prepared a condensation product of p-XDC with naphthalene, taken in stoichiometric ratio, which exhibits both an apparent specific surface area of $1000\,m^2/g$ and an almost twofold volume increase when contacting acetone or methanol. Also, one could equally well use bis-chloromethylated diphenyl, instead of XDC, and carry out its self-condensation or co-condensation with some amounts of unmodified benzene or diphenyl [258].

The ease of preparing hypercrosslinked microporous polymers according to the polycondensation protocol attracts attention of specialists looking for inexpensive and efficient materials for hydrogen storage. Wood and coworkers [259] prepared a large series of microporous materials by the above self-condensation process of o-, p-, and m-isomers of XDC, as well as of 9,10-bis(chloromethyl)anthracene and 4,4′-bis(chloromethyl)-1,1′-biphenyl in anhydrous ethylene dichloride in the presence of 0.5–2 mol $FeCl_3$ based on the monomers. The use of o-XDC was consistently detrimental to the generation of surface area, both when used individually or in mixtures with the m- and p-isomers. As determined by solid-state nuclear magnetic resonance (NMR), the o-isomer produces materials with the lowest degree of condensation (i.e., crosslinking). The p-isomer provides the highest surface area values, up to $1430\,m^2/g$. Still better products were achieved from the bis(chloromethyl)-biphenyl. The most promising polymer displayed a specific surface area as high as $1900\,m^2/g$ as measured by N_2 adsorption and a hydrogen storage capacity of 3.68 wt% at 15 bar and 77.3 K, the highest yet reported for an organic polymer. Authors extensively discuss the structure of pores in the synthesized materials by analyzing the shape of nitrogen and hydrogen adsorption isotherms and by atomistic simulation of the network structure with the assumption that each aromatic ring in the poly-p-XDC, on average, receives three substituents. The calculations show the majority of the pore channels falling into the range between 0.3 and 2.0 nm. A sufficient portion of pores remains inaccessible to molecules of N_2 and even H_2.

3.5 Hypercrosslinked polyaniline and polypyrrole

Polyanilines contain a number of amine and/or imine functionalities that can enter the reaction of alkylation with alkyl halides [260]. A scheme below shows the microwave-assisted crosslinking of emeraldine base polyaniline, dissolved in dimethylformamide, with diiodomethane.

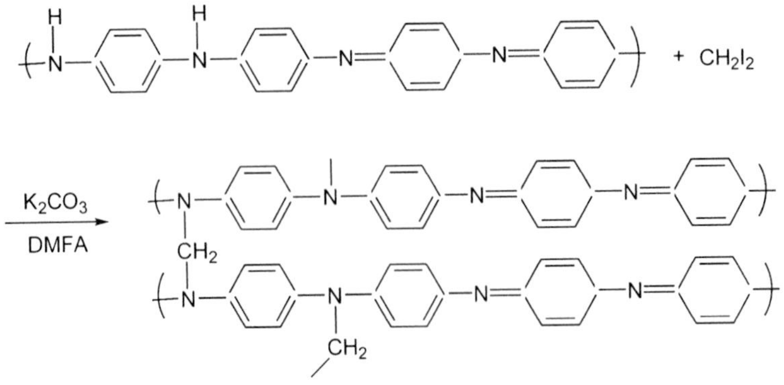

Interestingly, no quaternization reaction of amino groups takes place on introducing methylene bridges between the polymeric chains; the polymer network contains no detectable amounts of iodine anions. The thus-obtained hypercrosslinked material exhibits a high surface area of 630 m^2/g, total pore volume of 0.94 cm^3/g, and nanopore volume of 0.29 cm^3/g (the pore volumes being estimated from adsorption isotherms for nitrogen at relative pressures of 0.99 and 0.25, respectively). This polymer was tested as a medium for hydrogen storage, with the loading capacity achieving 2.2 wt% at 3.0 MPa and 77 K.

The porosity of polyaniline networks was found to rise as the concentration of the starting polymer solution increases, and, therefore, to enhance the solubility of starting linear polymers, another representative of polyanilines, leucoemeraldine (containing only amine functionalities), was deprotonated by using disodium salt of dimethyl sulfoxide. This salt of polyaniline, highly soluble in dimethyl sulfoxide, was then crosslinked with diiodomethane. However, the porosity of the final polymer network was pretty low. Alternatively to the above reaction, the microwave-assisted process of crosslinking emeraldine with paraformaldehyde results in obtaining a solid polymer having a surface area of about 500 m^2/g. Still, this material came short of fuel storage requirements, as well.

Germain at al. [261] used a similar chemistry to convert linear chains of polypyrrol into hypercrosslinked materials. The product crosslinked with diiodomethane exhibits a pretty high surface area of 730 m^2/g, mainly formed by small nanopores of around 1.0 nm in diameter. It reversibly adsorbs 1.6 wt% hydrogen at 77 K and 0.4 MPa. The iodoform-crosslinked material demonstrates smaller surface area, $400\,m^2/g$, and lower retention of H_2 under the same conditions, 1.3 wt%. The third representative of hyper-crosslinked polypyrrole was obtained by treating the polymer precursor with boron iodide. When measured by N_2 adsorption, the specific surface area of the material proved to be as small as $19\,m^2/g$. This product contains pores that are too small for nitrogen adsorption but large enough for adsorption of hydrogen. Still, the boron-crosslinked polypyrrol retains only 0.63 wt% H_2 at the above temperature and pressure. The authors believe that the difference in the adsorption of hydrogen onto the tested hypercrosslinked polypyrrole results from the difference in the size of crosslinking units that are schematically reproduced below. This is not the case, of course. In the first place, one should look for the reasons of the above differences in the conditions of reactions, which, unfortunately, were not specified in the paper. Besides, we should point out once more that the conformation of polymeric chains in solution does not allow the formation of suggested ladder-like structures.

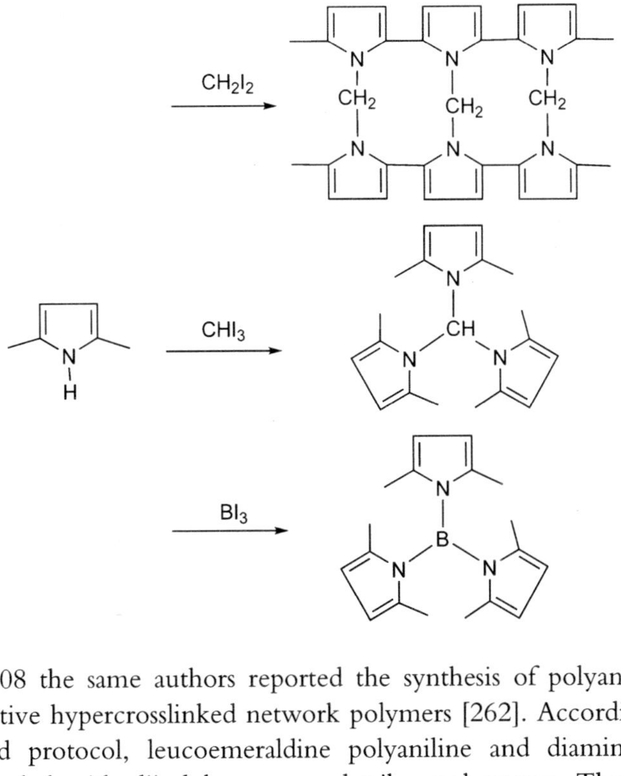

In 2008 the same authors reported the synthesis of polyaniline-type size-selective hypercrosslinked network polymers [262]. According to the developed protocol, leucoemeraldine polyaniline and diaminobenzene were coupled with diiodobenzene and tribromobenzene. The resulting polymers with porosity in excess of 35% consist of aromatic rings linked by trivalent nitrogen atoms. The pores in the material proved to be almost inaccessible for nitrogen molecules; that is why the surface area measured by means of this adsorbate did not exceed $100-300 \, m^2/g$. At the same time, the new hypercrosslinked polymers are capable of taking up 0.3–0.8 wt% of hydrogen at 77 K and 0.12 MPa.

3.6 Hypercrosslinked polyamide and polyimide networks

2,2′,7,7′-Tetraamino-9,9′-spirobifluorene was taken for the synthesis of polyamide (PA) and polyimide (PI) networks having an intrinsic micro-porosity, because the crosslike spatial orientation of spirobifluorene units provides loose packing of polymeric chains in final networks [263]. By reacting the tetraamine with terephthalic acid or pyromellitic acid dianhydride dissolved in N-methylpyrrolidone or o-cresol, respectively, microporous materials were obtained as depicted in the following scheme:

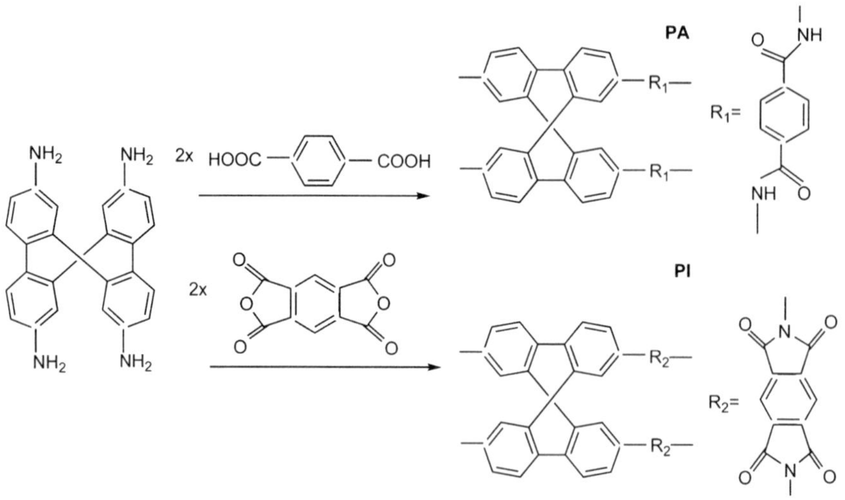

The dry PI network proved to be a highly porous material with an apparent specific surface area of about $1000 \, m^2/g$ and huge free volume of $0.62 \, cm^3/g$ (which was estimated by the volume of nitrogen adsorbed at $p/p_s = 0.99$). The PA network is more flexible and, therefore, its surface area does not exceed $50 \, m^2/g$. It is also worth noting that a strong hysteresis is characteristic of the sorption/desorption isotherm of nitrogen on the PI network, which the authors explained by swelling effects during the sorption of nitrogen.

3.7 Hydrophilic hypercrosslinked pyridine-containing polymers

Recently, Pavlova et al. [264] described the synthesis of the first hydrophilic hypercrosslinked polymers. These polymers were obtained in accordance with the basic principle of preparing hypercrosslinked polystyrene but through a mechanism fundamentally differing from the earlier studied processes of polymerization, polycondensation, or polymer-analogous transformations. The authors selected 4-vinylpyridine as the initial monomer which easily enters alkylation reaction on its nitrogen atom (the Menshutkin reaction) by means of halogen alkyls, particularly p-XDC. One of the interesting characteristics of 4-vinylpyridine is its inclination to a spontaneous polymerization after acquiring a positive charge via N-alkylation; a possible reaction mechanism has been described earlier by Kargin et al. [265]. The specific cationic polymerization is induced by the transformation of pyridine into a quaternary ammonium salt. The mechanism is further corroborated by the fact that the unmodified excess monomer does not take part in the polymerization. This chain of two successively parallel reactions has opened up a new original possibility to developing a hydrophilic hypercrosslinked ionic polymer. Indeed, the interaction of p-XDC with two molecules of 4-vinylpyridine activates two vinyl groups and results in the initiation of two independent polymeric chains which are automatically bound by the bridging $-CH_2-Ph-CH_2-$ fragment. The final product represents an anion-exchange resin with high density of positive charges, since every pyridine residue was converted into a quaternary ammonium salt:

The fast and sufficiently complete reaction proceeds in dimethyl sulfoxide, dimethylformamide, hexamethylphosphotriamide, and excess 4-vinylpyridine, but the best results were achieved in N-methyl-N-butylimidazolium tetrafluoroborate, an ionic liquid. The final polymers demonstrate a noticeable volume increase by taking up as much organic solvents, both polar and nonpolar, and even water, as a hypercrosslinked polystyrene crosslinked with the same crosslinking agent (Table 9.4). Still, the specific surface area of the hydrophilic hypercrosslinked polymer is smaller than that of the hypercrosslinked polystyrene, being less than $100 \, m^2/g$. In all probability, the rigidity of the hypercrosslinked vinylpyridine product is

Table 9.4 Swelling of hydrophilic hypercrosslinked polymers prepared in an ionic liquid (measured by weight of dry and swollen polymer, mL/g)

Sample	Benzene	n-Heptane	Acetone	Ethanol	Water
1	1.40	1.35	1.50	2.75	2.47
2	1.70	1.55	1.50	1.73	2.22
3	1.90	1.75	2.10	2.35	2.70
4	2.35	3.67	2.53	3.30	3.37

Source: After [264].

smaller, compared to that of the polystyrene material, because of the enhanced flexibility at the quaternary nitrogen atom. Note that except for the presence of these quaternary ammonium atoms, the structure of the poly (4-vinylpyridine) and polystyrene crosslinked with p-XDC are identical.

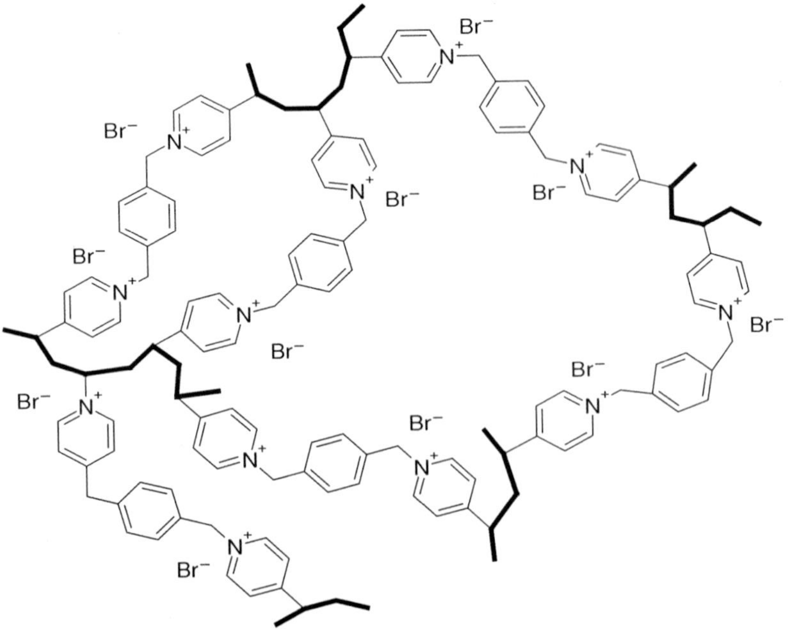

3.8 Other types of hypercrosslinked organic polymers

Several other types of hypercrosslinked networks can be found in scientific publications, although their authors did not recognize the unique nature of polymeric products obtained, and did not test their swelling in non-solvents. For example, Webster et al. [266] prepared in 1991 typical hypercrosslinked polymers by treating 4,4′-dilithiobiphenyl with dimethyl carbonate in tetrahydrofuran at −80°C:

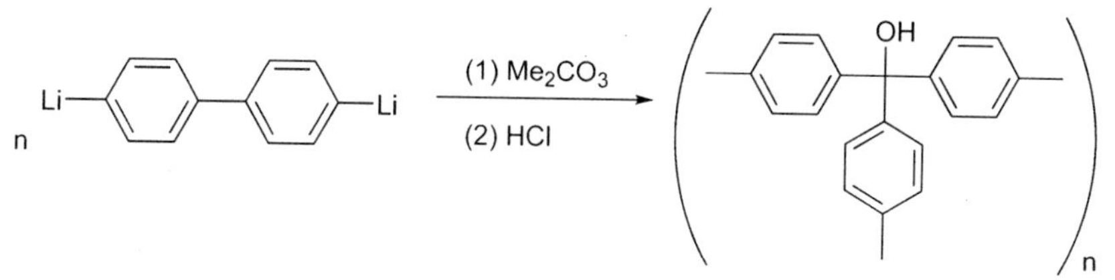

After slow warming to room temperature, washing, and drying, a product was obtained that exhibited a specific surface area of $1167 \, m^2/g$. The total specific pore volume of the product amounted to $0.5 \, cm^3/g$, while the pore diameters were quite small, about 3 nm. This polymer absorbed methylene dichloride (obviously, with an intensive swelling) up to 6 times its original weight, although each monomer unit in the chain was crosslinked.

Table 9.5 presents the structures of similar types of hypercrosslinked rigid polymers and values of their specific surface area.

In 2004, synthesis of the so-called polymers of intrinsic microporosity (PIMs) was reported [267]. It was found that microporosity can arise in non-crosslinked polymers, whose highly rigid and contorted molecular structure simply prohibits space-efficient packing of chains in the solid state. Rigid macromolecules are usually composed of fused rings some of which are spirocyclic. Thus, the spiro-center (i.e., a single tetrahedral carbon atom shared by two rings) in a monomer unit of the 3,3,3′,3′-tetramethyl-1,1′-spirobisindane provides for the site of contortion of the rigid chain having a ladder-type structure of condensed rings [268]:

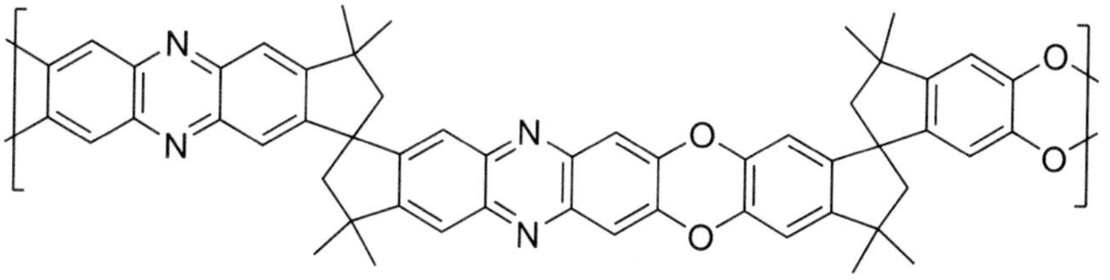

In this case, formation of a network of covalent bonds between the linear chains is not necessary for the generation and maintenance of

Table 9.5   Examples of hypercrosslinked rigid polymers

Monomers	Polymer	Specific surface area (m^2/g)
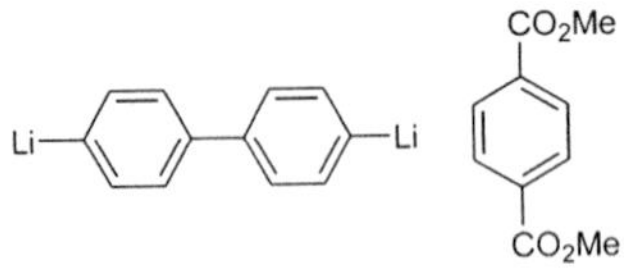	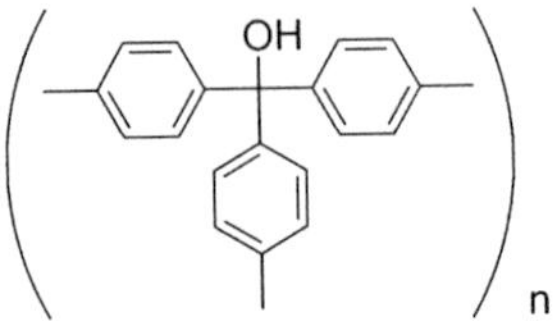	570
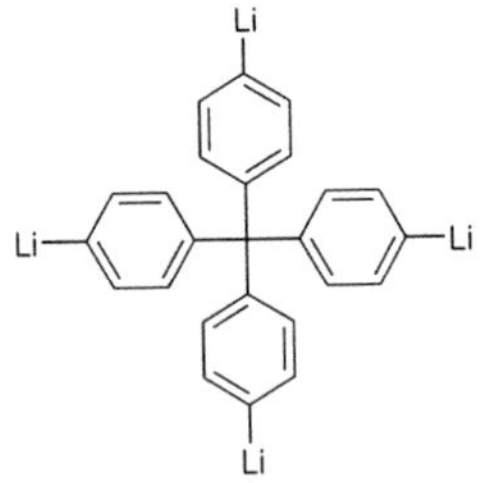	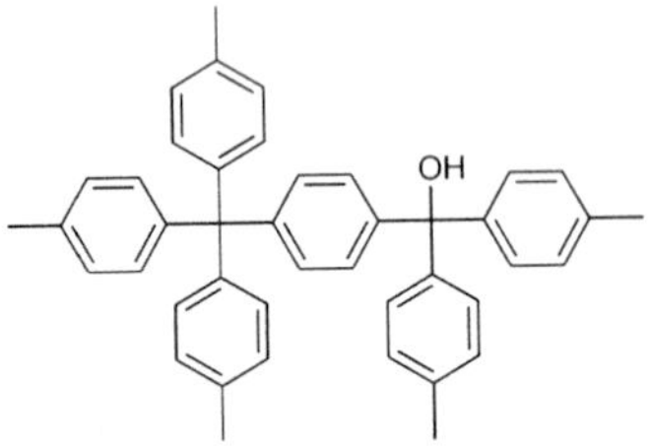	651
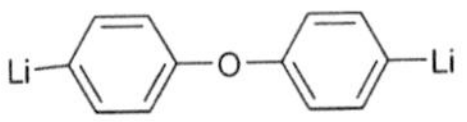	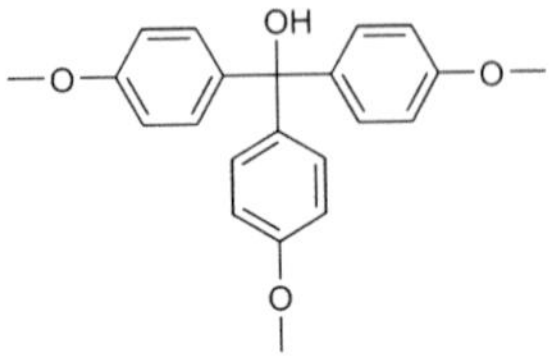	443
		570

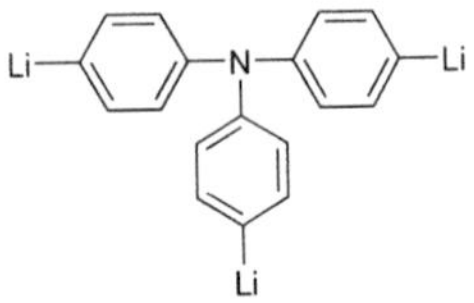

microporosity within the polymer material. Linear PIM materials are solution processible and can be used for casting semipermeable porous membranes, coatings, or bodies of desired shape, for example, for carrying nanodispersed metal catalysts (Budd et al., in review) [269].

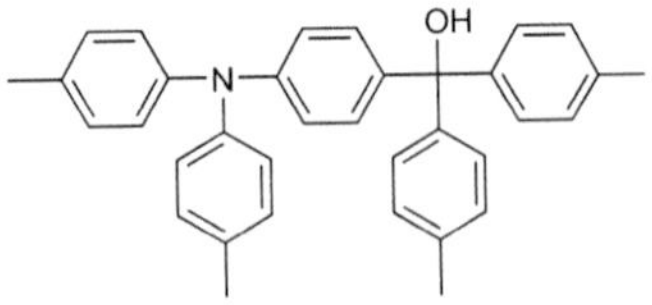

Still, compared to linear PIMs, 3D polymeric structures usually display higher values of specific surface area. Thus, a single-phase cyclotricatechylene-based network polymer with contorted bowl-shaped repeating units as shown below possesses large free volume and specific surface area exceeding $800 \, m^2/g$, which are provided by pores of average diameter less than 0.7 nm [269].

The polymer was tested for storage of hydrogen. The loading capacity proved to be not very high, 1.5–1.7% at 77 K and 5–10 atm, which, still, is comparable with that of activated carbons and carbon nanotubes.

Highly rigid and contorted aromatic spirobisindane-based linkers were shown to ensure a space-inefficient packing of phthalocyanine macrocycles, giving polymer networks [269] with gratifyingly high surface areas $(500–1000 \, m^2/g)$ and interconnected nanopores of diameters in the range 0.6–0.8 nm.

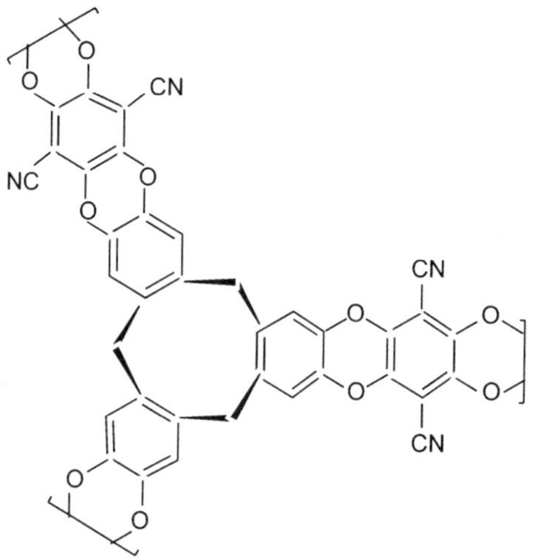

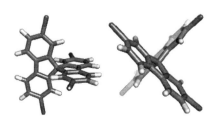

Weber and Thomas [270] prepared rigid poly(p-phenylene)-type polymeric networks by reacting tetrafunctional building blocks of tetrabrominated 9,9'-spirobifluorene with benzene-1,4-diboronic acid or 4,4'-biphenyldiboronic acid in a mixed organic solvent. Here, the contorted unit of spirobifluorene forces the four polyphenylene chains to grow in four different directions.

The materials thus obtained exhibit rather small specific surface area, 200–500 m^2/g, if measured by nitrogen adsorption. Interestingly, for the same samples, small-angle X-ray scattering method gives the surface area of 860 and 1030 m^2/g, respectively. The mismatch of data was explained by the presence of closed pores or pores inaccessible to nitrogen molecules.

Rigid microporous hypercrosslinked networks were also derived through Schiff base chemistry by condensation of trifunctional melamine with aromatic di- and trialdehydes [271]. If appropriate experimental conditions are chosen (heating at 180°C for 72 h in dimethyl sulfoxide under inert atmosphere), the imine double bonds of the initially formed Schiff bases are attacked by primary amines, resulting exclusively in the generation of aminals:

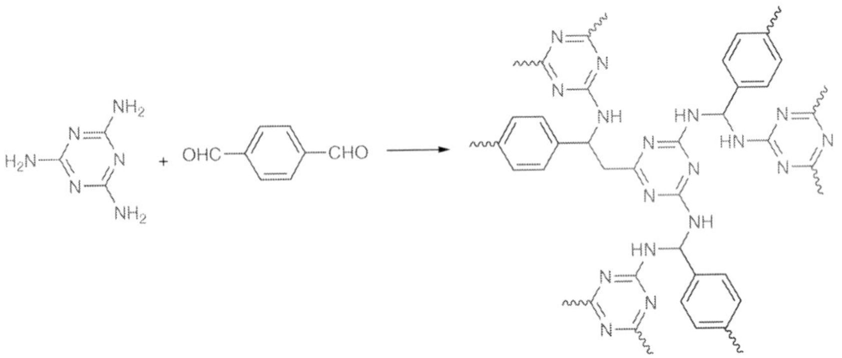

High surface area, 840–1400 m^2/g, and significant pore volume, up to 1 cm^3/g, make the inexpensive low-density materials with high nitrogen content (up to 40 wt%) promising candidates for potential applications in materials' science, for example, for the development of novel proton-conducting polymers.

Jiang, Cooper and coworkers [272–275] presented interesting and important results on the synthesis and properties of hypercrosslinked networks build of poly(arylene-ethynylene)-type struts, the first fully conjugated porous polymer. The networks were synthesized by palladium-catalyzed Sinogashira–Hagihara cross-coupling polycondensation of aryl halides and aryl ethynylenes. Both partners of the condensation reaction were either bi- or trifunctional. The group of aryl halides used consisted of 1,4-diiodobenzene, 4,4'-diiodobiphenyl, and 1,3,5-tris(4-iodophenyl)benzene. The other partners had similar structures, but bearing ethynyl groups

(–C≡CH) instead of iodine in the *para*-position of the aromatic rings. The coupling reaction resulted in topologically identical networks composed of rigid arylene-ethynylene struts of different lengths and trifunctional joints of the 1,3,5-substituted benzene, as exemplified below:

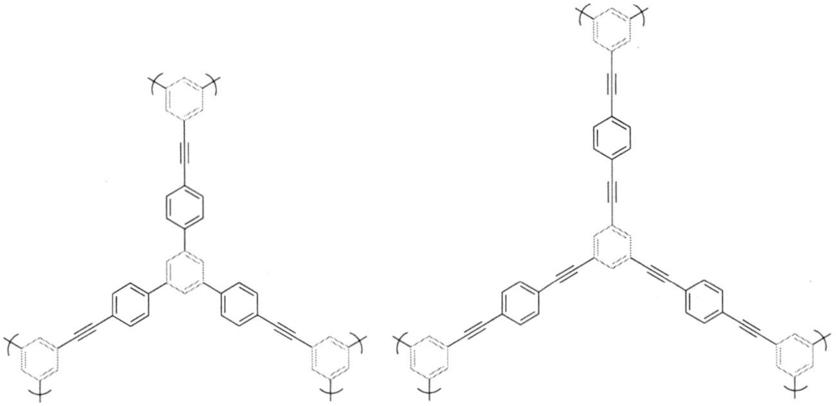

Depending on the combination of partners, the strut length could be varied in a wide range. The materials formed had specific surface areas between 500 and 1000 m²/g. The largest values of S_{BET}, 1018 m²/g, and pore volume, about 1 cm³/g, were characteristic of the network with shortest distance between the joints, prepared from coupling a pair of partners that were both trifunctional. Increasing the length of struts leads to an increase in the size of network meshes, but simultaneously enhances their flexibility and provides more conformational freedom for strut deformation and minimization of free volume. Accordingly, as the monomer strut length increased, the surface area and the overall micropore volume fall, while the micropore size distribution systematically shifts toward larger pore diameters. These parameters were found to be relatively unaffected even by large changes (× 10) in the monomer concentration during the polycondensation reaction. As discussed in Section 3.4, this feature is characteristic of the polycondensation processes between small molecules, in contrast to reactions of post-crosslinking of initial polymer chains where the porous structure of final networks is strongly affected by the concentration of chains in the solution. Since the micropore size of the poly(arylene-ethynylene)-type networks can be systematically controlled by selecting monomer pairs with appropriate molecular dimensions, the materials, though rater expensive, seem to be very promising for structural studies, development of materials with electrical properties, and specific separation technologies. The materials were found to be thermally stable and showed little decomposition under nitrogen gas below temperatures around 400°C.

The same group of Cooper [275] carried out a similar set of experiments with the only difference that the trifunctional components of the cross-coupling reactions derived from 1,3,5-tris(phenyl)benzene were now replaced by corresponding derivatives of triphenylamine (bearing three iodine atoms or ethynyl groups in *para*-positions). Again, by combining different coupling pairs, a series of networks topologically identical to the above-described poly(arylene-ethynylene)-type networks were prepared and examined. All above-made observations were found to hold for the new series of networks that now contained nitrogen atom instead of benzene ring as the junction point. The most rigid network obtained by condensation of tris-ethynyl-substituted benzene with tris-iodo-substituted triphenylamine was found to display the highest specific surface area of $1108 \, m^2/g$. Authors note that all materials are chemically stable and that the nitrogen atom does not break the conjugated structure of all chains.

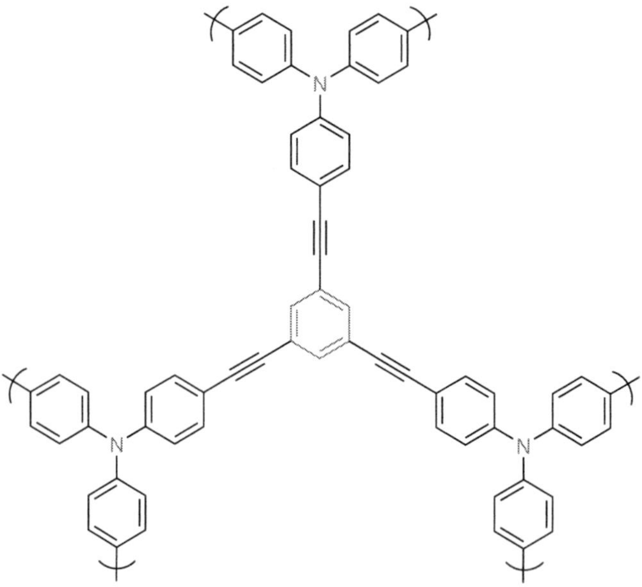

Though the above-described materials are amorphous polymers, they are most probably more regularly built than the products of post-crosslinking of linear polymers. This allows using the term "covalent organic frameworks" (COFs) for their description by analogy to crystalline "metal-organic frameworks" (MOFs) described later in this chapter. The well-known structure of the struts and network nodes also simplifies the computer simulation of their structure. Results of such atomistic simulations are presented in both the works [274, 275], visualizing the structure and connectivity of pores in the material. As can be expected, the pores are different in shape and size with the channels between them being rather small. We can imagine, however, that solvation and swelling of the materials with liquid media would significantly change the structure of the network and their permeability. Swelling may even make the structures with the longest struts more accessible to host molecules compared to the more rigid polymers having shorter struts and displaying higher surface area in dry state. Authors [275] make only a short note that the COFs investigated swell less readily than hypercrosslinked polystyrene polymers described in the literature.

In a short review Weder [276] compares perspectives of using PIMs, Davankov-type hypercrosslinked polystyrene, COFs, and the above-described rigid strut networks in the capacity of hydrogen storage media. While stating that it is important to fine-tune the structure of micropores, Weder admits that "the differences between these systems may not be as fundamental as previously thought. It is unclear to what extent the pore size distribution can be narrowed in disordered polymer networks, but the results appear to open the door for the further development of a broad range of materials with controlled pore dimensions." In a previous paper Weder and coworkers [277] pointed out another important prospective field of application of crosslinked polymers with conjugated molecular structure, namely that of semiconductors and materials with interesting optoelectronic and photoluminescent materials, the more that they can be easily prepared in the form of spherical milli-, micro-, and nanoparticles.

Finally, it could be mentioned here that the basic idea of hypercrosslinked network synthesis was also applied to the preparation of porous fibrous sorbents. Liu et al. [278] have suggested grafting of styrene–DVB copolymer to the surface of polypropylene fibers, followed by the post-crosslinking of the grafted copolymer with 4,4′-bis-(chloromethyl)-diphenyl to 100% degree of crosslinking in a mixture of ethylene dichloride, nitrobenzene, and cyclohexane. After being coated with several successive layers of such hypercrosslinked polystyrene, the final fiber material displayed a surface area of $360 \, m^2/g$ and pore volume of $0.345 \, cm^3/g$, formed predominantly by thin pores of $1.9 \, nm$ in diameter.

3.9 Hypercrosslinked polysilsesquioxane networks

Very interesting arylene-bridged polysilsesquioxane networks were described by Loy and Shea [279] and Shea et al. [280]. Aromatic hydrocarbons containing two (or more) triethoxysilyl groups were used as the starting compounds:

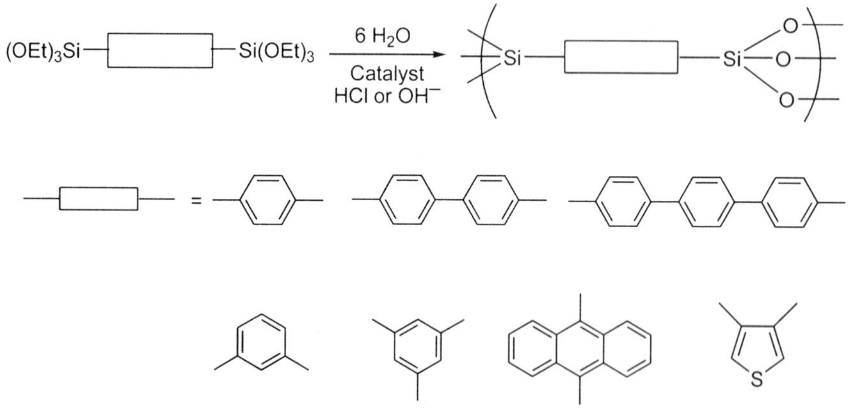

In the presence of excess water and an acidic, basic, or fluoride catalyst, triethoxysilyl functions of the monomers were hydrolyzed, yielding first silanol groups and then siloxane bonds. As the condensation proceeds, the initial solution of the monomers in ethanol or tetrahydrofuran (0.2–0.4 M) transforms into a monolytic gel of arylene-bridged polysilsesquioxane. No signs of phase separation were observed when the reaction was initiated with an acidic catalyst. The swollen transparent gel was then subjected to air drying (in this case the material obtained was referred to as a xerogel). The removal of large amounts of the initially present diluent, together with ethanol released in the condensation reaction (six molecules of ethanol per one monomer unit), leads to a 90–95% decrease in the gel volume. It is at the end of this stage where the major portion of siloxane-type crosslinks is formed. One can largely avoid the shrinking of the primary gel by replacing supercritical carbon dioxide for the original solvent and then slowly venting the CO_2. The opaque material thus obtained (aerogel) was close in its volume to that of the initial swollen gel. Obviously, final condensation of residual silanol groups in the carbon dioxide media takes place within a larger volume of the gel.

Arylene-bridged polysilsesquioxanes are highly porous materials. For example, xerogels based on 1,4-bis(triethoxysilyl)benzene exhibited specific surface area values as high as 600–1200 m^2/g, depending on the synthesis protocol. The specific pore volumes ranged from 0.3 to 0.9 cm^3/g. The average pore diameter was 10–50 Å. The same phenylene-bridged aerogel was found to have a specific surface area of 1880 m^2/g, which was supplied largely by 20 Å diameter micropores together with a certain portion of mesopores with diameters ranging from 20 to 500 Å. Interestingly, both xerogel and aerogel were found to be sufficiently rigid and did not change noticeably their volume on rewetting. Obviously, hypercrosslinked networks of the phenylene-bridged polysilsesquioxanes, similar to microporous silica gels and zeolites, are too rigid to allow volume changes. Quite important is the observation that introduction of two methylene groups instead of the phenylene ring caused a decrease in the specific surface area of the xerogels to 600–700 m^2/g, whereas the porosity disappeared completely when the bridge contained 6 (acid catalysis) or 10 (basic catalysis) methylene groups.

Shea et al. [281] have also succeeded in making a low-density aerogel from tetramethoxysilane and p-di(triethoxysilyl)benzene in one step using supercritical CO_2 as a solvent. The highly porous monolithic samples exhibited a specific internal surface area over 700 m^2/g.

Han and Zheng [282] reported the synthesis of microporous networks by hydrosilylative polymerization of 12- and 24-membered macrocyclic oligomeric silsesquioxanes with the general formula of $[vinylSi(OSiMe_2H)O]_n$, $n = 6, 12$. Each of the above 6 or 12 repeating units of the macrocycles contained both a reactive Si–H and $CH_2=CH$–Si bonds able to react with each other in the presence of Pd-containing Karstedt catalyst. It was expected that the high rigidity of polysilsesquioxane networks obtained would prevent the collapse of micropores after elimination of the reaction solvent (toluene) from the crosslinked products. The expectations were fully justified in the case of polymerization of the 24-membered oligomeric cycle yielding a porous material with $S = 300 \, m^2/g$, pore volume of 0.21 cm^3/g, and average pore size of 2.8 nm.

3.10 Metal-organic frameworks (MOFs) and covalent organic frameworks (COFs)

Scientists have always been interested in highly porous materials, both organic and inorganic, to use them in catalysis, separations, gas storage, or other applications. In this connection, a novel type of hybrid materials called metal-organic frameworks (MOFs) should be shortly mentioned here. The term was coined by Yaghi et al. [283] in 1995 to generally denote low-density crystalline salts formed by polyvalent transition metal cations with polyvalent organic acids, for example, terephthalic acid. Although the salts are not typical polymeric compounds, the MOFs were obtained, in fact, along the same principle of formation of rigid spacious 3D networks characteristic of hypercrosslinked polystyrene. The difference is that the network joints in MOFs are formed by electrostatic (and/or coordination) bonds between the functional groups of the organic spacer and transition metal cations.

Among these materials MIL-101 represents a truly striking product [284, 285]. It has been obtained by heating terephthalic acid (a bifunctional ligand), chromium nitrate, hydrofluoric acid, and water at 220°C for 8 h. MIL-101 consists of inorganic chromium carboxylate clusters linked by rigid organic (aromatic) moieties. The key building block is a tetrahedral supercluster consisting of four smaller clusters, held together by 1,4-benzene dicarboxylate groups. The superclusters form an extended framework containing cages of two different sizes, 29 and 34 Å. The unit cell of this material has an unprecedented high volume of 720,000 $Å^3$. This implies that after the removal of the solvent initially filling the pores, 90% of the interior proves to be empty space. For that reason, the apparent specific surface area of the solid amounts to a very large value of 5900 m^2/g.

MOF-177, another representative of MOF compounds, is a porous crystal [286]. It is composed of octahedral $Zn_4O(CO_2)_6$ clusters, each being attached to six triangular 1,3,5-benzenetribenzoate units. The material is constructed entirely of six-membered rings and in each formula unit there are 84 exposed edges and only 6 fused edges. This material is stable, robust, and highly porous; its specific surface area is also large, 4500 m^2/g, and the diameter of the pores is 1.1 nm. MOF-177 readily accommodates brominated naphthalene, anthracene, fullerene (C_{60}), or large polycyclic organic dye molecules. It takes up 140 wt% CO_2 at room temperature and moderate pressure of about 30 bar [287]. The trapped gas can be easily released by gentle heating.

More recently, Koh, Matzger, and coworkers [288] described a new uniformly structured type of MOF called UMCM-1 (University of Michigan Crystalline Material-1). This material contains Zn_4O clusters linked by dicarboxylate and tricarboxylate spacer groups, giving rise to six microcages surrounding a central "mesoporous" hexagonal channel. The diameter of the latter is about 3.2 nm, while the size of micropores amounts to about

1.4 nm in diameter. The specific surface area of UMCM-1 even exceeds that of MOF-177 and reaches 4730 m^2/g.

A series of 16 highly porous "isoreticular" (having the same cubical 3D network topology) MOFs were constructed from Zn–O–C clusters connected by 1,4-aromatic dicarboxylate links of different lengths [289]. The pore size of crystals can be expanded up to 28.8 Å using long molecular struts between the clusters, such as biphenyl, tetrahydropyrene, pyrene and terphenyl dicarboxylate. Open space can be boosted up to 91.1% of the crystal volume. Densities as low as 0.21 g/cm^3 were measured. IRMOF-6 proved to be especially appropriate for methane storage. At room temperature and 36 atm, regarded as a safe and cost-effective pressure, 1 cm^3 of this material takes up 240 cm^3 of methane (calculated for standard temperature and pressure conditions).

The above-described few examples of the MOF family provide only a superficial insight into the astonishing structure and properties of these appealing new materials. Nowadays, hundreds of scientific publications in this area appear every year around the world. Only one BASF pilot plant in Ludwigshafen, Germany, manufactures 100-kg-per-day batches of various MOFs [290]. Among the most interesting application areas of the materials are hydrogen and methane storage [291, 292]. The outstanding progress toward this goal was described by Dinca et al. [293]. The authors prepared an MOF incorporating Mn^{2+} coordination sites and 1,3,5-benzenetribenzoate ligands, along with other building blocks. Crystals of the compound remain intact upon desolvation and show a total H$_2$ uptake of 6.9 wt% at 77 K and 90 bar, which at 60 g H$_2$/L provides a storage density 85% that of liquid hydrogen. Due to direct binding at coordinatively unsaturated Mn^{2+} centers, the material exhibits a maximum isosteric heat of H$_2$ adsorption of 10.1 kJ/mol. Ma et al. [294] reported in 2008 that a new MOF compound prepared from dicopper units and anthracene-isophthalate derivatives and referred to as PCN-14 takes up unprecedented quantities of methane at 290 K and 35 bar, 230 eq. vol. (brought to normal conditions) per volume of solid. This value exceeds by 28% the key Department of Energy target for methane.

In 2009 the Yaghi group [295] first described crystalline interpenetrating MOFs. Their synthesis was based on the approach developed for obtaining the primitive cubic topology of the archetypical MOF-5, in which benzene (dicarboxylate) struts are connected by Zn$_4$O(CO$_2$)$_6$ joints. Extending this principle to longer and more complex structures, the authors managed to build twice, triply, or even fourfold interpenetrating MOFs. When the struts additionally contain crown ether fragments (e.g., 34- or 36-membered cyclic polyethers), as is the case with MOF 1001 and MOF 1002, the materials exhibit high affinity to electron-deficient guest molecules. Since the free volume within these crystals is exceptionally high (in MOF 1001 nearly 87% of space is not occupied by its atoms), large guest molecules like paraquat dications freely diffuse into the crystal interior and finally dock there, forming colored charge-transfer complexes.

The majority of MOFs are known to be rigid crystalline materials with permanent porosity. However, they were shown to display capability of reversibly expanding their structure on accommodation of "guest" molecules within the interior cavities. In this case MOFs are said to "breathe." The breathing was observed, for instance, on absorbing branched alkanes into MOFs composed of metal oxide–based coordination polymers with spherical periodicity in which 12 pentagonal $\{(Mo^{VI})Mo^{VI}_5O_{21}(H2O)_6\}^{6-}$ "ligands" are connected by 30 Cr^{3+}, Fe^{3+}, or $Mo^V_2O_4(acetate)^+$ "spacers." The structure of the MOF with $Mo^V_2O_4(acetate)^+$ spacers contains 20 Mo_9O_9 rings that are 3 Å in diameter and serve as openings to the material interior [296]. By migration through these openings, acetate anions can be exchanged against other carboxylate anions. The exchange rate strongly depends on the size of the alkyl carboxylates. Most interesting is the fact that even isopropyl and *tert*-butyl carboxylates, species whose size is larger than the pore dimensions, migrate through the pores. Obviously, small cooperative changes in the lengths and angles in the large ensemble of Mo_9O_9 rings lead to considerable fluctuations in the size of the pores. We may note here that the crystalline structure of MOFs does not necessarily prevent their swelling in liquid media, so that the sizes of swelled network meshes may differ from those determined by X-ray crystallography.

Indeed, as Trung et al. [297] reported, X-ray diffraction analysis unambiguously confirms the expansion of MNLL-53(Al, Cr) that accompanies absorption of normal alkanes into its interior. The breathing also manifests itself on the sorption isotherm as an unusual stepwise increase in the MOF capacity at rather low relative pressures. The same type of stepwise sorption isotherm for carbon dioxide at 196 K and clear-cut sorption–desorption hysteresis was also observed [298] for a flexible MOF formed by zinc carboxylate units joined by 1,4-diazobicyclo[2,2,2]octyl groups in one direction and by 2-amino-1,4-benzenedicarboxylates in the other direction. The amino functions of the MOF were then converted into amide by modification with alkyl anhydrides of various lengths. Short alkyl chains thus introduced narrow the pores and the longer chains hold the pores open, while the medium-size chains (1–2 C) are just right, creating a bistable framework conformation and facilitating breathing. In another example, twofold interpenetrating microporous MOFs synthesized with a flexible tetrahedral linker and Zn_2 clusters were found to take up a larger volume of CO_2 compared to that of N_2 or H_2, which may be explained only by a certain framework expansion on absorbing carbon dioxide [299]. All above findings suggest that the breathing of MOFs in fact is nothing but their swelling with vapors absorbed or with liquids, which is characteristic of all hypercrosslinked network materials.

In view of the above-discussed general flexibility of networks of MOFs, it is appropriate to answer here the question why these materials are crystalline while all organic hypercrosslinked polymers are amorphous. The latter holds even for the case where the exact structure of both the network nodes and struts between them is strongly controlled by the chemical structure of starting monomeric components. The answer follows from the basic difference in the mechanism of formation of the 3D structure of the two types of networks under consideration. Organic hypercrosslinked networks are formed through chemical reactions of active functional groups of starting components, resulting in closure of stable covalent bonds. Therefore, the whole process is kinetically controlled. Once formed, the nodes of the network cannot be changed or rearranged. The distribution of nodes through the polymer phase as well as the length of meshes thus closed follow the laws of statistics. Topology of microdomains of the network remains irregular and the material amorphous. Contrariwise, the nodes of networks in MOFs are formed by relatively weak electrostatic and coordination bonds in a reversible manner. The whole process is thermodynamically controlled so that the system tries to attain a state of minimal free energy under conditions of synthesis. Any erroneous bonds and structures will break or rearrange to finally produce a regular network and crystalline hypercrosslinked material. The final MOF material may well be stable, unless exposed to chemicals capable of disrupting electrostatic and coordination bonds of the network nodes.

In the year 2005 the group of Yaghi introduced another type of porous materials named covalent organic frameworks (COFs) [300]. They are built of rigid organic struts of various length and trifunctional nodes of planar boroxine rings, $(RB)_3O_3$. The latter are formed by condensation of three aromatic boronic acid moieties. Though this six-membered ring is formed by covalent B–O bonds, it is much less stable than a benzene ring and it can reversibly form and decay under synthesis conditions. This makes it possible, by starting from bifunctional or trifunctional aromatic boronic acids, to form networks under thermodynamic control and arrive at regular, even crystalline structures. Thus, 1,4-benzenediboronic acid moieties slowly condense in solution to the parent material COF-1, low-density crystals made of 2D sheets of the following structure:

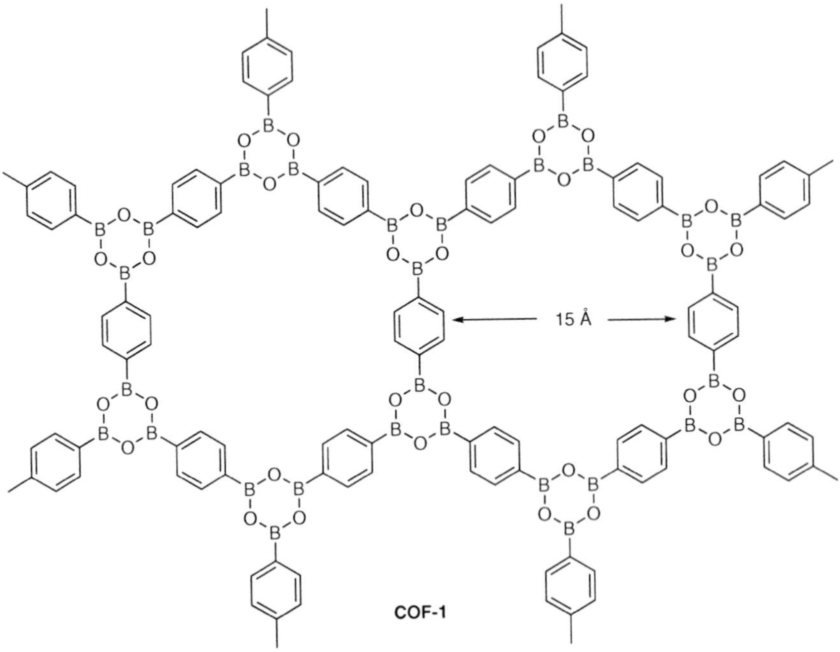

COF-1

More elongated bifunctional monomers condense to other hexagonal 2D COF layers with openings as large as 3 nm in diameter [301]. Aromatic boronic acids also readily condense with *ortho*-bis-phenols. By involving tri- or tetrafunctional monomers into polycondensation, e.g., tetra (4-hydroxyborylphenyl)methane (or -silane) and hexahydroxytriphenylene, 3D frameworks can be designed [302].

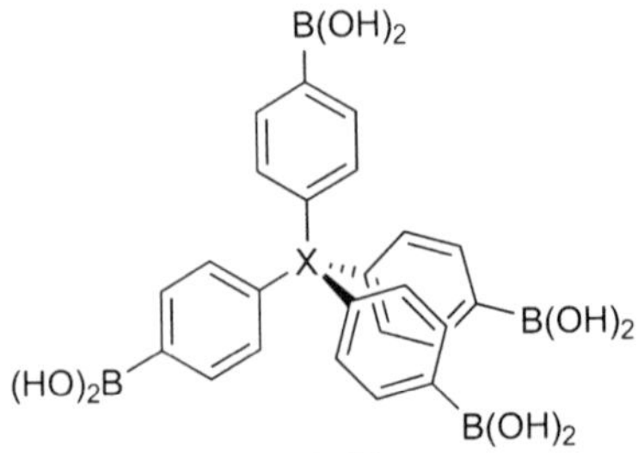

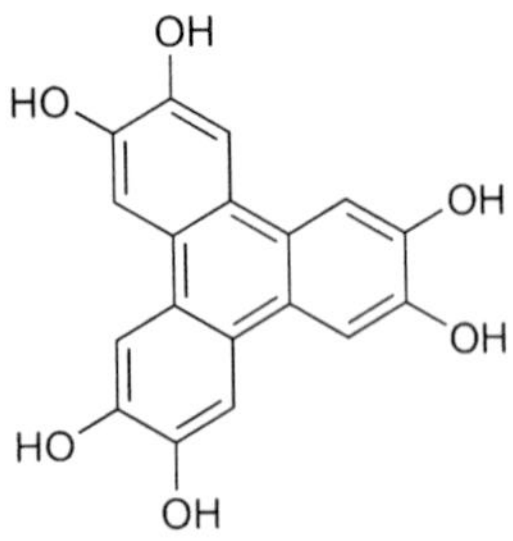

Very impressive 3D covalent organic frameworks were constructed by using three molecular building blocks, two of which are tetrahedral molecules and one is planar triangular molecule. The tetrahedral building blocks are four-armed structures where each arm is tipped with a boronic acid, $-B(OH)_2$, group. Three of the above building blocks condense in solution, with loss of three water molecules, to form a B_3O_3 ring that serves as a triangular core of what might be called a tri-tetrahedron unit. The latter undergoes further condensation reactions. The most striking feature of these materials, constructed of light elements such as C, O, and B, is their low density. For example, one member of this family, COF-108, has density as low as $0.17 \, cm^3/g$, which is the lowest-density crystal known so far [302]. A diamond, for comparison, has the density of $3.5 \, g/cm^3$. Being constructed from strong covalent bonds, C–C, C–O, C–B, and B–O, the materials have high thermal stability, 400–500°C.

One can imagine the enormous prospects offered by modern synthetic-organic and element-organic chemistry for the design and preparation of manifold scaffolds of MOFs and COFs. Indeed, the number of publications in this field initially (from 1995 to 2000) was low, constantly below 15 each year. Then, the number grew exponentially, and in 2007 more than 500 reports were devoted to ordered porous frameworks [303]. It is not possible to review them within the book that is concentrated on polymeric organic hypercrosslinked networks, all the more because metal and boron atoms are not considered to be organogenic elements and chemistry of these elements is quite different from conventional polymerization and polycondensation reactions.

Still, we believe to have convincingly demonstrated that, beside polystyrene, any other conformationally rigid highly crosslinked polymers or MOFs will always possess low density (high porosity), provided that it was formed in a solvating medium (porogen). At a certain level of overall rigidity, these expanded networks fail to achieve dense packing upon removing the porogen and develop inner stresses on shrinking. Whether amorphous or crystalline, the hypercrosslinked materials exhibit high specific surface area and enhanced sorption capacity toward any kind of host molecules. When in dry state, the stressed amorphous networks will always tend to return back to the initial volume in which they were formed and which is characterized by the smallest deviations from relaxed conformations of constituent meshes. Amorphous materials swell therefore stronger in liquid media compared to ordered crystalline frameworks. Naturally, each new initial monomer or polymer and each new method of preparing hypercrosslinked networks have to render the material its own special characteristics. Here, we did not discuss them in detail. The main goal of this chapter was only to demonstrate the porosity and tendency to swelling in any solvent as the main distinguishing features of hypercrosslinked networks as well as point at the practical usefulness of this new class of network materials. Hypercrosslinked polystyrene is not an exception; it just happened to be the first example of a new broad class of materials that can be referred to as hypercrosslinked networks.

4. COMMERCIALLY AVAILABLE HYPERCROSSLINKED POLYSTYRENE RESINS

Hypercrosslinked polystyrene networks not only present an academic interest because of their unique set of physico-chemical properties but they also acquire great practical importance as excellent adsorbing materials. Indeed, one of the basic distinguishing features of the single-phase highly porous hypercrosslinked matrix is that its polymeric chains are separated by homogeneously distributed, long, and rigid bridges, better to say, struts, and therefore remain permanently exposed to gaseous or liquid environment. Being deprived of close contacts to one another, these chains generate a strong uncompensated force field. For this reason, hypercrosslinked polystyrene attracts and retains a wide variety of organic compounds from the surrounding medium and concentrates them in the interior of the network. In contrast to common macroporous styrene–DVB adsorbents in which only the surface of pores is accessible to sorbate molecules, the hypercrosslinked polystyrene resin sorbs compounds by its entire network volume. The inevitable consequence of this fundamental difference in the structure and properties of these two network types is the exceptionally high sorption power and loading capacity of hypercrosslinked polystyrene resins. Notably, the hydrophobic polymer absorbs both nonpolar as well as highly polar organic compounds, such as phenol and the degradation products of triazine-type pesticides. It will be shown later that the naked aromatic rings of the resin readily enter π–π-type interactions with polar functional and aromatic groups of the sorbates. This π-selectivity, in combination with some other unusual rules of sorption onto hypercrosslinked sorbents, permits them to be considered as a new, third generation of polymeric adsorbents.

While it took a rather long time for many chemists to understand and accept the scientific ideas concerning hypercrosslinked networks, the practical usefulness of the new sorbents became evident rather early. In 1977 Krauss et al. [28] received a patent in the German Democratic Republic, and three years later patents were granted to Rohm and Haas in the USA [35, 43] for the synthesis of highly porous adsorbents by post-crosslinking of swollen styrene–DVB copolymers with the involvement of preliminary introduced chloromethyl groups or by a direct reaction with chloromethylated bifunctional reagents in the presence of Lewis acids. Actually these patents followed in our footsteps and utilized our idea defined 10 years earlier in a USSR patent and in patents of 16 other countries, including the USA, Great Britain, the two German states, Japan, France, Italy, and others [1]. Only Mitsubishi Chemical Industries Ltd. acquired in 1985 a patent [63] that protects a new technique for the post-crosslinking of swollen styrene–DVB copolymers by involving residual pendent vinyl groups of DVB into alkylation of phenyl groups in the presence of Friedel–Crafts catalysts. As stated, the main goal of the patent was to increase the surface area available for sorption.

Later on, in the 1990s a number of application patents were issued to the Dow Chemical Company [31, 39–42, 44–46]. In 2005 the company's Web site advertised five hypercrosslinked products of the Dow Optipore^{TM} series – V493, V503, SD-2, L493, and L323 – all of them being designed to remove organics from water or air [304]. All of these sorbents have a specific surface area above $1100 \, \text{m}^2/\text{g}$, with V493 and L493 being biporous products with a pore volume of $1.16 \, \text{cm}^3/\text{g}$ and an average pore size of 46 Å. Another company, Bayer AG in Germany, also offers two hypercrosslinked sorbents, Lewatit VP OC 1163 [305] and Lewatit S 7768 [306], having both a macroporous and microporous structure with specific surface area values of over 1500 and $1300 \, \text{m}^2/\text{g}$, respectively. The pore volume of both the versions is similar, $0.6–0.8 \, \text{cm}^3/\text{g}$, and the pore diameters are indicated as 0.5–10 nm. Importantly, the resins have a monodisperse bead size distribution of 0.45–0.55 mm in diameter.

It should be mentioned that neither these patents nor the advertising materials imply real large-scale production of the resins. Usually an intensive study of the adsorbent's structure and properties precedes its industrial production, and the results of these studies often appear in scientific publications. For example, in China, Jiangsu N&G Environmental Technology Co. Ltd. quite recently started manufacture of the hypercrosslinked adsorbents NG-99 and NG-100; indeed, several reports appeared earlier in the literature describing the adsorption activity of these Chinese hypercrosslinked sorbents (their properties will be discussed later in Part III). However, we are not aware of any scientific literature concerning the above-mentioned sorbents of either Dow Chemical or Rohm and Haas.

Hypercrosslinked polystyrene sorbents appeared on the world market in 1994 when Purolite International Ltd. (UK and the USA) began to produce, with our active participation, the Hypersol-Macronet[TM] (MN) series of resins on an industrial scale. At present, a broad assortment of hypercrosslinked adsorbing materials are available (Table 9.6). The list of sorbents includes neutral (non-functionalized) resins and resins containing sulfonic, tertiary amine, or carboxylic functional groups. The sorbent matrices are also different: some have a solely microporous structure, while the others represent biporous morphology. In addition to the micropores typical of hypercrosslinked polymers based on linear polystyrene or gel-type copolymers, the matrix of the biporous products also contains large transport pores of 800 or 300 Å in diameter.

Table 9.6 Hypersol-Macronet[TM] polystyrene adsorbing materials produced by Purolite International [308, 309]

Sorbent	Functional groups	Total capacity (eq/L)	S (min.) (m²/g)	W_o (cm³/g)	Pore size (Å)	
					d_{50} micro	d_{50} macro
MN-100	WBA	0.1–0.3	900	1.0–1.1	14	800
MN-102	WBA	0.15–0.3	750	1.0–1.1	15	600–900
MN-150	WBA	0.2	800	0.65–0.85	14	300–550
MN-152	WBA	0.2	750	0.55–0.75	14	200–450
MN-170	WBA	0.2–0.5	1200	0.2–0.4	25	–
MN-200	None	–	800–1000	1.0–1.1	15	800
MN-202	None	–	700	1.0–1.1	15	600–900
MN-250	None	–	750	0.6–0.8	14	300–400
MN-252	None	–	450	0.6–0.8	14	200–450
MN-270	None	–	1200	0.7–0.8	15	na
MN-500	SAC	0.5–0.8[a]	500	1.0–1.1	15	850–950
MN-502	SAC	0.5–1.0	700	1.0–1.1	15	600–900
MN-600	WAC	–	800	1.0–1.1	15	800

[a]eq/kg.
WBA = weak basic anion exchanger containing tertiary amine functions; SAC = strong acidic cation exchanger containing sulfonic functions; WAC = weak acidic cation exchanger containing carboxylic functions; d_{50} = size of 50% micro- or macropores; na = no data presented; S = specific surface area; W_o = specific pore volume.

Purolite constantly widens the assortment of hypercrosslinked sorbents. One new type of hypercrosslinked sorbents, now being manufactured, was named NanoNet: NN-381 and NN-781 are particularly suitable for the separation of mixtures of mineral electrolytes into individual components by the new chromatographic method of ion size exclusion (ISE). NN-781 is especially resistant to such aggressive media as H_2SO_4, HNO_3, HF, or oxidants.

Since the last decade, the hypercrosslinked polystyrene sorbents have also come into routine use in analytical chemistry. They are widely used as excellent solid-phase extraction (SPE) media for the pre-concentration of trace amounts of organic contaminants in the environment, food products, biological fluids, gases, aquatic pools, etc., in combination with various chromatographic analytical techniques. By now many companies offer fine beads or granular particles of hypercrosslinked sorbents for SPE. Their main characteristics are presented in Table 9.7.

Purolite is also developing a series of microbeaded resins named "Chromalite" [307] for chromatography, solid-phase synthesis, and bio-separation applications. These products, neutral or functionalized, have either gel-type, macroporous, or hypercrosslinked structure. Several experimental batches of monodisperse hypercrosslinked beads have been successfully tested as stationary phase for high-performance liquid chromatography (HPLC).

Table 9.7 Commercial suppliers of hypercrosslinked sorbents for SPE applications

Sorbent (Company)	Particle size (μm)	Surface area (m²/g)	Pore volume (cm³/g)	Pore diameter (nm)	Ref.
Purosep 200 *Purolite International, UK*	40–140	1000	1.1	2–3 and 85–100	[310]
Isolute ENV *International Sorbent Technology, UK*	40–140	1100	1.1	2–3 and 85–100	[311]
LiChrolut EN *E. Merck, Germany*	40–120	1200	0.75	3	[312–314]
HYSphere-1 *Spark Holland, The Netherlands*	5–20	>1100	–	–	[315]
HR-P *Macherey Nagel, Germany*	40–120	1300	–	2.5	[316]

It is appropriate to emphasize here that all hypercrosslinked polystyrene products designed for use in columns, both analytical and industrial, are prepared in such a special manner that their bead size remains nearly constant in any liquid media. Most variants of the sorbents are of biporous morphology. The size constancy of their beads is essential for avoiding column blocking or channeling. This property, however, does not mean that the hypercrosslinked polymer phase of a biporous product does not feel the nature of the microenvironment and does not change its volume at the expense of the volume of macropores.

Part III describes in detail the use of hypercrosslinked sorbents, both industrial resins and laboratory-made polymers, for the sorption of organic compounds from air, water, and biological liquids and for the pre-concentration of analytes in SPE, as well as the use of hypercrosslinked resins as matrices for ion-exchange resins and as separating media in gas and liquid chromatography.

[1] Rogozhin, S.V., Davankov, V.A. Tsyurupa, M.P. Patent USSR 299165 (1969); Patents: USA 3,729,457, Austria 312929, Australia 448487, Great Britain 1315214, Argentina 190867, Belgium 756082, DDR 85644, Holland 140869, Canada 909442, Italy 916194, France 2061341, FRG 2045096, Czechoslovakia 164371, Switzerland 542254, Japan 982759; Chem. Abstr. 75 (1971) 6841b.

[2] Torkelson, J.M., Gilbert, S.R. Macromolecules 20 (1987) 1860–1865.

[3] Fleischer, G. Macromolecules 32 (1999) 2382–2383.

[4] Davankov, V.A., Rogozhin, S.V., Tsyurupa, M.P. Vysokomol. Soedin. B15 (1973) 463–465.

[5] Davankov, V.A., Rogozhin, S.V., Tsyurupa, M.P. J. Polym. Sci. Symp. 47 (1974) 95–101.

[6] Grassie, N., Gilks, J. J. Polym. Sci.: Polym. Chem. Ed. 11 (1973) 1531–1552.

[7] Grassie, N., Gilks, J. J. Polym. Sci.: Polym. Chem. Ed. 11 (1973) 1985–1994.

[8] Krauss, D., Popov, G., Schwachula, G. Plaste Kautsch. 26 (1979) 214–215.

[9] Krauss, D., Popov, G., Schwachula, G. Plaste Kautsch. 24 (1977) 545–548.

[10] Jovine, C.P., Ray-Chaudhuri, D.K., Patent USA 4,448,935 (1984).

[11] Tsyurupa, M.P., Lalaev, V.V., Davankov, V.A. Dokl. AN SSSR 279 (1984) 156–159.

[12] Tsyurupa, M.P., Davankov, V.A., in: Progress in Science and Technology, Chromatography, K.I. Sakodynskii (Ed.), VINITI, Moscow, 5 (1984) 32–66.

[13] Popov, G., Krauss, D., Feistel, L., Schwachula, G. Patent DDR 150218 (1981).

[14] Feistel, L., Popov, G., Schwachula, G., Krauss, D. Patent DDR 220051 (1981).

[15] Rogozhin, S.V., Korshak, V.V., Davankov, V.A., Maslova, L.A. Vysokomol. Soedin. 8 (1966) 1275–1278.

[16] Pepper, K.W., Paisley, H.M., Young, M.A. J. Chem. Soc. (1953), 4097–4105.

[17] Kolarz, B., Stepien-Firkowicz, A., Marciniak, A., Prace Naukove Institutu Technologii Organicznej i Tworzyw Sztucznych Politechniki Wroclawskiej, N13, Konferencje N2 (1973) 105–118.

[18] Fiestel, L., Popov, G., Schwachula, G. Plaste Kautsch. 30 (1983) 250–251.

[19] Hauptman, R., Schwachula, G. Z. Chem. 8 (1968) 227–228.

[20] Corte, H., Netz, O. Patent USA 3,417,066 (1968).

[21] Galeazzi, L. Patent France 2,272,108 (1975).

[22] Dow Chemical Co., Brit. Patent 677,350 (1952).

[23] Warshawsky, A., Deshe, A., Gutman, R. Br. Polym. J. 16 (1984) 234–238.

[24] Roland, L.R. Millar, J.R. Chem. Ind. (4 January 1993) 10–13.

[25] Bacak, N., Koza, G., Yagci, Y. Polym. Mater. 8 (1991) 189–192.

[26] DeHaan, F.P., Djaputra, M., Grinstaff, M.W., Kaufman, C.R., Keithly, J.C., Kumar, A. et al. J. Org. Chem. 62 (1997) 2694–2703.

[27] Feistel, L., Schwachula, G., Reuther, H., Klinkmann, H., Falkenhagen, D. Patent DDR 249274 (1987).

[28] Krauss, D., Popov, G., Schwachula, G., Feistel, L., Hauptmann, R. Patent DDR 125,824 (1977).

[29] Xu, M., Shi, Z., He, B. Chin. J. React. Polym. 2 (1993) 119–128.

[30] Reed, S.F. Patent USA 4,263,407 (1981).

[31] Stringfield, R.T., Goltz, H.R., Norman, S.I., Bharwada, U.J., LaBrie, R.L. Patent USA 4,950,332 (1990).

[32] Maroldo, S.G., Kopchik, R.M., Langenmayr, E.J. Patent USA 5,037,857 (1991).

[33] Feistel, L., Popov, G., Schwachula, G. Plaste Kautsch. 30 (1983) 496–498.

[34] Feistel, L., Popov, G., Schwachula, G. Plaste Kautsch. 30 (1983) 548–549.

[35] Reed, S.F. US Appl. 927,221,24 (1978); Chem. Abstr. 93 (1980) 48118S.

[36] Rubner, J., Schwachula, G., Feistel, L., Schilf, R., Kruger, H. Patent DDR 278592 (1990).
[37] Stringfield, R.T., Goltz, H.R., Norman, S.J., Bhawada, U.J., Labric, L.R. Int. Appl. WO 8908,718 (1989); Chem. Abstr. 112 (1990) 79845S.
[38] Schneider, H.P., Görlach-Doht, Y.M., Kümin, M.A.M. Eur. Pat. Appl. 388140 (1990); Chem. Abstr. 114 (1991) 8006Z.
[39] Schneider, H.P., Görlach-Doht, Y.M., Kümin, M.A.M. Patent USA 5,079,274 (1992).
[40] Demopolis, T.N. Intern. Pat. WO 94/05724 (1994).
[41] Stringfield, R.T. Patent USA 5,460,725 (1995).
[42] Stringfield, R.T., Ladika, M. Patent USA 5,683,800 (1997).
[43] Reed, S.F., Pinschmidt, R.K. Patent USA 4,191,813 (1980).
[44] Norman, S.I., Stringfield, R.T., Gopsill, C.C. Patent USA 4,965,083 (1990).
[45] Dawson-Ekeland, K.R., Stringfield, R.T. Patent USA 5,021,253 (1991).
[46] Stringfield, R.T., Ladika, M. Patent USA 5,519,064 (1996).
[47] Häupke, K., Schwachula, G., Pientka, V. Plaste Kautsch. 32 (1985) 48–51.
[48] Negre, M., Bartholin, M., Guyot, A. Angew. Makromol. Chem. 80 (1979) 19–30.
[49] Negre, M., Spitz, R., Bartholin, M., Guyot, A. 9th Europhysics Conference on Macromolecular Physics, 1979, Warsaw, Abstr. I.4.
[50] Macintyre, F.S., Sherrington, D.S., Tetley, L. Macromolecules 39 (2006) 5381–5384.
[51] Liu, Q.-Q., Wang, L., Xiao, A.-G., Yu, H.-J., Tan, Q.-H. Eur. Polym. J. 44 (2008) 2516–2522.
[52] Xie, Z. Materials Lett. 63 (2009) 509–511.
[53] Tsyurupa, M.P., Lalaev, V.V., Davankov, V.A. Acta Polym. 35 (1984) 451–455.
[54] Tsyurupa, M.P., Davankov, V.A., Lyustgarten, E.I., Pashkov, A.B., Belchich, L.A. Patent SSSR 948110 (1982).
[55] Peppas, N.A., Valkanas, G.N. Polym. Prepr. 17 (1976) 510–514.
[56] Peppas, N.A., Valkanas, G.N., Diamanti-Kotsida, E.T. J. Polym. Sci.: Polym. Chem. Ed. 14 (1976) 1241–1247.
[57] Regas, F.P., Papadoyannis, C.J. Polym. Bull. 3 (1980) 279–284.
[58] Peppas, N.A., Staller, K.P. Polym. Bull. 8 (1982) 233–237.
[59] Peppas, N.A., Valkanas, G.N. Angew. Makromol. Chem. 62 (1977) 163–176.
[60] Regas, F.P., Papadoyannis, C.J. Polym. Bull. 3 (1980) 279–284.
[61] Peppas, N.A., Bussing, G.R., Slight, K.A. Polym. Bull. 4 (1981) 193–198.
[62] Ando, K., Ito, T., Teshima, H., Kusano, H., in: M. Streat (Ed.), Ion Exchange for Industry, SCI, Ellis Horwood, Chichester, 1988, 232–238.
[63] Itagaki, T., Ito, T., Ando, K., Teshima, H. Patent USA 4,543,365 (1985).
[64] Zhou, C., Yan, J., Cao, Z. J. Appl. Polym. Sci. 83 (2002) 1668–1677.
[65] Hao, D.-X., Gong, F.L., Wei, W., Hu, G.-H., Ma, G.-H., Su, Z.-G. J. Colloid Interface Sci. 323 (2008) 52–53.
[66] Soukupová, K., Sassi, A., Jeřábek, K. React. Funct. Polym. 69 (2009) 353–357.
[67] Nyhus, A.K., Hagen, S., Berge, A. J. Polym. Sci. Part A: Polym. Chem. 38 (2000) 1366–1378.
[68] Christy, A.A., Nyhus, A.K., Kvalheim, O.M., Hagen, S., Schanche, J.S. Talanta 48 (1999) 1111–1120.
[69] Gawdzik, B., Osypiuk, J. Chromatographia 54 (2001) 323–328.
[70] Hradil, J., Králová, E. Polymer 39 (1998) 6041–6048.
[71] Zhang, Q., Yan, H., He, B. Ion Exch. Adsorption (Chinese) 2 (1986) 8.
[72] Samsonov, G.V., Ion Exchange. Sorption and Preparative Chromatography of Biologically Active Molecules Consultants Bureau (a Division of Plenum Publishing Corporation), New York, 1986.
[73] Trostyanskaya, E.B., Samsonov, G.V., El'kin, G.E. Ion Exchange. Sorption of Organic Compounds, Nauka, Leningrad, 1969.

[74] Moskvichov, B.V., Samsonov, G.V. Zh. Phiz. Khimii 47 (1973) 941–943.

[75] Ergozhin, E.E., Zhubanov, B.A., Kushnikov, Yu.A., Prodius, L.N. Izv. AN KazSSR, Chem. Series N13 (1970) 44–47.

[76] Kurmanaliev, M., Ergozhin, E.E., Rafikov, S.R. Izv. AN KazSSR, Chem. Series N3 (1972) 49–57.

[77] Kurmanaliev, M., Ergozhin, E.E. Vestnik AN KazSSR, N9 (1971) 53–55.

[78] Musabekov, K.B., Ergozhin, E.E., Shapovalova, L.P. Izv. AN KazSSR, Chem. Series N6 (1972) 59–64.

[79] Kurmanaliev, M., Ergozhin, E.E. Izv. AN KazSSR, Chem. Series N5 (1972) 67–72.

[80] Ergozhin, E.E., Prodius, L.N., Rafikov, S.R. Izv. AN KazSSR, Chem. Series N2 (1972) 53–55.

[81] Prusova, V.N., Ergozhin, E.E., Zhubanov, B.A., Rafikov, S.R. Izv. AN KazSSR, Khim. Ser., N2 (1972) 59–64.

[82] Ergozhin, E.E., Rafikov, S.R., Imanbenkova, S.M., Zhubanov, B.A. Izv. AN KazSSR, Chem. Series N11 (1972) 2609–2612.

[83] Prusova, V.N., Ergozhin, E.E. Izv. AN KazSSR, Chem. Series N5 (1972) 62–66.

[84] Prusova, V.N., Ergozhin, E.E. Izv. AN KazSSR, Chem. Series N1 (1972) 51–54.

[85] Ergozhin, E.E., Zhubanov, B.A., Kushnakov, Yu.A., Prusova, V.N. Izv. AN KazSSR, Chem. Series N1 (1972) 44–47.

[86] Ergozhin, E.E., Zhubanov, B.A., Prodius, L.N., Musabekov, K.B., Nurkhodzhaeva, V.A. Izv. AN KazSSR, Chem. Series N5 (1971) 54–58.

[87] Trushin, B.N. PhD Thesis, Mendeleev Institute of Chemical Engineering, Moscow, 1968.

[88] Trushin, B.N., Davankov, A.B., Korshak, V.V., Chemistry and Technology of Organic Substances and High Molecular Weight Compounds, Transactions of the D.I. Mendeleev Institute of Chemical Engineering, issue 57, Moscow, 1968, 119–123.

[89] Trushin, B.N., Davankov, A.B., Korshak, V.V. Vysokomol. Soedin. A9 (1967) 1140–1143.

[90] Wiley, R.H., Allen, J.K., Chang, S.P., Musselman, K.E., Venkatahalam, T.K. J. Phys. Chem. 68 (1964) 1776–1779.

[91] Yamashita, T. Bull. Chem. Soc. Jpn. 45 (1972) 195–198.

[92] Sangalov, Yu.A., Yasman, Yu.B., Gladkikh, I.F., Minsker, K.S. Dokl. AN SSSR, 265 (1982) 671–674.

[93] Run, R., Pei, W., Jia, X., Shen, J., Jang, S. Kexue Tongbao (Chinese) 32 (1987) 388–394.

[94] Rabek, J.F., Lucki, J. J. Polym. Sci.; Part A: Polym. Chem. 26 (1988) 2537–2551.

[95] Theodoropoulos, A.G., Bouranis, D.L., Valkanas, G.N. J. Appl. Polym. Sci. 46 (1992) 1461–1465.

[96] Davankov, V.A., Tsyurupa, M.P. React. Polym. 13 (1990) 27–42.

[97] Joseph, R., Ford, W.T., Zhang, S., Tsyurupa, M.P., Pastukhov, A.V., Davankov, V.A. J. Polym. Sci.; Part A: Polym. Chem. 35 (1997) 695–701.

[98] Streat, M., Sweetland, L.A. React. Funct. Polym. 35 (1997) 99–109.

[99] Law, R.V., Sherrington, D.C., Snape, C.E., Ando, I., Kurosu, H. Macromolecules 29 (1996) 6284–6293.

[100] Grassie, N., Schoff, C., Cunningham, J.G. Eur. Polym. J. 12 (1976) 647–650.

[101] Nakanishi, K., Infrared Absorption Spectroscopy. Practical, Nankodo Company Ltd., Tokyo, 1962.

[102] Critchfield, F., Analysis of Basic Functional Groups in Organic Compounds, Mir, Moscow, 1965.

[103] Mohanraj, S., Ford, W.T. Macromolecules 18 (1985) 351–356.

[104] Rogozhin, S.V., Davankov, V.A. Vysokomol. Soedin. 9A (1967) 1286–1992.

[105] Bootsma, J.P.C., Eling, B., Challa, G. React. Polym. 3 (1984) 17–22.

[106] Patel, G.R., Patel, S.R. J. Macromol. Sci. A19 (1983) 653–662.
[107] Patel, G.R., Patel, S.R. J. Macromol. Sci. A19 (1983) 663–672.
[108] Patel, G.R., Amin, P.T., Patel, S.R. J. Macromol. Sci. A18 (1982) 939–947.
[109] Krakovyak, M.G., Anan'eva, T.L., Nekrasova, T.N., Klenin, S.I., Krivobokov, V.V., Skorokhodov, S.S. "Conference on Chemistry and Physics of High Molecular Weight Compounds" Leningrad, 1983, Abstr., 102–103.
[110] Grassie, N., Meldrum, J.G., Gilks, J. J. Polym. Sci. Polym. Lett. 8 (1970) 247–251.
[111] Grassie, N., Meldrum, J.G. Eur. Polym. J. 4 (1968) 571–580.
[112] Grassie, N., Meldrum, J.G. J. Polym. Sci. Symp. N30 (1970) 147–156.
[113] Grassie, N., Meldrum, J.G. Eur. Polym. J. 5 (1969) 195–209.
[114] Grassie, N., Flood, J., Cunningham, J.G. Eur. Polym. J. 12 (1976) 641–645.
[115] Iovu, M., Tudorache, E. Makromol. Chem., 166 (1973) 51–56.
[116] Peppas, N.A., Valkanas, G.N. J. Polym. Sci.: Polym. Chem. Ed. 12 (1974) 2567–2579.
[117] Barar, D.G., Staller, K.P., Peppas, N.A. J. Polym. Sci.: Polym. Chem. Ed. 21 (1983) 1013–1024.
[118] Collette, C., Lafuma, F., Audebert, R., Leibler, L., in: O. Kramer (Ed.), Biological and Synthetic Polymer Networks, Elsevier Applied Science, London & New York, 1986, 277–290.
[119] Krakovyak, M.G., Anan'eva, T.D., Anufrieva, E.V., Nekrasova, T.N., Klenin, S.I., Krivobokov, V.V. et al. Vysokomol. Soedin. A26 (1984) 2071–2076.
[120] Häusler, K.-G., Popov, G., Krauss, D., Schwachula, G. Plaste Kautsch. 28 (1981) 260–265.
[121] Jenny, R. Compt. Rend. 246 (1958) 3477–3480.
[122] Jenny, R. Compt. Rend. 248 (1959) 3555–3556.
[123] Jenny, R. Compt. Rend. 250 (1960) 1659–1661.
[124] Tsyurupa, M.P. "Hypercrosslinked polystyrene – a new type of polystyrene networks", Dr.Sci. Thesis, Inst. of Element-Organic Compounds, Moscow, 1985.
[125] Rogozhin, S.V., Davankov, S.V., Tsyurupa, M.P., Ermakova, I.P., Misyurev, B.I. Patent SSSR 434757 (1974).
[126] Yoshizaki, K., Urakawa, O., Adachi, K. Macromolecules 36 (2003) 2349–2354.
[127] Lipatov, Yu.S., Nesterov, A.E., Gritsenko, T.M., Veselovskii, R.A., Handbook on the Chemistry of Polymers, Naukova Dumka, Kiev, 1971.
[128] Davankov, V.A., Rogozhin, S.V., Tsyurupa, M.P. Angew. Makromol. Chem. 32 (1973) 145–151.
[129] Tsyurupa, M.P., Andreeva, A.I., Davankov, V.A. Angew. Makromol. Chem. 70 (1978) 179–187.
[130] Davankov, V.A., Tsyurupa, M.P., Rogozhin, S.V. Angew. Makromol. Chem. 53 (1976) 19–27.
[131] Davankov, V.A., Tsyurupa, M.P. Angew. Makromol. Chem. 91 (1980) 127–142.
[132] Tsyurupa, M.P., Davankov, V.A., React. Funct. Polym. 53 (2002) 193–203.
[133] Davankov, V.A., Tsyurupa, M.P., in: S. Aharoni (Ed.), Synthesis, Characterization and Theory of Polymeric Networks and Gels, Plenum Press, 1992, 179–200.
[134] Martsinkevich, R.V., Tsyurupa, M.P., Davankov, V.A., Soldatov, V.S. Vysokomol. Soedin. A20 (1978) 1061–1065.
[135] Wei, J., Bai, X.Y., Yan, J. Macromolecules 36 (2003) 4960–4966.
[136] Tsyurupa, M.P., Volynskaya, A.V., Belchich, L.A., Davankov, V.A. J. Appl. Polym. Sci. 28 (1983) 685–689.
[137] Tager, A.A., Kargin, V.A. Kolloid. Zh. 24 (1952) 367–371.
[138] Wilks, A.D., Pietrzyk, D.J. Anal. Chem. 44 (1972) 676–681.
[139] Reshet'ko, D.A. "On Relation Between the Processes of Physical Adsorption and Dissolution in the process of Interaction of Polymers with the Vapors of Low-Molecular Weight Liquids", PhD Thesis, Sverdlovsk, 1976.

[140] Belyakova, L.D., Shevchenko, T.I., Davankov, V.A., Tsyurupa, M.P. Adv. Colloid Interface Sci., 25 (1986) 249–266.
[141] Rozenberg, G.I., Shabaeva, A.S., Moryakov, V.S., Musin, T.G., Tsyurupa, M.P., Davankov, V.A. React. Polym. 1 (1983) 175–186.
[142] Alfrey, T., Jr., Lloyd, W.G. J. Polym. Sci. 62 (1962) 159–165.
[143] Lloyd, W.G., Alfrey, T., Jr., J. Polym. Sci. 62 (1962) 301–316.
[144] Millar, J.R., Smith, D.G., Marr, W.E., Kressman, T.R.E. J. Chem. Soc. 1962, 218–225.
[145] Rabelo, D., Coutinho, F.M.B. Macromol. Symp. 84 (1994) 341–350.
[146] Huxham, I.M., Rowatt, B., Sherrington, D.C. Polymer 33 (1992) 2768–2777.
[147] Roe, S.P., Sherrington, D.C. React. Polym. 11 (1989) 301–308.
[148] Poinescu, Ig.C., Garpov, A. Rev. Roum. Chim. 34 (1989) 1061–1073.
[149] Yan, J., Xu, R., Yan, J. J. Appl. Polym. Sci. 38 (1989) 45–54.
[150] Yan, J., Wang, X., Yang, Y. React. Funct. Polym. 43 (2000) 227–232.
[151] Davankov, V.A., Pastukhov, A.V., Tsyurupa, M.P. J. Polym. Sci. Part B: Polym. Phys. 38 (2000) 1553–1563.
[152] Tsyurupa, M.P., Lalaev, V.V., Bel'chich, L.A., Davankov, V.A. Vysokomol. Soedin. A28 (1986) 591–595.
[153] Lee, K.Y., Rowley, J.A., Eiselt, P., Moy, E.M., Bouhadir, K.H., Mooney, D.J. Macromolecules 33 (2000) 4291–4294.
[154] Tikhonov, V.E., Radigina, L.A., Yamskov, Y.A. Carbohydr. Res. 200 (1996) 33–41.
[155] Tsyurupa, M.P., Oslonovich, Yu.V., Nistratov, A.N., Radchenko, L.G., Davankov, V.A. Polym. Sci. A37 (1995) 595–599.
[156] Popov, G., Häusler, K.-G., Krauss, D., Schwachula, G. Plaste Kautsch. 27 (1980) 614–616.
[157] Veverka, P., Jeřabek, K. React. Funct. Polym. 41 (1991) 21–25.
[158] Tsyurupa, M.P., Pankratov, E.A., Davankov, V.A. Vysokomol. Soedin. B 28 (1980) 755–757.
[159] Pastukhov, A.V., Davankov, V.A., Sidorova, E.V., Shkol'nikov, E.I., Volkov, V.V. Izvestia Akad. Nauk Ser. Khim. 2007, 467–476.
[160] Flory, P.J., Rener, J. J. Chem. Phys. 11 (1943) 521–526.
[161] James, H.M., Guth, E. J. Chem. Phys. 15 (1947) 669–683.
[162] Hermans, J.J. Trans. Faraday Soc. 43 (1947) 591–600.
[163] Dušek, K., Prins, W. Adv. Polym. Sci. 6 (1969) 1–102.
[164] Peppas, N.A., Merrill, E.W. J. Polym. Sci.: Polym. Chem. Ed. 14 (1976) 459–464.
[165] Seno, M., Yamabe, T. Bull. Chem. Soc. Jpn. 37 (1964) 754–755.
[166] Cejnar, F. Kolloid-Z. Z. Polym. 250 (1972) 977–979.
[167] Borchard, W. J. Polym. Sci.: Polym. Symp. 44 (1974) 153–162.
[168] Flory, P.J. Macromolecules 12 (1979) 119–122.
[169] Dušek, K. Collect. Czech. Chem. Commun. 27 (1962) 2841–2853.
[170] Okay, O. Makromol. Chem. 189 (1988) 2201–2217.
[171] Sarin, V.K., Kent, S.B.H., Merrifield, R.B. J. Am. Chem. Soc. 102 (1980) 5463–5470.
[172] Erukhimovich, I.Ya. Vysokomol. Soedin. A20 (1978) 114–118.
[173] Flory, P.J. Polymer 20 (1979) 1317–1320.
[174] Tsyurupa, M.P., Papkov, M.A., Davankov, V.A. Vysokomol. Soedin. C51 (2009) 1339–1345.
[175] Tsyurupa, M.P., Davankov, V.A. React. Funct. Polym. 66 (2006) 768–779.
[176] Khirsanova, I.F., Soldatov, V.S., Martsinkevich, R.V., Tsyurupa, M.P., Davankov, V.A. Kolloid. Zh. 40 (1978) 1025–1029.
[177] Gregg, S.J., Sing, K.S.W. Adsorption, Surface Area and Porosity, second ed., Mir, Moscow, 1984.
[178] Kenyon, A.S., Gross, R.C., Wurstner, A.L. J. Polym. Sci. 40, (1959) 159–168.
[179] Slonimskii, G.L., Askadskii, A.A., Kitaigorodskii, A.I. Vysokomol. Soedin. A12 (1970) 494–512.

[180] Xu, M., Shi, Z., He, B. Chin. J. React. Polym. 2 (1993) 119–128.
[181] Askadskii, A.A., Matveev, Yu.I. Chemical Structure and Physical Properties of Polymers, Khimiya, Moscow, 1983, 248p.
[182] Karnaukhov, A.P., Adsorption. Texture of Dispersed and Porous Materials, Nauka, Novosibirsk, 1999.
[183] Tsyurupa, M.P., Davankov, V.A. J. Polym. Sci.: Polym. Chem. Ed. 18 (1980) 1399–1406.
[184] Halasz, I., Martin, K. Angew. Chem. 17 (1978) 901–908.
[185] Werner, W., Halacz, J. J. Chromatogr. Sci. 18 (1980) 277–283.
[186] Halasz, J., Vogtel, P. Angew. Chem. Int. Ed. Engl. 19 (1980) 24–28.
[187] Gorbunov, A.A., Solovyova, L.Ya., Pasechnik, V.A. J. Chromatogr. 448 (1988) 307–332.
[188] Jeřabek, K., Setinek, K. J. Polym. Sci. Part A: Polym. Chem. 27 (1989) 1619–1623.
[189] Jeřabek, K. Anal. Chem. 57 (1985) 1598–1602.
[190] Ogston, A.G. Trans. Faraday Soc. 54 (1985) 1754–1757.
[191] Casassa, E.F. J. Polym. Sci. Polym. Lett. 5 (1967) 773–778.
[192] Shantarovich, V.P., Suzuki, T., He, C., Davankov, V.A., Pastukhov, A.V., Tsyurupa, M.P. et al. Macromolecules 35 (26) (2002) 9723–9729.
[193] Shkol'nikov, E.I., Volkov, V.V. Dokl. Akad. Nauk 378 (2001) 507–510.
[194] Pastukhov, A.V., Babushkina, T.A., Davankov, V.A., Klimova, T.P., Shantarovich, V.P. Dokl. Akad. Nauk 411 (2006) 216–219.
[195] Nightingale, E.R. J. Phys. Chem. 63 (1959) 1381–1387.
[196] Tsyurupa, M.P., Davankov, V.A., Rogozhin, S.V. J. Polym. Sci. Symp. N47 (1974) 189–195.
[197] Davankov, V.A., Tsyurupa, M.P. J. Chromatogr. A 1087 (2005) 3–12.
[198] Tsyurupa, M.P., Pankratov, E.A., Tsvankin, D.Ya., Zhukov, V.P., Davankov, V.A. Vysokomol. Soedin. A27 (1985) 339–345.
[199] Authors acknowledge Prof. A.N. Ozerin for the measurements and discussion of the results obtained.
[200] Hughes, L.J., Brown, G.L. J. Appl. Polym. Sci. 5 (1961) 580–588.
[201] Sperling, L.H., George, H.F., Huelck, V., Thomas, D.A. J. Appl. Polym. Sci. 14 (1970) 2815–2824.
[202] Pastukhov, V.A., Tsyurupa, M.P., Davankov, V.A. J. Polym. Sci. Part B: Polym. Phys. 37 (1999) 2324–2333.
[203] Pastukhov, A.V., Davankov, V.A. Dokl. Akad. Nauk Russ. 410 (2006) 767–770.
[204] Askadskii, A.A., Litvinov, V.M., Kazantseva, V.V., Tsyurupa, M.P., Belchich, L.A., Davankov, V.A. et al. Vysokomol. Soedin. A28 (1986) 281–288.
[205] Vasserman, A.M., Kovarskii, A.L., Aleksandrova, T.A., Buchachenko, A.L., in: G.L. Slonimskii (Ed.) Modern Physical Methods of Investigation of Polymers, Moscow, Nauka, 1982, pp. 121–155.
[206] Pastukhov, A.V., Davankov, V.A., Aleksienko, N.N., Belyakova, L.D., Tsyurupa, M.P. in: Structure and Dynamics of Molecular Systems, Kazan, 10, (2003) No, 3, 29–32.
[207] Pastukhov, A.V., Aleksienko, N.N., Tsyurupa, M.P., Davankov, V.A., Voloshchuk, A.M. Russ. J. Phys. Chem. 79 (2005) 1371–1379.
[208] Neely, J.W., Isacoff, E.G., Carbonaceous Adsorbents for the Treatment of Ground and Surface Waters, Marcel Dekker, New York, 1982.
[209] Pastukhov, A.V., Ginzburg, S.F., Davankov, V.A. Izv. Akad. Nauk Ser. Khim. No. 5, (2006), 824–831.
[210] Aleksienko, N.N., Pastukhov, A.V., Davankov, V.A., Belyakova, L.D., Voloshchuk, A.M. Russ. J. Phys. Chem. 78, (2004) 1992–1998.
[211] Seiler, M. Fluid Phase Equilibria 241 (2006) 155–174.
[212] Arnautov, S.A., Davankov, V.A. Mendeleev Commun. 16 (2006) 79.
[213] Voit, B.I., Lederer, A. Chem. Rev. 109 (2009) 5924–5973.

[214] Rosen, B.M., Wilson, C.J., Wilson, D.A., Peterca, M., Imam, M.R., Percec, V. Chem. Rev. 109 (2009) 6275–6540.

[215] Allen, G., Burgess, J., Edwards, F.R.S., Edwards, S.F., Walsh, D.J. Proc. Roy. Soc. London, A, 334 (1973) N1599, 453–463.

[216] Forget, J.L., Booth, C., Canham, P.H., Duggleby, M., Kimg, T.A., Price, C. J. Polym. Sci.: Polym. Phys. Ed. 17 (1979) 1403–1411.

[217] Funke, W., Bauer, H., Joos, B., Kaczun, J., Kleiner, B., Leibelt, U. et al. Polym. Int. 30 (1993) 519–523.

[218] Antonietti, M., Basten, R., Lohmann, S. Macromol. Chem. Phys. 196 (1995) 441–466.

[219] Antonietti, M., Bremser, W., Schmidt, M. Macromolecules 23 (1990) 3796–3805.

[220] Antonietti, M. Angew. Chem. 100 (1988) 1813–1817.

[221] Okay, O., Kurz, M., Lutz, K., Funke, W. Macromolecules 28 (1995) 2728–2737.

[222] Staudinger, H., Husemann, E. Ber. Deutsch. Chem. Ges. 68 (1937) 1618–1634.

[223] Downey, J.S., McIssac, G., Frank, R.S., Stöver, H.D.H., Macromolecules 34 (2001) 4534–4541.

[224] Allen, G., Burgess, J., Edwards, S.F., Walsh, D.J. Proc. Roy. Soc. London A, 334 (1973) N 1599, 465–476.

[225] Allen, G., Burgess, J., Edwards, S.F., Walsh, D.J. Proc. Roy. Soc. London A, 334 (1973) N 1599, 477–491.

[226] Antonietti, M., Sillescu, H., Schmidt, M., Schuch, H. Macromolecules 21 (1988) 736–742.

[227] Jiang, J., Thayumanavan, S. Macromolecules 38 (2005) 5886–5891.

[228] Mecerreyes, D., Lee, V., Hawker, C.J., Hedrick, J.L., Wursch, A., Volksen, W. et al. Adv. Mater. 13 (2001) 204–208.

[229] Funke, W., Okay, O., Joos-Müller, B. Adv. Polym. Sci. 136 (1998) 139–234.

[230] Tsyurupa, M.P., Maslova, L.A., Mrachkovskaya, T.A., Davankov, V.A. Vysokomol. Soedin. A33 (1991) 2645–2651.

[231] Tsyurupa, M.P., Mrachkovskaya, T.A., Maslova, L.A., Timofeeva, G.I., Dubrovina, L.V., Davankov, V.A. et al. React. Polym. 19 (1993) 55–66.

[232] Kim, Y.H., Webster, O.W. J. Am. Chem. Soc. 112 (1990) 4592–4593.

[233] Kratochvil, P. Pure Appl. Chem. 54 (1982) 379–393.

[234] Lange, H. Kolloid, Z. Z. Polym. 240 (1970) 747–755.

[235] Lange, H. Kolloid, Z. Z. Polym. 250 (19732) 775–781.

[236] Davankov, V.A., Ilyin, M.M., Tsyurupa, M.P., Timofeeva, G.I., Dubrovina, L.V. Macromolecules 29 (1996) 8398–8403.

[237] Eskin, V.E., Light Scattering of Polymer Solutions, Nauka, Moscow, 1973, p. 353.

[238] Davankov, V.A., Ilyin, M.M., Timofeeva, G.I., Tsyurupa, M.P., Dokl. Akad. Nauk Russ. 352 (1997) 494–496.

[239] Davankov, V.A., Timofeeva, G.I., Ilyin, M.M., Tsyurupa, M.P. J. Polym. Sci. Part A: Polym. Chem. 35 (1997) 3847–3852.

[240] Davankov, V.A., Ilyin, M.M., Timofeeva, G.I., Tsyurupa, M.P., Yaminsky, I.V. J. Polym. Sci. Part A: Polym. Chem. 37 (1999) 1451–1455.

[241] Svedberg, T., Pedersen, K.O. The Ultracentrifuge, Clarendon Press, Oxford, UK, 1940.

[242] Tsvetkov, V.N., Eskin, V.E., Frenkel, S.Ya. Structure of Macromolecules in Solutions, Nauka, Moscow, 1964, p. 394.

[243] Vargaftik, M.N., Zagorodnikov, V.P., Stolyarov, I.P., Moiseev, I.I., Likholobov, V.A., Kochubey, D.I. et al. J. Chem. Soc. Chem. Commun. 937–939 (1985).

[244] De Young, L.R., Fink, A.L., Dill, K.A. Acc. Chem. Res. 26 (1993) 614–620.

[245] Kim, Y.H., Webster, O.W. Macromolecules 25 (1992) 5560–5572.

[246] Frechet, J.M.J., Hawker, C.J. React. Funct. Polym. 26 (1995) 127–136.

[247] Kumar, A.K., Ramakrishnan, S. Macromolecules 29 (1996) 2524–2530.

[248] Funke, W., Bauer, H., Joos, B., Kaczun, J., Kleiner, B., Leibelt, U. et al. Polym. Int. 30 (1993) 519–523.

[249] "International Conference Nano-Structures and Self-Assemblies in Polymer Systems" St. Petersburg, Moscow, 1995, Abstracts.

[250] Zhong, X.F., Eisenberg, A. Macromolecules 27 (1994) 1751–1758.

[251] Guo, A., Liu, G., Tao, J. Macromolecules 29 (1996) 2487–2493.

[252] Nikolaev, A.V., Yakhin, V.S., Yur'ev, G.S., Bogatyrev, V.L., Theory and Practice of Sorption Processes (Russian), Voronezh State University, Russia, No 12 (1978) 9–22.

[253] Pastukhov, A.V. "Physico-chemical properties and structural mobility of hypercrosslinked polystyrenes", Dr.Sci. Thesis, Moscow, 2009.

[254] Sperling, L.H., Taylor, D.W., Kirpatrick, L.M., George, H.F., Bardman, D.R. J. Appl. Polym. Sci. 14 (1970) 73–78.

[255] Sperling, L.H., George, H.F., Huelck, V., Thomas, D.A. J. Appl. Polym. Sci. 14 (1970) 2815–2824.

[256] Frenkel, M.D., Ratner, S.B., Encyclopedia of Polymers, Soviet Encyclopedia, Moscow 3 (1977) 846–847.

[257] Grubhofer, N. Makromol. Chem. 30 (1959) 96–108.

[258] Yan, S.G., Zhang, S.H., Zou, W.H., Zhou, Y.H., Zhou, X.H. Chin. Chem. Lett. 19 (2008) 611–614.

[259] Wood, C.D., Tan, B., Trewin, A., Niu, H., Bradshaw, D., Rosseinsky, M.J. et al. Chem. Mater., 18 (2007) 2034–2048.

[260] Germain, J., Fréchet, J.M.J., Svec, F. J. Mater. Chem. 17 (2007) 4989–4997.

[261] Germain, J., Fréchet, J.M.J., Svec, F. Chem. Commun. (2009) 1526–1528.

[262] Germain, J., Svec, F., Frechet, J.M.J. Chem. Mater 20 (2008) 7069–7076.

[263] Weber, J., Antonietti, M., Thomas, A. Macromolecules 41 (2008) 2880–2885.

[264] Pavlova, L.A., Pavlov, M.V., Davankov, V.A. Dokl. Akad. Nauk Russ. 406 (2006) 200–202.

[265] Kargin, V.A., Kabanov, V.A., Aliev, K.V., Rasvodovski, E.F. Dokl. AN 160 (1965) 604–607.

[266] Webster, O.W., Gentry, F.P., Farlee, R.D., Smart, B.E. Polym. Prepr. 32 (2) (1991) 74–75.

[267] Budd, P.M., Ghanem, B.S., Makhseed, S., McKeown, N.B., Msayib, K.J., Tattershall, C.E. Chem. Commun. (2004) 230–231.

[268] Ghanem, B.S., McKeown, N.B., Budd, P.M., Fritsch, D. Macromolecules 41 (2008) 1640–1646.

[269] Budd, P.M., Makhseed, S.M., Ghanem, B.S., Msayib, K.J., Tattershall, C.E., McKeown, N.B. Mater. Today April (2004) 40–46.

[270] Weber, J., Thomas, A. J. Am. Chem. Soc. 130 (2008) 6334–6335.

[271] Schwab, M.G., Fassbender, B., Spiess, H.W., Thomas, A., Feng, X., Müllen K. J. Am. Chem. Soc. 131 (2009) 7216–7217.

[272] Jiang, J.-X., Su, F., Trewin, A., Wood, C.D., Campbell, N.L., Niu, H. et al. Angew. Chem. Int. Ed. 46 (2007) 8574–8578.

[273] Jiang, J.-X., Su, F., Trewin, A., Wood, C.D., Campbell, N.L., Niu, H. et al. Angew. Chem. Int. Ed. 47 (2008) 1167.

[274] Jiang, J.-X., Su, F., Trewin, A., Wood, C.D., Niu, H., Jones, J.T.A. et al. J. Am. Chem. Soc. 130 (2008) 7710–7720.

[275] Jiang, J.-X., Trewin, A., Su, F., Wood, C.D., Niu, H., Jones, J.T.A. et al. Macromolecules 42 (2009) 2658–2606.

[276] Weder, C. Angew. Chem. Int. Ed. 47 (2008) 448–450.

[277] Hittinger, E., Kokil, A., Weder, C. Angew. Chem. Int. Ed. 43 (2004) 1807–1811.

[278] Liu, F., Yuan, S.G., Wang, X.L., Polikarpov, A.P., Shunkevich, A.A. Chinese Chem. Lett. 18 (2007) 588–590.

[279] Loy, D.A., Shea, K.J. Chem. Rev. 95 (1995) 1431–1442.

[280] Shea, K.J., Loy, D.A., Webster, O.W. Chem. Mater. 1 (1989) 572–574.
[281] Chem. Tech., April 1998, 3.
[282] Han, J., Zheng, S. Macromolecules 41 (2008) 4561–4564.
[283] Yaghi, O., Li, G., Li, H. Nature 378 (1995) 703–706.
[284] Férey, G., Mellot-Draznieks, C., Serre, C., Millange, F., Dutour, J., Surblé, S. et al. Science 309 (2005) 2040–2042.
[285] Férey, G., Mellot-Draznieks, C., Serre, C., Millange, F. Acc. Chem. Res. 38 (2005) 217–225.
[286] Chae, H.K., Siberio-Pérez, D.Y., Kim, J., Go, Y.-B., Eddaoudi, M., Matzger, A.J. et al. Nature, 427 (2004) 523–527.
[287] Millward, A.R., Yaghi, O.M. J. Am. Chem. Soc. 127 (2005) 17998–17999.
[288] Koh, K., Wong-Foy, A.G., Matzger, A.J. Angew. Chem. Int. Ed. 47 (2008) 677–680.
[289] Eddaoudi, M., Kim, J., Rosi, N., Vodak, D., Wachter, J., O'Keeffe, M. et al. Science 295 (2002) 469–472.
[290] Chem. & Eng. News, August 25, 2008, 13–16.
[291] Zhou, W., Wu, H., Yildirim, T. J. Am. Chem. Soc. 130 (2008) 15268–15269.
[292] Chem. & Eng. News, March 23 (2009) 37.
[293] Dincă, M., Dailly, A., Liu, Y., Brown, G.M., Neumann, D.A., Long, J.R. J. Am. Chem. Soc. 128 (2006) 16876–16883.
[294] Ma, S., Sun, D., Simmons, J.M., Collier, C.D., Yuan, D., Zhou, H.-C. J. Am. Chem. Soc. 130 (2008) 1012–1016.
[295] Li, Q., Zhang, W., Miljanic, O.Š., Sue, C.-H., Zhao, Y.-L., Liu, L. et al. Science 325 (2009) 855–859.
[296] Ziv, A., Grego, A., Kopilevich, S., Zeiri, L., Miro, P., Bo, C. et al. J. Am. Chem. Soc. 131 (2009) 6380–6382.
[297] Trung, T.K., Trens, P., Tanchoux, N., Bourrelly, S., Llewellyn, P.L., Loera-Serna, S. et al. J. Am. Chem. Soc. 130 (2008) 16926–16932.
[298] Wang, Z., Cohen, S.M. J. Am. Chem. Soc. 131 (2009) 16675–16677.
[299] Thallapally, P.K., Tian, J., Kishan, M.R., Fernandez, C.A., Dalgarno, S.J., McGrail, P.B. et al. J. Am. Chem. Soc. 130 (2008) 16842–16843.
[300] Côté, A.P., Benin, A.I., Ockwig, N.W., O'Keeffe, M., Matzger, A.J., Yaghi, O.M. Science 310 (2005) 1166–1170
[301] Côté, A.P., El-Kaderi, H.M., Furukawa, H., Hunt, J.R., Yaghi, O.M. J. Am. Chem. Soc. 129 (2007) 12914–12915
[302] El-Kaderi, H.M., Hunt, J.R., Mendoza-Cortés, J.L., Côté, A.P., Taylor, R.E., O'Keeffe, W. et al. Science 316 (2007) 268–272.
[303] Mastalerz, M. Angew. Chem. Int. Ed. 47 (2008) 445–447
[304] DOWEX Ion Exchange Resins and Adsorbents, Optipore L493 and V493, Product Information, 1998.
[305] Bayer Chemicals, Levatit VP OC 1163, Product Information, 2003.
[306] Bayer Chemicals, Levatit S 7768, Product Information, 2004.
[307] Chromalite, Purolite; www.purolite.com, Product information.
[308] Hypersol-Macronet™ Sorbent Resins, Purolite, 1999.
[309] Cation Exchangers, Anion Exchangers, Mixed Beds, Nuclear Grade, Special Products, Purolite, Catalog, 2002.
[310] Purosep SPEs – Solid Phase Extractants, Purolite, 1997.
[311] Isolute ENV+, International Sorbent Technology, Catalog.
[312] LiChrolut®, Merck, 1996.
[313] Puig, D., Barceló, D. J. Chromatogr. A 733 (1996) 371–381.
[314] Gun'ko, V.M., Turov, V.V., Zarko, V.I., Nychiporuk, Y.M. et al. J. Colloid Interface Sci. 323 (2008) 6–17.
[315] Less, M., Schmidt, T.C., von Löw, E., Stork, G. J. Chromatogr. A 810 (1998) 173–182.
[316] Masqué, N., Galià, M., Marcé, R.M., Borrull, F. J. Chromatogr. A 771 (1997) 55–61.

PART **Three**

Application of Hypercrosslinked Polystyrene Adsorbing Materials

CHAPTER 10

Sorption of Gases and Organic Vapors

1. POLYMERIC ADSORBENTS VERSUS ACTIVATED CARBONS

Numerous industrial chemical processes, food technologies, manufacturing of pharmaceuticals, petrochemistry, etc., have had to deal with the problem of separating complex mixtures. In many cases minor components have to be removed from large volumes of bulk solution. For example, such problems are faced in the decontamination of industrial wastewaters, potable water processing, or exhaust gas treatment. Here, various regular separation techniques are of little use on the industrial scale or are too cumbersome, for example, distillation of large volumes of liquids, or condensation of organic vapors from air streams on a cooled surface. In this respect, adsorption of impurities from aqueous solutions or air on the surface of porous solids is the technique of choice. The effectiveness of the separation process is largely determined by the quality of the adsorbent used. The latter must combine high loading capacity with good kinetic characteristics in order to realize its full potential at high flow rates. Good permeability of the adsorbent is especially important when relatively large molecules of organic contaminants have to be adsorbed. Naturally, a most important characteristic of the adsorbent is the simplicity and completeness of its regeneration for subsequent reuse. High thermal stability and high hydrolytic and osmotic resistance of the material are also required. In addition, the adsorbent should display sufficient mechanical strength to withstand large pressure drops in industrial columns. Finally, the price of the adsorbent should not prohibit its practical use.

In many respects, activated carbons meet these requirements and today a vast assortment of microporous and macroporous activated carbons are commercially available. These sorbents are very attractive due to their low cost, particularly if prepared by pyrolysis of natural raw materials. The main advantage of carbonaceous materials consists in their high adsorption capacity with respect to a variety of organic compounds.

Comprehensive Analytical Chemistry, Volume 56
ISSN 0166-526X, DOI 10.1016/S0166-526X(10)56010-1

At the same time, some inherent drawbacks of activated carbons strongly restrict their use in many technologies. First of all, the removal of adsorbed matter from the carbons is rather burdensome. Thus, heat regeneration of activated carbons is highly energy consuming and accounts for up to 70% of all expenses in their usage. For example, when applied on a large scale in the sugar industry for decolorizing sugar syrups, burning of the adsorbed colored bodies at high temperatures proved to be the solely acceptable method for the regeneration of activated carbons. Most often, the used active carbon adsorbents just have to be combusted.

Activated carbons, as a rule, contain a large portion of fine micropores, causing a slow diffusion of many sorbate molecules. Although largely hydrophobic, carbonaceous materials display a certain amount of oxygen-containing functional groups. These represent active sites for the adsorption of water vapors. Therefore, high humidity may reduce the adsorption capacity and further slow down the sorption process from a gaseous phase. For example, activated carbons take up only slowly the vapors of ethanol, benzene, or hexane from humid air, particularly at low sorbate concentrations.

One of the principal shortcomings of activated carbons is their inflammability. Even the heat of adsorption may provoke ignition of carbons. For example, formation of pyrophoric compounds, such as FeS and Fe_2S_3, during the adsorption of carbon disulfide vapors from the off-gases of viscose production [1] in a corroding environment may facilitate the ignition of carbon column packings. Finally, high catalytic activity of the carbon surface, combined with the warming up, may cause the chemical transformation of unstable organic sorbates. Therefore, subsequent utilization of the sorbates often requires their additional purification.

Polymeric adsorbents, particularly macroporous styrene–divinylbenzene (DVB) copolymers, are free from these drawbacks of activated carbons. The heat of adsorption onto the polymeric adsorbents is significantly lower and, accordingly, the regeneration of polymers proceeds under much milder conditions. As a rule, the copolymers have no functional groups (apart from those specially introduced) that are capable of catalyzing chemical transformation of an adsorbate. However, in contrast to activated carbons, the specific surface area of most polymeric adsorbents is not very high; therefore, their sorption capacity is lower, often making the use of macroporous polystyrene-type adsorbents unprofitable. This circumstance impelled scientists to develop new types of polymeric materials with an enhanced porous structure, and hypercrosslinked polystyrenes are the result of efforts in that direction.

Up to now four main groups of hypercrosslinked sorbents have been developed and intensively tested. The first group, Styrosorb 1, incorporates laboratory samples of nanoporous (microporous) single-phase sorbents prepared by intensive post-crosslinking linear polystyrene of about 300,000 Da molecular weight, dissolved in ethylene dichloride, with monochlorodimethyl ether or p-xylylene dichloride. The irregular particles of these sorbents have pores with a diameter of about 20–30 Å and display an apparent specific surface area as high as 1000–1500 m²/g. The pore volume of Styrosorb 1 materials usually amounts to 0.4–0.5 cm³/g.

The second group, Styrosorb 1BP, represents the products of crosslinking the same linear polystyrene with monochlorodimethyl ether in cyclohexane solution. These sorbents manifest a bimodal pore size distribution: in addition to pores of 20–30 Å, they also have larger pores of up to 250–350 Å in diameter. Compared to Styrosorb 1, the apparent specific surface area of Styrosorb 1BP is lower, mostly 350–650 m²/g, only occasionally approaching 1000 m²/g, while the pore volume is considerably larger, 0.7–1.2 cm³/g. Similarly to Styrosorb 1, Styrosorb 1BP resins are obtained by converting the initial polystyrene solution into a rigid polymer block which – after crushing and sizing – yields irregularly shaped particles.

The third group, Styrosorb 2, represents nanoporous single-phase polymers derived from spherical beads of gel-type styrene copolymers with largely 0.7% DVB, post-crosslinked in swollen state with monochlorodimethyl ether. The size of the micropores is approximately 10–30 Å, and the apparent specific surface area reaches very large values of 1000–1900 m²/g, which is comparable to the range of the best activated carbons. On the other hand, the pore volume of these materials is rather small, 0.2–0.3 cm³/g.

The fourth group of hypercrosslinked polystyrene sorbents represents the commercially available resins of the Hypersol-Macronet, MN series (Purolite, UK) and has already been described in Chapter 9, Section 4.

The present chapter deals with several aspects of adsorption of gases and organic solvent vapors onto Styrosorbs 1 and 2. Most of the adsorption experiments were carried out shortly after the development of hypercrosslinked polystyrene materials. The choice of sorbates was largely dictated by practical needs. Sorption of hydrocarbons and their chloro and fluoro derivatives as well as vapors of organic solvents was thought to be important from the environmental viewpoint, while sorption of water vapors and carbon dioxide was of interest for solving problems encountered in closed spaces. These investigations immediately demonstrated distinguishing characteristics of hypercrosslinked polystyrenes and their distinct advantages over other known types of organic and inorganic sorbents and activated carbons.

2. ANALYSIS OF ADSORPTION ISOTHERMS ON HYPERCROSSLINKED POLYSTYRENES

We already mentioned the non-trivial fact that hypercrosslinked polystyrenes retain large amounts of gases at low temperatures, while being single-phase materials and exhibiting no real interfaces in the interior of the sorbent beads. Figure 10.1 demonstrates some typical sorption isotherms for nitrogen at −196°C on Styrosorb 1. All isotherms belong to type IV according to the classification of Brunauer, Deming, Deming, and Teller (BDDT), and are characteristic of conventional rigid mesoporous adsorbents. Usually the S-shaped isotherms are considered to be suitable for calculating the inner surface area, provided that they produce a straight line in the coordinates of the Brunauer, Emmett, and Teller (BET) equation (Eq. [3.1]) (Chapter 3). Indeed, all these isotherms are linear in the usual relative pressure (p/p_s) range of 0–0.35 (Fig. 10.1). This is consistent with the common interval of the BET equation's validity [2, 3] and can serve as a formal sign of applicability of the BET equation to calculations of the specific surface area of hypercrosslinked polystyrenes. However, as the polymer's porous structure is of a special nature and shows some peculiar properties, the adsorption isotherms require a more rigorous analysis.

Let us consider the adsorption of water vapor, nitrogen, and carbon dioxide at 20°C, as compared to the sorption of argon and nitrogen at −196°C on samples of hypercrosslinked polystyrenes prepared by crosslinking a linear polymer with p-xylylene dichloride [4]. Figures 10.2 and 10.3 demonstrate the sorption isotherms for water and nitrogen, respectively, at 20°C. At any crosslinking degree of the polymers, the isotherms have an S-shaped profile with distinct adsorption–desorption hysteresis. The isotherms for water present straight lines in the coordinates of the BET equation up to the relative pressure of 0.35:

$$\frac{p/p_s}{a(1-p/p_s)} = f(p/p_s) \qquad [10.1]$$

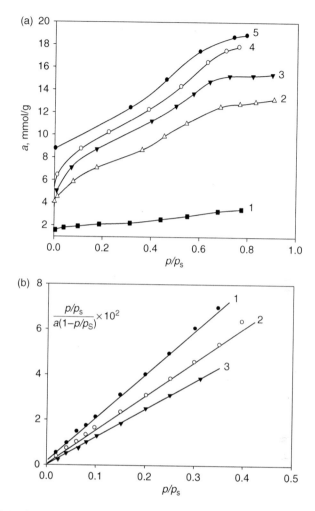

Figure 10.1 Sorption isotherms for nitrogen at -196°C on Styrosorb 1 (monochlorodimethyl ether, MCDE) with the crosslinking degree of (1) 25, (2) 43, (3) 66, (4) 80, and (5) 100% presented (a) in usual coordinates and (b) linear coordinates of BET equation.

On the other hand, the isotherms for N_2 and CO_2 are linear only until $p/p_s = 0.018$ and 0.01, respectively; this very narrow range of the validity of the BET equation is strange in itself. Furthermore, if one specifies the monolayer capacity a_m on the isotherm calculated by Eq. [3.1], then a_m must correspond to the position of the so-called B-point. This point on the isotherm indicates that the formation of a monomolecular adsorption

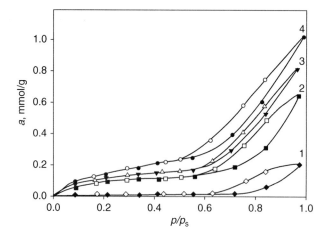

Figure 10.2 Sorption isotherms for water vapors at 20°C onto Styrosorb 1 (*p*-xylylene dichloride, XDC) with the crosslinking degree of (1) 25, (2) 43, (3) 66, and (4) 100%. *(After [4].)*

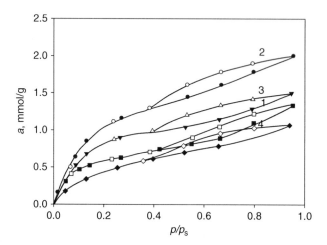

Figure 10.3 Sorption isotherms for nitrogen at 20°C on Styrosorb 1 (XDC) with the crosslinking degree of (1) 25, (2) 43, (3) 66, and (4) 100%; p_s was assumed to be equal to 760 mmHg. *(After [4].)*

layer is completed and a rectilinear section of the isotherm starts [5]. In the case of the CO_2 adsorption isotherm, the B-point shows up rather clearly and is located in the normal relative pressure range, but the sorption capacity a at the B-point does not coincide with the a_m value calculated from the BET equation. This is not typical for common rigid adsorbents.

One can assume that the chemical nature of hypercrosslinked polystyrenes changes only insignificantly with the increase in their degree of crosslinking. Correspondingly, the constant C in Eq. [3.1] would have to depend only slightly on the number of crosslinking bridges. Nevertheless, if one considers the sorption of carbon dioxide on several polymer samples, the C constant increases by a factor of 7 with the crosslinking degree X increasing from 25 to 66%. At the same time in the case of water adsorption, the constant C decreases by 2 orders of magnitude with X increasing from 25 to 100% (Table 10.1). In a number of cases this constant acquires rather unrealistic values. These irregular and even opposite tendencies, combined with the above-mentioned observations, can only imply the inapplicability of the BET theory to the adsorption of gases and vapors on hypercrosslinked polystyrenes.

It is of no surprise that the calculations of specific surface areas from the sorption isotherms for water and carbon dioxide also give totally different results. Thus, when calculated from the adsorption of water vapors, the specific surface area of the sample with 100% crosslinking degree amounts barely to 15 m^2/g, while the adsorption of carbon dioxide on the same polymer gives a 6 times greater value, 100 m^2/g.

Sorption of inert gases at low temperatures appears to produce more habitual data. Now, the values of the C constants do not significantly differ from those for conventional adsorbents; also, they are sufficiently high to allow a reliable estimation of the surface area (Table 10.1). The calculated monolayer capacities, a_m, are consistent with the position of the B-points. Also, the surface areas calculated from the adsorption isotherms for inert gases at −196°C prove to be many times greater than those calculated from the isotherms for water or carbon dioxide at room temperatures.

The sorption isotherms for nitrogen at 20°C are worthy of note, because the hysteresis loops are located in the relative pressure range corresponding to the capillary condensation of N_2 vapors in mesopores at low temperatures. However, it is obvious even to the naked eye that the capillary condensation of nitrogen at room temperature is impossible; therefore, fundamentally different reasons should be responsible for both the S-shaped form of these isotherms and the presence of hysteresis loops.

Table 10.1 Sorption parameters of various sorbates on hypercrosslinked polymers prepared by crosslinking linear polystyrene with p-xylylene dichloride

X (%)	Adsorbate	Temperature (°C)	S (m²/g)	B-point capacity (mmol/g)	Constants of BET equation	
					a_m	C
25	H_2O	20	2	–	0.025	801
25	CO_2	20	72	0.52	0.547	554
25	N_2	–196	2	0.02	–	–
25	Ar	–196	0	–	–	–
43	H_2O	20	7	–	0.081	32
43	CO_2	20	90	0.50	0.687	301
43	N_2	–196	258	2.10	2.640	95
43	Ar	–196	530	–	–	–
66	H_2O	20	9	–	0.103	19.5
66	CO_2	20	92	1.30	0.701	3601
66	N_2	–196	442	3.40	4.530	111
66	Ar	–196	823	–	–	–
100	H_2O	20	14	–	0.16	7.88
100	CO_2	20	100	1.40	0.761	2168
100	N_2	–196	673	5.70	6.90	146
100	Ar	–196	956	–	–	–

Source: After [4].

When considering the sorption of water vapors, we could indeed assume that capillary condensation starts at p/p_s larger than 0.35. However, the amount of water adsorbed by the sample with a 100% degree of crosslinking at the beginning of the hysteresis loop amounts to no more than 0.4% of the polymer's weight. We cannot assume that such small amounts of water can cover all accessible polymeric chains by a monomolecular layer before the capillary condensation begins that causes the hysteresis loop. Notably, adsorption isotherms for water on neutral hydrophobic hypercrosslinked polystyrenes display a convex initial part, while it is well known that adsorption of water on hydrophobic surfaces, such as those of activated carbons, gives concave isotherms. Isotherms which are convex to the ordinate axis are characteristic of hydrophilic surfaces. In our case the convex form of the isotherms could indicate a definite affinity of small amounts of water to the hypercrosslinked aromatic polymer.

Sorption of water vapors by hydrophobic hypercrosslinked polystyrenes is of particular interest because it points to a complex interaction mechanism between the sorbate molecules and the hypercrosslinked network. As explained in Chapter 9, Section 2, the network obtained by fixation with many rigid bridges of the initially strongly solvated polystyrene chains is free of serious stresses when in the swollen state. Inner stress arises during shrinkage of the material while removing the initial solvent, since all constituent network meshes are forced to acquire unfavorable conformations. Possibly, some strongly distorted and strained network meshes display a tendency of hosting sorbate molecules, thus representing the primary adsorption sites with a relatively high affinity even to water molecules. Restoration of more favorable conformations of the most distorted network cycles due to positioning of a guest molecule into the host cycle, accompanied by a partial relaxation of local strains, is likely to cause a certain amount of rearrangement of network fragments and an increase in the total network volume, that is, its swelling. We believe that these changes in the microstructure of the hypercrosslinked network are the reason for the adsorption–desorption hysteresis, rather than the capillary condensation of water vapors.

Due to the high mobility of the hypercrosslinked network and the reversibility of the volume changes of the material, we can expect an adsorption equilibrium to be obtained for any sorbate, for any particular pressure. However, the network structure and the magnitude of its inner stresses at any equilibrium position will then depend on the route of approaching the equilibrium state, always causing sorption–desorption hysteresis loops. This mechanism must also be valid for the sorption hysteresis of nitrogen at room temperature. Naturally, the sorption magnitude is strongly influenced by the nature of the sorbate and its affinity to polystyrene. Thus, the low cohesion energy between the nitrogen molecules, compared with water molecules, is responsible for the higher adsorption of nitrogen.

Sorption isotherms for nitrogen at −196°C, as can be seen in Fig. 10.1, tend to level off at high relative pressures. For common adsorbents exhibiting this type of isotherms, the adsorption capacity at a relative pressure close to unity is considered to equal the pore volume of the sorbent. The ultimate quantity of nitrogen sorbed by the hypercrosslinked sample with 100% crosslinking degree was found to be 0.68 cm^3/g, while the total pore volume calculated from the true and apparent densities of the material amounts to 0.51 cm^3/g, only. The explanation for this discrepancy can only be found in the swelling of the polymer with sorbed nitrogen at −196°C (see also Table 7.7). Naturally, the marked swelling completely distorts all calculations of the specific surface area.

Considering all the above-discussed data, we conclude that the linear dependence in the coordinates of the BET equation does not yet imply that the general conclusions of the polymolecular adsorption theory can be applied to the description of sorption on hypercrosslinked polystyrenes. These cannot be regarded as rigid sorbents with a constant heterophase structure. Accordingly, calculation of the pore size distribution from the desorption branch of the isotherms for carbon dioxide using the Kelvin equation (Eq. [3.4], Chapter 3) results in the unrealistic pore dimensions of about 0.1 Å. The isotherms for water produce more sensible data for pore diameters, 20–40 Å, but one should not assume that the plots presented in Fig. 10.4 correspond correctly to the real pore size distribution of the material.

3. SORPTION OF ORGANIC VAPORS UNDER STATIC CONDITIONS

For organic solvent vapors the type of isotherms describing sorption onto the hypercrosslinked sorbent Styrosorb 2 strongly depends on the nature of the sorbate molecules (Fig. 10.5) [6, 7]. Hydrocarbons exhibit a reasonable affinity to the adsorbent and are characterized by a convex initial part of the isotherms and a general shape typical of mesoporous

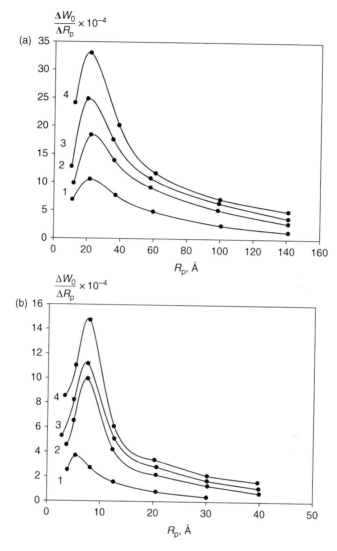

Figure 10.4 Pore volume distribution over effective pore radii calculated from the sorption isotherms for water vapors at limiting wetting angles of (a) 0 and (b) 61.3° for the networks obtained by crosslinking linear polystyrene with *p*-xylylene dichloride to (1) 25, (2) 43, (3) 66, and (4) 100%. *(After [4].)*

polymeric adsorbents. The inflection points on the isotherms are located within the relative pressure range of 0.35–0.45. The differences in the sorption values for *n*-alkanes are not great, although at low and moderate relative pressures the loading capacity decreases slightly in the

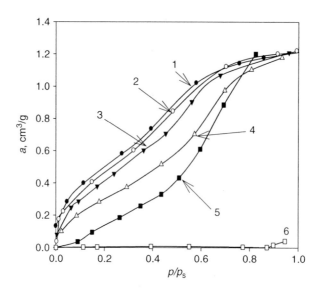

Figure 10.5 Sorption isotherms for (1) *n*-heptane, (2) *n*-hexane, (3) *n*-pentane, (4) acetone, (5) methanol, and (6) water vapors at 20°C on Styrosorb 1 (MCDE, $X = 100\%$). *(After [7].)*

sequence *n*-heptane > *n*-hexane > *n*-pentane. The more polar acetone adsorbs in a markedly smaller amount in the same pressure interval. The sorption isotherm for methanol looks quite different because of strong interactions between its molecules. In this case, the concave side of the isotherm faces the ordinate axis. With increasing relative pressures, the difference between the amounts of adsorbed solvent vapors decreases, so that the ultimate sorption volumes, V_∞, at $p/p_s = 1$ become practically equal ($1.208 \pm 0.021\,\text{cm}^3/\text{g}$) for the different solvents (Table 10.2) [7]. It should be remembered here that hypercrosslinked polystyrenes are hydrophobic materials and they take up no more than 2 wt% water from saturated water vapors; therefore, the sorbent capacity for solvents should not depend on the humidity of their vapor streams. Indeed, direct experiments showed that the presence of water vapors did not affect the sorption of trichloroethylene vapors on the hypercrosslinked sorbent prepared by post-crosslinking chloromethylated styrene–4-(*tert*-butyl)styrene–DVB copolymer [8].

Note that the ultimate amount of organic solvent vapors adsorbed by Styrosorb 2 is several times larger than the total pore volume of the dry sorbent ($0.22\,\text{cm}^3/\text{g}$), implying strong swelling of the sorbent during

Table 10.2 Ultimate volume of adsorbed solvents

Sorbate	Molar volume of solvent (cm³/mol)	Ultimate sorption at p/p_s 1 (mmol/g)	Volume of adsorbed solvent (cm³/g)
n-Pentane	115.22	10.5	1.210
n-Hexane	130.69	9.3	1.215
n-Heptane	146.54	8.2	1.202
Acetone	73.61	16.0	1.178
Methanol	40.51	30.5	1.236

Source: After [7].

sorption. However, the ultimate solvent volume sorbed from the gaseous phase never reaches the swelling capacity of the polymer in the corresponding liquid. In the case of hexane, these values amount to 1.21 and 1.36 cm³/g, respectively. Usually this phenomenon is explained by the sorbent having a certain portion of very large pores where no capillary condensation of vapors occurs. While this explanation might be acceptable for conventional rigid adsorbents, it is not valid for Styrosorb 2, since the latter does not contain any large pores. It is more reasonable to assume [7] that the surface tension on the interface between the saturated vapor and the swollen polymer prevents the volume of the polymeric beads to expand to their maximum, whereas in the case of beads immersed into the corresponding liquid no such compressive forces operate (the surface energy at the liquid/gel interface is markedly lower than that at the vapor/gel interface).

It is interesting to note that the distinct swelling in the organic solvent vapor starts at very low relative pressures. In the case of sorption of n-pentane, it already reaches 5% at p/p_s as low as 0.05. The swelling process starts almost simultaneously with the sorption, long before the moment when a considerable portion of the initially present voids (0.22 cm³/g) becomes filled with the sorbate (Fig. 10.6). The molecules of pentane not only occupy the guest positions in the stressed network meshes, thus filling out the smallest initial voids in the polymer, but also cause an immediate expansion of the polymeric network, leading to the formation of a new sorption space henceforth accessible to the sorbate. (It is appropriate to mention here that pentane, like water, fails to cause any swelling of linear polystyrene.) In other words, sorption of pentane (and other vapors) represents the superposition of two processes: dissolution of pentane in the polymer and classic adsorption.

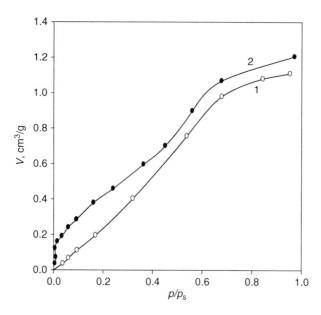

Figure 10.6 Increase in volume of Styrosorb 1 (MCDE, 100%) beads on swelling in the vapors of *n*-pentane (1) and the volume of *n*-pentane absorbed (2). *(After [7].)*

The cooperative structural rearrangement of the network associated with the continuous swelling of the hypercrosslinked sorbent with all the examined organic vapors is indicated by the hysteresis loops always extending throughout the entire range of relative pressures, down to zero (Fig. 10.7). This basically differs from the hysteresis loops caused by capillary condensation in the mesopores of adsorbents with rigid structure; in the latter case, the loop closes toward low vapor pressures.

The temperature dependence of the amount of adsorbed solvents also turns out to be very unusual. Figure 10.8 shows sorption isotherms for *n*-hexane vapors at temperatures varying from 0 to 60°C. Unexpectedly, in the low relative pressure range, the sorption at 0 and 10°C was found to be smaller than that at higher temperatures. (Normally, sorption decreases with increasing temperature.) Obviously, the mobility of the hypercrosslinked network increases in the above low-temperature range, facilitating the conformational rearrangement and swelling of the polymer during the sorption process.

It is relevant to note here that changes in the structure or behavior of polymers at a certain temperature can be revealed using the so-called inversed gas chromatography. In this technique, the polymer is tested in

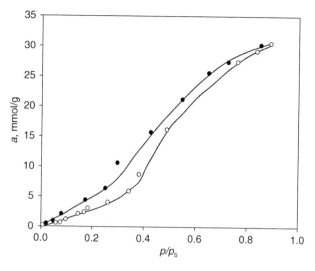

Figure 10.7 Sorption (o) and desorption (•) isotherms for methanol vapors on Styrosorb 1 (MCDE, 100%) at 20°C. *(After [7].)*

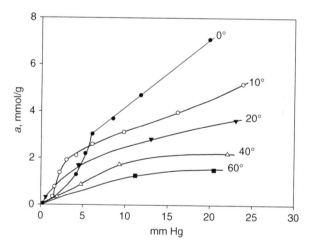

Figure 10.8 Sorption isotherms for vapors of *n*-hexane on Styrosorb 1 (MCDE, 100%) at different temperatures. *(After [7].)*

the capacity of an adsorbing material packed in a gas chromatographic column. Retention volumes (or times) of a test compound on the column are measured at various temperatures. The plot of the logarithm of the retention volume against inversed temperature usually presents a straight line. If, however, the behavior of the examined polymer changes at a certain temperature, for example, because of the transition from a glassy state to a viscous elastic state, the slope of the straight line changes at that temperature. Thus, macroporous adsorbing materials based on copolymers of styrene with DVB reveal clearly the glass transition temperatures between 105 and 125°C, depending on the DVB content. We applied the inversed chromatography technique for testing the hypercrosslinked polymers based on both linear polystyrene and styrene–DVB copolymers, post-crosslinked to 100% with monochlorodimethyl ether or xylylene dichloride. With methane as the probing adsorbate, the polymer glass transition temperature never exceeded 100°C. Instead, a small change in the straight line's slope was observed right above 0°C, which correlates with the aforementioned increase in the adsorption capacity of the hyper-crosslinked polymer for n-hexane in this temperature range. By using the inversed gas chromatography approach, Podlesnyuk et al. [9] also examined the retention of hexane, dichloroethane, tetrachloromethane, and pyridine on a series of hypercrosslinked polystyrene samples between 70 and 150°C with the result that no phase transition of the hypercrosslinked polymers occurs in this temperature range, contrary to conventional styrene–DVB copolymers.

The possibility of high sorption loading of hypercrosslinked sorbents with vapors of organic solvents, exceeding 1 mL/g, is very useful from a practical point of view. Figure 10.9 shows the sorption isotherm for n-hexane vapors on Styrosorb 2 in comparison with isotherms on typical conventional adsorbents. Styrosorb 2 exhibits exceptional adsorption capacity, superior to that of the macroporous styrene–DVB sorbent Polysorb-1 and activated carbon AR-3 (both Russian products) at relative pressures higher than 0.03, zeolite NaX at the value of p/p_s above 0.07, and semicoke in the whole relative pressure range.

In addition to the high loading capacity with n-hexane, the hypercrosslinked sorbent has the important advantage of easy regeneration, which can be accomplished sufficiently fast on evacuating or heating the sorbent at 80°C. In contrast, regeneration of activated carbons requires heating at 200–250°C and that of zeolite heating at 300–360°C. The low temperature of Styrosorb 2 regeneration correlates with the relatively low value of integral heat of sorption of n-hexane, amounting to 37.7 kJ/mol, while in the case of activated carbon and zeolite it is within the range of 58–67 kJ/mol [1].

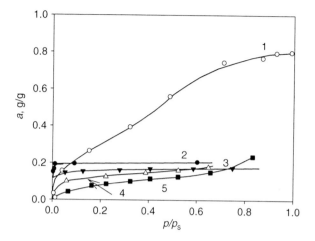

Figure 10.9 Sorption isotherms for vapors of *n*-hexane at 20°C on (1) Styrosorb 1 (MCDE, 100%), (2) zeolite NaX, (3) activated carbon AR-3, (4) semicoke, and (5) macroporous polystyrene sorbent Polysorb 1. *(After [7].)*

Perfluorocarbons exhibit low affinity to polystyrene-related materials and interact non-specifically with polystyrene [10–12]. In contrast to the S-shaped isotherms for *n*-heptane or *n*-hexane, *n*-perfluorooctane produces a Langmuir-type isotherm, which is more typical of microporous sorbents (Fig. 10.10) [13]. In the case of the large n-C_8F_{18} molecule, the relatively small pores of Styrosorb 2 slow down the diffusion into the sorbent beads, increasing the time required to achieve equilibrium. It is at times rather difficult to reliably register the establishment of equilibrium, which is exhibited as a discontinuance on the sorption branch of the isotherm on interrupting its registration. Large sorption–desorption hysteresis extends throughout the entire relative pressure range; this may be caused not only by the swelling of the polymer with n-C_8F_{18} vapors but also by the slow kinetics of the sorption and desorption processes. Still, the ultimate sorption of *n*-perfluorooctane, $0.6\,\text{cm}^3/\text{g}$, significantly exceeds the total pore volume of the dry sorbent, $0.2\,\text{cm}^3/\text{g}$. Although the swelling process alters the conformational structure of the hypercrosslinked network, after evacuating the sample at room temperature within 12 h, the sorption–desorption isotherms reproduce accurately, implying the full reversibility of the structural changes.

Interestingly, the adsorption isotherms for n-C_8F_{18} on the hypercross-linked sorbents Styrosorb 1BP have much in common with traditional

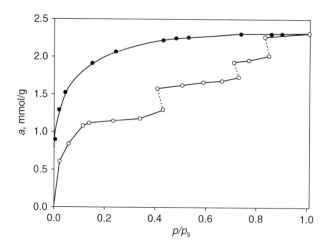

Figure 10.10   Sorption (o) and desorption (•) isotherms for n-C_8F_{18} vapors at 25°C on Styrosorb 2; discontinuities on the sorption branch correspond to keeping the sample at a constant p/p_s value for several hours. *(Reprinted from [13] with permission of Elsevier.)*

isotherms on mesoporous solids. Figure 10.11 shows that sorbents prepared in cyclohexane from linear polystyrene with 80–3000 kDa molecular weight display hysteresis loops belonging to type E according to de Bour's classification [13]: their desorption branch declines abruptly, while the adsorption branch rises smoothly. On the other hand, Styrosorb 1BP produced in cyclohexane by crosslinking a low-molecular-weight polystyrene, 13–17 kDa, has isotherms with a narrow, A–type hysteresis loop (Fig. 10.12), the latter being shifted to the range of higher relative pressures, as compared with the position of the loops for Styrosorb 1BP presented in Fig. 10.11. The formation of hypercrosslinked networks in cyclohexane was demonstrated to be accompanied by micro-phase separation. The shift in the position of hysteresis loops toward high vapor pressures testifies to the larger size of mesopores in Styrosorb 1BP formed during the phase separation in solutions of the low-molecular-weight polystyrene. The smaller deviation of the desorption branch from the adsorption branch of the isotherms in the low relative pressure range is a sign of sorbent swelling in n-perfluorooctane vapor. Indeed, the maximum quantity of n-C_8F_{18} vapor adsorbed by Styrosorb 1BP is larger than the total pore volume of the dry material. However, swelling of all hypercrosslinked samples in n-C_8F_{18} vapor

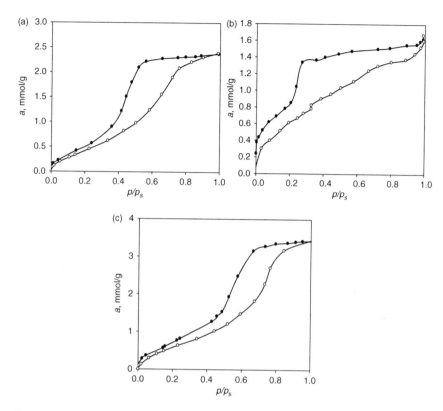

Figure 10.11 Sorption (○) and desorption (•) isotherms for n-C_8F_{18} vapors at 25°C on Styrosorb 1BP prepared on the basis of polystyrene with molecular weight of (a) 3×10^6, (b) 3×10^5, and (c) 8×10^4 in cyclohexane. *(After [13].)*

remains smaller than the swelling of the same sorbents with n–hexane or methanol vapors [13]. Most likely this is caused by some smaller pores in the hypercrosslinked Styrosorb remaining inaccessible to the large n-C_8F_{18} molecules as well as by the low affinity of perfluorocarbons to polystyrene.

4. KINETICS OF SORPTION OF HYDROCARBON VAPORS

For practical applications of adsorption materials an important value is the rate of approaching adsorption equilibrium. Provided that the adsorption isotherm is linear, and the diffusion front has not yet reached the center of the bead (which corresponds to the initial stage

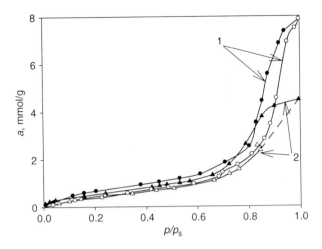

Figure 10.12 Sorption isotherms for n-C$_8$F$_{18}$ at 25°C on Styrosorb 1BP prepared on the basis of polystyrene with molecular weight of (1) 17×10^3 and (2) 13×10^3 Da in cyclohexane. *(After [12].)*

of adsorption when the degree of bead saturation, $\gamma \leq 0.6$), Fick's second law can be presented in the following simplified form:

$$\gamma = \frac{6}{R_\mathrm{b}} \sqrt{\frac{D_\mathrm{e} \tau}{\pi}}$$ [10.2]

where R_b is the radius of the spherical bead, τ is time, and D_e is the diffusion coefficient of the sorbate molecules within the sorbent bead.

A linear dependence of the degree of bead saturation γ on the square root of time (at $\gamma \leq 0.6$) is a sign of the constancy of the diffusion coefficient, that is, normal kinetics of the sorption process [14, 15]. Deviations from this linear dependence, that is, abnormal kinetics, have been observed for polymeric sorbents that swell during the sorption of vapors.

Figures 10.13 and 10.14 show the kinetic plots for the sorption of n-hexane vapors on Styrosorb 2 at 20 and 40°C [16]. The anomalous character of kinetics shown in the nonlinear dependence of γ versus $\sqrt{\tau}$ is clearly seen at the low and moderate extents of bead saturation. By replacing the curved kinetic plots with a set of short straight lines and calculating the effective diffusion coefficients, D_e, according to Eq. [10.2] within each linear interval, one arrives at the important conclusion that the diffusion rate of the sorbate molecules increases by more than 1 order of magnitude (Fig. 10.15) during the sorption process.

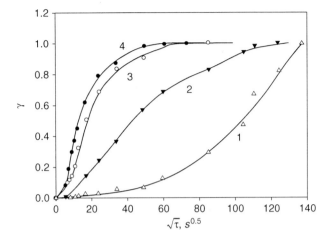

Figure 10.13 Kinetics of *n*-pentane vapor sorption on Styrosorb 2 at 20°C at different concentrations of the sorbate leading to final sorption value of (1) 10.0, (2) 27.5, (3) 50.0, and (4) 58.5 wt%.

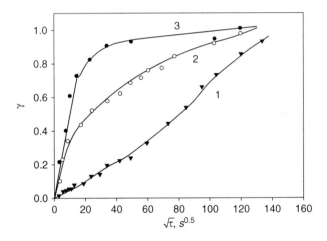

Figure 10.14 Kinetics of *n*-pentane vapor sorption on Styrosorb 2 at 40°C at different concentrations of the sorbate leading to final sorption value of (1) 6.2, (2) 10.7, and (3) 25.1 wt%.

Dependence of the diffusion coefficient on the degree of bead saturation is also known for porous adsorbents with a rigid structure, such as activated carbons [14, 15]. In this case, two processes – Knudsen diffusion (diffusion in porous materials) and migration of sorbate molecules along the pore

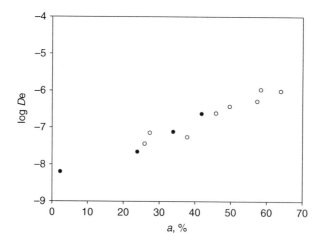

Figure 10.15 The dependence of effective diffusion coefficients on the degree of filling of Styrosorb 2 (MCDE, 100%) beads with (○) *n*-hexane and (•) *n*-pentane.

walls – determine the overall mass transfer. The occasional increase of the diffusion coefficient for that kind of sorbent has usually been related to the acceleration of surface diffusion because of a rapid consumption of the most active adsorption sites (the latter immobilize the sorbate molecules). In polymeric sorbents, swelling of the material, which increases the segmental mobility of polymeric chains, is thought to facilitate additionally the mass transfer of sorbate molecules [16].

Obviously this explanation is also valid for the hypercrosslinked polystyrene beads that readily swell with the vapors adsorbed. However, one has to consider the specific character of the swelling kinetics of hyper-crosslinked polymers, which is different from that of conventional gel-type and macroporous polymeric sorbents (see Chapter 7, Section 2). Because of the extremely high degree of crosslinking, the rigid network can signifi-cantly increase its volume only through a cooperative rearrangement of the whole structure. The outer layers of the bead cannot swell significantly before the first sorbate molecules arrive at the central part of the bead. From this moment on, the bead expansion accelerates, which logically facilitates the further diffusion of sorbate molecules. Therefore, the sorption kinetics on hypercrosslinked sorption materials has a peculiarly complex character.

It is important to note that the cooperative character of rearrangement of the hypercrosslinked network favorably accelerates sorbate diffusion during the regeneration process. The outer layers of the bead cannot shrink before the central part of the bead starts losing sorbed molecules. Therefore, regeneration of the material in vacuum largely proceeds by the sorbate molecules diffusing through an expanded network with very high diffusion coefficients, many times exceeding those of the sorption process. Indeed, the removal of adsorbed hydrocarbons proceeds easily with almost constant diffusion rates up to very small residual loading. At a 50% bead filling with n-hexane, the main portion of the sorbed hydrocarbon desorbs within 100 s. The desorption process slows down only when the residual portion of the hydrocarbons decreases to a few percent. It is this stage that is accompanied by a gradual reduction of dimensions of network meshes and emergence of strong inner stresses.

Thus, the cooperative character of the conformational rearrangement of hypercrosslinked networks causes a certain time lag in the expansion and shrinkage of the bead and explains the unusually strong hysteresis in the sorption–desorption kinetics. Generally, hypercrosslinked polystyrene sorbents display superb adsorption capacity, combined with high and steadily growing adsorption rates and very high desorption rates.

5. SORPTION OF HYDROCARBON VAPORS UNDER DYNAMIC CONDITIONS

Sorption of hydrocarbon vapors under dynamic conditions by hypercrosslinked sorbents also exhibits definite peculiarities. The process described below was examined in such a manner that along a column of 60 cm in length and 2 cm inner diameter (ID), packed with Styrosorb 2 beads (diameter: 0.5–1.0 mm), seven checkpoints were arranged, each supplied with a thermocouple for measuring local temperature and an outlet for sampling the gas flow in order to determine sorbate concentrations by gas chromatography. The checkpoints allowed observation of the movement of the sorption zone and the distribution of heat along the sorbent bed. Figure 10.16 shows typical breakthrough curves in each column section, generated by percolating concentrated n-pentane/air mixtures through the column. All the breakthrough curves have an asymmetric profile with long diffusive "tails"; the relative concentration of the sorbate in the mobile phase $C/C_o = 0.8$ is reached within 5–10 min, while in the C/C_o range of 0.8–1.0 the process slows down markedly, so that a much longer time is needed to reach the feed concentration of the sorbate in each section. This

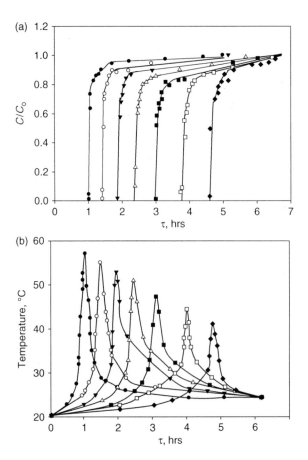

Figure   10.16 *n*-Pentane   breakthrough   curves   (a)   and   thermograms (b) corresponding to different checkpoints; pentane concentration: 52.9%. *(Reprinted from [17] with permission of the Society of Chemical Industry.)*

situation was found for all initial concentrations of *n*-pentane in the air and for linear flow velocities varying in the range 0.5–2.3 cm/s.

The sorption of hydrocarbons proved to be a strongly exothermic process. This is not surprising because both the adsorption (pentane–polystyrene dispersion interactions) and the relaxation of strong inner stresses of the network result in heat generation. The temperature increase of the packing affects the sorption process in two different ways. First, it results in a considerable acceleration of the sorption process due to the facilitated diffusion of the sorbate molecules into the bead interior (the effective diffusion coefficient increases by 1 order of magnitude with the temperature increasing from 20 to 40°C). On the other hand, the heating of the column bed displaces the adsorption equilibrium and reduces the adsorption capacity of the sorbent. Therefore, the heat distribution along the column and the movement of the heat wave determine the efficiency of the process in each section and in the column as a whole. Naturally, heat evolves at the moment when sorbate molecules interact with the sorption sites; hence, heat mainly generates at the front of the sorption zone. For this reason the highest heat increase in our experiment was observed to occur at the first checkpoint when the sorption zone reached it; there, the temperature increase amounted to 40°C if a 50% pentane/air mixture was sent through the column. Further on, the heat increase flattens and widens; in fact the carrier gas transports the generated heat ahead of the sorbate front and preheats the subsequent sorbent layers. The preheating effect reduces the sorption capacity of Styrosorb 2 in subsequent column sections and scales back the generation of new heat.

New portions of the feed mixture gradually cool down the hot section of the sorbent bed and soon complete its saturation process. The length of the heat wave with a significantly enhanced sorbent temperature amounts to no more than 7–10 cm, even in the case of percolating the concentrated pentane/air mixture. Therefore, only a small portion of the total column experiences the temporary reduction of its adsorption capacity and, at the moment of pentane arrival at the column outlet, the major capacity of the sorbent bed proves to be used up almost completely. The heat evolution is less significant and the capacity loss is much smaller in the case of lower concentrations of pentane in the feed.

When percolating a low-concentration gas mixture containing 11.5% pentane, both the heat wave and the sorption zone move along the column rather slowly. The preheating effect at the first checkpoint (17 cm from the inlet) becomes significant after 2 h of the process, while the breakthrough of pentane in this position occurs only after 3.3 h. Of course, the distance between the heat wave and the breakthrough front of the sorbate increases downward on the column. This distance in length and on the timescale was minimum in the experiment with the highest concentration (59%, v/v) of pentane in the feed mixture.

Two important conclusions follow from these experiments. The first relates to the sorption process with concentrated sorbate in the carrier gas. In this case, the column bed will require a special heat exchanger if the sorption capacity of the polymer is expected to be fully exploited. Second, a cooling jacket appears to be needed for a safe performance of the sorption process after a possible heat regeneration of the packing. However, cooling is not needed for processing gas mixtures with low sorbate concentrations. The process can safely be started immediately after the heat regeneration of the column because the excessive carrier gas will reduce the temperature of the whole packing long before the sorption front arrives at the end of the column.

One of the most important parameters characterizing the efficiency of the practical operation of an industrial adsorption column is the so-called protection time – the length of time the column can accept the sorbate without its breakthrough from the sorbent bed. This parameter is today also called the breakthrough time. As far back as 1929 Shilov [18] developed a simple model of frontal adsorption dynamics for adsorbents having permanent porosity. According to his model, one should distinguish two periods of adsorption. When the adsorbate and carrier gas pass through the first elementary layer of the adsorbent the degree of saturation of this layer progressively increases until its capacity becomes completely exhausted. Only this first layer operates under conditions where the inlet concentration of the sorbate is constant, while the outlet concentration gradually increases from zero to the maximum. Starting with the complete saturation of the first layer, the concentration profile of the sorbate along the column stabilizes. For all remaining sorbent layers the operating conditions with this constant concentration profile of the sorbate become identical. Then, "parallel mass transfer" will characterize the second step of the adsorption process and, therefore, the sorption zone starts to move along the column with a constant velocity. This leads to the linear dependence of bed protecting time, τ_p, on the bed depth, H:

$$\tau_p = KH - \tau_0 \qquad [10.3]$$

Here, τ_o is the time required for the formation of the constant concentration profile of the sorbate (it is called the loss of protecting time) and K is the protecting time coefficient (the Shilov constant) indicating the protecting time of a layer of unit length.

As shown by Fig. 10.17, the linear dependence of τ_p on H holds only for low concentrations of pentane in the feed mixture. When the pentane proportion in the feed does not exceed 10%, the value of τ_o remains in the range between 1 and 2 h. This is a small value compared to the total protecting time τ_p of a 50-cm-long column, which amounts to not less than 14 h.

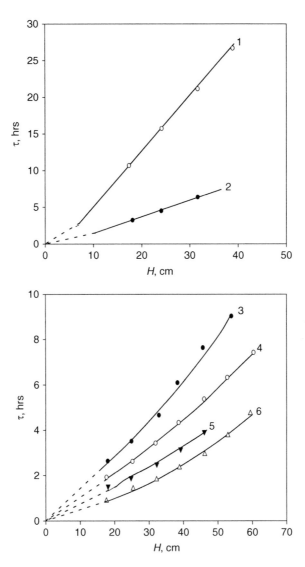

Figure 10.17 Dependence of protecting time on bed depth at the n-pentane concentration of (1) 2.09, (2) 11.53, (3) 14.16, (4) 28.9, (5) 34.18, and (6) 59.23%.

At higher concentrations of pentane in air, the Shilov dependence [10.3] is no longer linear. Analysis of the obtained data leads to the conclusion that the rate of propagation of the mass transfer zone gradually slows down with bed depth. Obviously the adsorption efficiency of the column packing's lower layers increases, compared to that of the first sorbent layers. This can find an explanation by taking into account the finding (Fig. 10.16b) that the heat wave flattens while the sorption zone moves down the column, gradually improving the temperature conditions for adsorbing the hydrocarbon by Styrosorb 2. In general, this Shilov model of adsorption has been developed for an isothermal adsorption process on rigid sorbents. It is not really applicable to hypercrosslinked polystyrene column packings, which swell with sorbate vapors and function with a significant heat evolution and an order-of-magnitude increase in the effective diffusion coefficient of sorbate molecules.

Isomeric pentanes represent the major constituents of the volatile fraction of gasoline. Usually the composition of the volatile fraction varies insignificantly for different types of gasoline. On average, the head space over gasoline at 30°C incorporates 0.79% n-propane, 2.23% isobutane, 6.66% n-butane, 7.48% isopentane, 6.45% n-pentane, 0.31% 2,2-dimethyl butane, 2.10% 2,3-dimethyl butane, 1.49% 3-methyl pentane, 1.40% n-hexane, 0.78% n-heptane, and traces of aromatic hydrocarbons, in total about 30% of hydrocarbons in air. When percolating this mixture through a 55-cm-long × 2-cm-ID column, packed with Styrosorb 2, at 30°C and a linear flow velocity of 1.0 cm/s, the breakthrough of hydrocarbons was observed to occur after 2.7 h. Due to the different sorption activity of Styrosorb 2 toward the constituent components of the gasoline sample, the hydrocarbons partially separate according to the principles of frontal chromatography. The order in which the components arrive at the column outlet corresponds to their boiling temperatures. The profiles of the breakthrough curves point to the displacement of light hydrocarbons by heavier components as the mixture moves along the column packing. Therefore, if desired, partial fractionation of hydrocarbons is feasible during the heat regeneration of the column.

6. DESORPTION OF HYDROCARBONS

Hydrocarbons that were adsorbed by the hypercrosslinked polymer Styrosorb 2 in the experiments described in Section 5 can easily be removed by treating the material with steam of 105°C passing through the column at a linear velocity of 0.015 m/s. Further increase in the flow rate of steam does not accelerate the desorption process, which confirms the limiting role of internal diffusion in the regeneration kinetics.

Usually the kinetics of desorption process can be described by a simple exponential equation for drying [19]:

$$\gamma = \exp\left[-\kappa\tau\right] \qquad\qquad [10.4]$$

where γ represents the relative value of residual loading, k is the desorption coefficient that is mainly determined by the diffusion coefficient of the sorbate molecules, and τ is the current time of the process. Figure 10.18a shows the actual kinetic plot for desorption of the volatile gasoline fraction, while in Fig. 10.18b the same experimental data are given in the

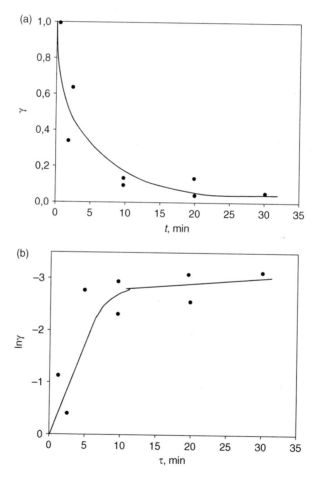

Figure 10.18   Kinetic curves for desorption of petrol vapors at 105°C and gas flow rate of 0.03 m/s: (a) experimental data and (b) those in the coordinates of Eq. [10.4].

coordinates of Eq. [10.4]. As a whole, the process of desorption does not obey this equation. However, with a fair degree of approximation one can distinguish two linear parts of the kinetic plot, each part being characterized by its own diffusion coefficient.

This circumstance finds a logical explanation in that the diffusion of sorbate molecules proceeds easily from the expanded network of beads swollen with gasoline vapors, but, as the hydrocarbons are removed, the network shrinks and the desorption slows down. Under the examined conditions the boundary between the first rapid and second slow stages of the process shows up at a low loading of $\gamma = 0.1-0.2$. The effective diffusion coefficient for the first stage amounts to $6.17 \times 10^{-3}\,cm^2/s$, but it drops to $1.67 \times 10^{-4}\,cm^2/s$ at the second stage. However, it is very important that the major portion of the sorbate, about 90%, is removed within 10–15 min, and a more thorough regeneration of the column is most likely not needed in practice.

As already mentioned, the adsorption of water vapors on the hydrophobic Styrosorb 2 is extremely low, so that no more than 5% residual moisture remains in the sorbent beads after steam regeneration. Therefore, it is not necessary to dry the sorbent additionally before conducting the next sorption cycle.

Finally, for practical use, it is very important to know how many sorption–regeneration cycles the sorbent tolerates without losing its sorption properties. In order to test Styrosorb 2, 100 sorption–desorption cycles were carried out, using n-hexane vapors as the sorbate and 105°C steam for regeneration. After the 1st and then 25th, 50th, and 100th cycles, the sorption isotherms for n-hexane were measured. The sorption capacities in the 1st and 25th cycles differed by less than 5%. Afterward, the change in the capacity did not exceed the experimental error. These results signify that Styrosorb 2 tolerates long-run operation without any mechanical destruction of the beads or deterioration of their properties.

The exceptional sorption capacity of Styrosorbs with respect to vapors of organic compounds combined with a high sorption rate, simplicity of the regeneration process, and durability of the material open up new opportunities for the practical application of hypercrosslinked polystyrene sorbents. In particular, they can be recommended for the protection of gasoline and oil storage and transportation tanks from natural emissions of volatile hydrocarbons into the atmosphere, as well as for the purification of exhaust gases of industrial plants from organic vapors and recuperation of valuable organics. In the latter case, however, the hypercrosslinked sorbent should not catalyze any chemical transformation of the recovered organic compounds.

7. PASSIVITY OF HYPERCROSSLINKED SORBENTS

Activated carbons are known to catalyze oxidation of many organic compounds, such as methyl ethyl ketone, diacetone, alcohols, or, especially, cyclohexanone. It is well documented in the literature that the heat of the oxidation reaction, in conjunction with the heat of adsorption, can result in developing the so-called hot spots in a sorbent bed. These hot spots can cause an increase of the local temperature up to the point when the bed ignites. The oxidative activity of many carbons appears to be caused by their functional groups or impurities of inorganic nature originating from the initial raw materials, or formed in the sorbent bed during its exploitation. There is no simple way to remove these impurities from activated carbons. In this respect, macroporous polystyrene resins have a serious advantage, because of the absence of catalytic functional groups and smaller thermal effects of adsorption. Still, it was felt advisable to examine the capability of hypercrosslinked sorbents to catalyze oxidation of some sensitive organic compounds.

We chose cyclohexanone (I) as a diagnostic probe. Oxidation of cyclohexanone yields easily 1,2-cyclohexanedione (II) and then adipic acid (III):

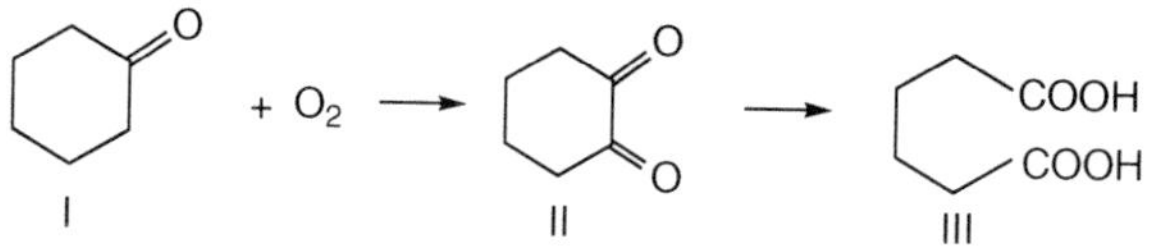

Two commercially available neutral hypercrosslinked polystyrene resins, MN-200 and MN-250, as well as the industrially used activated carbon SKT-6A (Russia) were loaded with cyclohexanone by a prolonged exposure of the materials to its vapors in a dessicator at ambient temperature (mostly, 25°C). The sorption capacity for both polymeric samples amounted to 400–500 mg/g. Afterward, the cyclohexanone-loaded sorbents were incubated at 85°C. At random intervals as indicated in Table 10.3, portions of each sorbent were washed with acetone and then a small dose of the acetone extract was injected into a gas chromatograph.

Table 10.3   Oxidation at 85°C of cyclohexanone absorbed on different sorbents

Sorbent	Incubation time (h)	Amount of analytes (mg/g)	
		I	II
MN-200	48	488	0
MN-200	360	474	0
MN-250	48	403	0
MN-250	360	388	0
Carbon SKT-6A	48	354	8.7
Carbon SKT-6A	96	405	12.7
Carbon SKT-6A	144	333	85.5

I = cyclohexanone; II = 1,2-cyclohexanedione.

Results given in Table 10.3 show that MN-250 and MN-200 are inert materials: even after 360 h of heating no traces of diketone II could be detected in the acetone extracts, which only contained cyclohexanone. Contrary to this, the activated carbon provoked an intensive transformation of cyclohexanone. After heating the preloaded carbon sample for 48 h at 85°C, the amount of diketone II formed was found to reach 9 mg/g. Naturally, on further heating of the sample, oxidation of the adsorbed cyclohexanone by the carbon becomes increasingly pronounced.

8. EVALUATION OF ADSORPTION ACTIVITY OF HYPERCROSSLINKED SORBENTS BY MEANS OF GAS CHROMATOGRAPHY

Gas chromatography represents an appropriate method for a rapid assessment of a sorbent's capability to take up vapors of various organic compounds differing in their chemical nature. Porous neutral adsorbents retain adsorbate molecules by various intermolecular interactions, among which non-specific interactions are general for all sorbent/sorbate combinations. Non-specific interactions comprise universal dispersion and inductive interactions. Adsorbents showing only such interactions are referred to as non-specific. The dispersion interaction between such adsorbent and adsorbate molecules is largely determined by the electronic polarizability of the latter. For members of a homologous series, for example, saturated hydrocarbons, the logarithm of retention volumes increases linearly with the increase in the polarizability of sorbate molecules.

Our first investigations with packed gas chromatographic columns containing Styrosorbs showed that at temperatures of 150°C or higher the polymers behave like non-swelling materials, giving symmetrical peaks for all polar or nonpolar, aliphatic or aromatic compounds. When the sample injected is really small, only equilibrium adsorbent/adsorbate interactions take place in the system, and, thus, the specific retention volume, V_g^T, becomes equal to the Henry constant. Hence, the specific retention volumes can serve as objective characteristics of the adsorption activity of an adsorbent. In addition, gas chromatography provides insight into the extent of chemical homogeneity of the sorbent surface, which is difficult to estimate by using any other method. Since polymeric adsorbents are widely used in gas chromatography as the separating media, it was interesting to find out how useful the hypercrosslinked sorbents may be in this field.

Hypercrosslinked polystyrenes strongly retain many organic compounds (Table 10.4); therefore, the chromatographic properties of Styrosorbs were investigated using a short, 30 cm × 1.5 mm ID column. The column was packed with spherical or irregularly shaped particles, with diameters of 0.16–0.20 mm [20].

Adsorption of organic compounds on hypercrosslinked sorbents proceeds largely due to dispersion interactions. This conclusion logically

Table 10.4 Specific retention volumes (mL/g) of organic compounds on two types of Styrosorbs and MN-100 measured at $T = 150°C$

Adsorbate	Mol. weight (kDa)	Boiling point (°C)	α_p (Å^3)	μ (D)	V_g^T (mL/g)		
					1	2	3
n-C_4H_{10}	58	−0.5	8.1	0	9	20	–
n-C_6H_{14}	86	69	11.8	0	50	260	1887
C_2H_5OH	46	78	5.1	1.7	12	20	–
C_3H_7OH	60	98	6.9	1.7	36	70	–
CH_3CN	41	82	4.2	4.0	21	30	–
CH_3NO_2	61	101	4.8	3.4	35	60	–
C_6H_6	78	80	10.6	0	69	260	1450
CH_2Cl_2	85	40	6.4	1.6	26	50	256
$CHCl_3$	119	60	8.2	1.1	54	160	813
$C_2H_4Cl_2$	99	84	$(8.2)^a$	1.5	74	200	1244
CF_2ClBr	165	−3.8	(7.9)	–	8	10	134
$C_2F_4Br_2$	260	47	(11.5)	0.6	21	40	774

α_p = polarizability; μ = dipole moment; 1 = Styrosorb 1BP (linear PS crosslinked with MCDE to 100% in cyclohexane); 2 = Styrosorb 2 (beaded styrene–DVB crosslinked with MCDE to 100% in EDC); 3 = MN-100.
[a]Calculated α values are given in parentheses.
Source: After [23].

follows from the direct dependence of the differential molar change in the energy of adsorption, $-\Delta U_1 = q_1$, on the polarizability of the adsorbate molecules (the heat of adsorption, q_1, was calculated from the plot of $\ln V_g^T$ versus $1/T$). However, $-\Delta U_1$ for benzene and polar nitrogen- and oxygen-containing compounds proved to be slightly higher than for n-hydrocarbons having the same polarizability [20]. Possibly some specific interactions between the π-electrons of polymeric phenyl rings and the lone electronic pairs of nitrogen- or oxygen-containing groups contribute to the total adsorption energy. One may estimate this contribution as

$$\delta q_{1,\mathrm{sp}} = q_{1,\mathrm{pol}} - q_{1,n-\mathrm{alk}} \qquad [10.5]$$

Here, $q_{1,\mathrm{pol}}$ and $q_{1,n-\mathrm{alk}}$ are the adsorption heats of a polar compound and a hypothetical n-alkane the polarizability of which equals the polarizability of the investigated polar compound, respectively ($q_{1,n-\mathrm{alk}}$ is usually determined from the dependence of q_1 for n-alkanes on their polarizability). Thus, in the case of adsorption on Styrosorb 2, the contribution from specific interactions, $\delta q_{1,\mathrm{sp}}$, amounts to 3 kJ/mol for C_6H_6, 4 kJ/mol for $(C_2H_5)_2O$, 11 kJ/mol for $C_nH_{2n+1}OH$, 11 kJ/mol for $(CH_3)_2CO$, 13 kJ/ mol for CH_3CN, 13 kJ/mol for CH_3NO_2, and 15 kJ/mol for C_5H_5N [20]. Therefore, strictly speaking, Styrosorb 2 should be considered a weakly specific sorbent [20, 21]. This conclusion is further corroborated by the data presented in Table 10.5, which shows values of $\delta q_{1,\mathrm{sp}}$ for various popular adsorbing materials. The first four sorbents, including Styrosorb 2, represent weakly specific polymers; they do not differ noticeably from each other. In contrast, the last three sorbents have polar functional groups in addition to π-electrons of benzene rings, and for them the contribution of specific

Table 10.5 Contribution of specific interactions to the total interaction energy for different polymeric adsorbents, $\delta q_{1,\mathrm{sp}}$, kJ/mol

Adsorbent	Adsorbate			
	C_6H_6	$(C_2H_5)_2O$	$(CH_3)_2CO$	C_2H_5OH
XAD-2	1.2	0.4	6	6
Polysorb-1	–	0.8	10	12
Porolas V-2T	2	0.4	6	13
Styrosorb 2	3	4	11	11
Tenax GC	–	13	12.5	8.8
Porolas SG-2T	3	10.5	16.7	20.9
Porapak T	6.3	8.8	18.0	17.6

Source: After [23].

interactions to the total adsorption energy is already noticeable. For example, the adsorption of ethyl alcohol on Porolas SG-2T (Russia) containing fragments of methacrylic acid proceeds specifically resulting in the high $\delta q_{1,sp}$ value of 21 kJ/mol. In the case of Styrosorbs their specificity might be caused by both the interactions of adsorbate molecules with the π-electron system of the aromatic polymer and the presence of traces of some polar groups in the polymer structure.

At present polymeric adsorbents, mostly Tenax, are widely used for the pre-concentration of harmful organic compounds present in air for their subsequent quantitative chromatographic analysis. It is interesting to compare the ability of various sorbents to accumulate volatile pollutants at room temperature. A much used simple manner of determining the V_g value at 20°C consists of the extrapolation to 20°C of the $\ln V_g^T = f(1/T)$ relationship established at higher temperatures. Although such extrapolation is not fully justified for some polymeric adsorbents (the decrease in temperature may lead, at a certain point called the glass transition temperature, to a sharp reduction in network mobility and a drop in the accessibility of sorption sites), these data may still provide insight into the sorbent's adsorption activity. According to the extrapolation results, Styrosorb 2 and MN-100 demonstrate at room temperature an exceptionally strong retention of n-hexane and benzene: the specific retention volumes reach 340 and 300 L/g on Styrosorb 2, and 3072 and 1495 L/g on MN-100, respectively. For comparison, the respective specific retention volumes of n-hexane and benzene on XAD-2 are 31 and 17 L/g.

It is quite interesting that both at high and low temperatures MN-100 retains n-hexane and benzene much stronger than does Styrosorb 2, although these two sorbents have practically identical specific surface areas: 1000 and 910 m^2/g, respectively [22]. Moreover, the MN-100 resin retains n-hexane twice as long as benzene. Evidently the value of the specific surface area of hypercrosslinked polystyrenes does not play a decisive role in these adsorption processes. More important is the size of the nanopores. Styrosorb 2 has a crosslinking degree of 100%, while the extent of bridging of the MN-100 resin exceeds significantly 100% (its phenyl rings are involved in the formation of more than one cross-bridge). Therefore, the size of nanopores in MN-100 resin must be smaller than that in Styrosorb 2. This assumption is in good agreement with the values of the intrinsic energy of nitrogen adsorption (calculated from low-temperature adsorption isotherms), which were found to be higher for MN-100 than for Styrosorb 2, namely 17.7 and 15.8 kJ/mol, respectively. This difference in the pore size basically explains the stronger retention of many organic substances on MN-100 (and MN-200 and Purosep 200, as well) [23, 24]. In addition, due to "reptation" movements, the flexible linear hydrocarbon molecules penetrate easier into the small pores of MN-100 and are retained there longer compared to rigid aromatic rings, which is opposite to the situation with wide porous polystyrene-type adsorbing materials.

Beside n-alkanes, MN-100 also shows high affinity to halogenated hydrocarbons (Table 10.6). Chlorocarbons are retained somewhat longer than the corresponding n-alkanes having approximately the same polarizability. The especially strong retention of trichloroethylene is peculiar, and is most likely caused by an additional contribution of specific interactions between the π-system of the adsorbate's double bond and the aromatic polymer to the total adsorption energy. In contrast, fluorocarbons are more weakly retained than n-alkanes with similar polarizability; most likely, the dispersion interactions of fluorocarbons with aromatic polymers are weaker.

The commercial hypercrosslinked sorbents MN-100 and MN-200 tolerate elevated temperatures well; the specific retention volumes of

Table 10.6 Specific retention volumes of halocarbons at 150°C on hypercrosslinked sorbents

Sorbate	α_p (Å^3)	μ (D)	V_g^T (mL/g)	
			Styrosorb 2	MN-200
C_3H_8	6.3	0	8	–
CF_3Br (Freon 13B1)	(6.5)[a]	–	1–2	–
$C_2H_2F_4$ (Freon 134a)	(6.0)	–	–	24
C_2HF_5 (Freon 125)	(6.3)	–	–	21
CH_2Cl_2	6.4	1.58	50	256
n-C_4H_{10}	8.1	0	20	–
CF_2ClBr (Freon 12B1)	(7.9)	–	10	134
$CHCl_3$	8.2	1.1	160	813
1,2-$C_2H_4Cl_2$	(8.2)	1.5	200	1244
C_2HCl_3	–	0.94	–	1723
n-C_5H_{12}	10	0	–	534
CCl_4	10–11	0	–	1499
n-C_6H_{14}	11.8	0	260	1887
$C_2F_3Cl_3$ (Freon 113)	(11.0)	1.44	40	–
$C_2F_4Br_2$ (Freon 114B2)	(11.5)	0.56	40	774

[a] Calculated α_p values are given in parentheses.
Source: After [25].

n-hexane, benzene, and acetone were found to remain fairly constant at a prolonged use of the sorbents at 250°C. The values of V_g^T for these compounds dropped slightly with time only at a higher temperature, around 300°C. However, this is not unexpected. When receiving a sufficient portion of heat energy, the loose and strongly stressed hypercrosslinked network rearranges to bring down the surface energy. It leads to shrinkage of the network (see also Chapter 7, Section 8.2), reduction of the adsorption surface area, and, as a result, a decrease in the retention volumes.

Thus, under gas chromatographic conditions hypercrosslinked polystyrenes retain many organic compounds (alkanes, esters, ketones, alcohols, halocarbons, etc.), with MN-100 and MN-200 resins demonstrating particularly strong retention. The retention on these two sorbents appears to be too long for using them as separating media in gas chromatographic columns. Nevertheless, these resins are the materials of choice as highly efficient sorbents for the pre-concentration of priority pollutants from air. The compounds absorbed can then be removed by thermal desorption and quantitatively analyzed by any appropriate method. Indeed, the high efficiency of hypercrosslinked polymers has been demonstrated for the pre-concentration of organic pollutants from ambient air in the home [26].

9. SORPTION OF *ORTHO–PARA*-SPIN ISOMERS AND ISOTOPE ISOMERS OF WATER

Russian spectroscopy specialists Tikhonov and Volkov succeeded in constructing a unique spectrophotometer that is capable of registering separately the *ortho-* and *para*-spin isomers of water at 36.605 and 37.137 cm^{-1}, respectively. According to the fundamental quantum property of a water molecule, it can exist in one of the two spin states, paramagnetic *ortho-* or nonmagnetic *para*-, with parallel or antiparallel proton spins. Under normal conditions, the amounts of gaseous *ortho-* and *para*-molecules are in a statistical 3:1 equilibrium. In 2002, the above authors reported [27] an intriguing fact: the two isomers demonstrate different adsorption properties on a solid surface. When a portion of water vapor was sent with a flow of nitrogen at a reduced pressure of about 200 Torr through a column packed with hypercrosslinked polystyrene MN-200, the very first part of the water zone at the column exit was found to be enriched in the *ortho*-isomer with the *ortho/para*-ratio as high as up to 10:1. The last water portions in the elution zone had an excess of *para*-isomers. While a similar problem of shifting the ratio of *ortho–para*-spin isomers of dihydrogen H_2 was resolved long ago [28] by cooling the gas to very low temperatures, no progress was made with respect to water spin isomers. Properties of separate *ortho-* and *para*-water are at present completely unknown, so that accessibility of the two forms of water could stimulate research in many branches of science. Several research groups, including that of the authors of the present book, tried to register, at least, a partial resolution of a water probe in gas chromatographic systems containing hypercrosslinked polystyrene, activated carbon, and some other adsorbing materials. Temperature of the column was varied from 0 to 120°C. However, without the isomer-specific detector no conclusive results were obtained. The very possibility of separating spin isomers of water was soon challenged by theoretical considerations [29].

Still, Tikhonov and Volkov repeatedly observed the above shift in the ratio of water spin isomers after contacting water vapor with the polymeric adsorbent both under dynamic (Fig. 10.19) and static conditions [30]. Activated carbons, zeolites, and silica gels generated similar effects. The adsorbed water was strongly *para*-enriched. Interestingly, water samples enriched in the *ortho-* or *para*-isomers were found to be stable in the form of ice, but equilibrated to the ratio of 3:1 in the liquid state at room temperature within 20–30 min. As the separation takes place at the

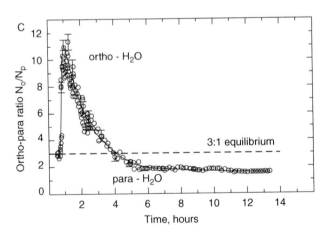

Figure 10.19   Ratio between the *ortho-* and *para*-spin-isomers within the elution zone of water vapors. Column 180 cm × 10 mm ID packed with MN-200 beads of about 0.5 mm in diameter. Ambient temperature. Points – experimental data. Lines – theoretical description of a non-equilibrium chromatography process. *(After [27].)*

partial water vapor pressures below 1 Torr, the authors suggest describing the process in terms of adsorption kinetics and non-equilibrium chromatography [31].

The enhanced interest to the separation of water isomers stimulated another group [32, 33] to test hypercrosslinked polystyrene in the capacity of column packing in gas chromatography of normal and heavy water. Heat of adsorption of D_2O on the polymer MN-272 was found to slightly exceed that of H_2O in the examined range of 50–70°C. Similarly, activated carbon retained heavy water slightly longer, as well. Specific interactions between water molecules and the polymer were estimated to contribute 53.8% for D_2O and 42.5% for H_2O, while the rest is contributed from dispersion interactions. These data render the aromatic polymer weakly specific with regard to water molecules that are not prone to enter dispersion interactions. Still, the resolving power of the packed column was insufficient for providing separation of these water isotope isomers.

CHAPTER 11

Sorption of Organic Compounds from Aqueous Solutions

Isolation of target organic compounds from aqueous solutions or removal of minor impurities present in technical or natural waters by their adsorption onto the surface of porous solids is one of the important constituents of many modern technologies, such as fine polishing of potable water, wastewater treatment, sugar syrup refining, and production of antibiotics. Traditionally, many of these technologies are based on exploiting activated carbons. However, in recent years, polymeric adsorbents, in particular, macroporous sorbents based on styrene–divinylbenzene (DVB) copolymers, have begun to gradually replace the activated carbons. Chiefly, this is accounted for by the simpler regeneration and reuse of macroporous polymeric adsorbents, though the latter are inferior to the activated carbons in terms of loading capacity. In this respect, highly permeable hypercrosslinked adsorbing materials are more promising, compared to the traditional macroporous resins, as the hypercrosslinked polystyrenes successfully combine both the high adsorption capacity and the ease of regeneration. This chapter considers the potential of hypercrosslinked sorbents in taking up a wide variety of organic compounds from aqueous media, pointing out those areas in which the practical application of hypercrosslinked sorbents would be very beneficial [34].

1. SORPTION OF ORGANIC SYNTHETIC DYES

It is well known that the porous structure of activated carbons, silica gels, zeolites, conventional macroporous styrene–DVB copolymers, and related rigid adsorbing materials remains unchanged when an organic solvent or water dislodges air from the internal porous space of the adsorbents. Therefore, similarly to adsorption from gaseous media, adsorption on the surface exposed to a liquid phase obeys a simple rule: the higher is the surface area of a sorbent, the higher is its adsorption capacity. Correspondingly, the tendency is to use in practice those adsorbing materials which have the largest surface area. The apparent specific surface area of dry

Comprehensive Analytical Chemistry, Volume 56
ISSN 0166-526X, DOI 10.1016/S0166-526X(10)56011-3

hypercrosslinked polystyrene measured by low-temperature sorption of nitrogen is very high, $1000-1500 \, m^2/g$. Moreover, when contacting organic solvents or water, the hypercrosslinked polystyrenes can strongly increase their volume. In this process, a cooperative conformational rear-rangement of constituent network meshes changes the network structure so that both the diameter of channels within the polymeric matrix and the number of network fragments exposed to the liquid phase increase con-siderably. Therefore, the ability of a hypercrosslinked material to take up solutes from aqueous solutions by no means correlates with the value of the apparent specific surface area measured for the dry network.

To confirm the above statement, a number of hypercrosslinked networks with 43, 100, and a nominal 200, 300, 400, and 500% degree of crosslinking were obtained by reacting 1 mol of styrene–0.5% DVB copolymer with 0.3, 0.5, 1.0, 1.5, 2.0, and 2.5 mol of monochlorodimethyl ether (MCDE), respec-tively. Theoretically, in the last three cases, all benzene rings of initial polystyrene must acquire in total four, five, and six substituents. However, this is not possible for steric reasons. Nevertheless, the conversion of the bifunctional MCDE taken in the amount of 2.0 and 2.5 molar excess is surprisingly high, and the content of residual chlorine in products with the nominal degree of crosslinking of 400 and 500% is relatively low (Table 11.1).

If one takes into account the $\pm 20\%$ accuracy of the single-point BET method of surface area measurement, the specific surface area of the above Styrosorbs 2 can be regarded as remaining largely invariable within the limits of $1300-1500 \, m^2/g$ with the degree of crosslinking rising from 43 to 400%. It decreases only a little for the last sample with nominal 500% degree of cross-linking. In contrast, the network free volume (pore volume) rises significantly

Table 11.1 Parameters of porous structure and water regain of hypercrosslinked Styrosorbs 2[a] with degree of crosslinking X rising from 43 to 500%

X (%)	MCDE (mol/ mol PS)	Residual chlorine (%)	S (m^2/g)	W_o (cm^3/g)	Water uptake (mL/g)
43	0.3	No	1350	0.03	0.32
100	0.5	<0.5	1300	0.21	0.62
200	1.0	0.8	1325	0.42	0.97
300	1.5	1.3	1480	0.63	1.31
400	2.0	2.3	1550	0.43	1.21
500	2.5	2.6	970	0.28	0.81

[a] Syntheses were conducted in EDC in the presence of 1 mol $SnCl_4$ per 1 mol of MCDE at 80°C within 10–12 h.

up to a maximum of 0.63 cm³/g at $X = 300\%$. The pore volume then starts to decrease with a further rise in the degree of crosslinking. Similarly, the water uptake of swollen polymers displays an extremal dependence on the degree of crosslinking with the highest value of 1.31 mL/g at $X = 300\%$. Bridging of polystyrene chains in the initial swollen styrene–0.5% DVB gel with MCDE is known to proceed in two consequent steps, chloromethylation and then alkylation of free positions of phenyl rings of a neighbor chain. Presumably, the highest rate of formation of a rigid hypercrosslinked network corresponds to the reaction composition where 1.5 mol of MCDE is allowed to react with 1 mol of styrene repeat units, to give a network with the degree of crosslinking of 300%. If more MCDE and more catalyst are involved in a reaction, the chloromethylation process may proceed too quickly, so that the second step gets retarded, because of the reduced concentration of remaining free positions in benzene rings to be suitably substituted in this bridging step. In any case, the samples thus prepared from the same initial copolymer with gradually increasing amounts of the post-crosslinking agent, MCDE, have comparable specific surface area measured for dry materials, but differ in their water uptake.

The above hypercrosslinked samples have been tested under identical conditions for their ability to extract from aqueous solutions three synthetic organic dyes: Malachite Green (MG), Methylene Blue (MB), and Direct Bordeaux (DB):

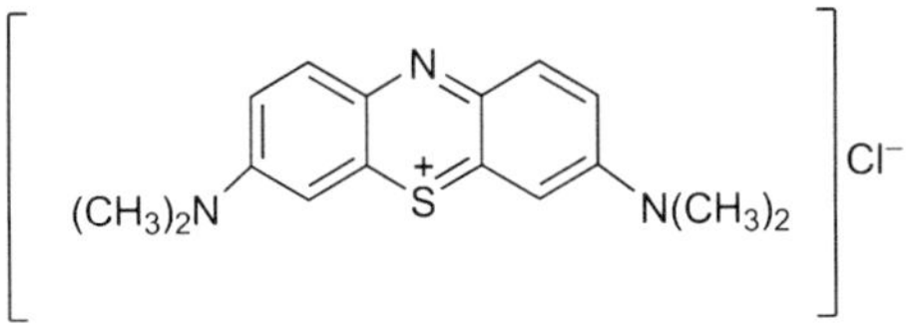

Methylene Blue $M = 318\,kDa$ 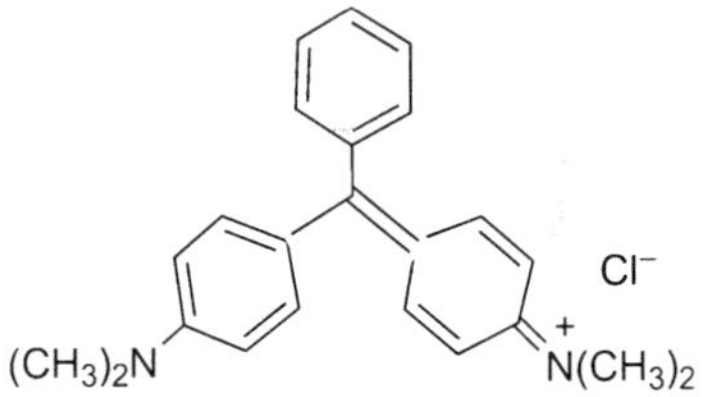   Malachite Green $M = 364\,kDa$

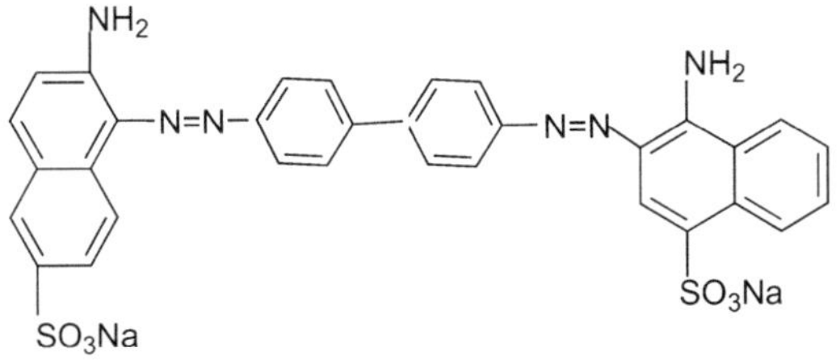

Direct Bordeaux $M = 576\,kDa$

Figure 11.1a demonstrates clearly the absence of any correlation between the sorption activity of the water-swollen polymers and their specific surface area measured in the dry state. On the contrary, the dye uptake obviously rises with the sorbent' water regain, reaching a maximum value at the highest swelling that corresponds to a 300% degree of crosslinking. The sorption capacity of Styrosorbs with a nominal 400 and

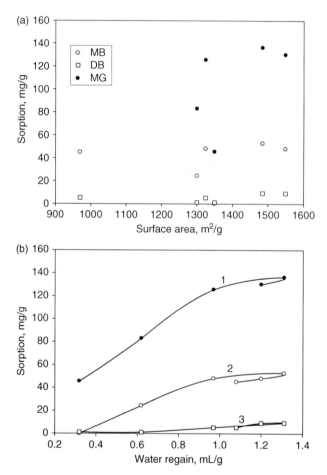

Figure 11.1 Sorption of (1) Malachite Green, (2) Methylene Blue, and (3) Direct Bordeaux onto Styrosorbs 2: (a) dependence on apparent specific surface area of the polymers and (b) dependence on their swelling in water. Experimental conditions: 20 mL solution of 0.3 g/L in concentration, 0.1 g dry polymer, 25°C, 24 h, spectrophotometric detection at (1) 615, (2) 665, and (3) 510 nm.

500% degree of crosslinking logically reduces, parallel with decreasing water regain of these samples.

All the above Styrosorbs 2 samples demonstrate the highest sorption capacity for Malachite Green, up to 140 mg/g at $X = 300\%$ (Fig. 11.1b). Methylene Blue, although smaller in molecular size, exhibits a strong tendency to aggregation at the concentration of the starting solution of 0.3 g/L and, therefore, adsorbs in smaller quantities [35]. Large non-planar molecules of Direct Bordeaux can hardly diffuse into the densely cross-linked microporous networks.

Other synthetic organic dyes, such as Rhodamine C (sorption up to 1.2 g/g [36]), Crezyl Violet, or Neutral Red, having a relatively small molecular size, are efficiently retained by both microporous and biporous Styrosorbs. Large transport pores in the biporous MN-100 and related neutral sorbents were found to be essential for the diffusion of such big molecules as Brilliant Blue R, which can be approximated by a species of 25 Å in diameter. Azo dye Acid Red 14 readily adsorbs onto both neutral MN-200 and weak basic anion exchanger MN-100 [37, 38]. A homemade representative of hypercrosslinked sorbents derived from a chloromethylated styrene–DVB copolymer in nitrobenzene with $ZnCl_2$ takes up comparable amounts, 50–60 mg/g, of both large molecules of Rhodamine B and rather small molecules of Methyl Orange [39], the latter dye being prone to form aggregates in aqueous solutions.

The high potential of hypercrosslinked resins for sorption of dyes indicates that the sorbents could be successfully used to recover various dyes from wastewaters of the textile industry, in order to return the valuable chemicals again into the dyeing processes and reduce pollution of the environment. Most dyes can be easily desorbed from hypercrosslinked polystyrene with alcohols or by appropriate acidic or basic aqueous solutions, which is hardly possible in the case of activated carbons.

2. SORPTION OF TRIBUTYL ESTER OF PHOSPHORIC ACID

Tributyl phosphate (TBP) is a widespread extracting agent generally used in processing transuranium elements and removing them from wastewaters of the nuclear industry. Although the solubility of the ester in water is not very high, its loss, nevertheless, is significant because of the large scale of application. As Fig. 11.2 demonstrates, Styrosorb 2 (MCDE, 100%) absorbs approximately 1 g/g of TBP from 0.5 N HNO_3 [40]. Note that this value is much higher than the total pore volume of the dry sorbent, 0.2–0.3 cm^3/g,

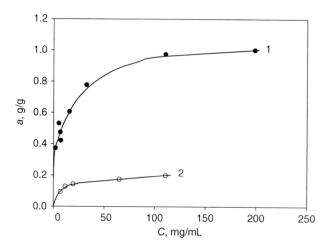

Figure 11.2 Sorption isotherms for tributyl phosphate from 0.5 N HNO$_3$ on (1) Styrosorb 2 and (2) Polysorb 40/100 at 20°C. *(After [40].)*

and even higher than the water uptake, 0.5 cm^3/g, of the polymer. Rather, the sorption capacity of Styrosorb 2 is very close to the swelling of this polymer with toluene, 1.0–1.2 g/g. It may only imply that in the course of sorption TBP not only actively displaces water from the sorbent interior but also causes additional strong swelling of the polymer with relaxation of inner stresses of its network.

Accordingly, since the total structure of the porous polymer changes in the course of absorption, all attempts of describing the sorption process by any theoretical or empirical approaches developed for solids with constant porosity seem to be meaningless. In the following, we will be guided by this idea when considering the sorption of a vide variety of organic substances onto hypercrosslinked sorbents.

Contrary to the hypercrosslinked adsorbing material, the conventional macroporous polystyrene sorbent Polysorb 40/100 absorbs the phosphate ester poorly, no more than 200 mg/g. Importantly, the sorption capacity of Styrosorb 2 with respect to TBP depends neither on the nitric acid concentration nor on the content of uranyl ions. Under dynamic conditions, one bed volume of the hypercrosslinked sorbent purifies, at the moment of breakthrough, more than 1100 volumes of the nitric acid solution containing 450 mg/L of TBP, while Polysorb 40/100 can treat only 350 vol. of the same solution.

3. SORPTION OF *N*-VALERIC ACID

n-Valeric (pentanoic) acid is considered to be a rather suitable model for a qualitative and quantitative examination of adsorption properties of porous sorbents, because, on the one hand, the solubility of valeric acid in water is sufficiently high and, on the other hand, it adsorbs readily on hydrophobic surfaces owing to its C_4-aliphatic chain. Besides, valeric acid does not form micelles even in rather concentrated aqueous solutions.

Sorption isotherms for *n*-valeric acid on hypercrosslinked microporous Styrosorbs 1 crosslinked with *p*-xylylene dichloride, biporous Styrosorb 1BP, and, for comparison, macroporous Amberlite XAD-2 are given in Fig. 11.3. Again, one cannot see any correlation between the specific surface area of dry hypercrosslinked sorbents and their capability to take up the acid. Indeed, Styrosorb 1 with $X = 25\%$ is a nonporous material when in the dry state, but it swells in water (0.15 mL/g) and, accordingly, retains a noticeable amount of the sorbate. On the other hand, the apparent specific surface area of dry Styrosorb 1 and Styrosorb 1BP, both having a 100% degree of crosslinking, are of the same order of magnitude, about $1000\,m^2/g$, but the biporous Styrosorb 1BP takes up twice as much water as Styrosorb 1 does (2.90 and 1.31 mL/g, respectively) and exhibits an exceptional sorption capacity at equal equilibrium

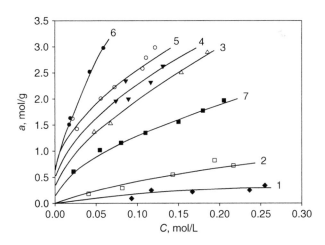

Figure 11.3 Sorption isotherms for *n*-valeric acid at 20°C onto Styrosorb 1 (*p*-xylylene dichloride, XDC) with crosslinking degree of (1) 11, (2) 25, (3) 43, (4) 66, and (5) 100%, (6) Styrosorb 1BP (100%), and (7) Amberlite XAD-2. *(After [41].)*

Table 11.2 Parameters of *n*-valeric acid sorption onto Styrosorbs and XAD-4

Sorbent	Bead size (mm)	D_c (cm)	S_F (cm/min)	D_{MTZ} (cm)	EF (%)
Styrosorb 2	0.2–0.4	0.2	0.038	1.77	100
XAD-4	0.3–0.5	2.1	0.063	2.33	83
MT-50[a]	0.3–1.0	1.5	0.027	3.67	90
XAD-4	0.3–1.0	2.4	0.058	4.89	84

[a]Laboratory analogue of MN-200.
Experimental conditions: 20×1.4 cm ID glass column, 5 g/L *n*-valeric acid solution, 1 ml/min.
Reprinted from [34] with permission of Elsevier.

concentrations of *n*-valeric acid. Although the surface area of XAD-2 is larger than that of dry Styrosorb 1 with a 43% degree of crosslinking, the latter takes up larger amounts of the acid [41].

Under dynamic conditions the capacity of microporous Styrosorbs at 5% breakthrough amounts to 0.4–0.5 g/g, which relates to 30–40 bed volumes of treated solution with a concentration of 5 g/L. In contrast, dynamic breakthrough volumes for Amberlite XAD-4 and XAD-1180 resins turned out to be lower by a factor of 2.

The superiority of hypercrosslinked sorbents is also evident from calculations of adsorption parameters [42, 43]. Depth of the mass transfer zone (D_{MTZ}), speed of the adsorption front (S_F), critical bed depth for which immediate breakthrough occurs (D_c), and efficiency (EF) were determined from breakthrough curves obtained for different bed depths of the sorbents. S_F was calculated from the straight-line dependence of t_b (5% breakthrough time) on bed depth as the slope $1/S_F$. This dependence also gives D_c value at $t_b = 0$. Then, $D = S_F(t_s - t_b)$, where t_s is the time required to saturate (95%) the column. Finally, $EF = 1 - D_c/D_p$, $D_p = 16$ cm being the highest bed depth in experiments. All these characteristics, calculated for XAD-4 and microporous Styrosorb 2, are given in Table 11.2. As compared to Styrosorbs, XAD-4 with both narrow and wide bead size distribution displays a considerably higher D_c value, wider mass transfer zone, and smaller efficiency.

4. CLARIFICATION OF COLORED FERMENTATION LIQUIDS

Large-scale production of optically active α-amino acids has usually been performed by microbiological synthesis or through hydrolysis of waste proteins. In both cases, a serious problem is the isolation of pure crystalline amino acids because numerous by-products are present in solutions. In particular, large amounts of colored bodies, such as melanoidins, prevent crystallization of amino acids even from concentrated solutions. This makes it necessary to purify amino acids by adsorption on a cation-exchange resin followed by desorption with ammonia [44]. Unfortunately, the procedure is accompanied by a significant dilution of amino acids. Moreover, some of the colored bodies adsorb on the ion exchanger too [45]; they partially co-elute with the target product, partially contaminate the resin, and gradually reduce its performance. Hypercrosslinked polystyrene sorbents can considerably simplify the isolation of amino acids from fermentation solutions, because they readily remove colored bodies but do not retain hydrophilic amino acids.

Melanoidins are a complex mixture of middle-molecular-weight proteins conjugated with polysaccharide fragments of a general approximate formula $(C_{10-11}H_{49}O_{20-21}N_4)_{40}$ [46]. Because of their large size, they only adsorb onto biporous Styrosorb 1BP. Study of the adsorption kinetics of melanoidins (isolated from a cultural liquid of L-lysine production) showed that the equilibrium adsorption is established within 30 min at pH 1.5 or in neutral medium and within 1.5 h at pH 11.9 [47]. The sorption isotherms for melanoidins approach a value of 650 mg per gram of the adsorbing material at low pH, while considerably smaller amounts of melanoidins adsorb from neutral and alkaline solutions (Fig. 11.4), thus

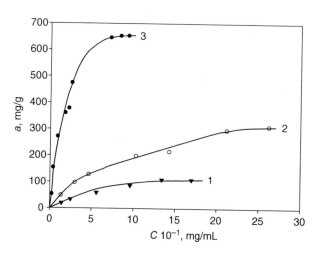

Figure 11.4 Sorption of melanoidins at 20°C on Styrosorb 2 at pH values of (1) 12.6, (2) 6.8, and (3) 2.7. *(After [47].)*

implying the possibility for regeneration of the sorbent with an alkaline washing. Infrared spectroscopy suggests conformational rearrangement of polypeptide chains in melanoidin molecules on changing from acidic media to alkaline one [46]. Another reason for the observed drop of sorption would be the increased negative electrostatic charge of the solute in alkaline solutions.

Under dynamic conditions the breakthrough of some colored bodies takes place rather soon, possibly that of the most hydrophilic or most voluminous fractions of melanoidins. However, even the last portion of 150 mL cultural liquid percolated through 3 g of the biporous sorbent shows a 70% degree of clarification; here the dynamic sorption capacity achieves a very high value of 670 mg/g. Of several regeneration tests, the best results were achieved when using mixtures of sodium hydroxide solution with isopropanol; 3–5 bed volumes were found to be sufficient to regenerate the sorbent comprehensively. Three consequent cycles of sorption–desorption do not deteriorate the performance of the hypercrosslinked Styrosorb 1BP.

Complex colored bodies present in L-tryptophan fermentation solutions readily adsorb onto industrial biporous MN-200. One bed volume of the latter clarifies approximately 30 volumes of the cultural liquid or 17–18 vol. of its concentrate. The sorption capacity of other industrial sorbents is lower.

A similar problem of clarification of cultural liquids also arises in the biotechnological production of nucleosides. In the Tripol'e Biochemical Plant in (Ukraine), MN-200 exhibited, on the whole, better performance than commonly used activated carbons [48]. A comparative study of MN-200, Styrosorb 1BP, Styrosorb 1 (MCDE, 100%), and granular activated carbon OU-A (Russia) was conducted by filtrating 200 mL cultural liquid of riboxin (inosine) production at 20°C or 250 mL aqueous saturated solution of technical crystalline riboxin at 60°C (to avoid nucleoside crystallization) through a 12 mL sorbent bed at a flow rate of 0.2–0.3 mL/min. In both solutions pH was between 3.3 and 3.6. After rinsing the sorbents with water, they were regenerated with 100 mL 2 N NaOH. The results obtained are given in Table 11.3.

As can be seen, all adsorbing materials unfortunately partially remove the target product, riboxin, together with the contaminating colored bodies. However, commercial MN-200 and biporous Styrosorb display the most beneficial ratio of the above two components. Besides, a crucial advantage of hypercrosslinked polystyrene is exhibited in the ease of its

Table 11.3 Clarification of cultural liquid of riboxin and a solution of its technical crystals by means of Styrosorbs and activated carbon

Sorbent	Cultural liquid		Solution of crystals	
	Fraction of colored bodies absorbed	Fraction of riboxin absorbed	Fraction of colored bodies absorbed	Fraction of riboxin absorbed
MN-200	0.55	0.26	0.66	0.18
Styrosorb 1BP	0.38	0.28	0.52	0.15
Styrosorb 1	0.33	0.30	0.14	0.21
OU-A	0.20	0.19	0.54	0.17

regeneration, which proceeds readily and exhaustively. MN-200 and biporous Styrosorbs can be used many times without any loss in capacity and breakdown of beads. In contrast, only 85% of colored substances and 32% of retained riboxin were recovered from the activated carbon. For that reason, after two or three runs the performance of activated carbon deteriorates so much that the column packing must be replaced with a new portion. Interestingly, a macroporous conventional resin Polysorb 100/60 shows no selectivity in the decolorization process and removes 99% colored bodies together with 93% of the target product.

A concentrated mixture of all L-α-amino acids is obtained through acid hydrolysis of natural waste proteins and also needs to be freed from contaminating colored bodies. The latter are readily removed by Styrosorb 1BP [49]. When percolating the protein hydrolysate through 12 g Styrosorb 1BP, colored bodies emerge in the effluent almost instantly, but the increase in the color intensity proceeds very slowly; even the 17th bed volume of the treated solution at pH 1.0 still has a degree of clarification as high as 80%. At pH 3.5, the same high degree of clarification is achieved even in the 25th bed volume of the effluent. Interestingly, a spontaneous crystallization of amino acids occurs sometimes from the thus purified concentrated solution, which has never been observed before. It should also be emphasized that adsorption of α-amino acids on Styrosorbs is negligible and even the loss of aromatic amino acids is very small during the purification of acidic hydrolysate solutions.

In short, the potential for using hypercrosslinked polystyrene adsorbing materials in biotechnology is very high, but still remains barely recognized.

5. SORPTION OF LIPIDS

Isolation and processing of plant proteins has become an important branch of the food industry. Often, the proteins have to be purified from residual fats and oils, which reduces the shelf time of isolated proteins. Hypercrosslinked polystyrene is the material of choice for the removal of all lipids from protein-containing extracts, as well as from isopropyl alcohol where the latter is used for the extraction of lipids from protein masses [50]. Figure 11.5 shows sorption isotherms on Styrosorb 2 for lecithin and oleic acid from pure isopropyl alcohol and its mixtures with water. The equilibrium of lecithin sorption on the microporous Styrosorb 2 from pure isopropyl alcohol is established within 4 h. Addition of water accelerates the sorption process; in a water–alcohol mixture of 1:1 (v/v), the equilibrium is established within 2 h. By reducing the thermodynamic affinity of the medium to both polystyrene and lipids, water also enhances the sorption capacity, so that from the 50% isopropyl alcohol solution as much as 600–700 mg/g of lecithin can be taken up by the hypercrosslinked polymer, compared to 400 mg/g taken by the XAD-4 resin under the same conditions [51]. Sorption of oleic acid by Styrosorb 2 is smaller than that of lecithin, but, still, it reaches 400 mg/g.

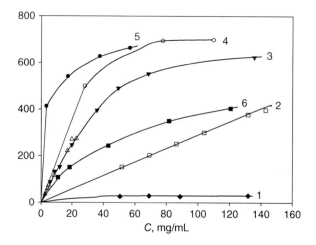

Figure 11.5 Sorption isotherms for (1–5) lecithin and (6) oleic acid at 20°C on (1) Polysorb 40/100, (2) XAD-4, and (3–5) Styrosorb 2 from (1, 2, 3) isopropyl alcohol and mixtures of the latter with water of (4) 4:1, (5) 1:1, and (6) 5.6:1 v/v, respectively, at 20°C; open and dark symbols on the plot 3 denote adsorption and desorption, respectively. *(After [51].)*

6. SORPTION OF GASOLINE

Contamination of surface waters, wastewaters, and soils by fractions of earth oil is an ever-growing environmental problem. The use of hypercrosslinked sorbents to remove oil traces dissolved or emulsified in water streams seems to be very promising. Table 11.4 demonstrates the superiority of Styrosorb 2 (MCDE, 200%) over XAD-4 when 1800 mg/L gasoline emulsion (not stabilized by surfactants) was pumped through 10 mL sorbent beds. XAD-4 retains gasoline weakly; even the first 100 mL of effluent contain 6.7 mg/L hydrocarbons. This value exceeds the concentration allowed for drinking water by one order of magnitude. On the contrary, one vol. of Styrosorb 2 is capable of purifying no less than 650 volumes of the initial emulsion, with the concentration of hydrocarbons in the effluent fully meeting the prescribed requirements.

Simple calculations show that, in this latter case, the uptake of hydrocarbons by Styrosorb 2 would have to approach 10 g/g, while the real swelling of Styrosorb 2 sample is much lower, only 0.7 g/g. The surprising fact, however, is that the sorbent beads cause an intensive coalescence and coagulation of emulsion droplets, resulting in the formation of a gasoline layer at the top of the column. Probably below the gasoline layer, only a true gasoline solution or its very fine emulsion contacts the polymer packing in the rest of the column. Taking into account the low solubility of hydrocarbons in water (approximately 30 mg/L), one can understand that the microporous Styrosorb functions in a different manner than the macroporous XAD-4 and can therefore purify a very large volume of the gasoline-containing aqueous emulsion.

Table 11.4   Extraction of hydrocarbons from a gasoline–water emulsion by Styrosorb 2 and XAD-4

Styrosorb 2		XAD-4	
Fraction volume (mL)	Hydrocarbon content (mg/L)	Fraction volume (mL)	Hydrocarbon content (mg/L)
1000	0.45	120	6.69
2000	0.30	250	5.98
3000	–	360	13.24
3600	0.15	490	13.50
4600	0.08	620	20.23
5600	0.50	760	21.98
6600	0.45	–	–

Reprinted from [34] with permission of Elsevier.

7. SORPTION OF PHENOLS

Removal of phenol and its chloro and nitro derivatives from wastewaters is a serious problem for many industry branches, in particular oil processing and organic syntheses. Contamination of aquatic pools with the toxic water-soluble phenolic compounds is inadmissible for ecological reasons; also, the large loss of the valuable feedstock material makes no sense from an economic viewpoint. In this respect adsorption and recovery of the above contaminants using porous polymers favorably distinguishes from other methods of waste treatment, for instance, electrochemical decomposition of phenols. However, not only the high adsorption capacity of the material toward phenols but also simplicity of the sorbent regeneration is equally essential for a feasible technology. Macroporous Amberlite XAD-4 resin demonstrates a good combination of the above properties. Indeed, column adsorption on XAD-4 yields around 30–40 bed volumes of pure water when starting from a 250 ppm phenol solution. Up to 430 bed volumes of water can be purified [52] from the stronger retained 2,4,6-trichlorophenol (513 ppm) as illustrated in Fig. 11.6. Two bed volumes of acetone or methanol are sufficient to regenerate completely the adsorbent exhausted with phenol.

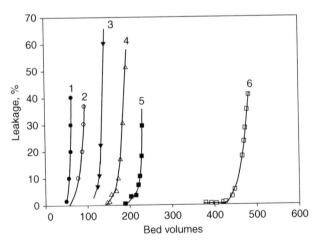

Figure 11.6 Column adsorption on Amberlite XAD-4 from aqueous solutions: (1) phenol, 249 ppm; (2) phenol + 13% NaCl, 252 ppm; (3) *m*-chlorophenol, 349 ppm; (4) *m*-chlorophenol + 13% NaCl, 349 ppm; (5) 2,4-dichlorophenol, 435 ppm; (6) 2,4,6-trichlorophenol, 513 ppm; experimental conditions: flow rate, 4 BV/h, 25°C. *(After [52].)*

It was found, however, that hypercrosslinked sorbents exhibit much higher adsorption capacities. Fig. 11.7 illustrates sorption isotherms for phenol on the biporous sorbents MN-200 and MN-150. The isotherms coincide completely, thus revealing no difference in the phenol adsorption between the two neutral resins that only differ in the size of their macropores [53]. Both MN-200 and MN-150 absorb about 0.4 g of phenol per gram of polymer without yet achieving full saturation of the resins. This value practically coincides with the amount of swelling water that is taken up by the hypercrosslinked polymeric phase of the biporous network.

When percolating a 200 mg/L phenol solution through a 100 mL column packed with MN-200, the position of sharp breakthrough curves proves almost independent of the flow rate, at least up to about 4 BV/h. One column bed extracts phenol from 100 volumes of the solution. A direct comparison of phenol extraction efficiency (Fig. 11.8) reveals the evident and expected superiority of hypercrosslinked MN-200 sorbent over the Amberlite XAD-4 and other macroporous resins of Porolas series (Russia).

Theoretically, the maximum value of the breakthrough volume is determined by the rate of migration of the solute chromatographic zone

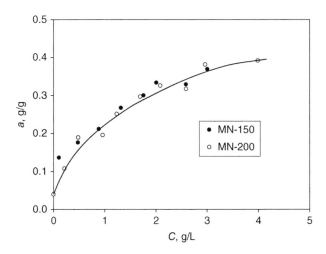

Figure 11.7 Sorption isotherms of phenol from aqueous solutions on Macronet MN-200 and MN-150 at 20°C. *(After [53].)*

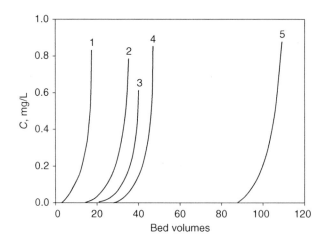

Figure 11.8   Breakthrough volumes for macroporous Porolas resins (1) V2T, (2) ST-2T, (3) XAD-4, (4) GM-2s, and (5) hypercrosslinked MN-200 on percolating 200 mg/L phenol solution at 2 mL/min flow rate and 20°C.

under conditions of linear equilibrium chromatography, that is, by a capacity factor k' for the solute of interest. Depending on whether or not saturation of the sorbent by that solute is reached under given frontal analysis conditions, the elution front propagates at a greater or smaller rate, the latter, however, remaining always faster than or equal to the intrinsic k'-limitation of the band movement rate. In the case of removing phenol from diluted aqueous solutions we most probably deal with this k'-limitation that corresponds to the relatively low affinity of the polar phenol molecules to the nonpolar hypercrosslinked polystyrene resin. Therefore, independent of the concentration of the initial phenol solution, the breakthrough volumes for the MN-200 and MN-150 sorbents were always observed to be around 100 bed volumes, and, fortunately, not much less at high phenol concentrations. In other words, purification of wastewaters with industrial resins MN-200 or MN-150 is rather predictable and safe, and a sudden surge in phenol concentration in the waste stream does not affect noticeably the operation time of the sorbent bed.

Streat and Sweetland [54] confirm the high efficiency of Hypersol-Macronet^{TM} polymers MN-200, MN-150, and MN-100 in the removal of low-level phenols and chlorophenols from water. They stress that weaker binding energies, in comparison with activated carbon, offer better regeneration characteristics of the polymers. They note that some exclusion effects for chlorophenols can be recognized, leading to a certain underutilization of the material's inner surface area that is available for nitrogen.

Other types of hypercrosslinked sorbents, namely commercial Lewatit EP63, laboratory samples obtained by post-crosslinking styrene–DVB copolymer with CCl_4 [55], by self-crosslinking macroporous copolymer of vinylbenzyl chloride with 20% DVB via Friedel–Crafts reaction [56], or by bridging preliminary chloromethylated macroporous styrene–DVB copolymer [57, 58], are all superior to XAD-2, XAD-4, and activated carbon Ambersorb XE340 with respect to phenol and nitrophenol removal. Material [20%-VBC(F.C.)] retains 2,4-dichlorophenol much stronger than unsubstituted phenol, and, therefore, the presence of phenol in the solution does not impact the retention of the chlorinated phenol. On the contrary, the sorption of phenol from the binary solution depends heavily on the concentration of 2,4-dichlorophenol.

Zhang et al. [59] found an interesting synergetic effect accompanying sorption of phenol and aniline from a mixed solution onto XAD-4 and hypercrosslinked NDA-100 resin (China). At high concentrations the sorption of both components from mixed solutions is larger than that predicted (by expanded Langmuir model) from sorption isotherms for the individual components. They suggest the synergetic effect to be conditioned upon acid/base-type interactions between phenol and aniline. Another pair of sorbates, 1-naphthol and 1-naphthylamine, demonstrates similar synergetic effects in adsorption onto NDA-100 and XAD-4 [60]. In both cases the synergy proved to be more pronounced for XAD-4, most probably due to larger pore dimensions in the latter sorbent. Recently, Valderrama et al. [61] also documented the synergetic effect on adsorbing phenol and aniline from aqueous solutions onto activated carbon F400 and hypercrosslinked resin MN-200. They observed both a larger uptake of the solutes from mixed solutions and also a faster extraction of phenol and aniline compared to their extraction from individual solutions. They explained the synergism of sorption by "cooperative effect between solutes on the surface of both adsorbents," which still cannot explain the acceleration of the sorption process. Obviously, formation of the acid/base- or hydrogen bond-type dimer between the two components doubles the effective molecular weight of the sorbate, thus shifting the phase equilibrium toward larger and faster adsorption.

Meng et al. [62] compared the ability of two hypercrosslinked resins to extract phenol from aqueous solutions. The resins possessed similar physical structures but carried different amounts of oxygen-containing groups (measured by Boehm titration [63, 64]). The experimental data show that the sorption of phenol rises with increasing total surface concentration of oxygen-containing groups; however, the latter do not influence the rate of sorption.

Sorption of very polar resorcinol and catechol on hypercrosslinked NDA-100 is relatively low, about 1 mmol/g, but it may be increased to 1.4 mmol/g by introduction of 3 mmol/g tertiary amino groups [65].

Finally, it should be noted that porous methyl methacrylate/DVB copolymers that exhibit some features of hypercrosslinked structures, but are directly wetted with water, retain no more than 50–60 mg/g phenol (at 250 mg/L equilibrium concentration), which is many times less than the adsorption capacity of MN-200 at the same equilibrium phenol concentration [66].

Regarding regeneration of hypercrosslinked sorbents loaded with phenol, various methods may be suggested. In spite of a relatively low phenol affinity to polystyrene, the increase in temperature from 20 to 80°C results in a twofold reduction of sorbent dynamic capacity, only. Presumably, steam stripping could remove phenol more efficiently. Of course, washing with a 2–3% sodium hydroxide solution at room temperature seems to be the simplest regeneration method. Besides, washing of the loaded sorbents with a few volumes of organic solvents, such as ethanol or acetone, results in complete recovery of all phenols.

8. REMOVAL OF CHLOROFORM FROM INDUSTRIAL WASTEWATERS

Since 1998 a chemical plant in Spain has used MN-200 for nonstop recovery of chloroform from industrial wastewaters. Figure 11.9 shows the functioning unit, which was designed and built specially for using the Macronet resin. High sorption capacity of MN-200, up to 100 g of chloroform per liter of the resin, together with a simple and complete regeneration by steam at a temperature as low as 100°C, allow recovery of up to one ton of chloroform using one adsorption column of 10 m^3 capacity. For the first year of operation there was approximately $1,170,000 saving in solvent use and waste management costs [67].

Figure 11.9 Operational chloroform removal plant. *(Reprinted from [58] with permission of Society of Chemical Industry.)*

9. SORPTION OF PESTICIDES

Pesticides are considered to be dangerous environmental pollutants. The US Environmental Protection Agency and European Union have established the maximum allowed concentration of pesticides in drinking water at the level of $0.1\,\mu g/L$ for an individual pesticide and $0.5\,\mu g/L$ for the sum of pesticides, including their main metabolites. It is clear, if the concentration of pesticides exceeds the above limit, water intended for human consumption has to be purified. Hypercrosslinked sorbents offer such an opportunity.

Isoproturon, atrazine, diuron, chlorotoluron, and simazine absorb well into MN-200. To the largest extent, this sorbent extracts isoproturon while the smallest loading was observed for simazine [68]. Nevertheless, the capacity of MN-200 even for this pesticide is pretty high. From the Freundlich adsorption isotherm obtained under static conditions, it follows that the loading capacity for simazine reaches 440 μmol/g (approximately 88 mg/g) at the equilibrium concentration of 1 μmol/L (0.2 mg/L) (note that the validity of the Freundlich equation as applied to hypercrosslinked sorbents is still debatable).

Under dynamic conditions the initial breakthrough of simazine and isotoluron to the legal level of 0.1 μg/L occurs at 100,000 and 180,000 BV, respectively, when the mixed solution of the four afore-mentioned pesticides in ultrapure water spiked with 20 μg/L of each component was pumped through a mini–column packed with 0.65 mL MN-200 beads having 53–70 μm diameter (Fig. 11.10) [69]. After per-colating 500,000 BV of the feed solution, the loading capacity for MN-200 sorbing simazine, chlorotoluron, atrazine, diuron, and isopro-turon was 15.7, 32.7, 39.7, 44.8, and 46.2 mg/g, respectively. In a frontal chromatographic process the more strongly retained isoproturon and diuron displace less strongly bound pesticides as they pass down the column, thus raising the concentration of atrazine and chlorotoluron in effluent up to 28 and 25 μg/L, respectively. After initial breakthrough the concentration of pesticides in the filtrate rises slowly, so that 200,000 bed volumes of the solution can be treated before the concentration of atrazine and simazine reaches the allowed tolerance limit of 0.1 μg/L.

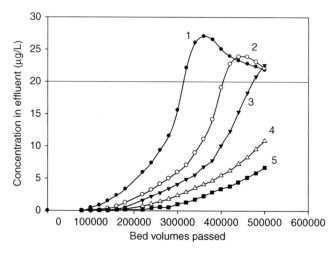

Figure 11.10 Breakthrough curves for MN-200 sorbing (1) simazine, (2) chlorotoluron, (3) atrazine, (4) diuron, and (5) isoproturon from 20 μg/L solutions in ultrapure water. *(After [69].)*

In contrast to the above triazine pesticides, sorption of highly water-soluble herbicides, namely benazolin, bentazone, imazapyr, and triclopyr, onto hydrophobic MN-200 is small [70]. When passing the feed solution of 20 µg/L of each component through a 0.70 mL mini-column with MN-200, the breakthrough of the herbicides occurs almost instantly, which is reflected in the low equilibrium adsorption capacity of only 0.48, 0.42, 0.26, and 1.28 mg/g for benazolin, bentazone, imazapyr, and triclopyr, respectively.

As compared to MN-200, activated carbon Chemviron F-400 exhibits better sorption for all the above pesticides. The breakthrough of atrazine and isoproturon at the legal level occurs after passing more than 800,000 BV of the pesticide solutions in ultrapure water through the above mini-column. However, if fulvic acid is present in the feed at the level of 10 µg/L, the breakthrough of the pesticides occurs sooner on both the sorbents. For MN-200 0.1 µg/L concentration is reached already at 75,000 BV for simazine and between 100,000 and 110,000 BV for other pesticides. The influence of fulvic acid is even more pronounced for F-400: in this case the 0.1 µg/L breakthrough of atrazine and isoproturon decreases to 160,000 BV.

The aforementioned herbicides dissolved in ultrapure water break through the column with F-400 in the following order: imazapyr (80,000 BV), bentazone (105,000 BV), benazolin (140,000 BV), and triclopyr (160,000 BV). The adsorption capacity of the carbon for these herbicides reached in the column experiment was 15.22, 22.84, 34.20, and 38.01 mg/g, respectively. However, the presence of fulvic acid in the feed solution reduces drastically the adsorption capacity of F-400 and the breakthrough of all the herbicides occurs at once.

Apart from the fact that in the presence of the natural organic matter in water MN-200 loses only 10% of its capacity for the above triazine pesticides, the regeneration of MN-200 (exhausted with simazine, chlorotoluron, isoproturon, atrazine, and diuron) proceeds easily by simple washing of the polymer with 3–5 bed volumes of acetone, methanol, ethanol, or 1-propanol [71]. Complete removal of atrazine, benazolin, bentazone, imazapyr, and triclopyr requires about 8 BV of aqueous ethanol adjusted to pH 12 at 50°C. Regeneration of carbon F-400 presents great problems; 200 BV of the above most efficient reagents comprehensively removes only benazolin and triclopyr, while the recovery of other herbicides remains below 50% [70].

Chang et al. [72] studied the adsorption of a more hazardous pesticide, methomyl, that is used worldwide in the cultivation of vegetables, corn, tobacco, cotton, etc. As was found, neutral MN-150 takes up approximately 35 mg/g methomyl at the equilibrium concentration of 12 mg/L in pure water. Importantly, in the range of low equilibrium concentrations (up to 6 mg/L), the loading capacity of MN-150 does not depend upon the presence in solution of humic acid. The sorption of methomyl onto sulfonated resin MN-500 is not of practical significance.

10. EXTRACTION OF CAFFEINE FROM COFFEE BEANS

Dawson-Ekeland and Stringfield [73] suggested using a hypercrosslinked sorbent for the decaffeination of aqueous extract from coffee green beans. The sorbent was prepared by post-crosslinking chloromethylated gel-type styrene–DVB copolymer via Friedel–Crafts reaction, followed by treating residual chloromethyl groups with dimethylamine. A comparative study of the synthesized polymer with Amberlite XAD-4 and phenol-formaldehyde sulfonate, Duolite S761, traditionally used for decaffeination, has demonstrated an obvious advantage of the hypercrosslinked adsorbing material. By equilibrating 1.5 g sorbent (in anticipation of dry material) with 75 mL extract containing 7–10 mg/mL caffeine and 65–95 mg/mL chlorogenic acids at 60°C within 16 h under static conditions, the authors observed a removal of 25–35% caffeine, accompanied by a small loss of chlorogenic acids (Table 11.5). Such treatment of the coffee bean extract was repeated

Table 11.5   Caffeine and chlorogenic acid removal from coffee beans extract

Sorbent		Cycle number				
		1	2	3	4	5
				Caffeine		
Initial concentration (mg/mL)		7.02	6.8	8.53	9.13	7.02
% removal	Hypercrosslinked resin	34	38.1	25.2	28.5	35
	XAD-4	17.5	29.4	12.2	22.9	17.7
	Duolite S761	7.7	19.1	10.1	15.2	13.7
				Chlorogenic acid		
Initial concentration (mg/mL)		72.3	75.4	91.3	93.4	67.4
% removal	Hypercrosslinked resin	14.2	15.2	12.2	11.4	16.2
	XAD-4	5.3	24.2	21.1	12.9	19.9
	Duolite S761	18.6	51.2	10.3	16.8	21.9

Source: After [73].

several times, with regeneration of the sorbents between the sorption cycles by shaking with 9:1 (v/v) methanol–water mixture. As can be seen, the adsorption capacity of the hypercrosslinked sorbent remains pretty high from cycle to cycle.

Unlike other decaffeination processes, a product treated with the hypercrosslinked sorbent was rich in color and very flavorful. The ion exchanger and the neutral polystyrene macroporous adsorbent failed to produce similar good results.

11. DECOLORIZING OF AQUEOUS SUGAR SYRUPS

Decolorizing of sugar syrup is the most important stage in sugar refining. For this purpose, activated carbons or ion-exchange resins have usually been employed [74]. Both types of the adsorbents exhibit serious disadvantages. Thermal regeneration of activated carbons, which is a power-consuming process, significantly raises the cost of the target product, while insufficient adsorption capacity with respect to colored bodies in sugar and incomplete regeneration with chemicals are the negative characteristics of ion-exchange resins. The hypercrosslinked sorbents were found to be more beneficial for removing colored bodies from sugar syrup. The following example described in [75] illustrates the superior properties of hypercrosslinked adsorbent resin in comparison with those of the conventional anion-exchange resin Dowex MSA-1. The hypercrosslinked sorbent was obtained by traditional post-crosslinking of swollen chloromethylated macroporous styrene–DVB copolymer via Friedel–Crafts reaction; the residual chloromethyl groups were reacted with trimethylamine. High-fructose corn syrup was pumped through a column loaded with 5 mL resin in hydroxide form at 50°C and a flow rate of 2 mL/min. The increase in percent of color leakage is summarized in Table 11.6.

Regeneration was conducted by rinsing the resin with a fivefold-diluted mixture of 2% NaOH, 1% NaCl, and 0.5% Na_2SO_4 (this mixture is a representative of regenerant effluent from anion-exchange resins employed for demineralization of the sugar syrup) with a flow rate of about 1.2 BV/h. The total volume of the solution passed was 15 BV. The obtained results are given in Table 11.7 as the ratio of total amount of color bodies in the regenerant solution to that of bodies adsorbed from sugar syrup. One can see that 3–4 bed volumes of the above solution are sufficient to remove color bodies almost completely.

Table 11.6   Decolorizing of high-fructose corn syrup under dynamic conditions

Sorbent	Breakthrough of colored bodies (%)					
	1 h	2 h	3 h	4 h	5 h	6 h
Dowex MSA-1	8.7	16.0	19.8	25.0	29.2	–
Hypercrosslinked resin	2.5	4.8	7.5	12.4	17.7	21.5

Source: After [75].

Table 11.7   Regeneration of the hypercrosslinked sorbent loaded with sugar color bodies

Bed volume of regenerant	Percent of colored bodies desorbed
1.2	41.2
2.4	88.2
3.6	91.2
6.0	93.4
9.6	96.5

Source: After [75].

12. REMOVAL OF BITTERNESS FROM CITRUS JUICE

Norman et al. [76] described the application of a hypercrosslinked biporous sorbent for removing bitter components from citrus juices. The basic representatives of bitter compounds are flavanoids, mostly naringin and limonin, which are initially present in fresh juices or result from non-bitter precursors, such as limonoic α-ring lactone, in acidic medium on processing, concentration, heat pasteurization, or storage of juices. The major part of the population detects the bitter taste when the concentration of limonin exceeds 6 ppm, whereas 20% of the population can detect levels down to 2 ppm. That is why the removal of bitterness is a problem of great importance. Weak basic anion-exchange resins used to extract bitterness, though absorbing the latter components, increase the juice pH markedly. It leads to denaturizing proteins, which, in its turn, has a negative effect on the taste of juices. MN-100 and MN-150 do remove limonin, but their sorption capacity is smaller than that of the "reference resin," Amberlite XAD-16. On the contrary, the hypercrosslinked sorbent MN-102 having larger micropores and a more flexible matrix than MN-100 [67] takes up the bitter compounds much better. As Table 11.8 indicates, the breakthrough of limonin at the level close to allowed

Table 11.8   MN-102 in orange juice de-bittering (12 m^3/h flow rate)

Time (h)	Brix[a]	Limonin (ppm)	Polyphenols (ppm)	Removal efficiency (%)	
				Limonin	Polyphenols
0	6.54	38.27	38.24	–	–
1	5.97	0.09	0.240	99.88	99.37
2	6.08	0.19	0.512	99.63	98.66
4	6.30	0.80	1.896	98.63	95.05
6	6.23	1.93	2.718	95.96	92.90
7	6.29	2.38	2.836	94.37	92.59

[a] The units representing percentage by weight of sugar in the solution.
Source: After [67].

3 ppm occurs only after 7 h of percolating orange juice through the column with MN-102. However, together with limonin, MN-102 extracts polyphenols representing useful natural antioxidative agents.

The regeneration of exhausted hypercrosslinked sorbents does not require the use of concentrated acids and alkalis. As was said in [76], a simple washing with 1 N sodium hydroxide and water is sufficient for removing the absorbed components. The sorbent maintains constant adsorption capacity, at least within 30 cycles of adsorption–regeneration.

13. SORPTION OF CEPHALOSPORIN C

It is hardly necessary to mention the important role antibiotics play in health care. Cephalosporin C belongs to the basic class of β-lactam-type antibiotics produced by microbiological methods on an industrial scale. The isolation of cephalosporin C from fermentation liquids is a multistage procedure, the last stage of which involves adsorption of the antibiotic onto such neutral sorbents as Amberlite XAD-2, XAD-4, XAD-16, or XAD-180. Therefore, it was interesting to test the commercial hypercrosslinked materials, namely unfunctionalized MN-200, functionalized MN-100, MN-150 (both are weak basic anion exchangers), MN-400 (strong basic anion-exchange resin), and sulfonate MN-500 for final polishing of cephalosporin C solutions [77]. As can be seen from Fig. 11.11, MN-200 exhibits the highest sorption capacity, compared to MN-100, MN-150, MN-400, MN-500, and XAD-4. In column experiments MN-200 showed excellent capacity to breakthrough: one column bed treats 200 bed volumes of 1 g/L solution

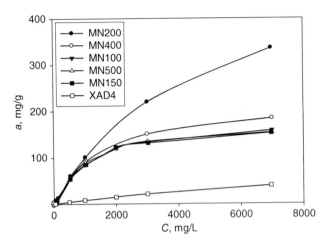

Figure 11.11 Sorption isotherms for cephalosporin C onto commercial hypercrosslinked sorbents at 25°C. *(After [77].)*

when the latter passes with a flow rate of 4 BV/h. The washing of exhausted sorbent with 3.5 BV of methanol results in complete recovery of cephalosporin C and obtaining of highly concentrated solution, 50 g/L, of the antibiotic.

14. SORPTION OF MISCELLANEOUS ORGANIC COMPOUNDS

Many other organic compounds readily adsorb onto hypercrosslinked polystyrene materials. For instance, CHA-111 ($S_{BET} = 1646\,m^2/g$) sorbent produced by China [78] takes up pretty well, 50–100 mg/g, such aromatic acidic substances as benzoic, o-phthalic, benzenesulfonic, and 2-naphthalenesulfonic acids at pH values lower than acids' pK_a. Sorption capacity of CHA-111 markedly rises after introducing 1.53 mmol/g tertiary amino groups by reacting residual chloromethyl functions with dimethylamine. In this case, in addition to dispersion interactions dominating in adsorption on the neutral resin, the interaction between the nitrogen free electron pair with acid protons contributes to enhance sorbent loading.

Jeřabek et al. [79] reported the adsorption of furfural onto the hypercrosslinked sorbents prepared by post-crosslinking both chloromethylated gel-type styrene–5% DVB copolymer and chloromethylated macroporous copolymer of styrene with 8% DVB via Friedel–Crafts reaction. The authors corroborated our early observations [80] that, on the whole, the post-crosslinking does not change the morphology of initial polymers, generating, nevertheless, a large amount of micropores which are responsible for the appearance of high specific surface area (940 and 720 m^2/g for the above samples, respectively). Sorption of furfural onto the biporous hypercrosslinked sorbent was found to be lower than that on the macroporous adsorbent XAD-4 in the entire range of the sorbate concentrations used. Adsorption of furfural on the microporous hypercrosslinked sorbent based on the gel copolymer markedly exceeded that on XAD-4 (having a surface area of the same order of magnitude) only at the equilibrium concentration less than 2%. It seems that for preparing hypercrosslinked samples authors used initial copolymers that contained too much DVB. They swell insufficiently during the post-crosslinking reaction resulting in a material with low permeability. Commercial hypercrosslinked sorbents are expected to take up furfural much better.

A comparison of Chinese hypercrosslinked sorbents, non-functionalized NG-100 and NG-99 containing tertiary amine groups, with XAD-4 in terms of their ability to retain phenylhydrazone derivatives showed that the capacity of XAD-4 yields by 20–70% to that of both NG-99 and NG-100 [81], although the specific surface area of the hypercrosslinked sorbents is by no means higher. The authors attributed this difference to uniform micropore structure of NG-99 and NG-100 resins, their partial polarity, enhanced π–π interactions, and better compatibility of sorbate molecules with the sorbent matrices.

A hypercrosslinked material derived from macroporous chloromethylated poly-DVB by additional bridging with isocyanuric acid well absorbs tannin from aqueous solutions [82]. Experimental isotherms are consistent with the empirical Freundlich equation, which is thought to be a sign of an energetically heterogeneous surface. It was suggested that multiple hydrogen bonds, π–π interactions, and hydrophobic interactions are responsible for the retention of tannin on the sorbent prepared.

In the search for sorbents that effectively remove polyaromatic hydrocarbons (PAH), ranging from two to four condensed rings, from aqueous solutions, Valderrama et al. [83, 84] studied the kinetics of sorption for naphthalene, fluorene, anthracene, acenaphthene, pyrene, and fluoranthene onto MN-200 and a granulated activated carbon. The better kinetic properties of MN-200 are transduced in lower hydraulic residence time and consequently in lower barrier thickness. A simulation of the barrier thickness to treat a PAH-polluted flow showed that 1.0 m of sorption media is enough even for high groundwater fluxes such as $2 \, m^3/m^2/day$.

Adsorptive immobilization of inulase *Aspergillus awamori* on Styrosorb 1BP or MN-500 sulfonate from aqueous solution at pH 4.7 results in introducing about 25 mg/g enzyme in each sorbent. At that, the enzyme retains 80–97% of its initial activity in disintegrating inulin to fructose [85]. Early on, the best results were achieved with a mixture of starch and β-cyclodextrin, which took up, however, only 4.8 mg/g of inulase with retention of only 74.4% its activity. Similarly, by treating Styrosorb 1BP with an aqueous solution of glucoamylase, it was possible to load into the sorbent approximately 10 mg/g enzyme [86]. The thus immobilized glucoamylase is more resistant to the change in pH and more thermostable as compared to its native form. Therefore, one may expect better performance of the immobilized enzyme in the hydrolysis of starch, which is widely used in the industrial production of ethanol.

It should also be mentioned that microporous and biporous Styrosorbs based on linear polystyrene adsorb papaverine [87], pectin (12.6 kDa molecular weight) [88], and phosphatidylcholine [89] in amounts of 30–40 mg/g and diprazine [90] and aminosine [91] in larger amounts, up to 120–140 mg/g.

In order to impart hydrophilicity to neutral hydrophobic hypercrosslinked adsorbing resin CHA-101 (also obtained by post-crosslinking chloromethylated styrene–DVB copolymer), the latter was subjected to heating at 100°C within 12 h in nitrobenzene in the presence of FeCl3 (6 g per 30 g of sorbent beads) [92]. The resulting resin, NDA-702, was stated to be directly wetted by water. A comparative study of dimethyl phthalate adsorption at 10°C onto NDA-702, XAD-4, and AC-750 activated carbon showed that the sorbents take 500, 400, and 300 mg/g of the ester, respectively. The loading capacity of the sorbents decreases by 50–100 mg/g with temperature rising to 40°C.

The hypercrosslinked sorbent NDA-150 prepared according to a similar protocol of treating initial material with nitrobenzene and FeCl3 was tested for adsorption of 1-naphthylamine [93]. On a direct contact of dry NDA-150 beads with the sorbate aqueous solution, this sorbent extracts more than 4 mmol/g of the aromatic amine, while the pre-wetted XAD-4 absorbs only 3 mmol/g of 1-naphthylamine. Under dynamic conditions the breakthrough capacity and total capacity were found to be 1.49 and 1.82 mmol per cm^3 of the NDA-150 resin (after percolating 100 and 350 bed volumes of diluted feed solution), respectively. A 100% recovery of

1-naphthylamine was achieved by eluting the exhausted NDA-150 with ethanol at 40°C.

Finally, the hypercrosslinked polystyrene adsorbing materials derived from low-crosslinked styrene–DVB copolymers and MCDE or chloromethyl ethyl ether [94] take up many volatile organic solvents (toluene, chlorobenzene, carbon tetrachloride, etc.) both from aqueous solutions and vapor phase. The sorption capacity of the in-house made polymer samples exceeds that of the commercial hypercrosslinked sorbent XUS 43493 (Dow Chemical) or Amberlite XAD-4 [95]. Interestingly, the sorption of the organic solvents from saturated aqueous solutions was observed to be somewhat higher than that from the vapor phase. This finding was attributed to the wetting procedure, that is, to the sample swelling, for instance, in sorbed toluene, especially at high concentrations of the latter in water. Although the polymeric sorbents tested in the studies [94, 95] do swell in the chosen solvents (the volume of solvents absorbed is well above the pore volume of the dry materials), the above-expressed supposition does not provide any explanation for the different sorption from aqueous and vapor phases. Meanwhile, as far back as 1903, Shroeder [96] described a 25-fold smaller equilibrium swelling of gelatin gel in water vapors compared to that in liquid water. Over the last one hundred years this phenomenon was observed for various polymers and even named "Schroeder's paradox," but no comprehensive elucidation of this phenomenon exists so far.

15. HYPERCROSSLINKED SORBENTS VERSUS AMBERLITE XAD-4

Almost all of the works discussed in this chapter as well as many other publications [97–106] describe the adsorptive properties of hyper-crosslinked resins in comparison with those of popular Amberlite XAD-4. Until recently, XAD-4 was the best sorbent for extracting various organic compounds from aqueous solutions. As Kunin [52] reported, especially good results were achieved by using XAD-4 for removal of phenols from wastewaters as well as for purification of vitamins and antibiotics produced by microbiological fermentation. All publications accentuate the much higher potential of hypercrosslinked adsorbing materials, although the sorbents under comparison have a surface area of the same order of magnitude and are similar in their chemical nature. In this regard, it is worthwhile to discuss more elaborately the basic reasons for this difference.

XAD-4 represents a macroporous resin obtained by free radical polymerization of technical-grade DVB in the presence of porogens. Judging by the high specific surface area of XAD-4, 700–900 m^2/g, and its ability to swell a little not only in toluene but also in methanol (see Chapter 9, Section 3), at least one of the components in the porogen is a thermodynamically good solvent, for example, toluene. The latter solvent reduces the density of the polymeric phase and generates free space in the final rigid DVB network, stimulating formation of micropores in this phase. In addition to macropores that appear on aggregation of primary crosslinked nodules, the fine pores strongly contribute to the enhanced surface area of XAD-4. It is quite possible that, due to these micropores, both XAD-4 and typical hypercrosslinked polystyrene materials take up similar amounts of small nitrogen molecules, thus demonstrating comparable specific surface area values for the dry samples. However, it is easy to see the principal difference between XAD-4 and hypercrosslinked polystyrene when the adsorption of slightly larger organic compounds from aqueous solutions is the case in point. XAD-4 does not swell in water and, therefore, the adsorption of organic molecules is restricted to the surface of its mesopores of up to 100–150 Å in diameter [106, 107], with the majority of micropores remaining inaccessible to these molecules. (Still, the adsorption capacity of XAD-4 has always been larger than that of a typical macroporous XAD-2 that is obtained in the presence of a precipitant and contains few micropores.) On the contrary, hypercrosslinked polystyrene, being manufactured in the presence of solvating porogens, swells in water. This property undoubtedly enhances the accessibility of the entire network to the molecules of organic sorbates. Also, it is quite possible that the size of nanopores in the hypercrosslinked polystyrene is larger than the size of micropores in XAD-4. (Note here that reliable methods for measuring dimensions of micropores in polymeric sorbents still remain to be developed.)

On the other hand, really large organic molecules may prove to be totally excluded from microporous hypercrosslinked polystyrene sorbents if they were prepared from gel-type copolymer precursors. In these cases, only biporous sorbents can compete with the macroporous Amberlite XAD-4.

16. SORPTION OF INORGANIC CATIONS

Separation of cations and anions is commonly brought about by means of ion-exchange resins. These resins, mostly based on styrene–DVB copolymers, contain chemically bound functional groups with negative or positive charge, which can exchange their own protons or hydroxyls for metal counterions or anions, respectively, from a surrounding solution. Another efficient and selective means of separating metal ions capitalizes on their ability to form complexes with chelating functional groups. Not long ago Bicak and Sherrington [108] demonstrated an interesting example of extracting mercury ions from aqueous solutions with crosslinked poly (acrylamide). The amide functional groups of this polymer were found to readily form complexes with Hg^{2+}, the capacity of the polymer being extremely high, up to 1.5 g of mercury per gram of the dry resin. The uptake of mercury is reversible; washing with hot acetic acid permits quantitative stripping of Hg^{2+} ions. Since the poly(acrylamide) does not absorb Ni^{2+}, Cu^{2+}, Co^{2+}, Cd^{2+}, Zn^{2+}, Fe^{2+}, and Fe^{3+}, this polymer presents the unique example of a resin truly selective toward mercury ions.

Unmodified hypercrosslinked polystyrene contains no specially introduced polar functional groups, although many publications [109, 54] report the presence of hydroxyl and carbonyl (ketone) groups in these sorbents. Still, the amount of oxygen-containing groups is quite small, less than 1.5 mmol/g, since the value of 2.5% oxygen calculated from the elemental analysis of MN-200 sample as $O\% = 100 - (C\% + H\% + Cl\%)$ also includes noticeable amounts of adsorbed oxygen, carbon dioxide, water, as well as residues of the catalyst. Besides, incomplete combustion of the polymer that is prone to carbonization can easily contribute to the overestimation of the oxygen content.

In any case, the oxygen-containing groups cannot account for the retention of many ions of Hg^{2+} (Fig. 11.12) [110], which is characteristic of neutral hypercrosslinked polystyrene sorbents. The uptake of mercuric salts from weakly acidic aqueous solutions by the unfunctionalized materials proceeds very slowly. For Styrosorb 2 final equilibrium is established not earlier than within 8 days, leading to the incorporation of up to 200 mg of mercury per gram of the dry polymer. The approach to equilibrium is even slower on MN-200. Introduction of the same amount of mercury into this polymer would take no less then 12 days.

Interestingly, the process of sorption was found to markedly accelerate at elevated temperatures. Whereas at room temperature it has been possible to introduce 50, 110, and 200 mg/g of mercury into Styrosorb 2 within 30, 70, and 140 h, respectively, the sorption capacity of the sorbent at 55°C rises to 360 mg/g in 20 h and 420 mg/g within 60 h. This intensive increase in both the sorption rate and capacity suggests that the uptake of Hg^{2+} ions presents a process differing from a trivial adsorption.

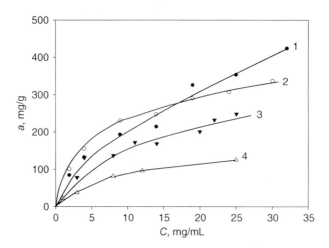

Figure 11.12 Sorption isotherms for mercury acetate from acetate buffer of pH 4 onto (1) Styrosorb 2 with $S = 1500 \, m^2/g$, (2) MN-200, (3) Styrosorb 1 (XDC, 100%), and (4) Styrosorb 2 with $S = 400 \, m^2/g$. *(Reprinted from [110] with permission of Taylor & Francis Group.)*

In order to understand whether or not the mercury uptake is accompanied by a chemical reaction, Styrosorb 2 and MN-200 were subjected to mercuration by treating the polymers with an excess of mercury acetate in glacial acetic acid in the presence of $HClO_4$ as a catalyst. MN-200 incorporates 30% of mercury while it proved to be possible to introduce as much as 55% of the metal into Styrosorb 2. This amount corresponds to 1.2 g of mercury per gram of the initial polymer or one Hg atom per each two phenyl rings. When using nitrobenzene as the solvent for the reaction of mercuration, the degree of substitution on benzene rings of the latter polymer decreases a little, but the polymer still takes up 47% of the metal. It should be emphasized that, under the same conditions, only 8% of mercury can be introduced into a gel-type styrene–2% DVB copolymer highly swollen with nitrobenzene, and about 30% of the metal into Amberlite XAD-2 [111]. Undoubtedly, porous sorbents with a very high content of chemically bonded mercury are of particular interest for the selective sorption of halogen ions or S-containing organic substances, for example, S-containing peptides. More important, however, is the fact that neither 1 M solution of hydrogen chloride nor 0.2 M solution of ethylenediamine tetraacetic acid (EDTA) can break the covalent chemical C–Hg bond, so that the above reagents fail to elute mercury from the mercurated polystyrene resins.

Hypercrosslinked polymers retain absorbed mercury ions rather strongly too. Exposure of the sorbents to a fresh portion of acetate buffer for 2 weeks results in no leaching of mercury ions. However, a complete and fast desorption of the retained mercury ions occurs on treating the sorbents with solutions of either hydrochloric acid or EDTA. Consequently, the sorption of mercury by the hypercrosslinked polystyrene was not accompanied by the formation of chemical covalent C–Hg bonds.

When considering the above results, we arrived at the conclusion that the acetate and nitrate salts of the bivalent mercury form charge transfer or π-complexes with solvent-exposed phenyl rings of polystyrene chains. In the open-framework and permeable structure of hypercrosslinked polystyrene most phenyls should be accessible to the interaction with the metal ions that display an affinity to aromatic systems. Obviously, at least two phenyl groups participate in the formation of the complex with one ion of mercury. If the coordination stoichiometry were 1:1, the complex formation would proceed much quicker, because no significant steric obstacles exist for the diffusion of small inorganic ions into relatively large meshes of the hypercrosslinked network. On the contrary, the formation of a 2:1 complex has to proceed slowly, because it needs time for two phenyl rings to fit into the metal ion coordination sphere, which, in its turn, requires a cooperative conformational rearrangement of the rigid network. The slow approach to the metal coordination equilibrium thus reflects the slow shift of the whole system toward the global minimum of energy of two coupled processes, the formation of coordination bonds and the conformational rearrangement of highly crosslinked polymeric ligand. This situation differs fundamentally from the generally very fast adsorption of organic compounds on the hypercrosslinked polystyrene materials.

Similarly to hypercrosslinked polystyrene, Amberlite XAD-4 is also capable of slow chemisorption of mercury ions in the amount of up to 300 mg/g. The latter are easily removed by hydrochloric acid, but not with fresh acetate buffers, thus pointing to the formation of π-complexes with aromatic rings.

Besides mercury, some other transition metal ions were found to bind to hypercrosslinked polystyrene, though less effectively. One gram of Styrosorb 2 takes up approximately 170 mg of silver(I) from a dilute nitric acid solution and about 80 mg of lead(II) from an acetate buffer (Fig. 11.13). Bismuth(III) ions bind to a somewhat smaller extent, 60 mg/g. Contrary to mercury ions, the sorption of silver, lead, and bismuth ions is reversible, so

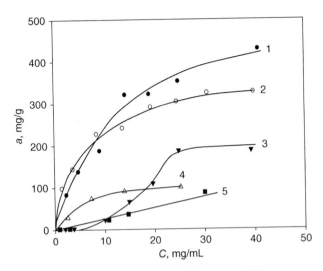

Figure 11.13 Sorption of ions of (1, 2, 4) Hg^{2+}, (3) Ag^{+}, and (5) Pb^{2+} from (1, 2, 4, 5) acetate buffer and (3) nitric acid solutions at pH 4. Surface area of Styrosorb 2 samples: (1, 3, 5) 1500, (2) 1200, and (4) 400 m^2/g. *(Reprinted from [110] with permission of Taylor & Francis Group.)*

that they can be easily removed by washing the polymer with fresh portions of the buffer or a solution of nitric acid.

Sorption of Cd^{2+}, Cu^{2+}, Ni^{2+}, Zn^{2+}, Co^{2+}, and Fe^{3+} was found to be small from solutions of corresponding nitrates at room temperature. These metal ions definitely display higher affinity to water molecules than to π-electrons of the aromatic rings.

CHAPTER 12

Nanoporous Adsorbing Materials in Ion Size-Exclusion Chromatography

The many years' experience in the preparation and examination of hypercrosslinked polystyrene materials resulted in the development of industrial manufacturing technologies for shape-resistant sorbents, which maintain their rigid nanoporous structures both in gaseous and aqueous media. The maximum of the rather narrow pore size distribution can be shifted down to about 2 nm, the range characteristic of zeolites and activated carbons. Indeed, Purolite International started manufacturing such hypercrosslinked polystyrene resins in their new NanoNet series. In looking for some specific application areas of these first polymeric nano-porous adsorbents, their adsorption and separation capacities were examined with respect to the smallest possible species, such as gas molecules and ions of mineral electrolytes. The latter class of compounds presents an important industrial challenge, in particular, with the advances in hydrometallurgy.

The average pore size of NanoNet polymers appears to be comparable with diameters of hydrated mineral ions, varying in the range of 0.6–1.0 nm. Therefore, larger ions can be expected to experience steric hindrance in passing into or through the smallest pores of the sorbent, which could provide a base for their separation from smaller ions according to the principles of size-exclusion chromatography (SEC). Previously, processing of mineral electrolytes was thought to be only the subject of ion exchange and extraction technologies. Nevertheless, we have been able to demonstrate the unexpected ability of hypercrosslinked polystyrene to efficiently separate mineral salts, acids, and bases, in spite of the absence of any polar, complexing, or ion exchanging groups in the chemical structure of the aromatic and hydrophobic sorbent material. Owing to its simplicity and extremely high productivity, the developed separation process, called ion size exclusion (ISE), representing a variation of SEC, proved to be of great interest to hydrometallurgy and other large-scale processing

Comprehensive Analytical Chemistry, Volume 56
ISSN 0166-526X, DOI 10.1016/S0166-526X(10)56012-5

techniques for mineral electrolytes. The separation was found to capitalize on the size exclusion phenomena, and the impassability of certain channels of the open-network structure of the nanoporous hypercrosslinked polystyrene to larger hydrated ions.

Let us first consider the accumulated literature concerning separation of mineral ions by chromatographic techniques other than stoichiometric ion exchange and complexation.

1. DEVELOPMENT OF THE CHROMATOGRAPHIC SEPARATIONS OF MINERAL ELECTROLYTES UNDER CONDITIONS EXCLUDING ION EXCHANGE; RELATED WORK BY OTHERS

As early as 1950, A.B. Davankov (not the present author) et al. applied for and, 5 years later, received a USSR patent [112] for "A method of removing salts from aqueous solutions by using amphoteric ion-exchange resins." For the first time, the amphoteric polymeric material incorporated both cationic and anionic functional groups and, therefore, simultaneously retained both anions and cations of a dissolved salt. Along these lines, a real success in the removal and/or separation of electrolytes by the "ion retardation" process was achieved with amphoteric "snake-cage poly-electrolytes" suggested in 1957 by Hatch et al. [113] and defined as a "crosslinked polymer system containing physically trapped linear polymer." Soon Dow Chemical Co. commercialized these materials under the trade name "Retardion" [114]. Retardion-550WQ2 (water content 41.6%), a typical snake-cage amphoteric polyelectrolyte, was made by polymerizing (*ar*-vinylbenzyl)trimethylammonium chloride inside a sulfonated cation exchanger Dowex-50Wx2 and had strong basic and strong acidic functional groups. On a 100 mL column packed with this resin, a 20 mL sample 2.0 N in NH_4NO_3 and 1.6 N in HNO_3 was almost completely separated into its constituents [115]. Similarly, 1.98 N $FeCl_2$ was partially separated from 3.15 N HCl [115].

In 1958 in a theoretical study on the activity coefficients of electrolytes in the resin phase, particularly in anion exchangers, Nelson and Kraus [116] arrived at the conclusion that "because of the relatively low activity coefficients of HCl in the examined anion exchanger Dowex-1x10, separation of HCl from concentrated halide solutions is possible." Notably, they report that in the presence of salts, the HCl form of that anion exchanger *retains* HCl, instead of rejecting it from the resin phase.

Thus, the breakthrough curves of 0.1 N HCl shift from *ca* 1.5 column volumes in 5 N LiCl to more than 8 column volumes in the concentrated (16 N) LiCl solution. A similar strong retention of HCl (because of the "reduced activity coefficient of HCl") was also characteristic of the 10 N solution of $MgCl_2$. Yet, retention of HCl in amphoteric resins and anion exchangers contradicts the concept of "ion exclusion" according to which *all* strong electrolytes should be effectively excluded from absorption into ion-exchange resins, because of the Donnan equilibrium effect [117, 118].

Five years later Hatch and Dillon rediscovered the fact that conventional ion-exchange resins efficiently separate acids from their salts under conditions that exclude normal ion exchange. The strong basic anion-exchange resin Dowex-1x8 (8% divinylbenzene (DVB), 40% water content) was found to function especially well and the investigators decided to introduce "acid retardation" as a new term. "Such separations can be defined as 'acid retardation' separations, since they are based on a preferential absorption of strong acids, which causes the movement of the acid on the bed to be retarded – i.e., slowed down – relative to the movement of the salt" [115]. Acid retardation, like ion retardation, can be done at high flow rates, especially at elevated temperatures. These processes have been optimized and since 1976 widely exploited by Eco-Tec Canada [119, 120], as well as by others [121], on the industrial scale.

Concerning the mechanism of acid retardation, Hatch and Dillon [115] note that various ideas or their combinations may be offered to explain the absorption of strong acids on anion-exchange resins:

(i)   Electrolyte interactions, including "salting effects," the HCl being salted out into the resin phase

(ii) Interaction of protons of the strong acid with the benzene rings of the resin matrix

(iii) Association of strong acids to form ion pairs and non-ionized molecules in the resin phase, due to the lowered dielectric constant of that phase

(iv) Possibly an entropy-increasing or energy-lowering effect of excess protons on the microstructure of the water inside the resin matrix

However, having examined many experimental data, Hatch and Dillon [115] concluded that "None of the ideas are proved by the data currently available. ... At this time, any of the theoretical explanations of the acid absorption certainly are far from proved."

During the 50 years that have passed since the publication of these classical studies, numerous research groups examined the salt and, in particular, acid retardation processes in ion exchanger, but no new ideas have been suggested for the explanation of the non-trivial phenomenon of electrolytes' discrimination. Interestingly, a more or less adequate mathematical description of experimental data is possible both in terms of self-association of acids in the homogeneous resin phase and association of acid molecules with the functional groups of the resin [122] and in frames of a two-phase model of the resin bead [123]. However, both approaches are based on the above ideas that have been formulated, considered, and rejected by the pioneers of the method. Indeed, mathematical models may well fit experimental data into equations, but they do not prove the reality of the concept they are based on.

An entirely different area of research presents analytical-scale separations of inorganic compounds on neutral hydrogels, such as Sephadex G-10 to G-50 (epichlorohydrine-crosslinked dextran) or Bio-Gel A (crosslinked agarose). From the very beginning these materials have been designed for gel filtration, later called gel permeation chromatography, so that the size exclusion separation mechanism of polyphosphates, molybdophosphates, silicic acid, and similar inorganic polymers has been early recognized. A detailed review by Yoza [124], "Gel chromatography of inorganic compounds," also deals with additional effects that complicate the SEC separation mechanism of inorganic electrolytes on hydrogels, such as adsorption, ion exclusion, and secondary complexation equilibrium in solution.

With respect to the simplest cations and anions, correlation has been noticed between their retention in the hydrogel column and the radii of the hydrated ions [125–127]: the larger ions emerged first from the column. Collins [128] examined thoroughly the chromatographic behavior of monovalent cations, including NH_4^+ and all halogen anions on Sephadex G-10 (in 0.1 M NaCl solutions, in order to suppress possible ion exchange interactions). The goal of his studies was to reveal the reasons for the selective transportation of K^+ and Rb^+ cations through some biological membrane channels. He concluded that the small monovalent ions of Li^+, Na^+, and F^- flow through the hydrogel with water molecules attached, whereas the larger ions K^+, Rb^+, Cs^+, Cl^-, Br^-, and J^- easily lose their hydration shell and are adsorbed to the nonpolar regions of the gel (irrespective of the aromatic or aliphatic nature of the gel) due to hydrophobic interactions. Therefore, the dehydrated "sticky" ions of K^+, Rb^+, and Tl^+

can "slide" down a narrow greasy hydrophobic channel of the cell membrane, while Cs^+ is too big for the channel and strongly hydrated Li^+ and Na^+ ions have no affinity with the interior of the channel. From his chromatographic experiments, Collins [128, 129] also concluded that anions retain water much stronger than cations having the same ion radius. Interestingly, an opposite opinion also exists, namely that cations have a higher degree of hydration than anions [130].

Computer simulations give an idea about the dimensions of pores that could provide size selectivity for ion transportation. In general, hydrophobic pores with a radius smaller than 3.5–4 Å contain water in the form of vapor only, though at least three water molecules could fit into the cross section of such a pore. A thermodynamically stable aqueous phase can form in pores having a radius of more than 5.5–6 Å. On the walls of such hydrophobic pores layered structures of water molecules must form, with a density higher than the bulk density. In pores around 4–5 Å the situation is unstable and rapid fluctuations from vapor to liquid water take place. These pores are not totally closed, since small water clusters or strings of hydrogen-bonded water molecules may slide rather effortlessly through the "greasy" hydrophobic pore [131]. In pores with a radius of 6 Å or more, liquid water is the only thermodynamically stable phase and all hydrated mineral ions are expected to easily diffuse through such pores. Still, in the range of smaller biological channels, K^+ is expected to selectively migrate through the pores, since its first hydration shell is very flexible and can be lost, whereas the larger channels favor Na^+ and Li^+ having a more stable and structured water surrounding [132]. Nanopores made in artificial membranes have been studied recently and it has been shown that size selectivity in ion transportation can also occur in these simple pores [133, 134].

All these studies on chromatographic size exclusion separations on neutral separation media and selectivity of transportation of ions through membranes were carried out with very dilute electrolyte solutions. Only Rona and Schmuckler [135] examined Dead Sea concentrated brine on a Bio-Gel P-2 (crosslinked polyacrylamide) column and obtained a lithium-enriched fraction free of calcium and magnesium. Bio-Gel P, however, is known to retain cations and probably enters hydrogen-bond interactions between the anions and the amide hydrogen. The elution order of chlorides was thus different from that expected for pure SEC, namely, K^+, Na^+, Li^+, Mg^{2+}, and Ca^{2+}, all, however, emerging before the hold-up (dead) volume of the column.

2. PREPARATIVE SEPARATION OF ELECTROLYTES VIA ION SIZE EXCLUSION ON NEUTRAL NANOPOROUS MATERIALS

Ferapontov et al. [136, 137] were the first to try non-functionalized hypercrosslinked polystyrene, along with ion-exchange resins, as column packing material in separations of rather concentrated mineral electrolyte solutions. They suggested that the neutral polymer "absorbed" electrolytes in different proportions, but in quantities comparable to those absorbed by ion exchanger.

However, very soon we recognized that the observed separation effects are due to size exclusion phenomena, rather than to different affinity of ions to the hypercrosslinked polymeric phase [138–141].

In the intensive studies that followed we have used the most simple and, therefore, the most convincing technique that permitted registering the large separation effects with a precision sufficient for principal phenomenological considerations. The columns used had a volume of about 30 mL and were about 20 cm long. If not stated otherwise, they were slurry-packed with NanoNet® NN-381 (Purolite Int.) hypercrosslinked polystyrene having a bead size of 0.3–0.5 mm and were pre-filled with water (by washing the material with ethanol and then with water). In some earlier experiments, the commercially available Macronet Hypersol resins MN-202 and MN-270, and also experimental microporous carbons prepared by the pyrolysis of Macronets, were examined. Note that residual chloromethyl and/or hydroxymethyl groups are present in these polymeric resins in minimal amounts, <0.4 meq/g, and they do not play any noticeable role in the processing of the investigated concentrated electrolyte solutions.

An aqueous binary mixture of a salt and an acid, or other combinations of electrolytes, was passed through the column at a flow rate of about 1 mL/min ("forward" experiment). The concentration profile of the electrolytes in the effluent (the breakthrough curves) was determined by collecting fractions of about 1.0–1.5 mL and analyzing their aliquots by conventional methods. After the column was equilibrated with the feed solution, the electrolytes were displaced from the column by water ("reverse" experiment) and their concentrations were determined in the fractions collected at the column outlet. This technique corresponds to the chromatographic frontal analysis and generally permits the isolation, in the forward experiment, of fractions containing the more or less pure component with smaller residence time in the column and, in the reverse experiment, the pure second component that resides longer in the column.

The only non-trivial change in the experiment was that the feed solution in the forward experiment was introduced upward from the bottom of the column, while the elution in the reverse experiment was carried out by downward water flow from the top of the column. The aim of this approach was to avoid the undesirable broadening of zone fronts due to eddy diffusion and convection between the rather large polymer beads, caused by a significant difference in the densities of water and the concentrated feed mixtures. Thus, a model solution (4 N $CaCl_2$ + 4 N HCl) often used in our experiments has a density of 1.332 g/cm^3, compared to 0.998 g/cm^3 for water at 20°C. The density of a mixture that is 4 N in both calcium chloride and calcium nitrate is even higher, 1.357 g/cm^3. Even a more dilute mixture such as 1.5 N $CaCl_2$ 1.5 N KCl is definitely heavier than water, 1.191 g/cm^3. As was mentioned, in the elution of the remaining components from the column in the reverse experiment, water was flowing from the top to the bottom. The reversal of the flow direction in this elution step had no adverse effect on the form of the chromatographic zones, because elution from a column that is completely equilibrated with the feed can be carried out in any direction. If both the column loading and elution of electrolytes were carried out in one direction, from the top downward, the leading fronts of the heavy chromatographic zones were observed to be widened. If both steps were carried out in the other direction, column upward, the trailing edges of the zones showed strong tailing. Naturally, the use of much smaller resin particles and higher flow rates, as in high-performance liquid chromatographic columns, would totally eliminate the negative influence of convection and make the direction of the flow fully irrelevant.

The concentration of the electrolytes in each fraction was determined by titration [139, 140]. Generally, the acid or the base was directly titrated in an aliquot of each fraction and then the metal cation or the anion was determined by selective titration. Thus, chloride anions were titrated with a 0.2 N solution of $Hg(AcO)_2$, and Ca(II) cations were determined by complexing with a 0.05 M EDTA solution. In other cases, when selective techniques were not available, the total concentration of the electrolytes in the fraction was established by passing an aliquot of the fraction through a bed of Dowex-50Wx8, hydrogen form, washing the resin with water, and titrating the total electrolyte in the wash with an

NaOH solution. Then the metal ion concentration in the fraction equaled the total electrolyte concentration minus the separately found acid concentration.

By using this simple chromatographic technique, we first registered the breakthrough curve of an aqueous (*ca* 4 N $CaCl_2$) solution from the 44 mL column packed with the beaded NN-381 material (forward experiment). Then, the elution curve of the salt from the column with pure water was also determined (reverse experiment). Analogus breakthrough and elution curves were further obtained for a *ca* 4 N HCl solution. The resulting concentration profiles for each electrolyte in the combined forward and reverse experiments show a typical trapezoidal shape (Fig. 12.1). However, it was impossible to neglect the fact that in both the forward and reverse experiments, $CaCl_2$ emerges from the column with smaller elution volumes than the acid. Still, the latter elutes from the column with a volume not exceeding the total volume of water in the column (the hold-up volume). These facts imply that (i) neither $CaCl_2$ nor HCl is retained by the neutral polymer and (ii) $CaCl_2$ is excluded from the porous space of the polymer to a greater extent than HCl is.

If now a mixed solution of the salt and the acid is sent through the same column, the salt will elute before the acid, allowing us to obtain a fraction

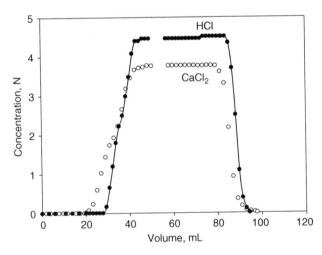

Figure 12.1 Elution profiles for HCl and $CaCl_2$ (measured separately) from the chromatographic column containing NN-381. Experimental conditions: 44 mL column, 0.8 mL/min flow rate. (*After [170].*)

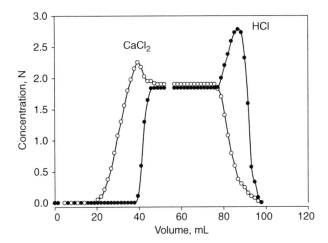

Figure 12.2 Elution profiles for HCl and CaCl₂ (taken as a mixture) from the chromatographic column containing NN-381. Experimental conditions: 44 mL column, 0.8 mL/min flow rate. *(After [170].)*

of pure CaCl$_2$ solution in the forward experiment and a fraction of the acid in the reverse experiment (Fig. 12.2). Notably, the breakthrough curves of the salt and acid diverge now by almost half of the column volume, much more than one could expect from the previous experiment presented in Fig. 12.1. Another unprecedented feature of the frontal chromatography of the mixture is that both the separated CaCl$_2$ and HCl fractions contain the electrolytes in concentrations significantly exceeding their initial concentration in the feed solution. Similar remarkable separations were also obtained by passing rather concentrated solutions of LiCl, KCl, or NaCl in hydrochloric acid solutions.

Several questions arise from these experiments, but the most important question to be answered is why the neutral hypercrosslinked polystyrene discriminates salts and acids.

Due to the local electroneutrality principle, the cations move along the column together with their anions. Since the salts and the acid in these systems have the same common chloride anion, the divergence of the salt and acid zones must result from the different behavior of the proton and the metal cations in the hypercrosslinked sorbent phase. This difference may be due to one of two reasons:

(i)   either polystyrene retains the proton stronger than it retains the alkali or alkaline-earth metal cation, or

(ii) different pore volumes are accessible for the proton and for these cations; that is, they are excluded to different extents from the sorbent phase.

Positive interactions between cationic species, including protons, with aromatic structures comprise an intensively examined and already well-documented phenomenon [142, 143]. In the hypercrosslinked polystyrene these interactions may well be enhanced by a possible presence of condensed aromatic systems. As was shown in Chapter 6, Section 4.4, anthracene-type structures may easily be formed by the condensation of two chloromethylated styrene repeating units, followed by a subsequent oxidation. However, the early elution of pure HCl in Fig. 12.1 does not imply any retentive interactions between protons and the polymer. The retention of HCl occurs only in the presence of a salt. But why would the properties of HCl in the polymeric phase change so dramatically in the presence of metal chlorides, while no association of HCl with LiCl or $CaCl_2$ takes place in solution? The version (i) of attractive interactions of protons with the polystyrene phase thus cannot be accepted without serious doubt.

Size exclusion phenomena (ii) can adequately explain the earlier elution of large hydrated metal cations (together with the equivalent amount of chloride anions), if one takes into consideration that, with the maximum of the pore size distribution of the material at about 2 nm, a substantial amount of channels or pores could prove to be inaccessible to the hydrated cations. Still, for version (ii) to be accepted as the main reason for the differentiation between a salt and an acid, one must assume that the effective size of protons in the system is substantially smaller than that of metal cations. In addition, several other unusual characteristics of the elution profiles shown in Figs. 12.1 and 12.2 have to be explained in frames of the SEC concept. First of all, the spontaneous concentration increase of both separated components has never been reported in SEC. In fact, it is neither characteristic of any other chromatographic technique. Only two exceptions exist: frontal adsorption chromatography, where the less retained component of the initial mixture emerges from the column in a more concentrated form, and the above-mentioned "acid retardation" process, where the stronger retarded acid undergoes self-concentration. Another important phenomenon that requires explanation is common for both the acid retardation process and the ISE on neutral polystyrene: this is the slowing down of the acid movement by the presence of a mineral salt.

In order to discuss the reasons for these unique features of the ISE process, let us first consider the special features of SEC, including those that have never been considered in analytical-scale separations.

3. REMARKABLE FEATURES OF SIZE-EXCLUSION CHROMATOGRAPHY

Size-exclusion chromatography (SEC), also called gel permeation chromatography (GPC) if the column packing is a polymeric gel, separates the analytes according to their hydrodynamic sizes in solution. Generally, the column packing is a porous material or a gel with pores or channels that are comparable in size to the size of the analytes. Very large analyte molecules do not enter the pores of the sorbent and remain in the interstitial volume. Actually, the latter is the only part of the mobile phase in the column that really moves between the sorbent particles toward column outlet. Therefore solutes that are totally excluded from the porous sorbent elute from the column with the interstitial volume, amounting to about 40% of the total column volume [144]. Very small analyte molecules freely diffuse into the porous space of the sorbent and reside for a while in the stagnant zones of the mobile phase. Therefore, they elute from the column with the column hold-up volume, which is the sum of the interstitial volume and the volume of pores. All other analyte species emerge from the SEC column in the rather small window between the interstitial volume and the hold-up volume of the column. For that reason, the resolution power of SEC is much more limited, as compared to adsorption chromatography, where the retention of analytes can amount to tens of the column volume.

SEC has been mainly used for the separation of large species on the analytical scale. Applying this technique to preparative separations requires closer consideration of some of the peculiar features of the technique, which may have escaped the attention of most chromatographers. Some of these features may even sound paradoxical.

Species separated by SEC are transported along the column with the mobile phase, but move faster than the mobile phase. As explained above, the totally or partially excluded analytes spend more time in the interstitial volume, that is, in the moving part of the mobile phase, while all solvent molecules often diffuse into stagnant zones in all accessible pores of the packing. As a result, *the analytes in SEC quickly depart from their "mother sample solvent."* The analytes that are excluded from the stationary phase constantly exploit the movement of those molecules of the column mobile phase which happen to locate in the interstitial space, and they leave behind these solvent molecules the moment the latter enter the porous space.

Let us consider the situation when a pump starts delivering a solution of analytes into a column packed with porous material. Large analytes arrive at the column outlet and emerge in the effluent before smaller particles and before the sample solvent. At the moment the front of the sample solvent arrives at the outlet, the column becomes equilibrated with the feed mixture. At that moment, certain portions of excluded particles have already left the column with the effluent. Therefore, *a size exclusion column, being equilibrated with the mixture under separation, always incorporates liquid with reduced concentrations of the excluded species.* Even at a prolonged pumping, concentration of these species within the column will never reach that of the feed. This fully corresponds to the name of the process: the species are *excluded* from the packing, that is, from the column. The larger are the molecules of the analyte, the lower is their relative concentration within the column.

Nevertheless, the situation is fully restored at the column outlet, that is, in the effluent. The effluent receives the larger species at an enhanced rate, since they move through the column with a velocity higher than the velocity of the mobile phase. The larger are the molecules of the species, the lower is their concentration in the column, and the faster are they delivered into the effluent. The same statement is valid for the case of elution of all species remaining in the "analyte-saturated" column. As a result, *in SEC, the concentration of all species under separation in all corresponding fractions of the effluent is equal to their concentration in the initial mixture.* It is important that *SEC does not cause dilution of the solutes.* of course, we assume here that all fronts are sharp and there is no noticeable dispersion at the borders of rectangular (or slightly trapezoidal) chromatographic zones.

Another special feature of SEC is that the process of *interaction between the solutes and the stationary phase is non-stoichiometric.* For this reason, the fundamental concept on exchange capacity or adsorption capacity of the column is no longer applicable, and the concentration of the sample solution can be selected arbitrarily. Only the solubility of sample components in the mobile phase can set a threshold for their concentrations. Naturally there is a limit for the volume of the sample that can be separated in one chromatographic cycle; theoretically, this is the total volume of porous space in the column packing (it will be shown later that in practice much larger sample volumes can be successfully applied).

It should be noted here, that all the above-emphasized features of the SEC principally distinguish this technique from the common variants of adsorption, distribution, or ion-exchange chromatography. In the latter techniques, the analytes move through the column *slower* than the mobile phase (as they are partially retained by the stationary phase); the total concentration of the analytes in the column equilibrated with the feed solution is higher than the concentration of the initial feed (as analytes additionally accumulate in the stationary phase); the analytes eluting with pure mobile phase from the "analyte-saturated" column appear in the effluent at concentrations lower than in the feed (as removing analytes from the sorbent requires additional amounts of the mobile phase); and the columns are easily overloaded (as the exchange capacity of every sorbent is a limited value).

4. SIZE OF HYDRATED IONS

To discuss separations of electrolytes in the framework of the ISE process, while accepting the idea that strong electrolytes completely dissociate into constituent ions, one needs to have information about the size of the individual ions in aqueous media. There has been a great deal of debate for more than a century on the solvation of ions and the effect that dissolved ions have on the tetrahedral network structure of water. A central theme of that discussion has been the notion that different ions have "structure making" or "structure breaking" effects, depending on the charge and size of the ion. While potassium is regarded as being rather neutral in its effect on the structure of water [145], the information collected thus far for other mineral ions, even for the simplest ions, is very contradictory and inconclusive.

In general, a group of ions ("cosmotrops") is discussed that bind water in their first coordination sphere more strongly than water is bonded to water. Typical for the group are SO_4^{2-}, HPO_4^{2-}, F^-, $Tris^+$, and, probably, Cl^-. They elute from Sephadex G-10 before tritiated water (THO). Interestingly, in aqueous solutions, they enhance the stability of proteins, contrary to the action of "chaotrops." The latter group, including, for example, Br^-, I^-, ClO_4^-, SCN^-, and trichloroacetate$^-$, binds water more loosely,

Table 12.1 Hydration numbers of halogen ions and methods of their determination

Method	F^-	Cl^-	Br^-	I^-
Stokes radius	5.5	3.9	3.4	2.8
Isothermal compression	4.5	2.0	1.8	1.5
Thermochemistry	–	4.4	4.8	–
NMR: time relaxation	9.9	13.2	16.2	21.8
X-ray diffraction	4.5	6	6	8.8
Neutron diffraction	–	5.8	–	–

Reprinted from [156] with permission of Elsevier.

compared to the water–water bonds. These ions easily dehydrate, elute from Sephadex after tritium-labeled water THO, and are said to be sticky [146].

With respect to the number of water molecules comprising the hydration shell of an ion, static and dynamic methods of investigation show incompatible results (Table 12.1). To the static methods belong thermochemistry [147, 148], nuclear magnetic resonance (NMR) time relaxation [149], infrared (IR) spectroscopy [150], X-ray diffraction [151–152], and neutron scattering [151–153]. They all provide information on the size of the hydration shell of ions in the bulk solution. Among these static methods, X-ray diffraction and neutron scattering require high electrolyte concentrations. Viscosimetry, determining the Stokes radius [154], and isothermal compression of solutions [155] are considered as dynamic methods, which characterize the moving ions.

Even within the group of dynamic methods, one can find in the recent literature entirely different hydration numbers, for instance, those presented in Table 12.2 for biologically important ions [157]. Surprisingly, the fundamental Stokes–Einstein relationship between the hydrodynamic radius and the diffusion coefficient of the ion is being used in several different manners in the calculation of the effective hydrated radius of an ion (compare [158] and [159]).

A direct mass-spectrometric measurement of hydrated halogen ions by the electrospray technique [156] showed that up to seven to nine water molecules can be detected with the ions after evaporation, but the most probable hydrates in the mass spectrum contain two or three water molecules associated with the F^- anion, two water molecules with Cl^-, one or two with Br^-, and only one with I^-. The HO^- anion strongly retains three water molecules. At the same time, the combined technique of neutron

Table 12.2 Apparent dynamic hydration numbers

Cation	Hydration number	Anion	Hydration number
Mg^{2+}	5.8	PO_4^{3-}	5.1
Ca^{2+}	2.1	HPO_4^{2-}	4.0
Ba^{2+}	0.35	$H_2PO_4^-$	1.9
H^+	1.93	SO_4^{2-}	1.8
Li^+	0.6	F^-	5.0
Na^+	0.22	HO^-	2.8
K^+	0	HCO_3^-	2.0
NH_4^+	0	Cl^-	0
Rb^+	0	Br^-	0
Cs^+	0	NO_3^-	0
–	–	ClO_4^-	0
–	–	I^-	0
–	–	SCN^-	0

Reprinted from [157] with permission of Elsevier.

diffraction with hydrogen isotope substitution showed recently [153] that the anion hydration number is fairly stable around 5.5–6 water molecules for all the halogen anions, in spite of a steady increase in the distance of the first hydration shell from the ion as we go down the halide series. However, there is a pronounced second shell with the fluoride ion, whereas it starts to overlap with the first shell for the larger anions.

Reliable, though relative, information about the size of ions in aqueous media can be obtained from data on the electrophoretic mobility of these ions [160], as the velocity of their movement in an electric field is directly proportional to their charge and inversely proportional to their hydrodynamic radius. According to these estimations (Table 12.3), the size of hydrated cations and anions decreases according to the following series:

$$Tris^+ \gg Li^+ > Na^+ \sim EtNH_3^+ > NH_4^+ \sim K^+ \ggg H_3O^+ \text{and}$$
$$SO_4^{2-} \sim Acetate^- > Formate^- \sim F^- \gg NO_3^- > Cl^- \ggg OH^-.$$

There is not much information available on the real size of hydrated ions. For a few simple alkali and alkaline-earth cations, the Stokes radius decreases from about 4.4 to about 3.3 Å according to the following series: $Mg^{2+} \sim Zn^{2+} > Ca^{2+} \sim Sr^{2+} > Ba^{2+} > Li^+ > Na^+$. Note that this sequence is largely opposite to that of the van der Waals radius (Table 12.4) of the cations themselves [161]. Obviously, *smaller cations exhibit larger charge*

Table 12.3 Electrophoretic mobility of ions, $\mu_e \, 10^5 \cdot (cm^2 \cdot v^{-1} \cdot s^{-1})$

H_3O^+	362.5	OH^-	205.5
K^+	76.0	Cl^-	79.1
NH_4^+	72.2	NO_3^-	74.1
$EtNH_3^+$	53.1	F^-	57.4
Na^+	51.9	Formate	56.6
Li^+	40.1	Acetate	42.4
$Tris^+$	29.5	SO_4^{2-}	41.5×2

Source: After [160].

Table 12.4 Ion radius, Å

Cation	Li^+	Na^+	Mg^{2+}	Zn^{2+}	Ca^{2+}	Sr^{2+}	Ba^{2+}	La^{3+}
Crystal	0.60	0.97	0.65	0.74	0.99	1.13	1.35	1.15
Hydrate	3.7	3.3	4.4	4.4	4.2	4.2	4.1	4.6

Source: After [161].

density on their surface and stronger binding of water molecules, compared to larger ions. This rule explains why ionic mobilities increase initially with ionic size, but with the largest ions such as Cs^+ and I^- that do not bind water, they decrease again [162].

Information accumulated thus far on the hydration of mineral ions has been critically analyzed in the recent review by Marcus [163] entitled "Effect of ions on the structure of water: Structure making and breaking."

It is important that definite changes have been noted in the water structure and in the structure of diffuse hydration shells with electrolyte concentrations. Neutron diffraction of $CaCl_2$ and $Ca(NO_3)_2$ solutions in D_2O has shown [164] a decrease of Ca^{2+} hydration number from 10 to 6 when the salt concentration increased from 1 to 4.5 M.

In addition to the concentration of the electrolyte solution, another barely examined factor influencing the structure and size of the hydration shells of ions is temperature. Investigations of spin-lattice relaxation rates of ^1H and ^7Li nuclei in aqueous LiCl solution as a function of concentration and temperature indicate that heating may destroy the tetrahedral water structure with inserted lithium cations that exists at temperatures below 30°C and enable the ions to construct a surrounding octahedral with the coordination number of 6 at temperatures above 40°C [165]. However, molecular dynamics simulations with current water structure models do not show this type of transition [166].

Finally, it can be noted that adding alcohols to an aqueous $CaCl_2$ solution was found to strongly affect the sizes of both the hydrated Ca^{2+} and Cl^- ions, when calculated from the measurements of the self-diffusion coefficients of $^{45}Ca^{2+}$ and $^{35}Cl^-$ as radioactive tracers [167]. Surprisingly the relationships are not monotonous and strongly differ for methanol and *tert*-butanol as compared to *n*-propanol. Small amounts of all these alcohols also demolish the Na^+ hydration shell.

This review of the literature suggests that both the size and stability of the hydration shell of an ion are equally important factors determining the ion's state and behavior; both may substantially change under different conditions.

Still, to have an orientation in the elution order of different ions under the SEC conditions, information on the relative sizes of the ions is essential. Therefore, while fully recognizing the argumentativeness of *the absolute* values of ion radii and noting that substantially different values can be found in the literature, we present here one of the most complete and self-consistent list of crystal radii and the effective radii of hydrated ions from the work of Nightingale [158]; the hydrated radii have been calculated from the limiting ionic equivalent conductance and limiting ionic diffusion coefficients (Table 12.5).

The sizes of hydrated protons and hydroxyl ions deserve special consideration. On the one hand, proton must have the highest possible surface charge density and, therefore, bind water molecules particularly strongly. The first water molecule converts the proton to the stable hydroxonium cation H_3O^+, but, most probably, the latter strongly coordinates at least three additional water molecules. Hydroxyl anion is also known to bind three water molecules [156]. From this point of view, the sizes of the two ions should be comparable to or larger than the sizes of other monovalent ions. On the other hand, acids and bases are known to cause especially high conductivity of aqueous solutions, pointing to extremely high mobility (i.e., small size) of the hydroxonium and hydroxyl ions (see Table 12.5.).

The obvious contradiction is resolved by taking into account that measuring electrophoretic ion mobility is a dynamic method, but protons and hydroxyls as such do not need to move in the aqueous phase under the action of an electric field. Instead, a rapid shift of electrons along "a hydrogen-bonded water wire" (the chain of hydrogen bonds between

Table 12.5 Crystal radii (r_{cryst}) and effective radii of hydrated ions (r_{hydr}) (25°C)

Ion	r_{cryst} (Å)	r_{hydr} (Å)	Ion	r_{cryst} (Å)	r_{hydr} (Å)
$(Me)_4N^+$	3.47	3.67	OH^-	1.76	(3.00)
$(Et)_4N^+$	4.00	4.00	F^-	1.36	3.52
$(n\text{-}Pr)_4N^+$	4.52	4.52	Cl^-	1.81	3.32
$(n\text{-}Bu)_4N^+$	4.94	4.94	Br^-	1.95	3.30
$(n\text{-}Pe)_4N^+$	5.29	5.29	I^-	2.16	3.31
H^+	(2.82)	–	NO_3^-	2.64	3.35
Li^+	0.60	3.82	ClO_3^-	2.88	3.41
Na^+	0.95	3.58	BrO_3^-	3.08	3.51
K^+	1.33	3.31	IO_3^-	3.30	3.74
Rb^+	1.48	3.29	ClO_4^-	2.92	3.38
Cs^+	1.69	3.29	IO_4^-	3.19	3.52
Ag^+	1.26	3.41	MnO_4^-	3.09	3.45
Tl^+	1.44	3.30	ReO_4^-	3.30	3.52
NH_4^+	1.48	3.31	CO_3^{2-}	2.66	3.94
Be^{2+}	0.31	4.59	SO_4^{2-}	2.90	3.79
Mg^{2+}	0.65	4.28	SeO_4^{2-}	3.05	3.84
Ca^{2+}	0.99	4.12	MoO_4^{2-}	3.23	3.85
Sr^{2+}	1.13	4.12	CrO_4^{2-}	3.00	3.75
Ba^{2+}	1.35	4.04	WO_4^{2-}	3.35	3.93
Ra^{2+}	1.52	3.98	$Fe(CN)_6^{4-}$	4.35	4.22
Mn^{2+}	0.80	4.38	–	–	–
Fe^{2+}	0.75	4.28	–	–	–
Co^{2+}	0.72	4.23	–	–	–
Ni^{2+}	0.70	4.04	–	–	–
Cu^{2+}	0.72	4.19	–	–	–
Zn^{2+}	0.74	4.30	–	–	–
Cd^{2+}	0.97	4.26	–	–	–
Pb^{2+}	1.32	4.01	–	–	–
Al^{3+}	0.50	4.75	–	–	–
Cr^{3+}	0.64	4.61	–	–	–
Fe^{3+}	0.60	4.51	–	–	–
La^{3+}	1.15	4.52	–	–	–
Ce^{3+}	1.1	4.52	–	–	–
Tm^{3+}	0.9	4.65	–	–	–
$Co(NH_3)_6^{3+}$	2.55	3.96	–	–	–

Source: After [158].

water molecules in an aqueous media) can immediately generate a positive or negative charge, that is, a hydroxonium or hydroxyl ion, respectively, in any position of that aqueous phase. In fact, the electron–proton-coupled transfer of a charge in an aqueous phase proceeds by electrons hopping along the "wire" with the protons in the chain of hydrogen bonds hopping in the opposite direction [168, 169]:

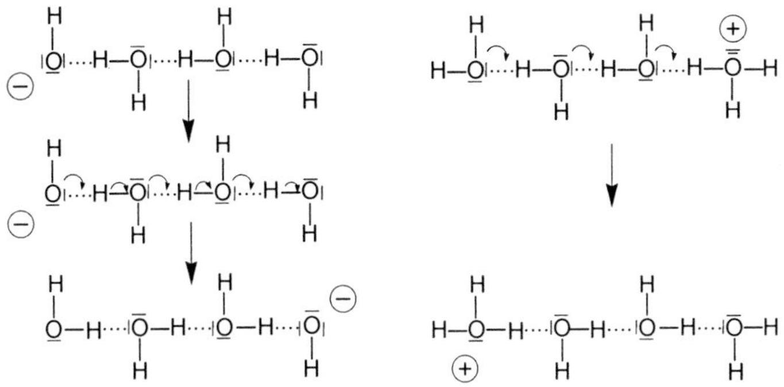

By analogy, we consider chromatography to be a dynamic system where protons and hydroxyl ions can appear at any point that accommodates a water molecule, provided that the latter is connected to the bulk aqueous phase by chains of hydrogen bridges. If a charge is required, protons or hydroxyls can be generated in any smallest pore of the nanoporous material that is accessible to water. Therefore, in the dynamic ISE process, mineral cations and anions should govern the separation, while protons and hydroxyls serve for the relocation of charges and maintenance of local electroneutrality, without occupying any eigenvolume. We thus assume that these water-borne ions do not endure any exclusion from the water-filled nanopores of the hyper-crosslinked polystyrene. This fully explains, in terms of SEC, the facts that acids or bases tend to effectively separate from their corresponding salts having a common anion or cation, respectively, and always elute from the column after the salt. For instance, all metal chlorides elute before hydrochloric acid, and all sodium salts elute before sodium hydroxide.

Finally, while comparing the sizes of hydrated ions, one must bear in mind that the hydration shells are labile and may react differently for different ions on the change of concentration and temperature. Some rather unpredictable phenomena may emerge with highly concentrated or hot solutions. The influence of temperature on the size of hydrated ions still requires a detailed examination, and chromatography may even provide unique information not accessible by other techniques.

5. SELECTIVITY OF SEPARATION IN ION SIZE-EXCLUSION CHROMATOGRAPHY

With the information available about the relative sizes of hydrated ions, one can test the basic conception of the size exclusion mechanism of electrolyte separations on the neutral nanoporous material.

According to the theory of SEC, all partially excluded analytes elute in a relatively narrow window between the interstitial volume and the hold-up volume of the column. The interstitial volume of a column packed with a beaded material of broad bead size distribution amounts to about 40% of the column volume [144]. In addition to this volume, the mobile phase (water) also occupies the porous volume within the sorbent. In the case of our polymeric packings, the total pore volume amounted to about half of the polymer volume. All analytes are thus expected to elute in the window between 40 and 70% of the column volume. (The size of the separation window equals the total pore volume in the column packing.) Each analyte must have a fixed position in this window corresponding to the portion of the pore volume that is accessible to its molecules. In analytical SEC, the hydrodynamic radius of a species thus can be directly read from the calibration plot showing the relationship between the analyte sizes and their elution volumes. Importantly, the distance between the elution volume of a totally excluded analyte and that of a small species of the size of a water molecule should not exceed the above one-third of the bed volume. This is the maximum separation selectivity that can be expected for the pure size exclusion mechanism of separation.

In the case of ISE of electrolytes, one has to take into consideration that in aqueous solutions strong electrolytes dissociate into constituent ions. If only one electrolyte is present in the sample, the cations and anions are forced to move together, even in the case that there is no tendency to form ionic pairs. In this manner the principle of local electroneutrality is abided by. Obviously, the larger ion that happens to be excluded from the porous space will govern to a greater extent the velocity of the salt zone migration in the column and the final elution volume. The smaller counter ion will probably not retard the zone to a noticeable extent.

If several electrolytes are present in the sample, the free ions will have the possibility to regroup. Then, the first eluting salt must be composed of the largest cation and the largest anion, while the last eluting salt (or acid, or base) must incorporate the smallest ions of the sample. Naturally, the stoichiometry of salts and electroneutrality of all fractions must be observed.

In the simplest case presented in Fig. 12.2, the initial mixture contained three ions, Ca^{2+}, H^+, and Cl^-, the chloride anion being common for the two electrolytes, $CaCl_2$ and HCl. Obviously, the separation of the mixture is caused by the difference in the sizes of Ca^{2+} and H^+, actually presenting the only difference between the two electrolytes. However, the rate of the movement of $CaCl_2$ must be determined by the exclusion of the largest ion, Ca^{2+} ($r_{hydr} = 4.12\,\text{Å}$), while the smaller Cl^- ions just follow their cations, probably without retarding noticeably the movement of the latter. Similarly, the rate of propagation of the hydrochloric acid zone must be determined by the size of the Cl^- ion ($r_{hydr} = 3.32\,\text{Å}$), rather than by the size of protons (we assume the latter to be close to zero). Protons of the acid must move, or emerge, in one way or another, in close vicinity of the moving chloride anions, in order to maintain local electroneutrality. From this consideration it logically follows that *the selectivity of separation of two electrolytes MeCl and HCl is mainly determined by the difference in sizes of the cation Me$^+$ and the anion Cl$^-$, the largest ions of the two electrolytes* under separation, and not by the size difference of those two ions (Me$^+$ and H$^+$) that actually make the electrolytes differing from each other [140].

To make possible some semiquantitative estimations of the separation selectivity of electrolyte mixtures in a frontal ISE process, we can determine in the forward experiments the breakthrough volumes, $BtV_{0.05}$, at the 5% level of the initial concentration of each of the two electrolytes under separation. From a theoretical point of view, more meaningful are the breakthrough volumes measured at the half of the height of the concentration waves, $BtV_{0.5}$. These values were estimated for both the forward and reverse experiments. Fig. 12.3 shows schematically the positions of these values on the elution scale. In the forward experiment, the difference Δ_1 between the $BtV_{0.5}$ values of the two separated electrolytes, when divided by the bed volume (30 mL), was thought to characterize the separation selectivity (expressed in bed volumes) with respect to the faster moving component. Similarly, selectivity Δ_2 of the isolation of the slower moving component results from the difference between the two $BtV_{0.5}$ values determined in the reverse experiment. Both sets of values are presented in Table 12.6 for a series of electrolyte mixtures examined on hypercrosslinked polystyrene-type sorbents. Note that in each electrolyte pair the first line of data in the Table 12.6 corresponds to the faster moving electrolyte whereas the second line of data characterizes the slower moving component.

When the size of the hydrated cation exceeds the size of its anion (as is the case for all chlorides and some sulfates), the breakthrough volume of the salt

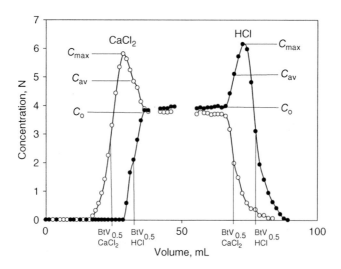

Figure 12.3 ISE separation of 4 N calcium chloride and 4 N hydrochloric acid on D4609 carbon. Column volume: 30 mL; flow rate: 0.6 mL/min. Feed solution (50 mL) delivered from the bottom upward, then the eluent (water) delivered from the top of the column downward. *(After [140].)*

correlates with the size of the cation. The largest cations appear in the effluent very early. Measured at the level of 5% of the zone height (BtV$_{0.05}$, mL), the elution sequence of salts on sorbents initially examined in [159] is as follows:

$$MN\text{-}202 : Ca^{2+}(18) < Li^{+}(19.5)$$
$$MN\text{-}270 : Al^{3+}(13) \sim Fe^{2+}(13) < Ca^{2+}(15) < Li^{+}(16)$$
$$Carbon\ D4609 : Al^{3+}(14) < Fe^{2+}(15) < Ca^{2+}(17) < Na^{+}(19).$$

Naturally, all metal chlorides elute before hydrochloric acid. When judging by the values of BtV$_{0.05}$ under similar conditions, the separation selectivity of HCl from metal chlorides increases slightly on all tested sorbents in the series $K^{+} < Na^{+} < Li^{+} < Ca^{2+}$, along with the increase of the radius of the hydrated cations. In combination with the larger anion SO_4^{2-} ($r_{hydr} = 3.79$ Å), only polyvalent cations have a chance to significantly contribute to the separation. Indeed, Cu(II), Fe(II), and Al(III) sulfates readily separate from sulfuric acid. Hydrated radii of these cations are estimated as 4.19, 4.28, and 4.75 Å, respectively, according to which aluminum has the smallest breakthrough volume.

   Even sodium sulfate slightly separates from sulfuric acid on nanoporous carbon, although the sulfate ion is a little bit larger than the sodium cation. In

Table 12.6 Separation of electrolytes on hypercrosslinked polystyrene resins

Electrolyte	MN-270						MN-202						MN-500				
	C_0 (N)	$BtV_{0.05}$ (mL)	$BtV_{0.5}$ (mL)	C_{max}/C_0	Δ_1 (BV)	P (BV)	C_0 (N)	$BtV_{0.05}$ (mL)	$BtV_{0.5}$ (mL)	C_{max}/C_0	Δ_1 (BV)	P (BV)	C_0 (N)	$BtV_{0.05}$ (mL)	$BtV_{0.5}$ (mL)	C_{max}/C_0	Δ_1 (BV)
$CaCl_2$ HCl	3.6	15	22.5	1.34	0.25	0.29	3.5	18	23.4	1.14	0.20	0.22	—	—	—	—	—
	3.8	28	30.0	1.50	0.25	0.31	3.9	27.5	29.5	1.41	0.15	0.18	—	—	—	—	—
$LiCl$ HCl	2.1	16.2	20.9	1.08	0.24	0.25	3.85	19.5	23.5	1.15	0.18	0.19	—	—	—	—	—
	2.1	25.8	28.0	1.36	0.20	0.24	4.0	26.3	29	1.45	0.14	0.17	—	—	—	—	—
KCl HCl	1.2	14	17.3	1.08	0.19	0.20	—	—	—	—	—	—	—	—	—	—	—
	4.2	21	22.9	1.08	0.17	0.18	—	—	—	—	—	—	—	—	—	—	—
H_2SO_4	4.0	17	22.4	1.07	0.14	0.14	4.0	20.5	24.9	1.05	0.06	0.06	4.5	18	23.2	1.00	0.05
HCl	3.9	23	26.5	1.12	0.05	0.05	3.9	23.5	26.8	1.07	0.05	0.05	3.5	19	24.6	1.06	0
Na_2SO_4	1.8	13.5	17.0	1.17	0.32	0.35	—	—	—	—	—	—	—	—	—	—	—
$NaOH$	0.95	23	26.5	—	—	—	—	—	—	—	—	—	—	—	—	—	—
$(Fe+Cu)$ SO_4	2.15	13	18.0	1.15	0.28	0.28	—	—	—	—	—	—	—	—	—	—	—
H_2SO_4	2.75	19	26.4	—	—	—	—	—	—	—	—	—	—	—	—	—	—
$Al_2(SO_4)_3$ H_2SO_4	0.9	13	18.0	—	0.26	0.26	—	—	—	—	—	—	—	—	—	—	—
	3.4	18	25.7	—	—	—	—	—	—	—	—	—	—	—	—	—	—
$CaCl_2$ $Ca(NO_3)_2$	—	—	—	—	—	—	1.8	—	23.4	1.01	0.05	0.05	4.0	19.5	23	1.00	0.05
	—	—	—	—	—	—	1.7	—	24.9	1.06	0.08	0.08	4.3	19.5	24.5	1.06	0.05
$NaCl$	—	—	—	—	—	—	1.9	—	23.3	1.0	0	0	—	—	—	—	—
$NaOH$	—	—	—	—	—	—	2.2	—	23.3	1.0	0	0	—	—	—	—	—

C_0 = initial electrolyte concentration; C_{max} = concentration at the maximum of breakthrough curve; $BtV_{0.5}$ = breakthrough volume at the middle of front; $\Delta = (Bt_1V_{0.5} - Bt_2V_{0.5})/30$ = selectivity of separation, 30 being the volume of column, 30 mL; $P = \Delta \times C_{av}/C_0$ = productivity of separation.
Reprinted from [139] with permission of Elsevier.

contrast to this, on the same carbonaceous material Na_3PO_4 is not separated from H_3PO_4 at all. Certainly, the very large phosphate and hydrogen phosphate ions determine the migration velocity of both electrolytes of the mixture, the salt and the acid, whereas the much smaller cations of Na^+ and H^+ are bound to follow the leading phosphate anions, without being separated at the end. Logically, there is no "acid retardation" in this system.

As can be expected, sulfuric acid separates well from hydrochloric acid, because the size of the two anions differs noticeably, $r_{hydr} = 3.79$ and $3.32\,Å$, respectively. Similarly, K_2SO_4 will separate from KCl with almost the same selectivity, since the K^+ ion, being the smallest in the mixture $(3.31\,Å)$, does not influence significantly the situation in the system.

Thus, the results obtained so far corroborate all qualitative predictions concerning the elution order of electrolytes, made on the assumption that the separation process is governed by size exclusion of the largest hydrated ions from the nanopores of the column packing.

However, an acceptable quantitative correlation between the ion size and the position of the breakthrough front exists only for the $BtV_{0.05}$ values of the first eluting component. Obviously, these values correspond to the travel velocity of the pure fast moving electrolyte at its rather low concentration. The three other zone borders, second one in the forward experiment and both fronts in the reverse experiment, were observed to shift dramatically along the volume axis, depending on the concentration of the electrolytes and their ratio. Accordingly, the selectivity values Δ_1 and Δ_2 vary in an extremely wide range. Also, the electrolyte separation selectivity Δ_1 in the frontal ISE process can amount to astonishingly high values, up to several column volumes, by far exceeding the theoretical limit, the total volume of pores (*ca* one-third of the column volume). Noteworthy is finally the fact that the Δ_1 and Δ_2 values for the two separated components are not equal in the majority of systems. These facts stand in a sharp conflict with the rules of SEC. Obviously, another powerful process interferes with the size exclusion separation mechanism, which up to now was not revealed in the common SEC practice.

6. PHASE DISTRIBUTION OF ELECTROLYTES AND THEIR MUTUAL INFLUENCE

Neither the special features of SEC discussed in Section 3 nor the absence of eigenvolumes of freely moving protons and hydroxyls in dynamic chromatographic systems, which was postulated in Section 4, can explain the following unique features of the ISE process (see, e.g., Fig. 12.2):

- the unprecedented self-concentration phenomena of both separated components; and
- the obvious retardation of acids in the presence of salts, which can enhance the selectivity of separation beyond the maximum theoretical value. (It will be shown later that the retardation is generally characteristic of any second-eluting electrolyte, not only acids.)

The first phenomenon has never been discussed in chromatography, while the second, although exploited in the acid retardation process, did not find an acceptable explanation.

Both the above phenomena are particularly striking when working with concentrated electrolyte solutions; they are definitely of a thermodynamic, rather than kinetic, nature and, therefore, must be valid not only under chromatographic (dynamic) conditions but also under static equilibrium conditions.

In order to reveal the origin of the apparent transition from exclusion mechanism to some kind of "adsorption" of smaller ions at high concentrations of the mixture, a model system, composed of calcium chloride and hydrochloric acid at varying concentrations, has been examined in detail [170–172]. This mixture is environmentally benign, allows preparation of highly concentrated solutions, and has the advantage that each of the three constituent ions, Ca^{2+}, H^+, and Cl^-, can be easily determined by titration. The test mixture has been studied in both static and dynamic experiments in combination with the hydrated nanoporous polystyrene NN-381.

In the first mode, known portions of the polymer were equilibrated with solutions of $CaCl_2$ and HCl, as well as with their mixtures of known concentrations. The final composition of the bulk solutions in equilibrium with the polymeric phase was determined by titrating the excess HCl acid with NaOH and by complexometric titration of the Ca^{2+} ions with ethylenediamine tetraacetate (EDTA). From these data the concentrations of the electrolytes within the porous space of the polymeric material were calculated and then the apparent phase distribution coefficients k of HCl and $CaCl_2$, defined as the ratio between the equilibrium concentrations of the corresponding electrolytes within and outside the polymeric beads. These calculations are strongly facilitated by the outstanding property of the neutral hypercrosslinked polystyrene sorbents, namely that their swelling does not depend on the electrolyte concentration, so that the volume of the porous space remains constant in all experiments. Thus, apparent phase distribution coefficients of HCl and $CaCl_2$, taken separately and in the mixture, have been obtained at varying concentrations of the electrolytes.

In the dynamic experiments, the corresponding solutions of HCl and $CaCl_2$, as well as those of their mixtures, have been sent through a chromatographic column packed with the same polymer NN-381 until the eluting solution became identical to the feed. Then, the liquid was displaced from the column with distilled water. In both loading and elution (forward and reverse) experiments, fractions of the effluent were collected and subjected to titration analysis.

Figure 12.4 presents the plots of the phase distribution coefficients of HCl and $CaCl_2$ taken separately as a function of their equilibrium concentration in the outer bulk solution. Only the middle part of the plots, corresponding to electrolyte concentrations between 0.5 and 4 N, can be considered to be trivial for size exclusion phenomena. In this concentration range, the apparent phase distribution coefficient for HCl remains constant at the level close to unity, indicating that this electrolyte is neither adsorbed on the polymer nor experiences any steric exclusion effects from the porous space. Thus, protons and chloride anions move freely within the space available to water molecules inside the polymeric beads. Nightingale [158] estimated the radius of hydrated proton as 2.82 and that of the chloride anion as 3.32 Å. These ions are obviously smaller than the pores of the polymer matrix. On the other hand, the apparent distribution coefficient for $CaCl_2$ in the middle concentration range is only 0.75, thus indicating

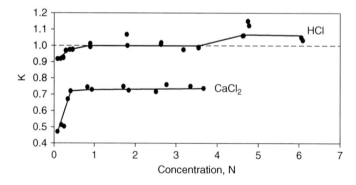

Figure 12.4   Phase distribution coefficients for HCl and $CaCl_2$ (determined separately) between the porous volume of NN-381 and the bulk solution as a function of the concentration of the latter at equilibrium. *(After [171].)*

partial exclusion of the large hydrated Ca^{2+} cation ($r_{hydr} = 4.12\,\text{Å}$) from about one-fourth of the porous volume.

Interestingly, at low concentrations the apparent distribution coefficients for both HCl and $CaCl_2$ decrease to about 0.9 and 0.45, respectively. This fact could indicate an increased hydration extent of the Cl^- and Ca^{2+} ions in very dilute solutions. It is more difficult to explain the unexpected increase in the values of the distribution coefficients over the threshold of $k = 1.0$ at electrolyte concentrations exceeding 4 N. Most likely the strongly dehydrated cations start interacting with the π-electronic system of the aromatic rings of the hypercrosslinked polystyrene. Such type of interactions is known to be observed between cations and aromatic amino acid residues in certain protein environments [142]. If this type of attraction really starts playing a noticeable role, it may cause some kind of retention or adsorption of the previously excluded electrolytes. In any case a stepwise decrease in the extent of hydration of the ions can be expected to proceed at increasing concentrations of the solutions, and ions with depleted hydration shells may exhibit some specific, sticky properties, not peculiar to stronger hydrated ions. The following observation may support this suggestion: normally, hydrochloric acid does not wet dry hypercrosslinked polystyrene beads, but a concentrated acid partially does.

The changing phase distribution coefficients shown in Fig. 12.4 fully agree with and largely explain the rather unusual breakthrough profiles for 3.75 N HCl and 4.5 N $CaCl_2$ solutions through the column containing the NN-381 sorbent and the elution of the electrolytes from the column with pure water, which were shown in Fig. 12.1. First of all, $CaCl_2$ emerges from the column with a smaller breakthrough volume than does HCl. Obviously, a part of the pore volume within the polymeric beads remains inaccessible to the large hydrated Ca^{2+} ions, thus causing a partial exclusion of the salt under chromatographic conditions.

Another interesting fact is the definite asymmetry of the chromatographic peaks in Fig. 12.1, with the trailing edge of the peaks being much sharper than the leading fronts. On elution with water, the concentration of the ions gradually drops to zero. If the hydration of ions increases with dilution, as follows from the static experiment, the ions must start to behave as species of gradually increasing size, tending to move with growing velocities through the size exclusion column. This must compress the trailing part of the elution zone. On the other hand, at the front of the peak, with the concentration of the electrolytes increasing to the high level of the feed solution, dehydration of the ions proceeds. The ions start behaving as smaller species that are now capable of entering smaller pores. The dehydrated ions move slower through the packing, thus stretching the front of the chromatographic peaks. At the end, they may even be weakly adsorbed on the material, which would flatten out the leading front of the zone even more.

Finally, definite bends can be noticed on the abnormally stretched front of the peaks. These bends can reflect the stepwise change in the hydration number of the ions with their concentration increasing on the leading edge of the peak.

Figure 12.5 presents the plots of apparent phase distribution coefficients of HCl and $CaCl_2$ for the case of their simultaneous presence in the mixture at equivalent concentrations. The plots coincide with those obtained for the individual components (Fig. 12.4) only in the range of very low concentrations. Here, the electrolytes must behave independently of each other and have maximum hydration numbers. In the middle concentration range, between 0.5 N and about 3.5 N for each component, the distribution coefficient for HCl quickly increases above the threshold level of 1.0 and reaches a value of 1.6, thus indicating the active accumulation of HCl within the porous volume of the polymer. On the other hand, the coefficient k for $CaCl_2$ never reaches its characteristic value of 0.75, but, instead, drops to 0.55 and then even to 0.45, thus indicating a stronger exclusion of the salt. These trends in the evolution of the distribution coefficients of the two electrolytes are oppositely directed, thus contributing additionally to

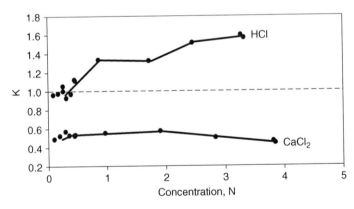

Figure 12.5 Phase distribution coefficients for HCl and $CaCl_2$ (taken as a mixture) between the porous volume of NN-381 and the bulk solution as a function of their concentration in the bulk solution at equilibrium. *(After [171].)*

the differentiation of the electrolytes with the increasing concentration of the mixture.

We are inclined to look for the reasons for this fortunate surplus differentiation in the mutual influence of the two electrolytes, which must increase with their concentration. The mixture investigated consists of two cations, Ca^{2+} and H^+, and a common anion Cl^-. According to the size exclusion principle, smaller pores are accessible only to HCl, while the bulk solution and the volume of large pores remain accessible to both HCl and $CaCl_2$. If no competition existed between the electrolytes for space in the system or, more correctly, for the available hydration water, small pores would contain only HCl, while the rest of space would accommodate both the electrolytes. The total concentration of the ions in the small pores would be just half of their concentration in the bulk and in the large pores. This situation is probably characteristic of diluted solutions. With the total concentration of the electrolytes increasing, the concentration gradients would become significant on the "borders" between the small and large pores. We may also speak of the sharp nonequivalence of the chloride ions' activities, or different activities of water, or different osmotic pressures of solutions in the two compartments of the aqueous phase, namely in the small pores and in the rest of the liquid. In any case, a driving force emerges that makes the chloride anions (and also equivalent protons) additionally migrate into the small pores and concentrate there. Still, the volume of fine pores is insufficient to accommodate the major part of HCl and eliminate the concentration gradients in the system. Therefore, calcium chloride is forced to additionally depart even from the middle–size pores of the polymer and concentrate in the large pores and bulk solution. This trend to the elimination of the concentration gradients equalizes osmotic pressure and the activity of water in all compartments of the aqueous phase, thus providing equal possibilities for hydration to all ions of the system. The redistribution of the electrolytes finally results in the self-concentration of HCl in the porous space (k increases up to 1.6). Formally this result simulates "retention" or "adsorption" of HCl by the polymeric sorbent. Notably, calcium chloride is not "adsorbed" from the mixed solution, but, instead, it is additionally excluded from the porous space (k drops to 0.45) and becomes self-concentrated in the bulk solution. To the best of our knowledge, the facts of a spontaneous differentiation of two components within one liquid phase under static conditions, with the obvious self-concentration of the components in two different compartments of the liquid, have not been described in the literature.

An analogous situation could also arise in the case of membrane separation of a concentrated mixture of $CaCl_2$ and HCl, if the selectivity of the membrane would permit to clearly reject the Ca^{2+} ions while easily transporting water, Cl^- ions, and protons. Note that such spontaneous separation of two components must proceed with a decrease in entropy of the aqueous system.

Experiments under static conditions with phase distribution of $CaCl_2$ and HCl taken at unequal proportions of the two components in the initial mixture support the above interpretation of the situation. Thus, small amounts of HCl, 0.6 N, cannot cause a significant additional exclusion of 4 N $CaCl_2$ from the porous space of the polymer into bulk solution. The distribution coefficient of the salt drops from 0.75 to 0.72, only, while HCl itself experiences a very strong influence from the major mixture component and its distribution coefficient increases from 1.0 to 2.6. Vice versa, if the initial mixture contains more HCl (3.84 N) than $CaCl_2$ (0.50 N), the k value for the former increases from 1.0 to 1.13, only, while the impact of HCl on the minor component, $CaCl_2$, is more significant: k value for the latter drops from 0.75 to 0.51.

The results presented in Fig. 12.2 of the chromatographic experiment with the concentrated $CaCl_2$/HCl feed mixture fully agree with the distribution coefficients found for the system under static conditions. Indeed, the distance between the elution fronts of HCl and $CaCl_2$ is now much larger than it was in the case of separately eluting the two electrolytes (Fig. 12.1). This fact reflects the increased difference between the phase distribution coefficients of the two components, that is, the increased separation selectivity in the concentrated feed solution.

It is also worth mentioning the much sharper breakthrough front for HCl in the mixture as compared to the chromatography of HCl alone. Indeed, the acid in the mixture elutes with the background of concentrated $CaCl_2$ solution, that is, under conditions of a constant (and rather small) solvation extent of the H^+ and Cl^- ions. On the contrary, the trailing edge of the $CaCl_2$ peak is now stretched, because it elutes with the background of concentrated HCl, so that the tail-compressing factor of increasing solvation number and the size of Ca^{2+} ions at their low concentration does not operate anymore. In general, the instability of the ion hydration shells under the conditions of changing concentrations favorably sharpens both the fronts of HCl, the second eluting component, but fails to do similarly for the first eluting component, $CaCl_2$ (see Fig. 12.2 as a typical example).

7. "ACID RETARDATION," "BASE RETARDATION," AND "SALT RETARDATION" PHENOMENA

The observation that the phase distribution coefficient of HCl in concentrated mixtures can exceed the threshold value of $k = 1.0$ (Fig. 12.5) implies that, when combined with a salt, the acid behaves in the chromatographic column as a "retained" component and moves through the column slower than the mobile phase does. This is against the rules of SEC, which have been generally formulated for diluted systems and do not consider phenomena related to mutual influence of the electrolytes and concentration gradients between different compartments of the aqueous phase. The deviation of concentrated systems from the rules has a practically important consequence, namely that the separation selectivity increases along with the concentration of mixtures under separation, so that in a single chromatographic run a volume of the feed solution can be successfully processed that exceeds the pore volume of the packing and even exceeds the total bed volume of the column. It has been mentioned in Section 1 that the "acid retardation" phenomenon on the anion-exchange resin Dowex-1x8 was developed to an industrially useful process [115, 119–121], without finding any acceptable theoretical explanation.

Our experiments reveal the concentration gradients (between the portion of the aqueous phase that is accessible to smaller ions, only, and the rest of the solution) as the driving force for squeezing these smaller ions into the fine pores and additional exclusion of the largest ions into the rest of the liquid phase. This process of spontaneous redistribution of electrolytes between the above two compartments of the aqueous phase continues until the total concentration of electrolytes in all parts of the liquid phase levels out. If this mechanism of the spontaneous separation of electrolytes corresponds to reality, it must operate in all nanoporous separation media and with all types of electrolytes.

To put to test the above suggestion, we examined nanoporous carbonaceous sorbent D4609 in combination with rather concentrated mixtures of NaCl and HCl or NaOH taken in equal proportions. In the first salt/acid pair the species that determine the rates of movement of solute fronts are hydrated Na^+ cation and Cl^- anion with ionic radii of 3.58 and 3.32 Å, respectively. The difference in their sizes is not too large, but obviously sufficient for a successful separation. As illustrated by Fig. 12.6, the fronts of NaCl and HCl diverge by one-third of the total bed volume ($\Delta = 0.33$) in both the forward and reverse experiments. Here, HCl behaves as a clearly

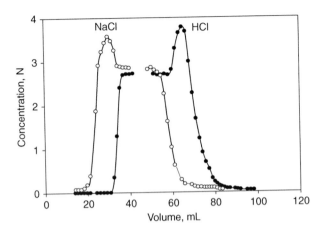

Figure 12.6   ISE separation of 2.8 N sodium chloride and 2.7 N hydrochloric acid on D4609 carbon. Experimental conditions: 28 mL column, 1 mL/min flow rate. *(After [140].)*

retained component that accumulates in the pores and emerges at the column outlet with an elution volume exceeding the total column volume. "Retention" of HCl in the "immobilized liquid phase" contributes to its separation from NaCl by the mere size exclusion effect.

In the NaCl/NaOH pair, the largest ion is Na^+ for both the electrolytes, and only a poor separation is observed ($\Delta_1 = 0.14$) on the same chromatographic column (Fig. 12.7). Here, NaOH is not in position to escape into pores that remain inaccessible to NaCl and the base elutes within the hold-up volume of the column. The separation is entirely due to the normal size exclusion process and reflects the fact that NaCl, as a whole, requires more space than NaOH, because hydroxyls do not exhibit any eigenvolume in the aqueous mobile phase. There is no additional "base retardation" in the NaCl/NaOH mixture. Though the finest pores of the sorbent receive no electrolyte and accommodate pure water, the concentration gradients in the system cannot be eliminated. The difference in the behavior of HCl and NaOH in their separation from NaCl could not be explained, if we try to compare the sizes of cations Na^+ and H^+ in the first pair and anions Cl^- and OH^- in the second pair, although exactly these ions distinguish the electrolytes in each pair and represent the target and the actual result of the separations.

Contrary to the above NaCl/NaOH mixture where NaOH behaves as a partially *excluded* electrolyte, NaOH turns into a *retained* component in its

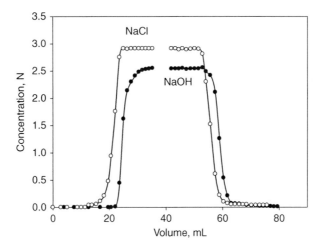

Figure 12.7   Separation of 2.9 N sodium chloride and 2.5 N sodium hydroxide on D4609 carbon. Experimental conditions: 28 mL column, 0.8 mL/min flow rate. *(After [140].)*

mixture with Na_2SO_4 on the same nanoporous carbon packing. Here, the electrolyte separation selectivity is governed by large sulfate anions $(r_{hydr} = 3.79 \, \text{Å})$ and the smaller sodium cations $(r_{hydr} = 3.58 \, \text{Å})$. NaOH is now in position to escape the concentrated mixture by occupying small pores that remain impassable for Na_2SO_4. With a very good separation selectivity of $\Delta_1 = 0.35$ from the salt, NaOH arrives at the column outlet well behind the front of the eluent. The trend to leveling out concentration gradients fully operates in this system, which actually presents the first case of "base retardation" chromatography.

Even more informative are results of examining mixtures of LiCl with HCl or LiOH, because, thanks to high solubility of LiCl, it is possible to prepare highly concentrated salt solutions. In these systems water is in deficiency and cannot fully hydrate dissolved ions. While in dilute solutions Li^+ cation is larger than Cl^- anion $(r_{hydr}$ 3.82 and 3.32 Å, respectively), the size ratio of the two ions may reverse on dehydration, since the van der Waals radius of Li^+ is much smaller $(f_{cryst} = 0.60$ and 1.81 Å, respectively). In the dehydrated form the electrolytes must tend to exist in the form of ionic pairs, where the size of LiCl is definitely larger than that of both HCl and LiOH (again, assuming negligible eigenvolumes of protons and hydroxyls in aqueous chromatographic systems). Both the above acid and base must be squeezed into the small pores and "retarded" there if the bulk solution is concentrated in LiCl [172].

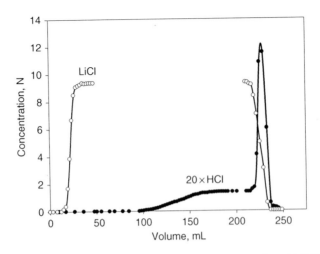

Figure 12.8 Elution profiles for a mixture of 9.4 N LiCl and 0.1 N HCl from the chromatographic column with NN-381. The concentration of HCl on the plot increased by a factor of 20. *(After [172].)*

Indeed, Fig. 12.8 demonstrates an example of extreme selectivity in "retaining" small amounts (0.1 N) of HCl from a rather concentrated (9.4 N) solution of LiCl. Here, more than 4 bed volumes ($\Delta_1 > 4$) of the feed solution could be processed, before the first traces of HCl arrived at column exit. Obviously, large volumes of the feed solution were required in this particular case, in order to increase the concentration of the common Cl^- ions in the small pores to the level comparable with their concentration in the bulk solution. On eluting with water the thus accumulated acid was recovered with the concentration of up to 0.85 N, which is 8.5 times higher than the initial HCl concentration of 0.1 N in the feed.

Two types of results can be obtained in the case of separation of LiCl from LiOH. In a mixture diluted to about 0.5 N, all constituent ions can be expected to exist in a fully solvated state. The largest hydrated ion of lithium that is common for the two electrolytes determines the migration rates of both LiCl and LiOH fronts. In accordance with the size exclusion principle, both fronts arrive at the column outlet before the hold-up volume with a minimal separation of only $\Delta_1 = 0.1$ bed volumes. This selectivity of salt/base differentiation rises by a factor of *ca* 20 when LiCl is taken at the high concentration of 3.5 N (Fig. 12.9). Now, the selectivity amounts to more than 2 bed volumes ($\Delta_1 > 2$) and LiOH behaves as a

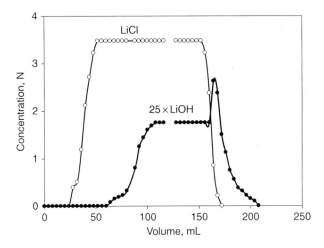

Figure 12.9 Separation of 3.5 N LiCl and 0.07 N LiOH. The concentration of LiOH on the plot increased by a factor of 25. *(After [172].)*

strongly retained component. Again, the accumulated base elutes with pure water in the form of a sharp concentrated peak.

The above-discussed "acid retardation" and "base retardation" in the "immobilized liquid phase" could be compared with the so-called "salting-out" effects. However, this term is hardly applicable to the case of "salt retardation," the first example of which was demonstrated by a successful removal of small amounts of NH_4Cl from a concentrated brine of $(NH_4)_2SO_4$. This practically important problem arises in the manufacture of caprolactam, where large amounts of sulfuric acid are converted into ammonium sulfate used for the preparation of the crystalline fertilizer. The new process of ISE on nanoporous NN-381 resin allowed an effective purification of very large volumes of concentrated sulfate brine, due to the fact that small ions of NH_4^+ and Cl^- are efficiently squeezed into and "retained" in the finest pores of the sorbent [172]. We consider this "salt retardation" process as a convincing proof of our interpretation of the mechanisms of the new electrolyte separation process.

More precisely, the basic size exclusion separation mechanism of concentrated mixtures obviously incorporates the concentration gradient-driven exclusion of smaller species from the mobile phase into stagnant zones in micropores combined with a steric and osmotic exclusion of larger species into the larger pores and moving parts of the mobile phase in the interstitial space. The porous material thus brings about a spontaneous size differentiation of the two electrolytes confined in the space of pores and, moreover, levels out their concentrations in corresponding compartments of the mobile phase. This process takes place both in static and dynamic (chromatographic) systems, indifferent of the chemical nature of the porous solid material. Though in all above presented examples the two electrolytes under separation had one common ion (Na^+, Li^+, Cl^-, or NH_4^+), the general formulation of the separation mechanism does not suggest the presence of a common ion as an essential prerequisite for the appearance of the "retardation" effect of the component composed of smaller ions. Indeed, the gradient-driven "retention" of one electrolyte can also be arranged in a mixture of four different ions. One example of this kind will be presented in Section 8.

Finally, it should be said that the above-discussed osmotic pressure–forced "retention" of one of the components of a mixture subjected to separation according to size exclusion mechanism does not exclude the possibility of true retention-type solute/sorbent interactions. Hypercrosslinked polystyrene may well enter attractive interactions with "soft" and lipophilic (chaotropic [163]) anions, such as sulfide, thiosulfate, rhodanide, perchlorate, tetrafluoroborate, and hexafluorophosphate. Also, some cations such as silver, copper, and mercury may interact with the aromatic π-systems of the polystyrene matrix, which will retard the movement of the ions. These interactions may contribute to the separation of the ions from their still larger competing ions, but they may also be counterproductive to the size exclusion effect and deteriorate the separation of the retarded ions from smaller species.

8. EXCHANGE OF IONS IN ELECTROLYTE MIXTURES

As an interesting example showing unique potentials of the new ISE technique we present [173] in Fig. 12.10 the results of chromatography of a mixture composed of four different ions. An initial mixture of two salts, $MgCl_2$ and K_2SO_4, was actually separated into fractions composed of $MgSO_4$ and KCl by using a column with water-swollen beads of NN-381. Fig. 12.10 demonstrates plots obtained by selective analysis of all four ions. The two largest ions of the mixture, Mg^{2+} and SO_4^{2-}, migrate through the column ahead of smaller ions K^+ and Cl^-. As the two initial salts were taken in equivalent proportions, the elution fronts of the larger pair of ions coincide; the same is true for the pair of smaller ions, in spite of

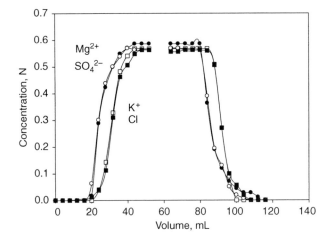

Figure 12.10   Elution profiles for MgCl₂/K₂SO₄ mixture from chromatographic column with MN-381. Experimental conditions: 44 mL column, 0.8 mL/min flow rate. *(After [173].)*

the fact that in each pair the ions do not need to be identical in their size. In this way, one pair of salts clearly converts into a new pair of salts.

In another experiment (Fig. 12.11), small ions of the above mixture were taken in deficiency compared to the two larger ions by preparing a mixed solution 4.6 N in MgSO₄ and 0.1 N in KCl. As can be expected, a

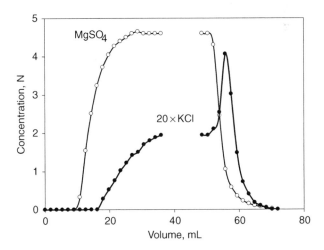

Figure 12.11   Elution profiles for a 4.6 N MgSO₄/0.2 N KCl mixed solution from the column containing NN-381. Experimental conditions: 36 mL column, 0.8 mL/min flow rate.

substantial increase in the total concentration of the electrolytes results in markedly improved selectivity of separation, owing to the operation of the concentration gradient factor. Importantly, the difference in the osmotic pressures between the water compartments in smaller and larger pores is now generated by entirely different ions, rather than by a concentration gradient of a common ion, as was the case in all previously considered systems. This example implies that the general thermodynamic origin of all above-discussed acid, base, or salt "retardation" phenomena consists of the size differentiation–caused difference in the osmotic pressures or, which is actually the same, difference in the water activity in different parts of the same aqueous mobile phase.

9. CONCEPTION OF THE "IDEAL SEPARATION PROCESS"

Chromatography is generally recognized as an unrivaled analytical tool, while, until recently, the preparative and industrial potentials of chromatography have been mainly exploited in the field of ion exchange and water treatment; recently the preparative potentials of the technique are increasingly utilized in the pharmaceutical industry, in connection with the chiral separations of enantiomers [174]. The benefits and disadvantages of chromatography in large-scale processes still remain to be fully recognized and compared with those of classical separation techniques, such as crystallization, distillation, flotation, sedimentation, and sieving.

In order to evaluate the practical utility of the new preparative ISE chromatography of mineral electrolytes, we have to pay attention to the striking effect of self-concentration of both the electrolytes in the final separated fractions, as shown in typical Fig. 12.3. From the viewpoint of general separation science, we would like to point out that the preparative ISE chromatography does not cause dilution of the initial sample with additional portions of the mobile phase. Let us use the term "ideal separation process" to define any separation that does not add any supplementary matter to the separated fractions of the initial mixture [138, 139]. An automatic consequence of isolating (removing) one component from the initial mixture in such an ideal separation process would be the inevitable concentration increase of all other components in the remaining mixture. When applied to distillation, crystallization, sieving, and other similar separation methods, this statement is so self-evident that it never needed to be formulated. On the other hand, in chromatography, transportation through the column with an adsorbent usually dilutes the initial sample with large amounts of the mobile phase, so that the very idea of observing a *self-concentration effect for all analytes because of their separation* seems unrealistic at the first sight.

Indeed, one usually deals with *retention* of analytes on the stationary phase, which reduces the concentration of the analyte in the moving zone of the mobile phase and requires additional amounts of the mobile phase to elute the retained portion of the analyte from the stationary phase. Cases of peak compression in chromatography are mostly coupled with the displacement of the adsorbed portion of the analyte (or analytes) by an auxiliary component of the mobile phase (a displacer or mobile phase modifier). In order to act like this, the latter must be adsorbed on the stationary phase even stronger than the displaced analytes. Only in frontal analysis can several weaker retained components of a mixture be obtained with an enhanced concentration, at the expense of the stronger retained component that functions as a displacer and remains in the column [175].

With this respect exclusion chromatography basically differs from all other modes of chromatography in that the analytes are *not retained* by the column packing and, therefore, do not need any special displacer or additional portions of the mobile phase, in order to be eluted from the column. Dilution of fractions separated in accordance with the size exclusion mechanism is no more unavoidable. (Dilution can be minimized to the diffusion effects at the front and tail of the analyte zone.) The absence of any supplementary matter in the frontal exclusion chromatography process relates ISE to the above-defined ideal separation process.

A starting solution that is 1 M in component A and 1 M in component B should be resolvable in a hypothetical ideal separation process in many different ways, for example, giving, from 1.0 L of the initial mixture, two fractions, 0.5 L each, one of them being 2 M in component A and the other being 2 M in analyte B, that is, resulting in the concentration enhancement of each of the components of the initial mixture by a factor of 2.

In one of our typical experiments (Fig. 12.2), the initial solution was 4 N in $CaCl_2$ and 4 N in HCl. A portion of 100 mL of that solution had a density of $1.332 \, g/cm^3$ and was composed of $22.20 \, g \, CaCl_2$, $14.58 \, g \, HCl$, and $96.42 \, g \, H_2O$. Removing HCl gas from that mixture would automatically enhance the concentration of the remaining $CaCl_2$ by 12%, while removing dry $CaCl_2$ from the mixture would enhance the concentration of HCl by 20%. In reality, the concentrations of the resolved fractions peaked up to 150% for $CaCl_2$ and up to nearly 170% for HCl. This is fully logical, as the faster-moving $CaCl_2$ molecules have also removed a sufficient amount of hydration water from the slower-moving zone of HCl. (Because of eddy diffusion on the sample borders in the low-efficiency column, the concentration enhancement was naturally less than 200%.)

Thus, the observed phenomena of self-concentration of the resolved CaCl$_2$ and HCl fractions are becoming self-evident from the close relationship between frontal exclusion chromatography and the above-defined ideal separation process. Indeed, the hydrated ions of calcium chloride, which move in the reverse experiment faster than the acid ions do, depart from the mixed-solution zone and liberate a substantial portion of space in this zone. In the ISE process, this portion of space is not filled with additional water that moves even slower; instead of this, the zone of hydrochloric acid, lagging behind the CaCl$_2$ zone, narrows down, resulting in the corresponding increase in acid concentration. In its turn, calcium chloride, which moves along the column at a higher rate than HCl does, gets rid of HCl and also becomes concentrated.

From the ideal separation concept it also logically follows that the greater the molar volume and the higher the concentration of the component being removed from the initial mixture are, that is, the greater the space liberated upon its removal is, the more pronounced must be the self-concentrating effect for the other remaining components. All experimental results that have been observed thus far fully correspond to this expectation.

Assume now that in our imaginary experiment 1 L of the initial mixture is 1 M in component A and only 0.1 M in component B. With the existing tendency of leveling out the concentration gradients, the ideal separation process could result in a 0.9 L fraction of component A, the concentration of which would be increased by 10%, and a 0.1 L fraction of component B, the concentration of which may rise to 10-fold. This is basically what happens when small amounts of an electrolyte having smaller ions, such as HCl or NH$_4$Cl, have to be removed from the dominating salt (see, e.g., Fig. 12.8).

An extreme value of the self-concentrating effect, $C_{max}/C_0 = 15$, thus far achieved was registered for HCl when separating 12 N LiCl from 0.06 N HCl on carbon D4609 [140]. The record-breaking selectivity and self-concentrating effect resulted from a combined action of three factors: low content of HCl, its smaller effective molar volume, and the above-discussed "retardation" of the smaller species in stagnant zones of small pores. An opposite result, stronger self-concentration of larger species, was demonstrated by separating (on MN-381) a solution 5.0 N in HCl and only 0.1 N in CaCl$_2$. Here, the concentration of CaCl$_2$ in the first eluting zone peaks up to 1.8 of the initial value, whereas the concentration of the last eluting major component, HCl, increases only insignificantly. The minor component in this system cannot be "salted out" into the stagnant pore zone, because $CaCl_2$ does not fit into small pores. In addition, the factor of molar volume difference of the two electrolytes acts here counterproductively. This is the reason for the much lower separation selectivity and smaller extent of the self-concentration of the minor component, $C_{max}/C_0 = 1.8$, which is entirely due to the fact of separation, that is, removing much HCl from the minor component, $CaCl_2$.

Still, the statement that *the separation must result in the self-concentration in separated fractions, provided that chromatography meets the requirements of the ideal separation process,* may not sound convincing to some chromatographer. To prove this statement, we have conducted a series of separations of a test mixture under conditions eliminating the diffusion-caused broadening of both the leading and trailing edges of the chromatographic zone. This can be done by applying the feed solution on a dry column packing (from the bottom of the column up) and then eluting the liquid from the column by using a water-immiscible organic liquid, *n*-butanol, directed downward. However, instead of the hypercrosslinked nanoporous polystyrene, we had to use nanoporous carbon, because only the latter is easily wetted by both the aqueous solution and *n*-butanol. The latter is barely soluble in water and does not interfere much with the separation process. As has been shown in Figs. 12.6 and 12.7, the neutral carbon D4609 behaves very similarly to resin NN–381 in all ISE separations.

Confined between two sharp fronts, air/water and water/butanol, the initial aqueous feed solution thus moves through the layer of the nanoporous carbonaceous column packing material, without experiencing any changes in the total volume of the probe. Only relocation of the two dissolved electrolytes, $CaCl_2$ and HCl, within the aqueous phase takes place in this chromatographic process, as the result of the faster movement of large calcium ions compared to HCl. According to the principles of the ideal separation process, such relocation of components within a constant volume of the sample mixture must lead to self-concentration of the components in the fractions finally occupied by the components. This effect of self-concentration because of separation must be independent of the initial concentration of the mixture and only depend on the ratio of the solvent volumes that the separated components finally occupy. Improving the quality of chromatographic systems is the only way to approach these theoretically predictable relations.

To demonstrate the validity of the notion on the ideal separation process, a rather diluted mixture of $CaCl_2$ and HCl, only 0.6 N with respect to each electrolyte, was sent through the column with dry nanoporous carbon and then eluted from the column with n-butanol. As presented in Fig. 12.12, the two components experienced relocation within the constant aqueous zone according to the tendency of $CaCl_2$ to move faster than HCl. The final self-concentration extent in the first and the last fractions of the aqueous phase amounted to 2.25 and 2.42 for $CaCl_2$ and HCl, respectively. Note that these large effects resulted entirely from the separation of the components due to the size exclusion mechanism, with no additional contribution from the electrolyte differentiation due to concentration gradients which are very small at concentrations that low. (Interestingly, a similar experiment with a more concentrated mixture, 4 N in each component, partially failed, because n-butanol proved to become miscible with HCl, whose concentration increased up to about 8 N in the last fraction.)

Still, Fig. 12.12 requires a more elaborated consideration. Removing slightly hydrated chloride ions in the concentration of only 0.6 N from the $CaCl_2$ solution, without 'physically' removing the corresponding amounts of protons, should not raise the concentration of $CaCl_2$ by a factor of 2.25. More realistic for $CaCl_2$ is the self-concentration coefficient of 1.25 that is characteristic of the following five fractions which are free of HCl, as well.

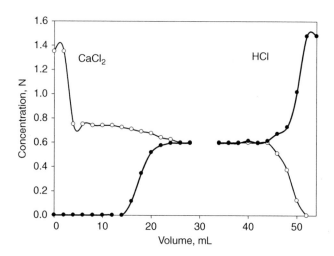

Figure 12.12 Redistribution of $CaCl_2$ and HCl within the aqueous solution passed through a column with dry carbon D4609 and then eluted with n-butanol. Initial concentration of each electrolyte in the feed, 0.6 N. *(After [171].)*

Nevertheless, the splash of $CaCl_2$ concentration is fully consistent with the size exclusion mechanism suggested. Indeed, large ions of the salt are trying to move faster than both the partner ions and the moving aqueous mobile phase. Being not in position to depart from the mobile phase on the dry column packing, the ions accumulate at the very front of the moving zone. This explanation was fully confirmed by a separate experiment with a 0.6 N solution of $CaCl_2$ alone. Here, too, the very first fraction on the front was found to contain the salt at an enhanced concentration, by a factor of 2.2, which was followed by a series of fractions with unchanged content of the salt, 0.6 N. Notably, two last aqueous fractions before the front of *n*-butanol were found to be free of the salt, thus confirming once more that analytes in SEC move faster than the mobile phase that carries them.

The above experiments on a dry column packing with no dilution of the initial sample unambiguously support the notion that any separation by SEC results in redistribution and self-concentration of all components of the initial mixture within the moving zone of the sample.

10. SIZE-EXCLUSION CHROMATOGRAPHY – A GENERAL APPROACH TO THE SEPARATION OF ELECTROLYTES

The above data, mainly accumulated from the experiments with $CaCl_2$/HCl mixtures at different concentrations and proportions, lead to a conclusion that separation of electrolytes on neutral nanoporous materials results from the discrimination of hydrated ions according to their size and is strongly enhanced by the tendency to even out the concentration gradients throughout the whole liquid phase. Moreover, the concept of ideal separation process must also be applicable to any system where elution of the analytes requires no excess eluent portions. If this interpretation of the experimental findings is conceptually correct, it must also be valid for any porous chromatographic material and any mixture of electrolytes differing in the size of constituent ions.

10.1 Use of other microporous column packings

The size exclusion mechanism of the differentiation of electrolytes should be applicable to separations on any microporous column packing material, regardless of its chemical nature. Of course, the chemistry of the material and pore walls can contribute in one way or another to the retention of ions, but the exclusion phenomena in fine pores can barely be eliminated. Indeed, the general validity of basic ideas developed for neutral microporous hypercrosslinked polystyrene was corroborated by the successful testing of several additional materials that can be expected to possess nano-sized pores.

One type is porous carbon, prepared, for example, by the pyrolysis of hypercrosslinked polystyrene sorbents under a protective atmosphere. The parameters of the porous structure of these new carbons, D4609 and D4610 (Purolite, UK), have not yet been disclosed. Our experiments on the pyrolysis of several hypercrosslinked polystyrene samples [176] showed that the process results in reducing by a half the bead size and weight, however, with a complete preservation of the overall morphology of the material. It would be logical to assume that the activated carbons prepared in this way have pores of 1–3 nm, similar to the parent polymer. Interestingly, as opposed to hypercrosslinked polystyrene, the carbon materials can be directly wetted by water. Some polar oxygen-containing groups can be suggested to exist on the surface of pores. Still, in all experiments on electrolyte chromatography, the carbons behaved very similarly to the hypercrosslinked polystyrene NanoNet 381. Typical chromatograms obtained on D4609 are presented in Figs. 12.6 and 12.7.

Another type of materials is poly-DVB, which receives microporous texture if the monomer is polymerized in the presence of an appropriate amount of a thermodynamically good solvent, such as toluene. However, in our ISE experiments, these were inferior to hypercrosslinked polystyrene sorbents.

Commercially available sulfonated hypercrosslinked polystyrene resin MN-500 (Purolite Int., UK) belongs to the group of biporous products such as MN-202 and, as a whole, has the same bimodal pore size distribution. Undoubtedly, large channels and transport pores are useless in the considered process, since only small pores can provide the separation of ions. Indeed, on both the neutral sorbent MN-202 and the cation exchanger MN-500 (used in the H^+ form), an almost identical, though reasonable, separation of H_2SO_4 from HCl was observed, with the acids emerging in this sequence. This implies that the sulfate and chloride anions can be recognized and discriminated according to their size even on these porous materials, although having a pore size distribution rather inappropriate for this particular application. On the other hand, this example demonstrates the total unimportance of the chemical nature of the polymer's surface in the ISE process, provided that the polymer functional groups do not cause retention of some ions. As can be expected, none of the two materials separates a mixture of $CaCl_2$ and $Ca(NO_3)_2$. Obviously, the large calcium ions determine the rate of migration of both the salts, while the difference between nitrate and chloride ions is small. Any possible interactions with the polar sulfonic groups of MN-500 (in the Ca^{2+} form) do not provide any discrimination between the chloride and nitrate anions.

As mentioned in Section 1, the acid retardation phenomena are known for nearly 50 years, but thus far did not find an acceptable explanation. We are convinced now that the size exclusion effect has been totally overseen as the main reason for the separation of salts from acids. The experimentally selected best separating media, anion exchangers Dowex-1x8 (Dow Chemical) and PCA-433 (Purolite Int.), are homogeneous gels with a reduced water uptake (*ca* 40%), which may well have "pores and channels" of most suitable size for the purpose of differentiating inorganic ions. Fig. 12.13 shows that there is no principal difference in the separation of the $CaCl_2$/HCl test mixture on the PCA-433 anion exchanger and on the hypercrosslinked polystyrene NN-381. This fact unambiguously points to the identity of the separation mechanisms and the nonparticipation of the quaternary ammonium functional groups in the separation. The polar groups only provide for the proper swelling of the gel-type polystyrene resin, which in the case of hypercrosslinked networks is secured by the rigidity of the expanded open-framework structure.

Sulfonated cation exchangers having the same 8% DVB as the crosslinking agent in the polystyrene matrix were found to be much less effective in the separation of electrolytes compared to the above anion exchangers. This fact can be related to the sufficiently larger swelling of the

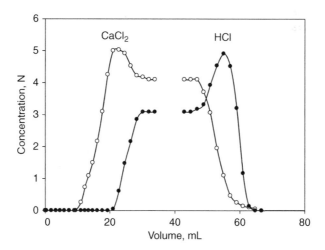

Figure 12.13 Elution profiles for a $CaCl_2$/HCl mixture from a chromatographic column with strong anion-exchange resin PCA-433 taken in Cl⁻ form.

cation exchangers and larger channel diameters in their matrix. Although prepared from the same styrene–DVB copolymer, the two types of resins may have different actual crosslinking densities, because the chloromethylation step in the preparation of anion exchangers is always combined with an additional intensive spontaneous post-crosslinking as a side reaction. The matrix of the gel-type 8%-DVB anion exchangers may thus approach the structure of macronet isoporous resins, whereas the structure of the similar 8%-DVB cation exchangers does not.

10.2 Productivity of the ion size exclusion process

The irrelevance of any stoichiometric relations between the sorbent and solute in the size exclusion mechanism of separation of electrolytes makes it possible to enhance the column productivity by simply increasing the concentration of the initial mixture. This situation favorably distinguishes the new ISE technique from all other types of adsorption chromatography and ion exchange and could be considered as a strong argument for preparative and industrial-scale applications of the new process.

Even more advantageous is the fact that with the concentration of the feed mixture increasing, the distance between the fronts of the two components under separation noticeably increases. This corresponds to an increase in the separation selectivity, which further enhances the productivity of the process. An analogous phenomenon was first observed by Nelson and Kraus [116] in 1958 in the separation of concentrated solutions of LiCl from HCl on the anion-exchange resin Dowex-1x10. The prolonged retention of HCl at increasing LiCl concentration was explained at that time by the authors as due to a drop of the activity coefficient of HCl in the resin phase (which, obviously, was not a correct explanation).

A third positive feature, unique of the ISE process, is that it results in obtaining products at concentrations higher than those in the initial feed mixture. Table 12.6 illustrated this self-concentrating effect by listing the ratio C_{max}/C_0 between the maximum concentrations of a component zone to the concentration of the component in the initial feed solution. Depending on the total concentration of the feed and the proportion of the two components, the value of C_{max}/C_0 varies in a broad range, observed up to 15 thus far [140]. According to the concept of the ideal separation process, the self-concentrating effect is particularly valid for the minor component of the mixture, regardless of the elution order of the components. The minor component occupies less space in the initial mixture and elutes after separation in the smaller portion of the effluent.

The practical value of any preparative chromatographic procedure is usually evaluated as the maximum volume of the sample that can still be tolerated by the column and the concentration of that sample. The two parameters usually counteract each other. From this viewpoint, ISE of electrolytes appears to be superior to all other chromatographic techniques in that the volume loading of a column increases with the increase in the concentration of the feed. This phenomenon results from the unique feature of the ISE technique that the phase distribution coefficients of two components fortunately change with increasing concentration: smaller ions are forced to additionally concentrate in stagnant fine pores, while large ions are additionally excluded into the interstitial volume.

The self-concentrating effect definitely enhances the practical value and the total productivity of the ISE process. Still, for estimating quantitatively its contribution, it is advisable to find concentration C_{av} graphically averaged over the total span Δ of the fractions containing the isolated component, and then the averaged self-concentrating coefficients C_{av}/C_0 for each of the two electrolytes (see Fig. 12.3). A product P of the Δ values with the corresponding averaged self-concentrating coefficients will then characterize the total productivity of the separation process; it is expressed in bed volumes (BV) of the initial mixture, from which the corresponding electrolyte can be isolated:

$$P = \Delta \times \frac{C_{av}}{C_0}$$

Notably, contrary to the selectivity values Δ and the self-concentrating coefficients for each of the two components, the corresponding process productivity P values are nearly equal. Such values can be found in Table 12.6, showing that the column productivity P can easily amount to 0.3–0.5 bed volumes for the comparable and not too high concentrations of two electrolytes. In extreme cases, more than 5–8 bed volumes of the feed solution can be processed, if traces of small ions (such as Cl^- or NH_4^+) are to be removed from very concentrated brines.

10.3 Ion size exclusion – green technology

When considering the practical aspects of the ISE process, some feasible modes of its realization should be compared.

We usually apply the technique originally developed for the "acid retardation" process, namely loading the column with the feed solution from the bottom up, and then eluting its content with water pumped in the opposite direction. This approach has helped to reveal all details and

potentials of the ISE process. However, from typical chromatograms such
as the one presented in Fig. 12.2, one can immediately note that the middle
part of the chromatogram, where the concentrations in the effluent are
equal to those in the feed, is not productive. Indeed, after determining the
process productivity by such type of preliminary experiments, this part of
the chromatogram can be easily minimized to no more than 1 bed volume;
one only has to stop introducing the feed solution at the moment that the
first eluting component becomes contaminated by the later eluting electrolyte. The mixed intermediate fraction obtained at the beginning of the
elution step can then be returned to the feed reservoir. Clearly, the process
can be easily automated.

Another approach for handling automatically electrolyte mixtures is a
pulse-type introduction of sample solutions and water in a constant direction, as in experiments by Hatch and Dillon [115]. Depending on the ratio
of the sampling and washing volumes, a more or less complete separation of
two components can be obtained. Then, the chromatographic zones have
to be automatically cut into fractions of desired concentrations and purity.
In order to achieve high flow rates and high productivity of this mode,
better quality of column packing is needed, and this can be provided by
using smaller bead particles with a narrower bead size distribution. In
general, reducing particle size is the common approach to enhance the
performance of any chromatographic process.

The potentially best way of separating mixtures of electrolytes in a
continuous procedure is provided by the simulated moving bed (SMB)
technique. It uses a series of packed columns connected into a cycle (or
carousel) and multiport valves. This method permits a continuous introduction of the feed solution and the eluent (water) into connecting tubes
between the columns and a continuous withdrawal of the two separated
fractions from two other connection points (Fig. 12.14). The input and
withdrawal positions are changed along a circle according to an optimized
program for opening and closing the valves. Results of a mathematical
simulation of the ISE process in the SMB mode have recently been
published [177, 178]. There, phase distribution coefficients obtained for a
model $CaCl_2/HCl$ mixture were used. This model predicts full feasibility of
a continuous SMB process that, for the first time, should permit to obtain
both components with concentrations higher than those in the initial
mixture, while securing the high purity of both products. Indeed, the
SMB process has become a mature large-scale technology in resolving
racemic compounds into constituent pairs of enantiomers [174], and it is

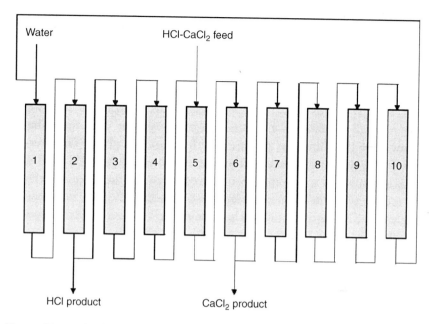

Figure 12.14 Configuration of the model SMB system. *(Reprinted from [177] with permission of Elsevier.)*

especially effective in the separation of binary mixtures. However, by involving a more complex scheme of valve switching, continuous separation of three-component mixtures is also possible [179].

The most obvious immediate advantage of ISE is that the elution process of electrolytes from the column does not require any displacing reagent, any acid or base. Pure water can easily carry all the non-retained ions through the column, so that the latter does not need any regeneration after the water wash. For this reason no waste streams are generated, which, in the case of regeneration of ion exchange columns, represent highly mineralized solutions that are expensive to dispose.

The concept of ISE thus sets new goals and new criteria for the evaluation of preparative chromatographic separations.

11. APPLICATION NICHE FOR SIZE-EXCLUSION CHROMATOGRAPHY OF ELECTROLYTES

For decades, SEC has served as an unrivaled technique for examining molecular weight distributions of polymers, that is, complex mixtures of large analytes that are difficult to operate by any other chromatographic technique. The method capitalizes on the size difference of the molecules and on their ability to penetrate into the pores of the column packing material. Although the theory relating the size of macromolecular coils to the diameters of the accessible pores is far from being mature, the practical usefulness of SEC in macromolecular research is beyond any doubt. However, SEC represents here an analytical technique, implying the injection of only a small portion of a diluted polymer solution into the column, followed by the elution of the probe with the mobile phase. This precondition provides an independence of the macromolecular coils from each other and prevents any competition between them for the space available to the mobile phase in the chromatographic column. Thus far, only micropreparative SEC separations have been described of macromolecular species that strongly differ in their size, for instance, separation of nanosponges from their clusters [180]. On a larger scale, SEC (the so-called gel filtration) operates in the purification of valuable enzymes and nucleic acids from inorganic salts, the former being excluded from hydrogel-type packing, while small inorganic molecules enter the gel phase and reside longer in the column. Because of the relatively low selectivity of SEC, it has never been considered to be a perspective approach to the separation of mineral electrolytes, especially on a preparative scale.

Preparative processing of concentrated solutions of mineral salts, acids, and bases is totally alien to SEC and other classical chromatography techniques, but presents an increasing problem in hydrometallurgy, plating and electrolysis processes, etc. This has been considered the application sphere of ion exchange or extraction technologies, but even they are economically feasible only for the selective isolation of some valuable metals or removal of toxic ions.

Processing rather concentrated mixtures of hydrochloric, nitric, or sulfuric acids with their metal salts, resulting, for example, from metal acid pickling, has been the successive niche of the "acid retardation" process [115]. The latter actually presents a very fortunate experimental finding, the physical sense of which remained obscure till our days. For this reason its application was limited to the partial recovery of excess acid from its mixtures with salts by frontal chromatography on strong anion exchanging resins.

The development of hypercrosslinked polystyrene, the first neutral microporous polymeric material, and the recognition of the size exclusion mechanism of the differentiation of ions migrating through the porous medium immediately opened new perspectives for SEC in processing mixtures of very small solutes, including mineral electrolytes. Three particular advantages of the technique should be mentioned.

First, the technique turned out to be applicable to any types of electrolyte mixtures, including mixtures of two acids, two salts, two bases, or a salt and a base, rather than just mixtures of a salt and its parent acid, as was the subject of the acid retardation process. The prerequisite of successful separation is the size difference of the largest ions of the two electrolytes to be separated. (Fractionation of mixtures composed of more than two components is also possible.)

Second, contrary to strong anion exchanging gel-type resins, nanoporous polystyrene is shape resistant and does not change its volume on an abrupt change from a concentrated electrolyte feed solution to pure water as the eluent. For this reason, much larger columns of most convenient shapes can be used with no danger of clogging, while anion exchangers require special shallow-bed short columns (75–60 cm long with about 120 cm diameter), as in the "Recoflow" process [120].

Third, the chemical inertness of hypercrosslinked polystyrene and such materials as nanoporous carbons allows the processing of such aggressive liquids as rather concentrated solutions of HF, HNO_3, and H_2SO_4, also containing H_2O_2 or chromates, while the much weaker crosslinked matrix of anion exchangers suffers from rapid oxidation resulting in increased water swelling and in the loss of separation selectivity with respect to electrolytes.

Thus, the ISE technique on neutral nanoporous materials can generally improve the productivity and/or efficiency of all processes in metal finishing applications where acid retardation is currently used, such as sulfuric acid aluminum anodizing, sulfuric and hydrochloric steel pickling, nitric/hydrofluoric stainless steel pickling, or copper, nickel, chromium, and aluminum etching involving nitric or sulfuric acid/peroxide mixtures. Many other acid-involving processes used in the metal industry, ore leaching, and hydrometallurgy could also be considered.

Additional separations on neutral nanoporous materials are also feasible, such as separation of phosphoric or sulfuric acid from hydrochloric, nitric, or hydrofluoric acids, separation of various salts from the corresponding bases (sodium aluminates from excess caustic, sodium chloride, or sodium sulfate from sodium hydroxide, etc.), or separation of salt mixtures. None of these separations could be performed on anion exchanging resins, because of the unavoidable ion exchange.

Separation of more complex electrolyte mixtures may also present great practical interest. Thus, in a very efficient separation of a mixture of iron and copper sulfates from excess sulfuric acid, the obviously larger iron ions were observed to elute before copper. Moreover, the metal-containing fractions are practically pH-neutral. Under these conditions, Fe(II) is easily oxidized by atmospheric oxygen to Fe(III) and precipitates in the form of brown iron hydroxide, leaving behind rather concentrated blue copper sulfate solutions. A simultaneous fractionation of metals thus becomes a new option in the recovery of excess sulfuric acid. Similarly, when processing mixtures containing chromium and nickel with nitric and hydrofluoric acids, the larger Cr(III) ions elute before Ni(II) ions, again offering an additional fractionation of the metals.

Another example of an important process that cannot be performed by any ion exchange or adsorption technique is the removal of traces of ammonium chloride from concentrated ammonium sulfate brines in the manufacturing of this commodity fertilizer. Chloride ions cause severe corrosion of the evaporation equipment and reduce the quality of the product. Due to the significant size difference of ammonium and chloride ions from the sulfate anion, and the large difference in the concentrations of sulfate and chloride (ca 40% against less than 2%), NH_4Cl is forced into the smallest pores, concentrating there from ca 3–5 bed volumes of the brine. It then elutes with water in the form of a sharp concentrated peak. If the ammonium sulfate is the by-product of caprolactam production, the brine must also be freed of the residues of caprolactam and other organic contaminants. This readily happens on the same hypercrosslinked polystyrene NN-381 that strongly retains organics due to the hydrophobic interactions combined with the salting-out effect of the salt. Caprolactam accumulates in the sorbent phase from up to 50 bed volumes of the feed. It must be finally displaced from the polymer with any water-miscible organic solvent. In the absence of the salt, caprolactam, being a rather polar compound, easily elutes in the form of a sharp peak (within 1.5 bed volumes) with ethanol. Due to the extremely high adsorption capacity of the sorbent with respect to caprolactam and to the relatively low concentration of organics in the feed, the column regeneration step can be introduced after many steps of the $(NH_4)_2SO_4/NH_4Cl$ separation. Notably, ethanol does not need to be fully removed from the column before the subsequent cycle of processing the ammonium sulfate feed solution, since the solvent does not interfere with the major ISE process of sulfate/chloride separation.

Finally, it should be noted that the shape resistance of neutral hypercrosslinked polystyrene permits conducting the size exclusion separation in any liquid medium, not just water. Thus, with the test mixture $CaCl_2/HCl$, almost identical resolutions were obtained by using water or aqueous methanol (1/1, v/v) as the mobile phase.

In conclusion, when comparing conventional adsorption chromatography and ion exchange with preparative SEC with respect to their convenience and productivity, one can formulate the following important advantages of the ISE process:

- The neutral column packing does not set any limit to the concentration of the feed solution and the mass of solutes in the sample.
- The resin is automatically regenerated when washing the column with water. No reagents are needed for the displacement of solutes from the column and no wastes are generated. ISE is a "green" separation process.
- As the analytes are not retained by the sorbent, no excess mobile phase is involved in the size exclusion process. Therefore, separation of the components automatically results in their self-concentration, according to the notion of the ideal separation process.
- Contrary to all retention-based chromatographic processes, the separation selectivity and overall productivity of ISE increase with the concentration of the feed solution.
- In general, the ISE process offers a totally new principle of fractionating complex mixtures of mineral ions, namely that based on the size of ions. This principle is complementary to fractionation according to their charge or ability to form coordination complexes.
- Finally, when dealing with a complex mineral/organic matrix, the combination of separations according to size exclusion and adsorption principles is possible on the same column packing.

12. CHROMATOGRAPHIC RESOLUTION OF A SALT INTO ITS PARENT ACID AND BASE CONSTITUENTS

One of the important assumptions in the above-outlined mechanism of ISE of electrolytes is that protons and hydroxyl anions have no eigenvolume in the dynamic system and do not experience exclusion from fine pores. If a charge is required at any point of the aqueous phase, a rapid shift of electrons along chains of hydrogen bonds between water molecules can easily convert any water molecule into a hydroxonium cation or hydroxyl anion. In this way electroneutrality can easily be preserved in all microdomains of the aqueous system.

From this understanding of the ISE chromatography process, it logically follows that cations and anions of a salt do not need to move through the porous space of a sorbent in the form of ionic pairs, since an occasional excessive local charge can be easily compensated by emerging protons or hydroxyls. If the sizes of the cation and anion of a given salt differ, they must tend to move through the nanoporous SEC column with different velocities. The incipient separation of the cations of a salt from their anions would be facilitated by the emerging protons and hydroxyls, preventing formation of charged micro-domains or charged zones in the aqueous phase. In other words, a salt in an SEC process can be expected to gradually resolve into the corresponding acid and base [181].

In order to test this prediction, 3 mL portions of different aqueous salt solutions were sent through conventional chromatographic columns packed with nanoporous hypercrosslinked polystyrene or poly(DVB). The latter was obtained by suspension polymerization of 80% DVB in the presence of 150% toluene. (The hypercrosslinked nature of the polymer reveals itself in the fact that the volume of dry beads increases in ethanol and water by a factor of 1.5 and 1.4, respectively.) The column effluent was conducted though a flow cell of a pH meter.

Figure 12.15 presents the plot of changing pH of the column effluent after injecting 3 mL portions of rather dilute potassium sulfate solutions. On moving through the non-functionalized polymeric medium, the smaller K^+ cations will experience more nanopores of the packing and will tend to fall behind the larger (hydrated) sulfate anions (the respective diameters are

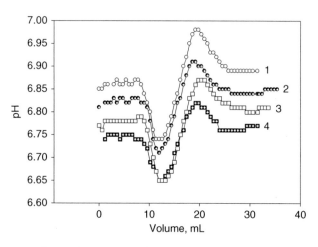

Figure 12.15 K_2SO_4 resolution on a poly-DVB column. Column, 28 mL; probe: 3 mL of (1) 0.015, (2) 0.025, (3) 0.035, (4) 0.05 M solutions; flow rate: 1.2 mL/min of the 0.015 M KCl background electrolyte. *(After [181].)*

6.62 and 7.58 Å). As seen from the initial drop of the effluent's pH, followed by its increase over the neutrality, this tendency indeed results in a partial separation of the zones of anions and cations, facilitated by the cooperation of protons and hydroxyls. The latter are always present in an aqueous phase, although in small concentration of 10^{-7} M (at pH 7 and room temperature), allowing the anions and cations to depart from each other, without the formation of electrically charged zones. Varying the concentration of the salt probe proves a consistency of the observed salt resolution phenomenon.

A similar partial resolution into acidic and basic fractions was also observed in chromatography of potassium iodide (and several other neutral salts). Obviously, the iodide anions, similar to sulfates, are excluded stronger from the nanoporous material, although the hydrated iodide anions have been reported to have a similar diameter to the hydrated potassium cations.

In order to achieve an opposite elution order, having the basic fractions before the acidic fractions (Fig. 12.16), a neutral salt composed of large tetrabutylammonium cations and small iodide anions (9.88 and 6.62 Å, respectively) was injected. In this case a typical hypercrosslinked poly-styrene material, Styrosorb 2, which is known to have only small nano-pores, was used as the sorbent. Indeed, as expected, the large organic cations were totally excluded from the polymer phase, while the smaller

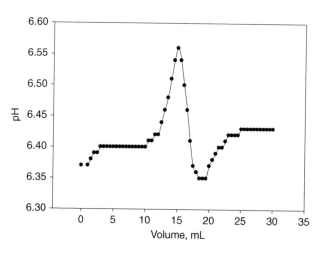

Figure 12.16 Resolution of tetrabutylammonium iodide on Styrosorb 2. Column: 28 mL; probe: 3 mL of a 1% solution; flow rate: 1.2 mL/min of the 0.015 M KCl background electrolyte. *(After [181].)*

I^- anions were in position to enter certain pores of the packing. Hence, the basic fractions emerged from the column ahead of the acidic fractions.

Naturally, such resolution of a neutral salt into constituent parent acid and base implies the creation of equivalent amounts of new protons and hydroxyls. Although the response of a pH electrode to the concentrations of these two ions is exponential, and the surface area under the negative and positive peaks in the chromatograms cannot be directly correlated with the amounts of the emerging acid and base (the deviation of the base line from pH 7.0 must also be taken into consideration), the rather symmetrical shape of the elution peaks in Figs. 12.15 and 12.16 suggests that the amounts of the acid and base are nearly equivalent.

Actually, size exclusion should not be the only possible separation mechanism resulting in a spontaneous chromatographic salt resolution. Any type of differentiated retention of the cation or anion must also generate acidic and basic effluent zones. Thus, in the case of ammonium acetate, simple dispersive (hydrophobic) interactions of the acetate anions overbalance their partial exclusion from the Styrosorb 2 polymer phase. Here, the ammonia-enriched alkaline fractions elute first, while the acetic acid–smelling acidic fractions elute from the column with a pronounced retention (Fig. 12.17). It should be noted that contrary to the bulky tetrabutylammonium cations, the acetate anions are barely excluded from the polymer phase and can experience hydrophobic retention in fine pores.

We can ask the question: what happens if one of the two ions becomes irreversibly retained on the neutral stationary phase by any of the possible mechanisms? Most probably, the remaining ion will elute in the form of a corresponding acid (or base). However, the eluting peak should be expected to be rather broad, since the "moving" ion must be retarded for a certain time by the "immobilized" counter ion. This ion-pair-type retention can explain the tailing of the ammonia peak in Fig. 12.17.

Finally, it should be mentioned that the resolution of a salt into the parent acid and base implies generation of additional protons and hydroxyls in the system through dissociation of an equivalent amount of water. This process can proceed spontaneously at room temperature, but requires a very substantial amount of energy (118 kcal/mol at 20°C). This energy can be taken from the kinetic energy of the molecules, which, however, must result in a certain decrease in the temperature within the moving chromatographic zone of the salt. Although up to now no quantitative measurements have been carried out, we see a conclusive indication that this is the mechanism by the fact that increasing the column temperature will result in

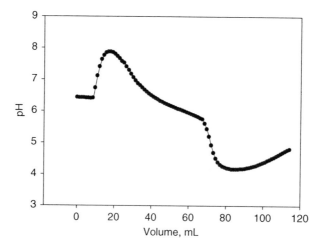

Figure 12.17   Resolution of ammonium acetate on Styrosorb 2. Column: 24 mL, probe: 3 mL of a 20% solution; flow rate: 1.2 mL/min of the 0.015 M KCl background electrolyte. *(After [181].)*

an increase in the deviations of the effluent's pH from neutrality. At 60°C these deviations were observed to amount to slightly more than one pH unit in each of the two directions.

One might think that the spontaneous chromatographic resolution of a salt is only of theoretical interest. Nevertheless, it turned out that the salt hydrolysis process facilitates the industrially important recovery of excess acids from titanium(IV)-containing solutions (or solutions of other polyvalent metals). In the ISE procedure, the initially eluting neutral Ti(IV) salts partially convert into almost neutral hydroxo or oxo complexes, thus liberating additional amounts of acid. The latter, being the actual target of the separation process, elutes later with a good selectivity. In this way, more acid can be recovered than what was present in the initial mixture in the form of free acid. Correspondingly, less alkali is now needed to precipitate TiO_2 from the first metal-containing fractions of the effluent.

CHAPTER 13

Hypercrosslinked Polystyrene as Column Packing Material in HPLC

1. MACROPOROUS POLYSTYRENE VERSUS SILICA-BASED HPLC PACKINGS

It was recognized by the end of the 1970s that the resolving power of column liquid chromatography can be substantially enhanced by using very small particles as the column packing materials, preferably with diameters in the 3–10 μm range. Additional optimization of the porous structure of these adsorbing particles allows reduction of the height equivalent to theoretical plate to two to three particle diameters, thus attaining high column efficiency of about 40,000–100,000 theoretical plates per meter column length. This was the beginning of the era of high-performance liquid chromatography (HPLC), which has become the leading analytical technique today.

Over 500 HPLC packings have been described in the literature. Nevertheless, as the result of years of development, only a limited number of types of stationary phases remain on the market. Most of the conventional HPLC separations today are performed using monodisperse silica gel 3 or 5 μm microbeads, especially those grafted with C4, C8, or C18 alkyl chains, as well as with cyano-propyl or amino-propyl groups. The last two bonded silicas and bare silica are used in normal phase (NP) HPLC, where the mobile phase (usually hexane with small amounts of isopropyl alcohol) is less polar than the stationary phase. Even more popular is the reversed phase (RP) mode, which uses polar eluents (mostly water or methanol with such additives as acetonitrile, methanol, or tetrahydrofuran (THF)) in combination with nonpolar alkyl-bonded stationary phases.

The drawbacks of silica-based materials are well known. Most serious among them are reduced hydrolytic stability in aqueous and aqueous-organic media, which practically eliminates the possibility of regenerating a contaminated column by an acidic or alkaline wash and dramatically reduces the lifetime of a column used outside of the 2–8 pH range.

Comprehensive Analytical Chemistry, Volume 56
ISSN 0166-526X, DOI 10.1016/S0166-526X(10)56013-7

Porous polymeric packings, in particular styrene–divinylbenzene (DVB) copolymers, though chemically resistant and more resilient mechanically, are too compressible. Besides, they show a tendency to swell in one type of solvent and collapse in the others, which dramatically reduces the palette of allowable mobile phases. Only macroporous highly crosslinked styrene–DVB microbeads prove to be suitable HPLC packing materials. It is not easy to prepare such small beads with a very narrow size distribution. Galia et al. [182] used suspended 4.1 μm polystyrene beads as shape templates. These polymer particles were swollen with a mixture containing dibutyl phthalate, styrene, DVB, and benzoyl peroxide. The resulting viscous droplets were polymerized in suspension to afford a series of uniformly sized 7.5 μm copolymer beads containing 30–80% DVB. Extraction of the porogens (dibutyl phthalate and also initial poly-styrene) provided porous packings with a specific surface area from 30 to 400 m²/g, respectively, and a broad pore size distribution. The 30%-DVB beads proved to be rather soft and deformed already during the packing procedure. The stronger crosslinked beads exhibited better flow properties and generated column efficiency of up to 28,000 and 14,000 plates/m for the peaks of (unretained in THF) toluene and polystyrene, respectively. These numbers for the low-molecular-weight analyte are rather poor, compared to those generated by silica-based columns, so that the polymeric material was of interest mainly for size-exclusion chromatography (SEC) of polymers and oligomers.

Indeed, microbeaded styrene–DVB copolymers, of which macroporous packings of the type PLRP-S (Varian, Polymer Laboratories) and PRP (Hamilton) are most popular, are predominantly used for SEC examination of macromolecules. For the calibration of the columns, polystyrene stan-dards of varying molecular size as well as dialkyl phthalate, butyrophenone, acetophenone, toluene, etc., dissolved in THF, are used [183]. Molecular sizes, Φ (Å), of the above test molecules are usually calculated from their molecular weights, M_w, according to the equation [184]:

$$\Phi = 0.62 M_W^{0.59}$$

Rather broad pore size distributions are generally found in this way for the polymeric SEC packings. This property widens the range of molecular sizes that can be determined by the polymeric column. In general, though, SEC columns based on macroporous silica or silica bonded phases function equally good.

For the chromatography of low-molecular-weight analytes, retention of those compounds that have solubility parameters comparable to that of polystyrene–DVB, that is, $18.6–19.0$ $(MPa)^{1/2}$, is observed to be particularly long [185]. They have a greater propensity to diffuse into the polymer matrix compared to those with very different solubility parameters. However, within the strongly crosslinked polymeric phase the analyte molecules experience steric hindrance for diffusion, and, moreover, are forced to spend much time in the most active small pores. Of the four commonly used components of mixed mobile phases in HPLC – water, methanol, acetonitrile, and THF – the last has the highest affinity to the stationary phase (solubility parameters 47.9, 29.7, 24.7, and 18.6 $(MPa)^{1/2}$, respectively) and, therefore, reduces significantly both the retention time and peak width of strongly retained analytes. Even an addition of $2–5\%$ THF to the MeOH–water mobile phase markedly improves the peak shape. This tail-suppressing effect of THF was explained by its molecules "blocking" the most active sorption sites within the polymeric phase of the packing, rather than increasing its volume by swelling effects [186]. Still, even with a very high content of acetonitrile, the most popular organic modifier of the mobile phase (up to $90/10$ acetonitrile/water), it was not possible to achieve column efficiencies over 6000 theoretical plates/m in the RP mode of chromatography of alkylbenzenes on monodisperse microbeaded macroporous polystyrene [187]. The situation is better with analytes having less affinity to the polystyrene phase. Here, the very high hydrophobicity of the macroporous polystyrene-type packings can facilitate analysis of compounds that are insoluble in water-rich media and, therefore, require high concentrations of organic modifiers, as is the case with the analysis of fat-soluble vitamins or acylglycerides (Fig. 13.1) [188].

Benefits of polystyrene-based HPLC columns best reveal themselves in aqueous–organic mobile phases with extreme pH values that cannot be tolerated by any conventional silica matrix. These conditions are often required for the direct analysis of some antocyanes, alkaloids, and other strong organic acids and bases. Analysis, purification, and preparative isolation of peptides and proteins also require sometimes very low or very high pH of the aqueous component of the mobile phase under size exclusion or RP chromatography conditions, as exemplified in Figs. 13.2 and 13.3 [189]. Protein chemistry seems to be the application area of macroporous styrene–DVB column packings with the best prospects.

Of course, through chemical modification, polystyrene-type materials can be easily converted into various ion exchanging, ligand-exchanging, or

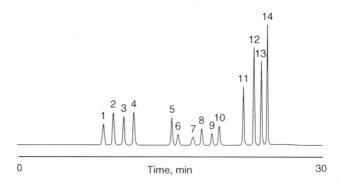

Figure 13.1 Fat analysis on a column (150 × 4.6 mm) with polymer PLRP-S (100 Å, 5 μm). Mobile phase: (A): 48% THF, 10% MeOH, 42% water; (B) 85% THF, 10% MeOH, 5% water; gradient 0–100% B in 20 min. (1–4) Monoglyceride, (5–10) diglyceride, (11–14) triglyceride. *(From [188].)*

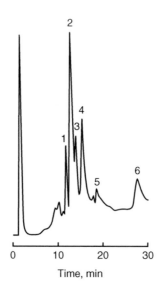

Figure 13.2 Separation of six proteins on a column (150 × 4.1 mm) with polymer PRP-3 (10 μm) at pH 2. Mobile phase: (A) water + 0.1% trifluoroacetic acid until pH 2, (B) acetonitrile + 0.1% TFA; gradient 0–50% B in 30 min. (1) Ribonuclease A, (2) insulin, (3) cytochrome C, (4) trypsin, (5) lysozyme, (6) myoglobin. *(From [189].)*

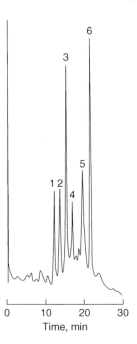

Figure 13.3 Separation of six proteins on a column (150 × 4.1 mm) with polymer PRP-3 (10 μm) at pH 12.2. Mobile phase: (A): 0.1% trifluoroacetic acid (TFA) in 50 mM sodium hydroxide pH 12.2; (B) 0.1% TFA in acetonitrile; gradient 0–60% B in 30 min. (1) RibonucleaseA, (2) insulin, (3) cytochrome C, (4) trypsin, (5) lysozyme, (6) pyruvate dehydrogenase. *(From [189].)*

biospecific phases, extremely useful in HPLC separations, but here we will restrict ourselves to unfunctionalized materials only.

2. HYPERCROSSLINKED POLYSTYRENE AS RESTRICTED-ACCESS ADSORPTION MATERIAL

Because of the diversity of applications of modern liquid chromatography, there is no universal column packing material. Still, a more or less "ideal" HPLC packing should be inert toward analytes, pH stable, and compatible with both nonpolar and polar organic solvents and even water, and allow fast diffusion of analytes in the interior of the sorbent bead. These conditions are best met by the new, third generation of polymeric adsorbent materials, hypercrosslinked polystyrenes. A rigid open-work-type hypercrosslinked network displays extremely high apparent specific surface area (up to

$1000–1500\,m^2/g$) and almost identical solvent uptake in both polar and nonpolar media, which explains the good compatibility of the material with all mobile phases, from hexane and THF to methanol and water. The whole interior of the hypercrosslinked polystyrene bead is accessible to analytes, as was the rather homogeneous network composed of small "pores" and channels of about 2.0–4.0 nm in diameter. (Biporous materials contain in addition large transport pores.) Hypercrosslinked polystyrene packings should equally well permit all separations that were carried out on macroporous polystyrene packings described in the previous section. But it is more interesting to find some specific applications that cannot be carried out on conventional silica-based materials and macroporous styrene–DVB packings.

One such very specific application possibility has been demonstrated rather early [190], which capitalizes on the unique feature of the hypercrosslinked network, namely its inherent microporosity. Pores in the 2.0–4.0 nm range should be easily accessible to the majority of low-molecular-weight analytes, but inaccessible to larger polymeric species. Analysis of biological fluids presents exactly this type of extremely important application where a direct chromatographic determination of small molecules is prevented by the presence of many proteins, polysaccharides, or other large molecules. Deproteination of such complex mixtures is cumbersome and often leads to the loss of target compounds that adsorb on the protein precipitate. Previously, Pinkerton et al. [191] suggested restricted-access materials (RAMs) for separations of such kind of samples. The RAM sorbent presented a silica matrix with a hydrophilic organic layer on the outer surface of the beads and a hydrophobic bonded phase within the pores. The proteins are not adsorbed on the accessible hydrophilic exterior of the beads and elute first from the column, ahead of small molecules that enter the hydrophobic pores.

Microporous hypercrosslinked polystyrene is a RAM material by definition. It rejects macromolecules by the size exclusion mechanism, but permits smaller analytes to interact with the whole hydrophobic interior of the bead. This property was convincingly demonstrated [190] by analyzing a mixture composed of the drug Amperozide®, its metabolite, and three related compounds, showing that macromolecules of plasma, serum, as well as individual proteins (ferritin and myoglobin) are not retained by the column. The packing was the microspherical (2–4 µm) Styrosorb whose residual chloromethyl groups were reacted with tris(hydroxymethyl)methylamine (Tris) to make the material surface more hydrophilic. It was assumed that residual chloromethyl groups are mainly situated on the surface of the beads, since there they have the smallest chance of finding a partner aromatic group to enter the conventional crosslinking Friedel–Crafts reaction. This would make the Tris-modified Styrosorb resemble the Pinkerton RAM packing in that both have a hydrophilic exterior and hydrophobic core. (It was found later that even without the Tris modification, the hypercrosslinked polystyrene does not adsorb proteins, contrary to conventional macroporous polystyrene packings. This outstanding property is explained by the open-network structure of the material not offering any solid hydrophobic surface on which the proteins can adhere.) In experiments with Amperozide®, it was found that at least 15% THF is needed for the aqueous-organic mobile phase to perfectly wet the interior of the hydrophobic RAM beads. The retention of Amperozide® strongly depends on the THF content, with the optimum separations resulting in the range of THF content between 38/62 and 60/40 (v/v) THF/phosphate buffer. Through SEC with polystyrene and polyethylene glycol standards in THF, the average pore diameter of the material was estimated as 1.45 ± 0.3 nm, which implies significant exclusion of solutes having molecular weights over 500 Da. Due to the high diffusion rates of the analytes within the hypercrosslinked network and small size of the sorbent particles, a baseline separation of the above five analytes of interest was achieved on a very short column of 29 mm in length within 15 min (Fig. 13.4).

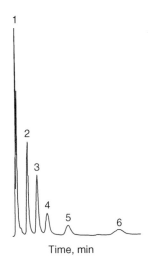

Time, min

Figure 13.4 Separation of Amperozide® (3) and related compounds with Tris-modified Styrosorb 2. Column: 29×4 mm; particle size: 2.5 μm; eluent: THF/15 mM phosphate buffer pH 8.2, 38/62 (v/v); flow rate: 0.5 mL/min; UV 262 nm. (*After [190].*)

In general, the size-sieving property of the inherently microporous hypercrosslinked polystyrene-based HPLC columns should be the most useful in the direct analysis of drugs and drug metabolites in blood and plasma matrices.

3. ION-EXCHANGING AND METAL-COMPLEXING ABILITY OF HYPERCROSSLINKED POLYSTYRENE

There are many techniques available for the analysis of trace metals in environmental samples. One of the most advanced techniques uses chelating dyes adsorbed on a hydrophobic HPLC packing that selectively retains complex-forming ions. Sutton et al. [192] have found hypercrosslinked polystyrene MN-200 (disintegrated to the mean particle size of 25 µm and packed into a 100×4.6 mm column) to be superior to all other available polymeric materials in the capacity of the dye carrier. It strongly retained 4-(2-pyridylazo)-resorcinol and showed the highest resolution of alkali-earth metals when eluted with lactic acid in potassium nitrate adjusted to pH 10 with ammonia. The elution order of the metal cations Me^{2+} (Ba, Sr, Ca, Mg) is opposite to that observed with simple ion exchange. This analytical technique is of value for samples such as milk, where using normal ion exchange would swamp the barium and strontium signals.

Immobilization on hypercrosslinked polystyrene of methyl orange (MO), a bipolar dye that has a strong acidic sulfonic group and a tertiary amine group (pI ~ 3), was also found to result in a very stable amphoteric ion exchange stationary phase, due to the high affinity of the matrix to two aromatic rings of MO. The bipolar nature of MO causes weak retention of both cations and anions and allows eluting them with very weak eluents and even water [193]. A series of cations were resolved on a 150×4.6 mm column eluted with cerium(III) nitrate of pH 5 in the order of $Na^+ < Li^+ \sim K^+ < Rb^+ < NH_3^+ < Cs^+$ and $Mg^{2+} < Sr^{2+} < Ca^{2+} < Ba^{2+}$, which again is different from the elution order from simple sulfonated polystyrene cation exchangers.

Rather unusual was the finding [194] that hypercrosslinked polystyrene as such displays some anion exchange properties in the pH 2.6–4.3 range. By measuring zeta-potential of small particles of neutral MN-200 polymer its surface was found to be neutral at pH 4.4 only. At higher pH it acquired a negative charge, while under acidic conditions it was charged positively, with the maximum zeta potential between pH 2.5 and 2.7. The authors explain the emergence of positive charges by protonation of small amounts of carbonylic groups resulting from some oxidation side reactions during the post-crosslinking process. We may add that some aromatic polymer systems may exhibit proton affinity, as discussed in Chapter 12 Section 6, thus also causing net positive charge of the matrix. In any case, in acidic media the polymer starts to discriminate and retain mineral anions in the following sequence (at pH 3.0):

$$SO_4^{2-} < Cl^- < IO_3^- < Br^- < NO_3^- < I^- << SCN^- \sim NO_2^-$$
$$<< IO_4^- \text{ for MN-200 and } F^- \sim H_2PO_4^- < Cl^- < IO_3^- < Br^- < BrO_3^-$$
$$< NO_3^- < ClO_3^- < NO_2^- < I^- < SO_4^{2-} \sim ClO_4^- << SCN^- <<IO_4^-$$

for a 10 μm spherical hypercrosslinked polystyrene. This retention order is different from that usually observed for common anion exchangers and shows a strong contribution from dispersion interactions between the polymer and ions prone to polarization. Another difference is the relatively weak retention of the double charged sulfate and the too strong retention of nitrate. Remarkably, conventional macroporous PLRP-S polymer does not exhibit any comparable anion-retaining properties, with the exception of nitrite for which a different retention mechanism is suggested (esterification of hydroxy groups). In the above ion chromatographic separation of anions on hypercrosslinked polystyrene, it is most appropriate to use dilute solutions of strong mineral perchloric, nitric, or sulfuric acids as the eluents. Herewith, the stronger retained perchloric and nitric acids are stronger eluents than the weaker retained sulfuric acid. The attained efficiency of the 3400 theoretical plates/m permits separation of up to seven anions on a 250 × 3 mm column packed with 10 μm beads.

Most interesting and unexpected was the observation that neutral underivatized hypercrosslinked polystyrene MN-200 (disintegrated to 20 μm particles) behaves similarly to chelating resins in that it shows definite affinity to polyvalent metal cations [195]. By eluting with 0.05 M KNO_3–0.05 M lactic acid adjusted to pH 4.4 with NH_3, it was possible to separate peaks of Zn^{2+}, Cd^{2+}, and Pb^{2+}, the latter being retained longer on the column than the first two metals (Fig. 13.5). The presence of high concentration of potassium nitrate ruled out any possibility of simple cation exchange occurring. Doubtless, the coordination of the metals took place with the participation of electronic systems of the aromatic moieties of the polymer exposed to the aqueous media.

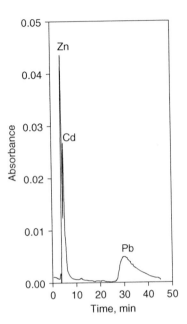

Figure 13.5 A 100 μL injection of 5 mg/L Zn^{2+}, 20 mg/L Cd^{2+}, and 20 mg/L Pb^{2+}. The eluent was 0.05 M KNO_3–0.05 M lactic acid adjusted to pH 4.4 with NH_3. A 15 cm unmodified 20 μm MN-200 column was used and post-column detection was with 4-(2-pyridylazo)-resorcinol (PAR) at 400 nm. *(After [195].)*

This π-donating property of the polymer will be treated in Section .4. It proved to be especially useful in developing a simple method for the direct determination of bismuth in lead–lithium alloys. These alloys are planned for use in the production of tritium by neutron bombardment in future nuclear fusion plants. Bismuth contamination of the alloys may lead to the formation of undesirable radioactive polonium isotopes and therefore needs to be frequently controlled. In high-acid-strength media (1.8 M HBr), lead is no longer retained by MN-200 and elutes with the solvent front, while bismuth emerges soon as a sharp peak when detected directly as a bromide complex at 370 nm. Nitric acid, which is used to digest the alloy, does not disturb the determination. A 5-cm-long column permits the determination of Bi with a detection limit better than 10 μg/L every 4 min. There was no indication of deterioration of the column after 2 months of operation with the strong hydrobromic acid. Obviously, several other analytical procedures for complex-forming metals can capitalize on the π-donating property of hypercrosslinked polystyrene.

4. π–π-INTERACTION SELECTIVITY IN HPLC ON HYPERCROSSLINKED POLYSTYRENE

Such important properties of hypercrosslinked polystyrene as excellent hydrolytic stability, pressure resistance, and ability to display an almost equal swelling in both polar and nonpolar solvents fully justify continuing attempts to prepare monosized microbeaded material and compare it with common silica-based materials for use as a general HPLC column packing. First of all, the performance of the polymeric material has to be evaluated under normal phase and RP conditions, the most widespread modes of separating organic compounds. This task was systematically accomplished by using principal component analysis in order to reveal basic factors that govern retention of a representative series of organic compounds eluted from a hypercrosslinked polystyrene column (Chromalite 5HGN, 5 μm, Purolite) with different mobile phases [196, 197].

4.1 Reversed phase chromatography

Under RP conditions, that is, in the combination of hypercrosslinked polystyrene with aqueous-organic mobile phases that are more polar than the hydrophobic polymer, the retention was found to be largely determined by dispersive type of interactions. In these chromatographic systems, the latter are also called hydrophobic interactions, and they have been well examined using C_{18}-modified RP silica. Accordingly, the two column packings were found to generate similar, though not identical, retention sequences, as shown in Fig. 13.6.

Comparing the separation results in Fig. 13.6, note that the experimental batch of hypercrosslinked Chromalite 5HGN still slightly yields to the well-optimized Zorbax SB-C_{18} silica RP packing with respect to the column efficiency (we may recall here that macroporous styrene–DVB packings are not usable under these conditions at all). In general retention on the polymeric phase is stronger than that on the RP silica, thus reflecting a markedly higher "hydrophobicity" of the polymer and its higher affinity to aromatic compounds. Earlier, retention increment for one phenyl group on hypercrosslinked polystyrene was found to be 2.5 times higher than on a macroporous styrene–60% DVB resin and 3.6 times higher than on RP alkyl silica [193]. Finally, some changes in the selectivity of analytes' retention can be noted from Fig. 13.6. Contrary to the situation with Zorbax, anisole and nitrobenzene retain longer than benzene on the polymeric phase. This phenomenon reflects some contribution from the π–π-interactions with

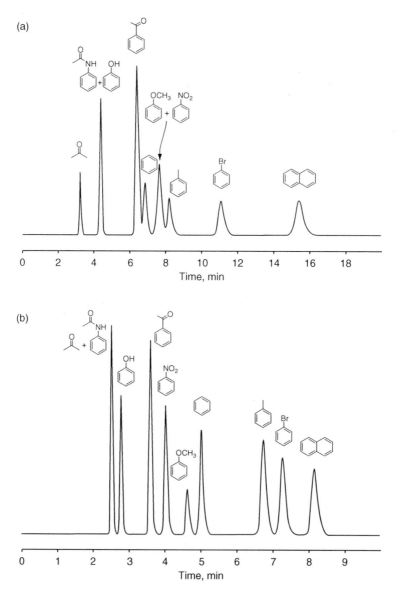

Figure 13.6 HPLC separation of a mixture of eight substituted benzenes, acetone, and naphthalene under reversed phase conditions. (a) Column: Chromalite 5HGN (250 × 4.6 mm); mobile phase: acetonitrile/THF/water (80:10:10). (b) Column: Zorbax SB-C, 5 μm (250 × 4.6 mm.); mobile phase: acetonitrile/water (60:40). Flow rate: 1 mL/min. *(Reprinted from [196] with permission of Elsevier.)*

the polymer to the dominating hydrophobic interactions characteristic of both packings (the general RP mechanism is evident from the fact that alkylbenzenes such as toluene retain longer than benzene).

Express affinity of hypercrosslinked polystyrene to aromatic rings of analytes makes it possible to successfully resolve under RP conditions complex mixtures of such hydrophilic aromatic compounds as dihydroxybenzenes, aminophenols, and phenylenediamines, which are scarcely retained on RP silica. These substances, though not really harmless, are conventional components of cosmetics such as sun creams, skin-toning creams, and hair-coloring dyes. Legislation in many countries prescribes the content of aminophenols to be less than 2%, and that of phenylenediamines less than 4%. A simple and rapid chromatographic analytical procedure was developed [198] using a column packed with Chromalite 5HGN. Cosmetics usually contain various herbal extracts and oils, lanolin, high-molecular-weight alcohols, monoglycerols, wax, etc., which all have to be preliminarily removed by adsorption on an RP silica. Most of the remaining polar analytes of interest are then separated under RP conditions as shown in Fig. 13.7. High concentration of the phosphate buffer in the mobile phase was found to improve peak shapes, probably by suppressing the influence of residual Friedel–Crafts catalyst metals in the polymer. The efficiency of the column was high, about 70,000 plates/m when calculated for the peak of resorcinol.

Enhanced hydrophobicity of hypercrosslinked polystyrene, combined with additional π-interactions with aromatic moieties, explains the unusually strong retention of phenol and its derivatives from aqueous solutions. This makes it possible to pre-concentrate traces of phenols and chloro- and nitrophenols from water samples directly on the top of the analytical HPLC column (with MN-200 disintegrated to 15 μm particles) [199] and then analyze the mixture on the same column with an aqueous-acetonitrile RP eluent. This approach increases the sensitivity of the determination because the whole amount of analytes initially present in the sample appears in the column and arrives at the detector, contrary to the situation with the off-line solid phase extraction (SPE) pre-concentration approach.

4.2 Quasi-Normal phase chromatography

Elution with nonpolar hexane modified with small amounts of methanol or propanol-2 is usually called normal phase chromatography. This is correct for systems with bare silica stationary phases that are definitely more polar

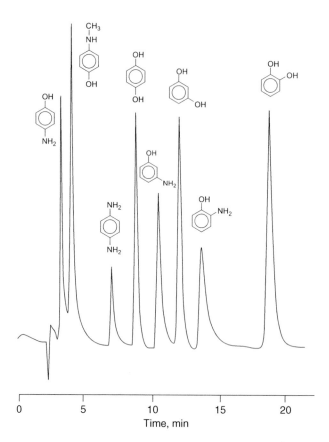

Figure 13.7 Separation of a mixture of dihydroxybenzenes, aminophenols, and p-phenylenediamine on a column (250 × 4.6 mm) with Chromalite 5HGN under RP conditions in acetonitrile/300 mM ammonium phosphate buffer of pH 5.15 (25/75); UV at 280 nm. *(Reprinted from [198] with permission of the Royal Society of Chemistry.)*

than the eluent. The term is not applicable to hypercrosslinked polystyrene columns, as they are by no means more polar than the above eluents. Therefore, we introduced the term quasi-normal phase [196, 197] for systems where both the stationary and mobile phases are nonpolar. Neither bare silica nor hypercrosslinked polystyrene retain nonpolar aliphatic compounds from hexane-reach eluents. Polar and aromatic compounds are retained on silica due to dipole-type interactions with surface silanols under the normal phase conditions. On the contrary, principal component analysis unambiguously indicates the π–π-interactions to be the governing retention mechanism on the polymer under the quasi-normal phase conditions. Indeed, retention strength of substituted benzenes strongly correlates with the Hammet–Taft constant σ_p^o. Accordingly, retention of aromatics generally enhances with the π-electron-donating or π-electron-accepting ability of the adsorbate, relative to the constant π-electron density of the polystyrene sorbent. The least difference in the π-electron densities of the analyte and polystyrene can be expected for benzene and toluene. Indeed, these analytes are found to be the least retained on Chromalite 5HGN under the quasi-normal phase conditions. Both electron-donating (e.g., hydroxy group) and electron-accepting (e.g., nitro group) substituents in the aromatic rings of analytes facilitate the retention, as can be inferred from Fig. 13.8.

Of all the organic solvents tested as the eluents in the quasi-normal phase chromatographic system, the strongest eluents are dichloromethane and chloroform, while hexane and pentane possess the weakest elution power.

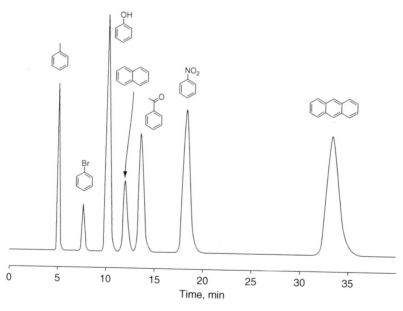

Figure 13.8 HPLC separation of seven aromatic compounds under quasi-normal phase conditions. Column: Chromalite 5HGN (250 × 4.6 mm). Mobile phase: pentane/CH$_2$Cl$_2$/isopropanol (75:5:20), flow rate, 1.0 mL/min. *(Reprinted from [196] with permission of Elsevier.)*

It is important to note that besides aromatic rings, π-type interactions are also peculiar to aliphatic compounds with double bonds, as well as functional groups that expose lone electron pairs at heteroatoms of O, N, and S. Therefore, numerous compounds incorporating polar functional groups can also be successfully separated under quasi-normal phase conditions on hypercrosslinked polystyrene packings.

4.3 Mixed-Mode chromatography

Alkyl substituents diminish the retention of aromatic compounds in nonpolar eluents (like hexane/chloroform mixtures), which is the opposite tendency to the retention under RP conditions in water/propanol-2 polar eluents. As hypercrosslinked polystyrene is compatible with both the above types of mobile phases, by just adding propanol-2 to a nonpolar mobile phase it is possible to mutually cancel the above contributions of alkyl substituents. As a result, only compounds having a different number of aromatic rings will separate, due to the difference in their susceptibility to π–π-interactions. This presents an opportunity to perform either a full resolution of a mixed aromatic sample or a group separation into mono-, bi-, and tricyclic aromatics on the same column by just running the analysis under RP, quasi-normal phase, or mixed-mode conditions (Fig. 13.9). A practical example of such sensitive group analysis of aromatics in real gasoline samples can be found in [197].

4.4 Other modes of HPLC separations

The outstanding property of hypercrosslinked polystyrene of being compatible with any kind of aqueous and organic mobile phase as well as any mixture of solvents allows many different solvent gradients to be designed in accordance with the needs of separation. It makes possible the separation of complex mixtures of analytes according to many different mechanisms of their interaction with the same stationary phase, dispersion, hydrophobic, π–π-interactions, complex formation, and size exclusion. Sychov et al. [197] present examples of different elution orders of components of the same analytes' mixture by changing the type of the mobile phase as well as by transition from one mode of chromatography to another in the course of gradient elution. In any case, the unique selectivity of the hypercrosslinked polystyrene largely depends on the π-activity of this packing. The authors suggest the selectivity of anthracene/phenol separation as the measure of the π-donating–accepting ability of a stationary phase under normal phase

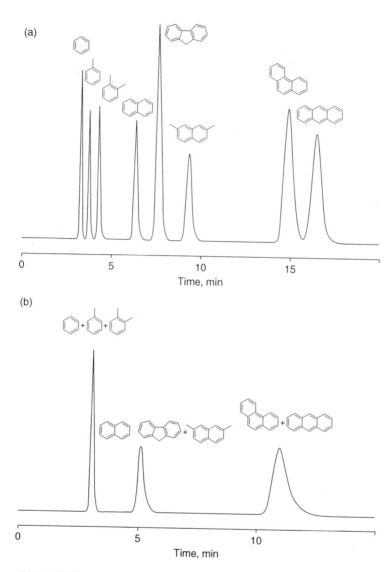

Figure 13.9 HPLC separation of a mixture of aromatic compounds under (a) reversed phase and (b) mixed-mode conditions. Column: Chromalite 5HGN (250 × 6 mm). Mobile phase: (a) acetonitrile/CH$_2$Cl$_2$/isopropanol (85:10:5), flow rate: 1.5 mL/min; (b) hexane/CH$_2$Cl$_2$/isopropanol (30:20:50), flow rate: 1.0 mL/min. *(Reprinted from [196] with permission of Elsevier.)*

chromatography conditions (in hexane/propanol, 93/7) and that of the nitrobenzene/benzene pair under RP conditions (in methanol/water eluents). In all cases the π-selectivity of hypercrosslinked polystyrene is markedly higher than that of other conventional HPLC packings, including phenyl-modified silica and even porous graphitic carbon Hypercarb. This may result from both the larger inner surface area of hypercrosslinked matrix and the fact that its phenyl groups are totally exposed to the analytes while they are part of a graphitic solid phase in Hypercarb or partially deactivated by interactions with residual silanols in the case of silica phenyl phase.

Another prospective area of application of the hypercrosslinked packing is modern high-temperature HPLC as well as chromatography under other extreme conditions. Enhanced temperatures, through facilitating the mass transfer, substantially increase the column efficiency and reduce analysis time. In gel permeation chromatography extreme conditions refer to either extremes of temperatures (typically between 80 and 220°C), extremes of solvent aggression (the use of highly aggressive or toxic solvents), or both [200]. Conventional silica-based materials, because of their insufficient chemical resistance, are not suitable for that purpose, while neutral hypercrosslinked polystyrene withstands temperatures of 220°C and aggressive solvents. An example of rapid and efficient HPLC separation at 60°C can be found in [196].

All the above benefits position hypercrosslinked polystyrene among the most prospective HPLC packings and stimulate efforts toward further optimization of its performance. In particular, the disadvantage of the column packing in that an occasional running dry will destroy the column efficiency still has to be dealt with.

A significant step toward expanding the use of hypercrosslinked polystyrene as HPLC separation media was made recently by the group of Svec [201]. They first prepared monolithic poly(styrene-*co*-vinylbenzyl chloride-*co*-divinylbenzene) precursor capillary columns and then hypercrosslinked the polymer to afford a monolith containing an array of small pores. This monolithic column exhibited a surface area of 663 m^2/g, more than 1 order of magnitude larger than measured for the precursor column. This monolithic column afforded good separation of uracil and alkylbenzenes in isocratic mobile phase mode and also proved useful for separations in size exclusion mode. A column efficiency as high as 73,000 plates/m was determined for uracil. In contrast, the presence of mesopores in this hypercrosslinked monolith had a detrimental effect on the separation of proteins. In fact, an amazingly wide variety of techniques enabling the preparation of porous polymer monoliths, discussed in detail in a recent review by Svec [202], offers many possibilities for further development of capillary HPLC.

Finally, it is worth mentioning a successful attempt to form a thin layer of hypercrosslinked polystyrene on the surface of macroporous (80 Å) silica particles, in order to obtain an HC–C_8 packing with an exceptional hydrolytic stability under aggressive acid aging conditions. Silica was first treated with dimethyl-chloromethylphenyl-chlorosilane and then reacted with styrene heptamer in the presence of $AlCl_3$. Four additional reaction steps followed, namely chloromethylation of the heptamer, its crosslinking via Friedel–Crafts reaction, derivatization of residual chloromethyl groups with octylbenzene, and, finally, benzene "end-capping" [203]. The stationary phase thus obtained was shown to behave as a typical RP material with a quite good (up to 100,000 plates/m for the 5 μm particles) chromatographic efficiency for neutral analytes. Again, compared to both alkyl and phenyl bonded silicas, the hypercrosslinked material offered unique chromatographic selectivity, because of express electron-donating properties. Besides, the silica-bonded hypercrosslinked polystyrene stationary phase was found to be extremely stable under highly acidic mobile phase conditions at a temperature as high as 150°C [204]. The wide pore (300 Å) material made in the above way allows ultrafast gradient elution of proteins at high temperatures and high flow rates. By reducing the time that proteins spend in the hot column, denaturation of thermally unstable proteins could be avoided.

CHAPTER 14

Solid-Phase Extraction of Organic Contaminants With Hypercrosslinked Sorbents

1. WHY IS PRE-CONCENTRATION NEEDED?

The dramatic growth of industry and agriculture in the later half of the twentieth century has led to a significant contamination of the Earth's aquatic pool with a wide variety of toxic organic compounds. Among them pesticides are the most widespread pollutants because of the enormous scale of their application around the world. In a practical illustration, in 1995 alone the amount of pesticides used worldwide reached 2.59×10^9 kg of active ingredients [205]. In European Union countries the annual scale of pesticides application increased from $291,895 \times 10^3$ to $327,280 \times 10^3$ kg between 1992 and 2001. Within the same period, annual pesticide usage in the USA declined from $426,377 \times 10^3$ to $402,790 \times 10^3$ kg [206]. Although there is a benefit of increase in crops, pesticides and their metabolites readily enter surface and groundwaters (the latter being a source of potable water in many countries), thus increasing the risk of harmful effects on human health.

The family of hazardous pollutants also includes phenol and its nitro and chloro derivatives. They enter the aquatic environment through wastewaters from many industries, such as petroleum processing and production of plastics, dyes, cellulose, pharmaceuticals, etc., or as the products of pesticides' decomposition. Phenols may also arise in drinking water from the reaction of natural humic and fulvic acids with chlorinating disinfectants. Even at non-toxic levels, they deteriorate the taste and odor of drinking water. To address the steady increase in water contamination with phenolic compounds and pesticides, the US Environmental Protection Agency (EPA) has included 26 phenolic compounds and 32 pesticides and their metabolites in the list of priority contaminants. In accordance with regulatory requirements, the allowed tolerance limit of these pollutants must not exceed $0.1\,\mu g/L$ for individual species and $0.5\,\mu g/L$

Comprehensive Analytical Chemistry, Volume 56
ISSN 0166-526X, DOI 10.1016/S0166-526X(10)56014-9

for the sum of pesticides or phenols in water intended for human consumption [207, 208].

A variety of other organic compounds contaminate soil, water, and air. Carcinogenic polyaromatic hydrocarbons, which are formed on combustion of organic matter, surfactants that are largely used in detergent preparations, chlorinated hydrocarbons employed in the chemical industry, freons in refrigerating engineering, and aromatic and heterocyclic compounds and others resulting from human activity pose a real or potential threat to human beings, flora, and fauna. Even non-toxic (for human beings) pharmaceuticals, when entering surface water and groundwater with wastes of sewage treatment plants, noticeably disrupt aquatic ecosystems. Thus, we now see that residues of contraceptive preparations in wastewaters of large cities dramatically reduce the reproductive ability of several valuable fish species in rivers of North America and Europe.

To control the extent of water contamination with substances of environmental concern it is necessary to develop rapid and sensitive analytical methods. Modern gas chromatography (GC) and liquid chromatography, especially in combination with mass spectrometry (MS), capillary electrophoresis (CE), and capillary electrochromatography are the most suitable techniques for the analysis of both an individual component and a complex mixture. However, as a rule, the concentration of solutes to be analyzed in water samples is too low, in the range of nanograms to micrograms per liter, to inject a probe directly into a chromatographic analytical column and, therefore, pre-enrichment of water samples is required.

Several methods for sample pre-concentration have been proposed: liquid–liquid extraction (LLE), solid-phase extraction (SPE), solid-phase microextraction (SPME), stir bar sorptive extraction (SBSE), and pervaporation (PV).

The greatest advantages of LLE, which is still employed in routine analysis, are its simplicity and numerous practical applications. At the same time, this method has many drawbacks, of which the major one is the low efficiency in extracting polar compounds. Even twofold extraction with methylene dichloride of water samples containing phenol and 4–nitrophenol results only in 61 and 44% recoveries, respectively [209]. Furthermore, prior to injecting an extract into an analytical column, the excess organic solvent should be evaporated. On the one hand, this may lead to partial loss of volatile compounds, and, on the other hand, it increases the risk of additional sample contamination. Additionally, large volumes of a very pure organic solvent required to extract a trace target compound significantly increase the cost of analysis. Usually, only a small portion of the extract is subjected to analysis while a major part of the solvent must be disposed of.

The performance of SPME is based on the adsorption of analytes onto a fused silica or polymeric fiber – uncoated or coated with various materials (polydimethylsiloxane, polyacrylate, Carbowax or its mixture with graphite, etc.) [210, 211] – which is housed inside the needle of a microsyringe. To conduct the analysis, the needle is placed into a contaminated water sample and the fiber is exposed to the solution for a controlled time. Since the adsorption equilibrium is established quickly on the thin layer of the absorbing material, long exposure of the fiber is not required. Then, the needle is placed into the injector of a gas chromatograph and the adsorbed compounds thermally desorb for subsequent separation and quantification. The method is simple and fast and does not require the use of solvents and so SPME is very popular in the analysis of volatile and semi-volatile compounds in food. However, sometimes poor reproducibility and linearity as well as the possibility to analyze only volatile and thermally stable compounds limit the application of the SPME method.

SBSE follows the same principle as SPME [212] but the sensitivity of this method is higher because the adsorbing polymeric film that covers the magnetic stir bar is thicker. This technique is particularly convenient for analyzing slurry samples and food homogenizates. Yet, the resulting extract is more contaminated compared to that obtained in SPME.

PV is a membrane-based separation technique designed for the accumulation of volatile compounds from solid or liquid samples before their analysis by GC. PV is performed in two configurations, either by simple collection of volatiles passed through a polymeric membrane followed by the injection of analytes into the analytical column, or by the additional pre-concentration of the analytes on such sorbents as Tenax, Chromosorb, and Carbowax (for more detail see [213] and references therein).

Over the last two decades, SPE has become a powerful alternative to the above methods because of its simplicity, flexibility, high efficiency and throughput, possibility of automation [214], and a relatively low laboratory cost. SPE owes its advanced evolution to the emergence on the world market of highly efficient polymer adsorbing materials, including hypercrosslinked sorbents, for pre-concentration of nonpolar and polar organic pollutants from different water matrices. In this Chapter, we will consider the modern analyses of numerous harmful organic compounds present in water, food, and biological liquids, placing the primary emphasis on the adsorbing properties of hypercrosslinked polystyrene sorbents in comparison with other packing materials for SPE cartridges.

2. BASIC PRINCIPLE OF SOLID-PHASE EXTRACTION

The basic principle of SPE enrichment consists in transfer of target analytes from a large volume of diluted contaminated water sample to the surface of a polymeric sorbent in a pre-column and subsequent elution of the adsorbed moieties with a small volume of an appropriate solvent. As a result of such operation, the concentration of the analytes in the eluate increases by several orders of magnitude compared to their concentration in the initial aqueous solution. Then, these analytes are separated, identified, and measured quantitatively by GC, high-performance liquid chromatography (HPLC), CE, or any other appropriate instrumental analytical technique.

In respect of the SPE performance involving chromatographic analysis, two modes have been used: off-line and on-line. Off-line trace enrichment is carried out separately from the subsequent chromatographic analysis. After sample loading on the porous sorbent cartridge and conducting the cleanup step to remove interferences from the complex sample matrix (washing of the sorbent), the adsorbed analytes are eluted with an appropriate liquid phase. The eluate can be evaporated to achieve a higher pre-concentration factor or replace the solvent with a more appropriate mobile phase. Finally an aliquot of the concentrate is injected into the analytical column. Undoubtedly, operational flexibility and the simplicity of the equipment required are great advantages of the off-line procedure [215]. The analyst can use small cartridges or syringes packed with any adsorbing material (or porous polymeric disks of *ca* 5 cm in diameter), employ any eluting solvent(s), and then provide the concentrate with any desired solvent composition and pH level. Eventually, the analyst can employ any convenient combination of analytical techniques, GC, liquid chromatography, isotachophoresis, capillary zone electrophoresis, planar chromatography, etc. However, numerous manipulations with the sample increase the risk of losing volatile moieties and/or additionally contaminate the sample under analysis.

On-line configuration permits the analyst to conduct the pre-concentration and subsequent chromatographic analysis within one set of analytical equipment. When HPLC is employed, a small sorbent-filled pre-column is connected as a sample loop to a multiport injector valve of the chromatograph. In the loop position disconnected from the mobile phase flow, a sample of water is percolated through the pre-column. Then, by switching the valve, the cartridge is directly coupled to the analytical column, and the flux of mobile phase (in back-flush mode) transports the analytes from the pre-column to the analytical one. The on-line SPE procedure is very handy for carrying out an analysis within a short period of time [216]. As compared to the off-line mode, this procedure is free from the risk of sample loss or contamination. It provides more accurate results and allows handling with much smaller sample volumes, because the entire body of the analytes of the initial water sample is taken into the analytical column. On the other hand, analytes that are strongly retained in the pre-column require a strong solvent to be transported into the main column. This strong solvent, however, adversely affects the final chromatogram, usually obtained on a C-8 or C-18 silica-based reversed phase analytical column (silica gel with grafted octyl or octadecyl aliphatic chains). In the optimal case, the analytical separating column packing should exhibit at least the same retention power as does the adsorbent in the pre-column. To this end, it is better to use the same sorbent for both the cartridge and analytical column. In this case, the same mobile phase elutes the analytes from the pre-column and performs analysis in the analytical one. Still, this on-line approach is rare in occurrence, since conventional HPLC packings are not very efficient in the capacity of SPE materials in the pre-concentration step.

Three key parameters characterize SPE efficiency: (i) breakthrough volume of the analyte; (ii) its recovery, R; and (iii) limits of its detection (LOD) and quantification (LOQ).

Breakthrough volume depends on the retention power of the SPE adsorbing material and determines the volume of the water sample, which can be percolated through the pre-column until the analyte arrives at the pre-column outlet. Thereby, the breakthrough volume determines the extent of analyte enrichment. Obviously, to achieve a high extent of pre-enrichment, the retention of the analyte should be a maximum at the sample loading step (but it should be a minimum at its elution step).

Generally, the breakthrough volume depends on the concentration of the analyte. In a practical approach, the breakthrough volume has usually been measured in one of two ways, either by passing through the pre-column different volumes of water sample containing a fixed amount of an analyte, or by passing constant volumes of water sample with rising concentrations of the analyte. In both cases the breakthrough of analyte is thought to occur when the recovery significantly decreases. Another approach was proposed in [53]. To measure the breakthrough volume of phenol employing SPE technique in an on-line configuration, a 16 μL cartridge packed with Styrosorb 2 was repeatedly loaded with 1 μg of phenol, followed by washing the sorbent with pure water, the volume of which was stepwise increased in each subsequent experiment, and then displacing phenol from the cartridge directly into the analytical column. The decrease in phenol peak height (or peak area) at a certain volume of water wash implies losing phenol at that breakthrough volume.

Recovery of an analyte is determined as the ratio of the chromatographic response to a known amount of the analyte that was spiked to the sample matrix and subjected to a full SPE procedure, on the one hand, with the response to an equal amount of analyte injected directly into the chromatographic column, on the other hand. Recovery is 100% when complete retention of the target compound by the sorbent in the SPE cartridge is followed by its complete elution with an appropriate liquid phase. One can expect 100% adsorption of a component under testing when the volume of the cartridge-percolated water sample is smaller than the breakthrough volume for that particular component. Incomplete sorption is observed frequently on extracting highly polar analytes with a nonpolar sorbent. When several analytes with a wide range of polarity are to be pre-concentrated simultaneously, the choice of sample volume is determined by a reasonable compromise between the loss of weakly retained components and the low extent of enrichment of more strongly retained components. In environmental monitoring, a recovery higher than 60% is considered to be acceptable for practical purposes.

Incomplete desorption of target compounds from the SPE material may also be the reason for low recovery. Off-line configuration offers many ways for avoiding this shortcoming. However, when working with an on-line SPE–HPLC system, the choice of the mobile phase is mostly dictated by the separation conditions in the analytical column, and this choice is not always beneficial for the compact and complete desorption of solutes from the pre-column.

The limit of detection (LOD) is the lowest concentration of a compound that can be detected in a given analytical procedure at the signal-to-noise ratio $s/n = 3$. In practice, for a reliable evaluation of the degree of contamination, it is generally acceptable to measure also the limit of quantification (LOQ) at s/n = 10. LOD and LOQ achievable in analysis are determined by both the retention capability of adsorbing material in the pre-column and the sensitivity of chromatographic detectors. Sometimes, to increase the sensitivity with respect to several analytes, two detectors are coupled in series [210].

At present, many automated systems have been suggested for off-line and on-line trace enrichment. One may find their description elsewhere [216–218]. Still, the important suggestion by Pocurrull et al. [216] must be mentioned here, since it elegantly solved, by applying a system of two six-port-valves, the problem of combining a strongly retaining hypercrosslinked polystyrene-type pre-concentration column with a weaker retaining conventional C18 analytical column. The system allows eluting the collected analytes from the pre-column with the organic component of the mobile phase, alone. The eluate is then mixed with the aqueous component of the mobile phase before the latter carries the probe into the analytical column. Researchers who followed this recommendation did not observe any peak broadening on the chromatograms. Otherwise, the elution of the analytes by the water-rich mobile phase from the pre-column proceeds slowly, thus reducing the hight and distorting the shape of all peaks on the final chromatogram.

3. PRE-CONCENTRATION OF PHENOLIC COMPOUNDS

Many inorganic and polymeric sorbents were tested for SPE pre-concentration of phenol and its derivatives before hypercrosslinked sorbents became commercially available. Early experiments revealed that both the end-capped silica-based C-8 and C-18 reversed phases and the phases with exposed hydroxyl groups retain polar phenolic compounds rather weakly [219–221]. On the one hand, dispersion interactions between aromatic rings and aliphatic chains of the above packings are rather weak, whilst, on the other hand, dipole–dipole interactions of the polar solutes with water molecules or formation of hydrogen bonds between them hold phenolic compounds predominantly in water.

Polymeric adsorbents, and, first of all, macroporous styrene–divinylbenzene (DVB) resins, retain phenols and polar pesticides substantially better, because in addition to hydrophobic forces, the π–π interactions between aromatic fragments of the solutes and the internal surface of the sorbents contribute to the analyte retention. Still, a set of macroporous Styragel resins with 3×10^3 to 1.5×10^4 Da exclusion limits (coupled in series), provided an unacceptably low recovery for phenol, <40% [222]. Retention of phenol on the typical macroporous PLRP-S resin having a rather high surface area,

$500 \, m^2/g$, and small particle size, $20 \, \mu m$, was also found to be weak [220]. Breakthrough of phenol on a $10 \times 3 \, mm$ ID pre-column occurs after pumping of only $1 \, mL$ of river water spiked at a $25 \, \mu g/L$ level, while the breakthrough volumes of more hydrophobic Cl and NO_2 derivatives of phenol exceed $50 \, mL$ [223]. Therefore, PLRP-S as such proves to be unsuitable for SPE enrichment of phenol-containing mixtures. Another macroporous polymer, Envi-Chrom P ($500 \, m^2/g$, $80-160 \, \mu m$ particles), retains phenol more strongly (breakthrough volume is $5 \, mL$ under identical conditions), but for the same reason Envi-Chrom P provides much broader peaks than PRLP-S, when the SPE cartridge is combined on-line with an HPLC C-18 analytical column. Nevertheless, recoveries of around 100% and good peak resolutions were attained for all tested pollutants including phenol by passing the water sample first through a PLRP-S pre-column to concentrate the moieties with $50 \, mL$ breakthrough volume, and then inserting in-line a cartridge with Envi-Chrom P to concentrate phenol from an additional $5 \, mL$ water sample. The limit of detection achieved in this experiment varied from $0.3 \, \mu g/L$ for phenol derivatives to $1.0 \, \mu g/L$ for phenol itself [223], the latter value being markedly higher than regulatory requirements, $0.1 \, \mu g/L$.

In order to reduce the LOD of the most polar EPA priority phenolic compounds to the required $0.1 \, \mu g/L$ level, Pocurull et al. [224, 225] suggested, first, to use tetrabutylammonium bromide as an ion-pair reagent enhancing the retention of phenols on PLRP-S and Envi-Chrom P extracting resins. Second, they coupled in series an ultraviolet (UV) and electrochemical detector to significantly enhance the detection sensitivity toward phenolic water contaminants, including those in tap and river water.

Amberchrom 161m polystyrene resins, chemically modified with hydroxyl and acetyl groups, were reported to extract many polar organic compounds, including phenol, with recoveries higher than 90%. The only exception was phloroglucinol (1,3,5-trihydroxybenzene), the recovery of which did not exceed 54% [211]. An unquestionable advantage of the modified Amberchrom 161m over unmodified styrene–DVB sorbents is their easier wetting with water, which allows shortening the time of analysis and decreasing the consumption of solvents in the off-line procedure.

Phloroglucinol strongly adsorbs from aqueous solutions onto porous graphitic carbon (PGC), the estimated retention factor k' for this sorbent being 1050 in contrast to 3 for polystyrene type PRP-1 resin and 0.3 for C-18 silica. Coquart et al. [226] believe that the powerful retention of 1,3,5-trihydroxybenzene is caused by some specific interactions between its hydroxyl groups and polar species present on the surface of the generally hydrophobic carbon. Unsubstituted phenol is retained by PGC four times stronger than by PRP-1 and 15 times stronger than by C-18. Trace-level enrichment of very polar phenolic compounds on the PGC pre-column and the on-line transfer of the enriched mixture into an HPLC analytical column packed with the same PGC enabled the authors to develop a simple and sensitive analysis of very polar pyrocatechol, resorcinol, phloroglucinol, and 2-chlorophenol at a level below 0.1 µg/L in a 50 mL sample. Most probably, strong retention of phenols points to the important contribution of π–π interactions between the aromatic compounds and the graphitic matrix.

A comparative study of Amberlite XAD-4 resin (900 m^2/g) and a laboratory-made porous copolymer of 1,4-di(methacryloyloxymethyl) naphthalene with DVB (*ca* 300 m^2/g) in off-line pre-concentration of 11 priority pollutant phenols showed that the retention characteristics of both polymers are generally the same [227]. On percolating 100 mL aqueous solution containing 0.4 µg/mL of each compound through 10 × 9 mm ID pre-columns packed with 200 mg of the above sorbents with particles of 40–50 µm in diameter, the recoveries obtained were around 100% for all the target analytes. The only exception was 2,4-dinitrophenol recoveries: 67% for the laboratory copolymer and 80% for XAD-4, as for both sorbents the breakthrough volumes were less than 100 mL. Also, the authors point out that the XAD-4 resin increases its volume in methanol by a factor of approximately 1.5, which complicates handling with XAD-4 cartridges.

The above brief review shows that conventional macroporous styrene-type sorbents, though meeting the needs of environmental monitoring with respect to many pollutants, still do not provide a desirable level of pre-concentration for the most polar phenols. Therefore, the benefits of hypercrosslinked polystyrene for SPE of polar organic pollutants were immediately recognized on the emergence of the sorbents on the world market.

Currently, several types of hypercrosslinked SPE sorbents are commercially available. Their characteristics and manufacturers are listed in Table 9.7. Purosep 200 [228] and Isolute ENV+ (Isolute ENV, IST ENV+) [229] are biporous products, the major portion of small pores having a diameter of 1–3 nm while the size of large transport pores being between 80 and 100 nm. LiChrolut EN, on the contrary, appears to be a microporous sorbent, the average pore size amounting to 3 nm [230, 231]. These three sorbents exhibit a high specific surface area of 1100–1200 m^2/g. Recently, a new material with small pores, HR-P, also intended for trace enrichment of organic pollutants, was described [232]. This polymer, as well as HYSphere-1 [233], with a specific surface area of more than 1000 m^2/g, is supposed to belong to the family of hypercrosslinked polystyrene–type sorbents.

However, the laboratory-made sample Styrosorb 2 with a 100% degree of crosslinking was the first hypercrosslinked resin examined for phenol trace enrichment [53]. To this end, 6 mg of Styrosorb 2 beads of 70–80 μm in diameter were placed in a 16 μL valve loop for the on-line SPE enrichment of the analytes and their subsequent HPLC separation and quantification using a C-18 analytical column and UV detection. Irrespective of the method used for cartridge loading, namely by injecting up to 2.0 mL of 0.5 mg/L phenol solution or by introducing 1 mL of solutions with concentrations ranging from 0.05 to 1 mg/L, the plot of chromatographic peak heights versus the amount of phenol introduced into the pre-column was found to be linear. It implies that 6 mg of Styrosorb 2 absorbs quantitatively at least 1 μg phenol injected in 2 mL of water and then releases it completely with an aqueous acetonitrile mobile phase into the analytical column. Moreover, peak broadening caused by the Styrosorb-packed SPE 16 μL pre-column was found to be less significant than that caused by the empty pre-column. By using the same approach, 100% recoveries were also obtained on concentrating 2-chlorophenol, 2,4- and 2,6-dichlorophenols, and 2,4,6-trichlorophenol.

Masque et al. [234] compared the SPE performance of PLRP-S (20 μm beads), Envi-Chrom P (80–160 μm), Amberchrom 161 (50–100 μm), acetylated Amberchrom GC-161 (50–100 μm), and LiChrolut EN (40–120 μm particles) in the on-line enrichment of phenolic compounds. By percolating a 10 μg/L standard phenol solution in Milli-Q water through 10 × 0.3 mm ID cartridges packed with the above sorbents, the largest breakthrough volume was found to be characteristic of LiChrolut EN, 35 mL, while PLRP-S demonstrated an eight times smaller breakthrough volume. On analyzing a 25 mL river water sample spiked with 4 μg/L of 11 priority phenolic compounds, the highest recovery for phenol, 80%, was also obtained with LiChrolut EN, in contrast to 50% recovery achieved with Envi Chrom P and 66% with the acetylated copolymer. Recoveries of other phenols were comparable and rather high, around 80–90%, for the latter three sorbents with the exception of pentachlorophenol, the recovery of which was only 40%.

When comparing the performance of PLRP-S, LiChrolut EN, Isolute ENV, and PGC in on-line pre-concentration followed by liquid chromatography analysis, Puig and Barceló [230] arrived at the conclusion that LiChrolut EN and Isolute ENV are the most suitable sorbents (Table 14.1) when the whole range of phenolic compounds has to be monitored, except for 2-amino-4-chlorophenol. The latter is protonated in acidic solutions and is not retained on neutral polymers. Both Isolute ENV+ and LiChrolut EN provide very good breakthrough volumes and identically acceptable

Table 14.1 Mean percentage recoveries ± standard deviations of phenolic compounds in groundwater using different sorbents and working with on-line 10 × 2 mm ID stainless steel SPE pre-columns; spiking, 4 μg/L of each phenolic compound; sample volume, 100 mL, except for phenol, catechol, and 2-amino-4-chlorophenol (50 mL), pH = 2.5

Analyte	Sorbent, particle size			
	PLRP-S 16–18 m	LiChrolut EN 15–40 m	Isolute ENV 40–140 m	PGC
Catechol	<20	55 ± 9	57 ± 8	61 ± 7
Phenol	34 ± 5	67 ± 7	62 ± 7	54 ± 6
4-Methylphenol	69 ± 6	75 ± 6	82 ± 5	52 ± 7
2,4-Dimethylphenol	81 ± 4	98 ± 4	92 ± 4	Nd
2-Nitrophenol	76 ± 5	88 ± 5	88 ± 5	Nd
4-Nitrophenol	78 ± 5	84 ± 6	100 ± 4	Nd
2,4-Dinitrophenol	100 ± 4	102 ± 5	98 ± 4	Nd
2-Amino-4-chlorophenol	<20	<20	<20	87 ± 6
4-Chloro-3-methylphenol	85 ± 5	92 ± 6	88 ± 5	Nd
2-Chlorophenol	76 ± 4	86 ± 6	81 ± 4	85 ± 7
3-Chlorophenol	78 ± 6	83 ± 5	79 ± 5	88 ± 6
4-Chlorophenol	85 ± 6	84 ± 5	80 ± 4	88 ± 5
2,4-Dichlorophenol	81 ± 3	94 ± 5	92 ± 3	Nd
2,4,6-Trichlorophenol	96 ± 4	103 ± 5	99 ± 5	Nd
2,3,5-Trichlorophenol	94 ± 5	96 ± 4	101 ± 6	Nd
2,3,4-Trichlorophenol	95 ± 5	101 ± 4	105 ± 6	Nd
3,4,5-Trichlorophenol	93 ± 5	99 ± 4	98 ± 4	Nd
Pentachlorophenol	100 ± 4	100 ± 3	99 ± 5	Nd

PGC = porous graphitic carbon; Nd = not detected.
Reprinted from [230] with permission of Elsevier.

recoveries, but the detection limits for the latter were found to be slightly better because of the smaller peak broadening, which undoubtedly corresponds to the smaller particle size of LiChrolut EN. Notably, many analytes were not eluted from the carbon under conditions of the on-line SPE–HPLC combination.

Surprisingly, contradictory results were reported in [235]. The introduction of tetrabutylammonium bromide as the ion–pairing reagent during the extraction procedure results in recoveries between 85% for 2-nitrophenol and 2,4-dinitrophenol and 103% for phenol with PLRP-S sorbent, while with LiChrolut EN recoveries were lower for most of the phenols, in particular for 2,4-dinitrophenol and 2,4,6-trichlorophenol, 66%. It is not an easy matter to interpret these results; however, partial size exclusion of voluminous ion pairs from the microporous hypercrosslinked LiChrolut EN could have played a role in reducing the breakthrough volume of the analytes.

As mentioned above, strong retention of the analytes on a hypercrosslinked sorbent in the pre-column and their substantially weaker retention on a conventional C-18 phase in the analytical column inevitably results in peak broadening when both columns are coupled to each other directly on-line. A logical idea arises from this observation, namely, to use the hypercrosslinked sorbent as packing material for both pre-concentration and analytical separation or, alternatively, to use a single analytical column first for the pre-concentration and then for separation of adsorbed compounds. The latter idea has been tried in [236] for the determination of phenols using Purosep 200 as column packing material. After percolating 2.0 mL of distilled water spiked with phenols at a level of 25–100 μg/L, subsequent separation of the on-column concentrated analytes was performed in an isocratic regime using a mixture of acetonitrile, water, and acetic acid as the mobile phase. An acceptable separation of phenol, 2-chlorophenol, 4-chlorophenol, and 2,4-dichlorophenol was obtained. Although this experiment provided an unsatisfactory high limit of detection, from 30 to 100 μg/L, it showed the principal applicability of hypercrosslinked polystyrene as the stationary phase for HPLC.

Fontanals et al. [237–239] synthesized macroporous copolymers of 4-vinylpyridine or N-vinylimidazole with DVB and a hypercrosslinked polystyrene sorbent through post-crosslinking of vinylbenzyl chloride–DVB copolymers. All these sorbents extract phenol and nitrophenols with good recoveries, comparable with those on the best commercial SPE materials. The hypercrosslinked sorbent exhibits even slightly better recovery for phenol than Oasis HLB (currently a very popular SPE sorbent, a copolymer of DVB with N-vinylpyrrolidone) [239]. The N-vinylimidazole-based copolymer did not show any noticeable advantages over LiChrolut EN in the determination of phenol, 4-nitrophenol, and 3,4-dinitrophenol [238]. Another laboratory-made sorbent, polypyrrol, showed no advantages over LiChrolut EN, either [240]; both sorbents take up comparable amounts of chlorophenols, but LiChrolut EN recovers phenol much better.

Hypercrosslinked biporous Isolute ENV+ was involved in the analysis of alkylphenolic pollutants [241]. Being the degradation products of nonionic surfactants, these compounds were found to induce a hormonal imbalance in fish. A high extent of enrichment achieved in the off-line pre-concentration from 1 L samples and recoveries close to 100% permitted the determination of 4-n-hexylphenol, 4-tert-octylphenol, 4-n-heptylphenol, n-nonylphenol, and 4-n-octylphenol at the 0.1 μg/L level.

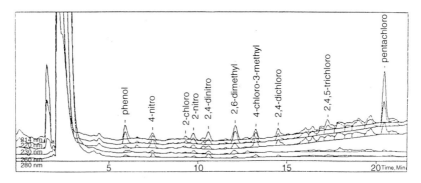

Figure 14.1 Gradient-mode HPLC separation of 18 priority phenolic contaminants off-line pre-concentrated on Purosep 200.

Also, this sorbent demonstrates exceptionally strong retention of chlorophenols [242]. No breakthrough of the analytes was registered after loading 200 mg Isolute ENV cartridge with a 2 L water sample containing 5 ng/L of chlorophenols, with a flow rate as high as 100 mL/min. Notably, complete elution of all the analytes from the loaded cartridge with a small volume of methanol (4 mL) was attained only in back-flush mode, thus implying the retention of the chlorophenols at the very top of the precolumn bed. The recoveries obtained were almost 100% while the limit of quantification was very low, ranging from 0.05 to 0.13–μg/L.

Thus, in contrast to well-known silica-based adsorbents and conventional macroporous styrene–DVB copolymers, hypercrosslinked polystyrene sorbents demonstrate a high potential for SPE trace enrichment of basically all phenolic compounds, widely differing in their polarity. Authors of several other publications also [243–246] arrived at the same conclusion. As one example, Fig. 14.1 shows a chromatogram of 18 EPA priority phenols off-line pre-concentrated on Purosep 200 with nearly 100% recoveries.

4. TRACE ENRICHMENT OF PESTICIDES

Atrazine and related triazines are the most widespread herbicides used throughout the world for the protection of crops from broadleaf weeds and for non-agricultural purposes, such as soil sterilization and road maintenance. After application, the herbicides experience degradation caused by oxidation, photolysis, and thermal or microbial impacts. Triazines and their metabolites are very resistant and may be found in soil several years after application. Concerning the global consumption of herbicides of over 3 million metric tons per year, it is not surprising that the control of water, soil, and food for contamination with these extremely toxic compounds has become of great importance. The European Commission regulations set the maximum allowed concentration of an individual pesticide in drinking water at 0.1 µg/L. Table 14.2 shows chemical structures of some widespread triazine pesticides and their metabolites. The parent atrazine, simazine, and propazine are more hydrophobic compounds compared to highly polar products of their degradation, especially didealkylatrazine, deisopropylatrazine (DIA), and hydroxyatrazine. Nevertheless, hydrophobic neutral hypercrosslinked adsorbing materials retain excellently both the herbicides and their metabolites listed in Table 14.2, as well as many other pesticides belonging to different chemical classes.

The comparison of LiChrolut EN (40–120 µm particles) performance in SPE trace enrichment of herbicides listed in Table 14.2 with that of a restricted-access material (RAM, 25 µm), alkyl-diol-silica (25 µm), reversed phase C-18 (10 µm), monofunctional C-18 (40–70 µm), and, finally, styrene–DVB Empore extraction disks revealed a decisive superiority of the hypercrosslinked sorbent in the retention of very polar DIA and DEA [247]. However, the strong retention power of LiChrolut EN causes significant band broadening when analysis is performed in on-line

Table 14.2 Chemical structure and trivial names of some s-triazine herbicides

Chemical structure of s-triazines	Compound name, abbreviation	R_1	R_2	R_3
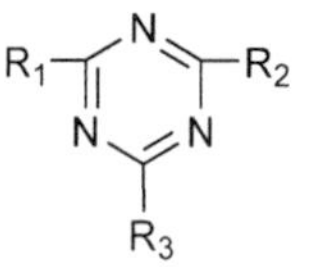 R_1—N=N—R_2, N=N, R_3	Didealkylatrazine, DDA	NH_2	NH_2	Cl
	Deisopropylatrazine, DIA	NH_2	NHEt	Cl
	Hydroxysimazine, HS	NHEt	NHEt	OH
	Desethylatrazine, DEA	NHPr	NH_2	Cl
	Hydroxyatrazine, HA	NHiPr	NHEt	OH
	Simazine, SIM	NHEt	NHEt	Cl
	Desethyl-terbuthylazine, DTBA	NHterB	NH_2	Cl
	Atrazine, ATR	NHiPr	NHEt	Cl
	Propazine, PROP	NHiPr	NHiPr	Cl
	Terbuthylazine, TBA	NHterB	NHEt	Cl

SPE–HPLC mode with conventional analytical columns. To lessen this shortcoming, the elution of analytes from the loaded pre-column must be conducted in the back-flush mode.

Another general problem that often emerges when analyzing natural waters, including seawater, is the marked sorption on the SPE material of humic and fulvic acids, which particularly increases at low pH. These matrix components cause problems with identification and quantification of the most polar analytes, which elute from the analytical column with the background of a wide interference band of humic and fulvic acids. Working at neutral pH of the water sample reduces the adsorption of natural acids [247]. Thus, using an LiChrolut EN pre-column it is possible to analyze for all herbicides in seawater and even in samples with salinity values up to 35%, but the high salt concentration was reported to shorten the lifetime of the sorbent [247].

A significant peak broadening during the on-line extraction of pesticides with a hypercrosslinked polystyrene sorbent (SDB-1, J.T. Baker), followed by their HPLC separation and quantification on a C-18 analytical column was reported [248]. The problem was solved by replacing the silica-based separating medium with PGC. This on-line combination provided trace determination of most polar pesticides such as clopyralid, methomyl, DIA, and picloram with detection limits at the low $0.1 \mu g/L$ level for $100 \, mL$ drinking water samples. The study shows that the hypercrosslinked sorbent retains the pesticides much better than the PLRP-S resin.

Curini et al. [249] tested a series of SPE sorbents for the extraction of 52 herbicides belonging to 12 chemical classes from dilute aqueous solutions. They concluded that graphitized carbon black is preferred in monitoring environmental waters for trace pesticides; recoveries of all analytes were close to 100%. The necessity to acidify water samples to enhance the sorption of pesticides on LiChrolut EN or Envi-Chrom P results in a partial loss of imazethapyr and haloxyfop, which are unstable in acidic solutions. Of 40 pesticides studied [250], LiChrolut EN does not recover only very polar desethyldeisopropyl atrazine. Also, the polymer poorly retains tris(2-ethylhexyl)phosphate (recoveries were between 40 and 50%). Recoveries of other compounds from various aqueous matrices were rather high and similar to those obtained with graphitized carbon black. Loos et al. [251] reported strong retention of very polar hydroxytriazine degradation products on LiChrolut EN at low pH values (when the analytes exist in a cationic form) and suggested the presence of anionic adsorption sites in the SPE material, without presenting additional

arguments for this statement. LiChrolut EN was also found to demonstrate good recoveries for many organophosphorous pesticides present in
various crop samples [252]. The applicability of LiChrolut EN and Isolute
ENV+ for trace enrichment of various pesticides is pointed out in the
literature [253–259] as well. When comparing the retention power of the
above hypercrosslinked sorbents toward pesticides, there is no marked
difference between Isolute ENV+ and LiChrolut EN [260]. The latter
publication, however, mentions a number of cases where Isolute ENV+
seemed to fail in adsorbing some pesticides, while almost quantitative
recoveries for the same compounds were obtained with LiChrolut EN;
however, such a dramatic difference in the performance of the two similar
materials is questionable. The retention power of LiChrolut EN, Isolute
ENV+, and Hysphere-q with respect to some polar pesticides, such as
oxamyl, atrazine, and diuron desethylatrazine, was reported to be comparable [261].

Frassanito et al. [262] focused on the analysis of glutathione–atrazine
conjugate, which plays an important role in developing plant resistance to
the herbicide. The adduct was synthesized, isolated by means of Isolute
ENV+ material, characterized by electrospray ionization mass spectrometry, and quantified by capillary zone electrophoresis. Recovery values of the
hydrophilic compound from fortified bacterial samples were between 85
and 100%.

Antifouling pesticides used in boat paints were found to severely
damage the population of bivalves and gastropods in the sea and reduce
oyster production even at extremely low concentrations. A simple and
sensitive method of analyzing some antifouling pesticides at the part-per-
trillion level [263] is based on passing 500 mL seawater samples through
cartridges packed with graphitized carbon black, LiChrolut EN, or Isolute
ENV. After the off-line pre-concentration, detection of the analytes was
carried out by HPLC coupled to atmospheric pressure chemical ionization
mass spectrometry. Both hypercrosslinked sorbents exhibited similar
retention characteristics; they extracted diuron, irgarol 1051, their by-
products, and Sea-nine 211 well, but better recoveries for chlorothalonil
have been obtained with graphitized carbon black. Nevertheless, by
employing on-line extraction on LiChrolut EN followed by aforementioned HPLC and detection systems, diuron and irgarol 1051, as well as
folpet and dichlofluanid (all are antifouling pesticides), could be determined
in 100 mL seawater samples in a fully automated manner at a level of
0.005–0.2 µg/L [264].

Recently, 4 µm monosized particles of an in-house made hypercrosslinked polystyrene–type sorbent were suggested for SPE pre-concentration of several very polar compounds, namely two pesticides, oxamyl and methomyl, phenol, guaiacol, 4-nitrophenol, 2,4-dinitrophenol, and (4-chloro-2-methyl-phenoxy) acetic acid [265]. The sorbents were obtained by post-crosslinking copolymers of vinylbenzyl chloride with 25, 50, and 75% DVB via Friedel–Crafts reaction. When percolating 100–500 mL water samples containing 0.2 µg of the above analytes through pre-columns incorporating approximately 30 mg of the 4 µm sorbent, the recoveries were between 85 and 95%. At the same time, recoveries between 80 and 90% were also obtained with large-particle LiChrolut EN (40–120 µm). In other words, the new sorbents did not demonstrate serious advantages over the commercial sorbent, with the sole exception for phenol, the recovery of which proved to be surprisingly low, 40–50% (on applying 300 or 500 mL samples). Good retention of pesticides and phenols by another commercial hypercrosslinked sorbent, HySphere-1, was also reported [266].

Disposable cartridges packed with the commercial hypercrosslinked sorbents LiChrolut EN and Isolute ENV were suggested for storage of collected pesticides and phenolic compounds. The latter were pre-concentrated by percolating 700–1000 mL water samples spiked at 5 µg/L and pH 2.5–3. Complete recoveries of 14 phenols from the priority pollutants' EPA list were obtained after storage of the cartridges at −20°C for 2 months and at +4°C for 2 weeks [267]. Four priority pesticides, desethylatrazine, fenamiphos, fenitrothion, and fonofos, can be stored at −20°C for 1 month after pre-concentration from 26 mL groundwater spiked at 10 µg/L [268]. Accordingly, in the course of environmental monitoring, it is not necessary to analyze the samples on-site immediately. After the pre-concentration of the pollutants, the loaded cartridges can be shipped within 2 weeks under refrigeration conditions from a sampling place to a central laboratory for analysis. One small cartridge substitutes a 1 L bottle in shipping, a decided advantage. Isolute ENV+ was also suggested for storage of extracted sulfonated benzene and naphthalene compounds [269].

5. TRACE ENRICHMENT OF PHARMACEUTICALS

Hundreds of residual pharmaceutical products and their metabolites enter groundwaters because of their incomplete elimination from wastewaters of

sewage treatment plants. Currently, traces of many carcinogenic, mutagenic, and reproductive toxic pharmaceuticals attack receptors of nontarget
aquatic organisms that are sensitive to individual pharmaceuticals or to their
combinations exhibiting synergetic effects. Though pharmaceuticals
unconditionally belong to the compounds of growing environmental concern, only a few publications describe the use of hypercrosslinked sorbents
for their determination in different waters.

Sacher et al. [270] developed methods for trace-level determination of
60 pharmaceuticals, including off-line SPE on reversed phase C-18 silica
and hypercrosslinked sorbents, followed by GC–MS or HPLC–electrospray
MS. It turned out that LiChrolut EN retains poorly very polar iodinated
X-ray contrast media, such as iopamidol or iomeprol, with the recovery
not exceeding 30%. Similarly, Isolute ENV+ poorly recovers sulfonamide
antibiotics. Nevertheless, the authors do not consider these results to be
negative, due to the high sensitivity of the developed analytical methods, 5
and 10 ng/L of the above-mentioned contrast media, respectively. On the
contrary, a good enrichment was achieved for macrolide and penicillin
antibiotics when using Isolute ENV+ extracting cartridges. Interestingly,
Klenner et al. [271] reported 75–85% recoveries of iodinated contrast
medium iohexol, when LiChrolut EN was involved into the analysis
of canine serum and rat urine. The application of Purosep 200 was also
found to be promising for the analysis of phenobarbital in aqueous
matrix [272].

The first studies on the determination of pharmaceuticals in human
blood can be traced back to the end of 1980s [273, 274]. These studies
dealt with pharmacokinetics of drugs in the blood of cardiac patients. As
commercial production of hypercrosslinked sorbents did not yet exist,
laboratory-made Styrosorb 2 was the sorbent of choice for drug pre-
concentration. This sorbent is attributed to the so-called RAMs, the
pores of which are open for low-molecular-weight compounds but
remain inaccessible even to relatively low-molecular-weight proteins.
Because of this, no sample pre-treatment is required prior to analysis;
even such complex matrices as plasma, serum, or urine can be directly
injected into the extracting cartridge. With this in mind, 200 μL plasma
sample was pushed through a 3 × 1 mm ID glass column by centrifugation. Then, the column was washed with 200 μL of acetonitrile, also
applying a centrifuge, and an aliquot of the extract was analyzed by
HPLC. Table 14.3 shows the chemical structure of 11 pharmaceuticals and
their metabolites, and their recoveries in comparison with the recoveries

obtained for the same drugs by SPE enrichment on reversed phase C-8 silica. The superiority of Styrosorb 2 is obvious. When applying a 1 mL sample on a Tris-modified Styrosorb 2 SPE cartridge, followed by HPLC separation with a highly selective detection system based on

Table 14.3 Recoveries (%) of some pharmaceuticals on Styrosorb 2 and reversed phase C-18 silica cartridges from human plasma

Formula and name	Synonyms	SPE	
		Styrosorb 2	RP-8
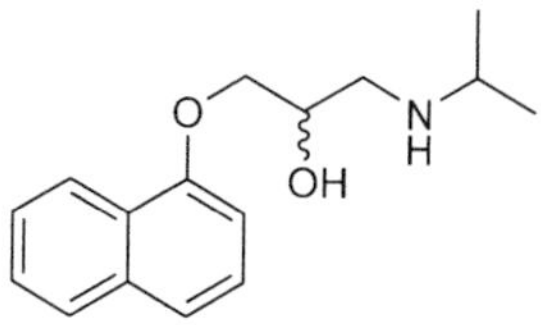 Propranolol	Alindol Avlocardil Betadren Obsidan Propral	99 ± 4	29 ± 12
S-Propranolol glucuronid (metabolite of propranolol)	–	89 ± 4	55 ± 8
R-Propranolol glucuronid (metabolite of propranolol)	–	92 ± 5	65 ± 15
4-Hydroxypropranolol sulfate (metabolite of propranolol)	–	98 ± 6	72 ± 9
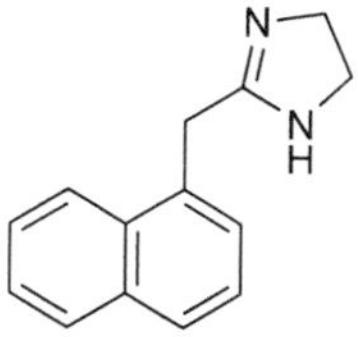 Sanorine	Naphazoline Privin Rhinazin	96 ± 3	94 ± 10
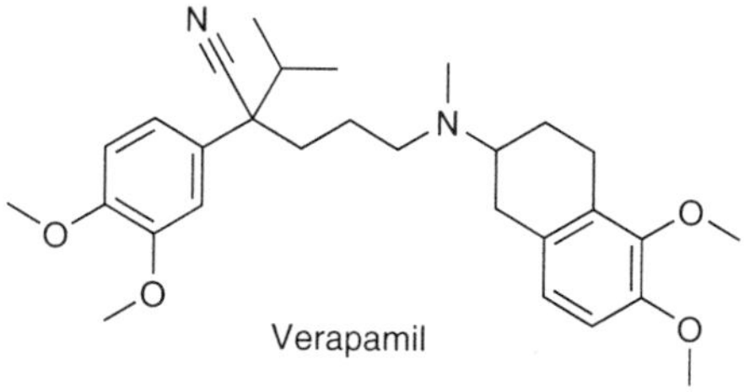 Verapamil	Finoptin Iproveratril Isoptin Isopine Vasolan Veramil	95 ± 4	–

(Continued)

Table 14.3   (Continued)

Formula and name	Synonyms	SPE	
		Styrosorb 2	RP-8
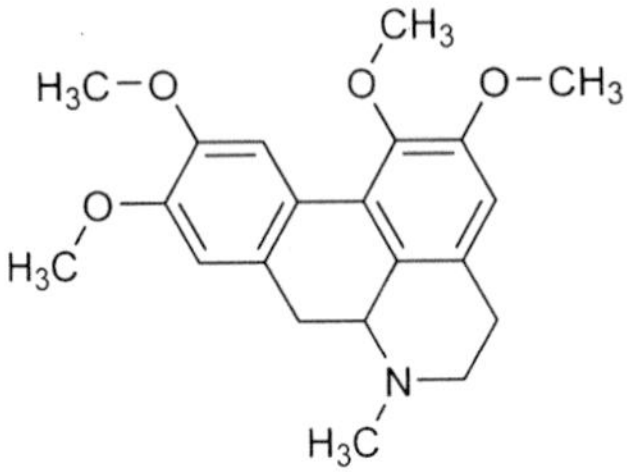 Glaucine	Tussiglaucin	90 ± 6	–
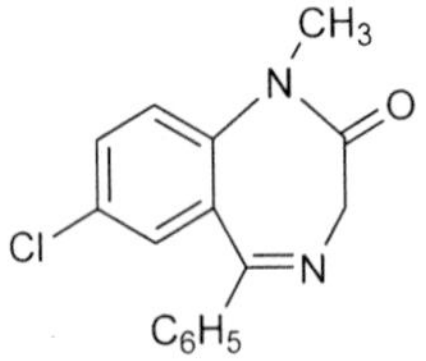 Seduxen	Ansiolin Apozepam Diazepam Valium Relanium	85 ± 3	–
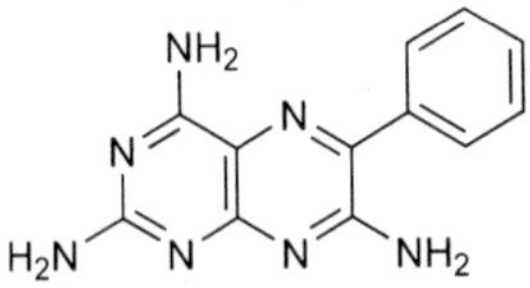 Triamterene	Anteren Diutac Dyrenium Fluxinar Noridyl	80 ± 3	–
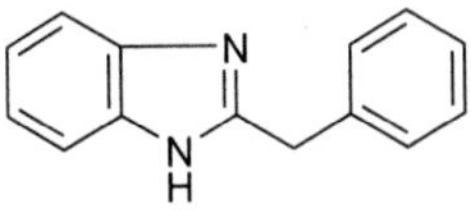 Bendazol	Bendazole Tromasedan	99 ± 5	–
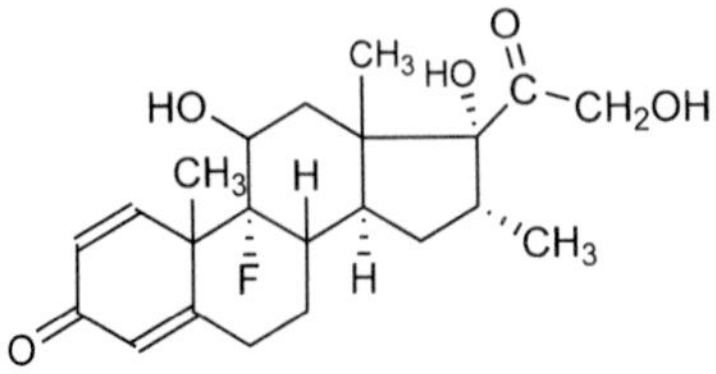 Prednisalon	Arcodexan Cortadex Deason Decardan Dexacort Resticort	72 ± 8	–

(Continued)

Table 14.3 (Continued)

Formula and name	Synonyms	SPE	
		Styrosorb 2	RP-8
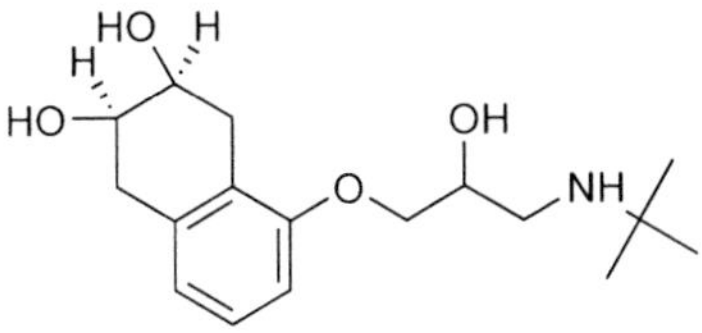 Nadolol	Anabet Betadol Corgard Nadic	48 ± 6	–
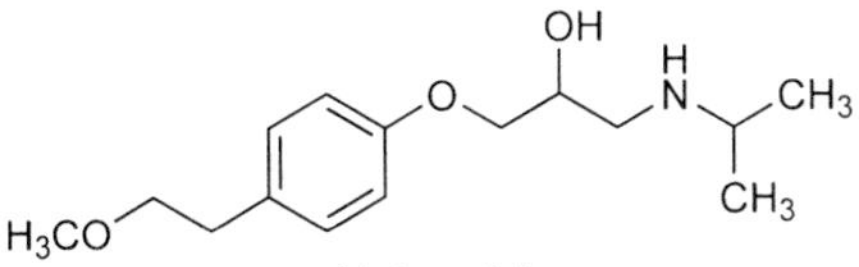 Metoprolol	Betalog Blocksan Neobloc Opresol Specicor	99 ± 5	–

SPE conditions: off-line configuration, 3 × 1 mm ID cartridge packed with 80–125 μm beads of Styrosorb 2, 200 μL of plasma, 200 μL acetonitrile for desorption;
HPLC conditions: 200 × 4 mm ID analytical column with Nucleosil 5 SA, acetonitrile as mobile phase, fluorimetric detection.

antibody–antigen interactions, the concentration of digoxin at the pg/mL level was determined in plasma of patients having taken orally a 1 mg dose of the medication [275].

6. EXTRACTION OF ORGANIC COMPOUNDS FROM BIOLOGICAL LIQUIDS

Polybrominated diphenyl ethers (flame retardant agents) were determined in human plasma of workers at an electronic dismantling facility via pre-concentration on either commercial 200 mg Isolute ENV+ cartridge or open extraction column with 27 g Hydromatrix (diatomaceous earth). Both sorbents provide recoveries of over 70% and low detection limits, ranging from 0.0076 to 0.13 ng/g, depending on the individual congener. However, Isolute ENV+ differs favorably from Hydromatrix in smaller consumption of solvents, fewer sample requirements, faster performance, and lower contamination risks owing to fewer operation steps [276].

Isolute ENV+ was also used in the determination of N-methyl-2-pyrrolidone and its metabolites in human plasma and urine [277]. At present, N-methyl-2-pyrrolidone replaces hexane and dichloroethane in many chemical processes because it is thought to be less harmful to human health. However, animal studies revealed harmful and reproductive toxic effects. Because of this, the monitoring of individuals exposed to the solvent is not unreasonable. After entering the human body, N-methyl-2-pyrrolidone quickly metabolizes as

An analysis of the above compounds comprising off-line pre-concentration on 50 mg ENV+ cartridge followed by liquid chromatography–tandem mass spectrometry showed that the metabolites reside to a larger extent in the urine of exposed individuals. Recoveries of all the analytes from urine, ranging from 50 to 100%, were found to be higher than from plasma. Lower recoveries of the most hydrophilic 5-hydroxyl-N-methyl-2-pyrrolidone were caused by the analyte loss at the washing step, but this drawback is partially compensated for by the sufficiently low detection limits of all analytes, ranging from 0.01 to 0.07 µg/mL (in urine).

Mendaš et al. [278] analyzed urine for chloro- and methylthiotriazine herbicides and some metabolites by HPLC or GC techniques using highly porous styrene–DVB copolymer (SDB-1, Baker, Netherlands) for off-line pre-concentration. Recoveries higher than 90% were achieved in ultrapure water. However, urinary matrix interferences adversely affect the quantification of polar analytes in HPLC analysis. Washing the cartridge with aqueous acetonitrile eliminates the interfering effect but leads to a complete loss of the most polar didealkylatrazine. To avoid the above complications, the effluent from the cartridge was evaporated, re-extracted with n-hexane, and analyzed by GC. However, the recoveries of monodealkylatrazine and, in particular, didealkylchlorotriazine were surprisingly low, 1–2%. The authors believe that the sorbent as such (rather than sample matrix) generates interferences which remain invisible in the GC chromatogram, but dramatically lower the response of the thermo-ionic sensitive detector.

7. SPE IN FOOD ANALYSIS

Not only useful compounds such as fats, proteins, carbohydrates, amino acids, vitamins, and salts but also many hazardous organic substances enter the human body with food and drinks and therefore food quality control is of growing concern. For instance, heterocyclic aromatic amines, in particular, 2-amino-1-methyl-6-phenylimidazo[4,5-*b*]pyridine (PhIP), a potentially mutagenic and carcinogenic compound, forms when meat or fish are fried or grilled at 150–250°C. Concentration of the heterocyclic amine in blood is too low to be analyzed even by applying SPE, while in urine it drops very soon after food intake. However, human hairs preserve the amines and may serve as a useful indicator of long- or medium-term exposure of individuals to the above toxicant. PhIP was isolated from an NaOH hydrolysate of male hairs by successive SPE on Isolute ENV+ and on a silica-based mixed-mode column with C-8 and −SO$_3$H functional groups. The use of GC separation and MS identification resulted in determining the above compound at a level between 0.25 and 25 ng per gram of hair [279].

In contrast to heterocyclic aromatic amines, resveratrol (3,5,4′-trihydroxystilbene) and its isomers, natural phenolic compounds present in grape and wines, are beneficial to human health for many reasons; primarily, there is a lower mortality from cardiovascular diseases observed in populations taking constantly a moderate amount of wine. SPE from red or white wine samples on LiChrolut EN and subsequent separation of the analytes by liquid chromatography provides selective and fast analysis of resveratrol and its isomers [280].

In addition to many useful compounds, wines may contain multi-residues of pesticides that are widely used in vineyards. Quantitative determination of pesticides in wine samples by means of SPE on LiChrolut EN followed by GC proved to be strongly complicated by the matrix effect. When using calibration plots obtained by standard solutions of pesticides in pure solvents, the recoveries for 37 pesticides and metabolites from real wine samples were found to be unrealistically high. This problem was explained by the incomplete transfer of the analytes from the injection port into the analytical capillary column and real sample matrix facilitating this transfer. Realistic calibration plots could only be prepared by using real wine samples spiked with different amounts of the pesticides. Another problem is the presence in the chromatogram of many co-eluting peaks. Still, washing of the loaded ENV+ (or Oasis) cartridge with water–isopropanol (9:1) mixture and conducting a further cleanup step of the ethyl acetate extract on a Florisil cartridge obtained much cleaner chromatograms. Recoveries over 80%, mostly between 90 and 100%, for all the compounds tested were finally achieved. Unfortunately, the LOD for several pesticides exceeded 1 μg/L [281].

Another problem of wine quality is connected to a possible presence of odor-active aldehydes. Aldehydes containing 8–10 carbon atoms are known to be strong odorants, of which (E)-2-nonenal provides a "sawdust" or "plank" off-flavor to the wine. All above aldehydes from a 200 mL wine sample are easily retained on 200 mg LiChrolut EN. After eliminating other matrix interferences by washing the loaded cartridge with an aqueous solution containing 49% methanol and 1% NaHCO$_3$, aldehydes are converted into O-pentafluorobenzyl-oximes directly in the cartridge within 15 min, and then eluted with dichloromethane and subjected to GC–MS analysis [282]. At concentrations of a few nanograms per liter, 2,4,6-trichloroanisole and 2,4,6-tribromoanisole are believed to be mostly responsible for moldy/musty off-flavor and cork taint of wines. A sample pretreatment protocol, very similar to the one described above, provided the basis for developing a fast, robust, and selective method of analyzing the halogenated anisole derivatives [283]. Sensitivity of the determination was largely enhanced by the large volume injection technique in the GC–MS analysis. The same group examined in a more detailed study [284] six different SPE materials for the pre-concentration of eight odor-active aliphatic lactones (with 8–12 carbon atoms) in wines. The authors asserted "LiChrolut EN and Isolute ENV+ have extraordinary extraction capacity and are, therefore, highly recommended sorbents for the extraction of polar compounds from wine or for the production of non-selective extracts." As for the selectivity of lactones' isolation, both hypercrosslinked sorbents are still inferior to Bond Elut-ENV.

We should also mention that LiChrolut EN was successfully used for SPE pre-concentration of 90 pesticides from water–acetone extracts of fresh fruits and vegetables with a subsequent pesticide analysis by GC–MS [285]. Only the most polar pesticides (methamidophos, acephate, and omethoate) could not be determined by this technique. Finally, a simple, fast, and sensitive method of determining a mycotoxine potuline (a metabolite of many moulds *Penicillium* and *Aspergillus*) in fruits and fruit juices was developed [286]. The method includes SPE pre-concentration of the toxin on Purosep 200 and its quantification by normal-phase HPLC and provides recoveries over 90%, with LOD five times lower than the allowed tolerance limit, 50 μg/kg.

8. EXTRACTION OF ORGANIC ACIDS

Benzene, naphthalene, anthraquinone, and stilbene sulfonates are widely used in many industrial and domestic processes including the production of pharmaceuticals, ion-exchange resins, dyes, textiles, paper, etc. Aromatic sulfonates are strongly acidic, highly hydrophilic, and biologically persistent compounds. Owing to their surface activity, the aromatic sulfonates assist desorption of toxic compounds from soil and their transfer into the aquatic environment. Sulfonates have regularly been found in wastes, groundwater, and even in tap and drinking water (at the μg/L level). For this reason, aromatic sulfonates belong to the so-called drinking water–relevant pollutants, though little is known about their toxicology and environmental impact.

Loos et al. [287] analyzed 14 representatives of aromatic sulfonates in Milli-Q, tap, and river waters, enriched at pH 2.0 on LiChrolut EN and subjected to a subsequent quantitative analysis by CE. Taking into account the high polarity of the analytes, recoveries were excellent, between 70 and 130% (Table 14.4). Only very hydrophilic 3-aminobenzenesulfonate,

Table 14.4 Recovery of aromatic sulfonates from 200 mL fortified water at 2 μg/L (for each compound) with 200 mg LiChrolut EN ($n = 6$)

Sulfonates	Milli-Q water, pH 1.0		Tap water, pH 2.0		River water, pH 2.0	
	R (%)	RSD	R (%)	RSD	R (%)	RSD
Diphenylamine-4-sulfonate	100.5	12.7	103.7	11.2	90.6	7.5
Anthraquinone-2-sulfonate	78.7	15.9	95.0	9.3	89.4	12.6
Diphenyl-4-sulfonate	100.6	8.4	96.4	9.0	96.4	8.2
1-Amino-5-naphthalenesulfonate	2.5	2.8	4.9	4.8	8.6	1.2
2-Amino-1-naphthalenesulfonate	70.9	10.1	65.0	5.9	131.6	12.8
Naphthalene-1-sulfonate	105.8	9.3	99.4	5.2	113.0	8.8
3-Aminobenzenesulfonate	–	–	–	–	–	–
4-Chlorobenzenesulfonate	96.3	9.4	80.2	13.1	72.2	8.6
Toluene-4-sulfonate	102.7	8.5	105.6	13.9	112.0	6.3
Benzenesulfonate	38.4	18.4	–	–	–	–
4,4′-Diamino-2,2′-stilbenedisulfonate	–	–	–	–	–	–
Naphthol-4-sulfonate	89.1	7.2	82.9	6.6	74.8	15.1
Naphthalene-1,5-disulfonate	–	–	–	–	–	–
2-Naphthol-3,6-disulfonate	–	–	–	–	–	–

R = recovery; RSD = relative standard deviation.
Reprinted from [287] with permission of Elsevier.

4,4'-diamino-2,2'-stilbenedisulfonate, naphthalene-1,5-disulfonate, and 2-naphthol-3,6-disulfonate were not extracted from water by LiChrolut EN. The detection limit of the combined SPE–CE method was approximately 0.1 µg/L for a 200 mL water sample. To achieve high recoveries (80–100%) for the most hydrophilic sulfonates, including naphthalenedisulfonates, by means of their on-line pre-concentration coupled to HPLC–UV determination, Gimeno et al. [288] exploited an ion-pair reagent, tetrabutylammonium bromide, added in a 3 mM concentration to the sample buffered with phosphates to pH 7.0. Analytes were then eluted from LiChrolut EN with methanol in a back-flush mode directly into the chromatographic C-18 column operated in a gradient mode and with a UV detector. According to Rao et al. [289], neither LiChrolut EN nor Isolute ENV+ retain sulfonic acids with two amino groups, such as *meta*-phenylenediamine-4-sulfonic acid, *para*-phenylenediamine-2-sulfonic acid, and *cis*-4,4'-diaminostilbene-2,2'-disulfonic acid, from groundwater spiked with 50 µg/L of the acids at pH 2.5. Recoveries for many other benzene- and stilbenesulfonic acids containing one or two amino groups along with chloro, acetyl, and nitro groups proved to be less than 58% at that pH value. In contrast to the SPE procedure, pre-concentration of these aromatic sulfonates by reverse osmosis or especially vacuum distillation was found to provide better recoveries [289].

An analytical method for ultra-trace determination (down to 0.1 µg/L) of 15 fluorinated aromatic carboxylic acids, used as water tracers, was described [290]. The method comprised off-line extraction of the acids with Isolute ENV+ at pH 1.5, elution with acetonitrile, and GC–MS quantification. The examination of the behavior of the fluorinated benzoic acids on the hypercrosslinked extracting material showed that with increasing acidity the breakthrough volume of the acid decreases, while increasing its molecular size increases the breakthrough volume because of more effective dispersion interactions with sorbent material.

It is also interesting to mention that SPE on Isolute proved to be a successful procedure for analyzing residual succinic acid in nucleoside derivatives to be used in nucleotide synthesis [291]. According to the developed protocol, almost all nucleoside derivatives are trapped by the sorbent at pH 7 whereas succinic acid is ionized at this pH value and does not bind the sorbent, so that examining the filtrate with GC–MS allowed safe determination of 0.18–0.24% of the disturbing admixtures of succinic acid in the nucleoside preparations.

It is only natural that weak organic acids are better adsorbed in their neutral (protonated) form, that is, from acidic aqueous solutions. They then elute together with all other accumulated neutral and basic compounds. To develop a selective SPE material for acidic compounds, Fontanals et al. [292] prepared microspherical anion exchanging hypercrosslinked polystyrene sorbents. Dispersion polymerization of vinylbenzyl chloride with 25% DVB in acetonitrile yielded 6 μm beads which were post-crosslinked under the action of $FeCl_3$ and finally treated with ethylene diamine or piperazine. The thus prepared weak anion exchangers had a specific surface area of $1000\,m^2/g$ and ion exchange capacity of 0.75–0.90 meq/g. They strongly retained from acidified river waters all acidic and neutral contaminants. The latter could then be eluted with methanol, while desorption of acids required ammonia-containing acetonitrile. Recoveries of all tested 12 compounds were found to be very high even from effluent waters of a water treatment plant.

9. TRACE ENRICHMENT OF MISCELLANEOUS COMPOUNDS

Many other organic compounds can be analyzed in water samples using extraction with hypercrosslinked adsorbing materials. Of 19 tested analytes [293] representing various classes of organic pollutants, excellent recoveries were observed for polar compounds such as caffeine, atropine, cyclohexanol, 2-ethyl-1-hexanol, and lindane, while for the most hydrophobic analytes, such as 2,4,6-trichlorophenol, n-butylbenzene, and n-octane, better recoveries have been found after pre-concentration on macroporous styrene–DVB copolymers or C-18 silica than after sorption on LiChrolut EN or Isolute ENV+. This result can only be attributed to the inefficiency of the extraction of the analytes from these SPE materials with the ethyl acetate/dichloromethane mixture (1:1 v/v) applied. The same LiChrolut EN and Isolute ENV+, as well as hypercrosslinked sorbent HR-P (Macherey-Nagel) cartridges containing 200 mg of each sorbent, were tested for the off-line SPE pre-concentration of 53 highly water-soluble aromatic amines from drinking and ammunition wastewaters adjusted to pH 9.0 [232]. The cartridges were then dried with air end-eluted with a 3 mL acetonitrile–methanol (1:1) mixture. Table 14.5 demonstrates good recoveries, >80%, with all the three hypercrosslinked SPE media. However, four other styrene–DVB type resins also showed acceptable recoveries. Only the most polar 1,2-phenylenediamine,

2,4-diaminotoluene, and 3,4-dimethylaniline were poorly extracted from drinking water. Surprisingly, some recovery values were found to be unrealistically high, up to 250–300%, when derivatization of the amines for their subsequent analysis by GC with an electron capture

Table 14.5   Comparison of the recoveries of 33 aromatic amines collected on polymer-based solid phases of different suppliers from a 100 mL sample of drinking water spiked at the level of 10–20 µg/L

Amine	Relative recovery (%) (RSD, %) after SPE with						
	HR-P Macherey-Nagel	Bond Elute ENV Varian	Envi Chrom P Supelco	SDB-2 Baker	Polymer Restek	Isolute ENV+ IST	LiChrolut EN Merck
Aniline	80 (7)	69 (4)	78 (8)	47 (4)	59 (9)	71 (18)	85 (21)
4AT	80 (3)	55 (2)	64 (8)	43 (2)	58 (12)	75 (25)	76 (17)
2A3NT	80 (9)	86 (4)	106 (10)	84 (5)	90 (3)	96 (3)	105 (11)
2A4NT	91 (5)	90 (1)	95 (2)	87 (5)	87 (3)	88 (1)	102 (19)
2A5NT	90 (6)	87 (2)	92 (2)	89 (7)	90 (3)	90 (2)	105 (17)
2A6NT	88 (9)	89 (1)	94 (2)	88 (6)	88 (2)	90 (1)	100 (19)
4A2NT	84 (5)	88 (4)	92 (2)	85 (6)	83 (3)	85 (2)	96 (17)
2,4DAT	52 (5)	30 (4)	46 (2)	62 (7)	60 (9)	51 (4)	50 (7)
2,6DAT	80 (7)	23 (3)	52 (4)	78 (8)	48 (6)	60 (2)	63 (10)
2A4,6DNT	81 (8)	63 (1)	68 (2)	96 (6)	89 (8)	90 (5)	77 (11)
4A2,6DNT	89 (8)	84 (2)	91 (1)	87 (7)	86 (3)	86 (2)	102 (16)
2,4DA6NT	95 (6)	88 (3)	90 (5)	92 (8)	89 (2)	90 (1)	107 (14)
2,6DA4NT	109 (6)	90 (4)	93 (7)	97 (8)	97 (2)	100 (2)	117 (14)
2,6DMA	111 (4)	55 (3)	58 (9)	101 (7)	108 (13)	107 (9)	91 (13)
3,4DMA	66 (4)	36 (4)	42 (5)	30 (2)	58 (14)	79 (28)	59 (10)
3,5DMA	99 (5)	46 (2)	51 (7)	84 (6)	98 (13)	99 (6)	81 (10)
6E2MA	179 (4)	65 (2)	69 (11)	113 (9)	119 (16)	113 (9)	138 (20)
2,6DEA	253 (5)	80 (2)	84 (12)	117 (11)	122 (18)	112 (10)	203 (34)
4IPA	116 (3)	65 (5)	87 (16)	46 (2)	70 (16)	85 (26)	111 (23)
3NA	89 (6)	87 (1)	93 (1)	85 (7)	84 (5)	83 (1)	101 (21)
4NA	97 (6)	92 (1)	94 (1)	89 (8)	88 (5)	87 (1)	108 (20)
2,6DNA	60 (8)	66 (5)	64 (3)	79 (3)	79 (1)	88 (5)	74 (14)
3,5DNA	83 (8)	70 (3)	70 (4)	98 (5)	96 (2)	100 (3)	88 (13)
1,3PhDA	35 (1)	9 (1)	35 (2)	58 (6)	15 (2)	33 (3)	38 (6)
1NphA	79 (5)	74 (5)	84 (8)	50 (4)	73 (12)	89 (19)	87 (17)
2NphA	82 (5)	72 (3)	88 (7)	49 (3)	71 (11)	83 (17)	87 (16)
2ABPh	80 (6)	83 (4)	99 (6)	61 (6)	81 (12)	90 (10)	93 (18)
4ABPh	94 (6)	87 (4)	96 (10)	74 (7)	89 (15)	97 (9)	104 (17)
Benzidine	69 (6)	81 (1)	91 (7)	76 (5)	76 (17)	78 (2)	89 (11)
4CNMA	120 (1)	122 (8)	128 (6)	106 (7)	112 (13)	116 (6)	132 (29)

(Continued)

Table 14.5   (Continued)

Amine	Relative recovery (%) (RSD, %) after SPE with						
	HR-P *Macherey-Nagel*	Bond Elute ENV *Varian*	Envi Chrom P *Supelco*	SDB-2 *Baker*	Polymer *Restek*	Isolute ENV+ *IST*	LiChrolut EN *Merck*
4C2MA	144 (3)	123 (3)	131 (12)	110 (6)	119 (17)	120 (6)	149 (32)
3,4DCA	100 (3)	103 (0)	105 (6)	97 (9)	99 (5)	100 (2)	112 (21)
3C4MOA	106 (5)	108 (4)	120 (4)	106 (6)	114 (8)	100 (16)	122 (23)

4AT: 4-aminotoluene; 2A3NT: 2-amino-3-nitrotoluene; 2A4NT: 2-amino-4-nitrotoluene; 2A5NT: 2-amino-5-nitrotoluene; 2A6NT: 2-amino-6-nitrotoluene; 4A2NT: 4-amino-2-nitrotoluene; 2,4DAT: 2,4-diaminotoluene; 2,6DAT: 2,6-diaminotoluene; 2A4,6DNT: 2-amino-4,6-dinitrotoluene; 4A2,6DNT: 4-amino-2,6-dinitrotoluene; 2,4DA6NT: 2,4-diamino-6-nitrotoluene; 2,6DA4NT: 2,6-diamino-4-nitrotoluene; 2,6DMA: 2,6-dimethylaniline; 3,4DMA: 3,4-dimethylaniline; 3,5DMA: 3,5-dimethylaniline; 6E2MA: 2-ethyl-6-methylaniline; 2,6DEA: 2,6-diethylaniline; 4IPA: 4-isopropylaniline; 3NA: 3-nitroaniline; 4NA: 4-nitroaniline; 2,6DNA: 2,6-dinitroaniline; 3,5DNA: 3,5-dinitroaniline; 1,3PhDA: 1,3-phenylenediamine; 1NphA: 1-naphthylamine; 2NphA: 2-naphthylamine; 2ABPh: 2-aminobiphenyl; 4ABPh: 4-aminobiphenyl; 4CNMA: 4-chloro-N-methylaniline; 4C2MA: 4-chloro-2-methylaniline; 3,4DCA: 3,4-dichloroaniline; 3C4MOA: 3-chloro-4-methoxyaniline.
Reprinted from [232] with permission of Elsevier.

detector was done in a drinking or wastewater matrix; however, the reason for this matrix effect could not be determined.

LiChrolut EN was used successfully to extract endocrine disruptors, manmade chemical compounds interfering with the normal functioning of estrogen, androgen, and thyroid hormones in human and animals and causing birth defects and metabolic disorders. The group of studied compounds comprised alkylphenols, estrogens, diuron (an antifouling agent), 2,4-dinitrophenol, and bisphenol A [294]. These species were sorbed on the polymer from 500 mL river and marine water or 100 mL sewage treatment plant effluent and then eluted with acetonitrile. By using an SPE–HPLC–MS system, endocrine disruptors were found to be present in waters at levels between 0.001 and 0.3 μg/L. A method for the determination of mutagenic polar aromatic compounds 1*H*-quinolin-4-one, 5*H*-phenanthridin-6-one, 10*H*-acridin-9-one, and 9*H*-fluoren-9-one was suggested [295]. The method comprising SPE and a cleanup step on LiChrolut EN and analytical determination by reversed phase HPLC was applied for the analysis of groundwater samples from a former gas plant. All the above four pollutants have been found in groundwater in concentrations up to 90 μg/L with recoveries from 76 to 100%.

The environmental analysis of estrogens and progestogens at physiologically active concentrations (low ng/L range) was performed in river water and extracts from river sediments. In the off-line procedures, all three cartridges, LiChrolut EN, Isolute ENV+, and LiChrolut C-18, showed equally good extraction capacity and elution efficiency for most of the analytes, but only the last phase was capable of satisfactorily extracting estriol. In the on-line mode, polymeric HySphere-GP, PLRP-S, and Oasis HLB were found to be more suitable for the coupling with HPLC analysis. In river sediments, high concentrations of ethynyl estradiol (22.8 ng/g) were observed [296].

Khryaschevsky et al. [297] developed a method of determining aliphatic alkylamines (from butylamine through tetradecylamine) by off-line SPE on a hypercrosslinked sorbent (Pyrosep 200), desorption with ethanol/1 M HCl (10:1), derivatization, and HPLC analysis. Compared to C-18 silica, the polymer displayed a much higher adsorption capacity.

It has also been demonstrated that styrene–DVB copolymer resins Sepabeads SP825, Sepabeads SP850, Optipore L-493, Diaion HP-20, and Amberlite XAD-4, of which the first three have a surface area of $1000 \, m^2/g$ and might be attributed to hypercrosslinked sorbents, accumulate lipophilic marine toxins okadaic acid and dinophysistoxin-1 from *Prorocentrum lima* cultures [298]. The procedure developed in the study represents a simple and efficient method suitable for solid-phase adsorption toxin tracking in both cultural liquids and natural waters. The toxins are eluted from the resins with methanol and subjected to LC–MS analysis.

10. HYPERCROSSLINKED POLYSTYRENE SORBENTS VERSUS OASIS HLB

Following the hypercrosslinked polystyrene SPE materials, another sorbent, Oasis HLB from Waters (USA), appeared on the market and soon became popular for pre-concentration of a wide variety of trace organic compounds present in aqueous matrices of different origin. Oasis HLB is a macroporous copolymer of DVB with *N*-vinylpyrrolidone, a hydrophilic component. It has a high specific surface area of $830 \, m^2/g$ and pores of an average 82 Å diameter [217]. The direct wetting of dry sorbent beads with water is said to favorably distinguish Oasis HLB from other neutral polymer adsorbents. This wettability of the material considerably simplifies the use of the SPE cartridges, provided that their pre-cleaning and conditioning proves to be superfluous in the routine analysis (which is seldom the case). Disposable cartridges incorporate 30, 60, or 200 mg of the sorbent beads of 30 μm diameter [299].

Oasis HLB was successfully used for SPE and trace enrichment in the determination of phenols in water [300, 301], tetracycline, macrolide, and sulfonamide antibiotics in aqueous media and agricultural soils [302–305], various pharmaceuticals in plasma and serum [306–308], herbicides in natural waters [309–311], polycyclic aromatic hydrocarbons (PAHs) in plants [312], aromatic amines in mainstream cigarette smoke [313], as well as many other types of compounds [314–319].

Many studies have weighed the properties of Oasis HLB cartridges against the properties of cartridges packed with LiChrolut EN or Isolute ENV+. From neutral aqueous solutions, Oasis HLB (60 mg, 30–100 μm beads) recovers nearly quantitatively 26 priority phenolic pollutants included in the 8041 US EPA method and the 76/464/EEC European Union directive. The only exceptions are phenol and 2,4-dinitrophenol, for which recoveries are 70 and 64%, respectively. Isolute ENV+ cartridge (200 mg, 40–140 μm particles) generated recoveries for these analytes below 60% under neutral conditions. Better retention of phenols requires the aqueous sample to be acidified to pH 2.5. However, the sorption of humic and fulvic acid then increases, causing interferences in the subsequent chromatographic analysis [320].

Oasis HLB (60 mg) was found to extract oxytetracycline and its three concomitant impurities from pig plasma better than does LiChrolut EN (200 mg), especially at neutral and basic pH [321]. Similarly, Oasis HLB (500 mg) and LiChrolut EN (200 mg) were tested for the extraction of four tetracycline and two quinolone antibiotics from different water matrices [322]. While the recoveries from a 100 mL sample were equally good for both SPE cartridges, they dropped for the 500 mL sample applied on Isolute ENV+, while still remaining larger than 80% for the pre-concentration of the analytes from 1000 mL samples on the Oasis HLB cartridge, and so the latter was the sorbent of choice for the analysis. Blackwell et al. [323] compared the polymers in the analysis in groundwater of veterinary antibiotics oxytetracycline, sulfachloropyridazine, and tylosin, which are the members of tetracycline, sulfonamide, and macrolide groups of antibiotics, respectively:

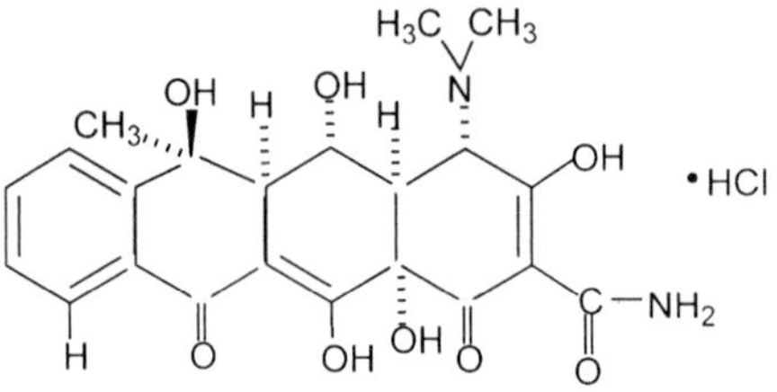

Oxytetracycline

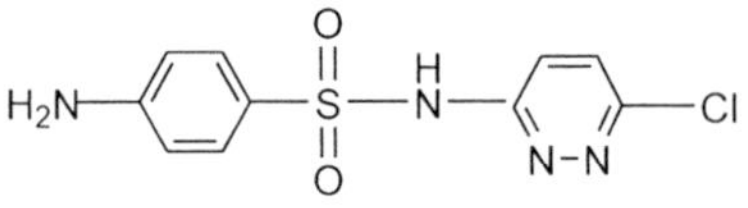

Sulfachloropyridazine

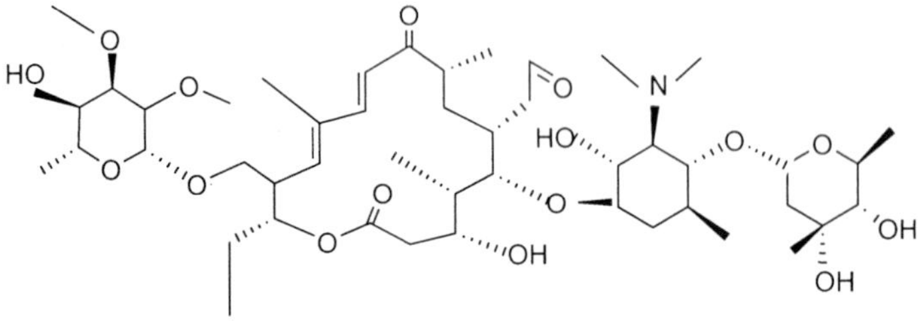

Tylosin

After applying 500 mL of a mixed aqueous solution (40 µg/L), the commercial Oasis HLB cartridge gave recoveries of over 90% for all the compounds, while Isolute ENV+ cartridge adsorbed 89% sulfonamide but did not recover oxytetracycline and tylosine. This result is in agreement with previous reports [321, 322, 324] that Isolute ENV+ is inefficient in the pre-concentration of tetracyclines. One may note the big difference in the size of the above three analytes.

Table 14.6 presents recoveries of some acidic and neutral pharmaceutical residues from 500 mL spiked water samples applied on 200 mg LiChrolut EN, 60 mg Oasis HLB, and 500 mg reversed phase silica commercial cartridges. On the basis of the chemical structure of the compounds under testing, the hypercrosslinked sorbent would have to retain all of them well. Surprisingly, the recoveries were found to be unsatisfactorily low for LiChrolut EN and very high for HLB. Even the reversed phase silica showed better results compared to LiChrolut EN. Only the incomplete wetting by water of the hydrophobic hypercrosslinked sorbent may explain these unanticipated results [325]. In fact, by using the Manifold equipment, it is easy to overlook the cartridge running dry after the

Table 14.6 Recoveries of selected pharmaceuticals obtained by different sorbents from spiked deionized water ($n = 4$)

Medicine	Recovery (%) (RSD)		
	RP C-18	LiChrolut EN	Oasis HLB
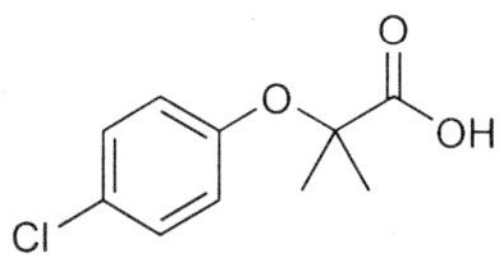 Clofibric acid	70 (9)	40 (10)	95 (1)
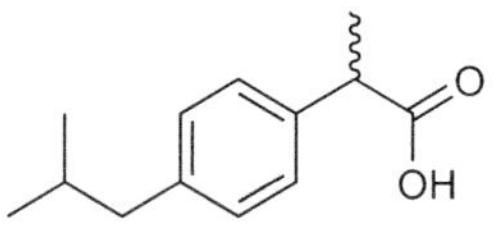 Ibuprofen	41 (18)	25 (8)	77 (5)
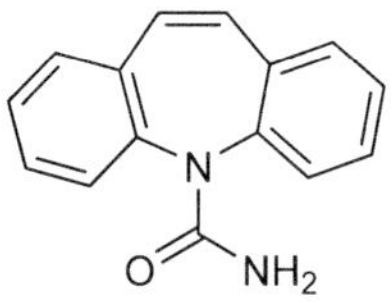 Carbamazepine	107 (11)	95 (10)	93 (6)
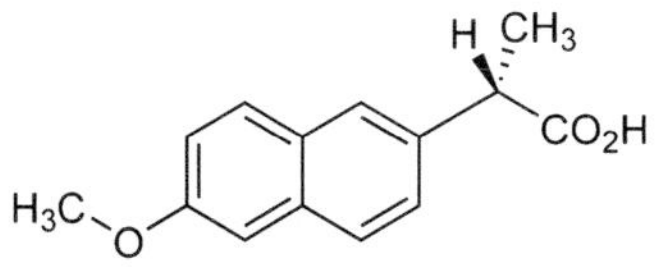 Naproxen	67 (10)	46 (9)	91 (10)
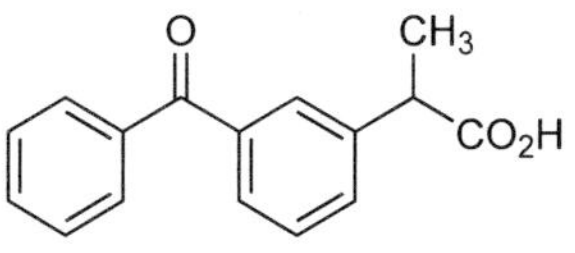 Ketoprofen	82 (10)	28 (20)	102 (6)

(Continued)

Table 14.6   (Continued)

Medicine	Recovery (%) (RSD)		
	RP C-18	LiChrolut EN	Oasis HLB
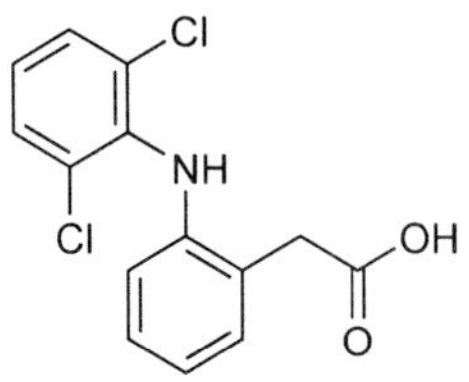 Diclofenac	56 (9)	30 (9)	92 (2)

Reprinted from [325] with permission of Elsevier.

polymer activation step with the organic solvent, which was acetone/methanol (3:2) in this particular case.

On the other hand, commercial Oasis HLB cartridges (60 mg) retain weaker mono-, di-, and trichloro- and -bromo-substituted acetic acids, compared to commercial cartridges packed with 200 mg LiChrolut EN [326]. Similar results were reported in [327] (Table 14.7), demonstrating an obvious tendency of reducing recovery of haloacetic acids in the sequence of sorbents LiChrolut EN > HR-P > Isolute ENV+ > Oasis HLB. Determination of the haloacetic acids is an important analytical task, since the sum of five regulated toxic compounds of this group should not exceed 60 μg/L in chlorinated tap water. As to the aromatic sulfonates, again, the above four SPE materials follow the same sequence, Oasis HLB being the least suitable material (Table 14.8), which yields to the first two polymers in the recovery values by 20% on average [328]. Note, however, that none of the polymers retains analytes with two or three sulfonate groups in one molecule.

With respect to extraction of atrazine herbicides and metabolites, the choice between Oasis HLB and LiChrolut EN cartridges proved to be in favor of LiChrolut EN [329] as well. As has been found, Oasis HLB does not retain the most polar metabolite, desisopropyl-2-hydroxyatrazine (DIHA), at all, and recovers only 10 and 18% desethyldesisopropylatrazine (DEDIA) and desethyl-2-hydroxyatrazine (DEHA), respectively. On the contrary, under the same conditions, LiChrolut EN exhibits 69% recovery for DEDIA and >90% recoveries for DEHA and DIHA.

Table 14.7   Recoveries of haloacetic acids from 50 mL spiked ground water at pH 1.8 (sulfuric acid) obtained with different SPE sorbents (elution with 0.5 mL water + 3.5 mL methanol/acetone 1:1; $n = 4$)

Acids	Conc. (mg/L)	Recovery (%) (RSD (%)			
		LiChrolut EN	HR-P	Isolute ENV+	Oasis HLB
Monochloroacetic acid	300	25 (3)	27 (2)	21 (2)	18 (3)
Monobromoacetic acid	200	42 (6)	57 (3)	28 (15)	29 (9)
Dichloroacetic acid	300	55 (5)	54 (3)	50 (4)	42 (7)
Bromochloroacetic acid	200	60 (8)	53 (4)	44 (3)	38 (5)
Dibromoacetic acid	100	56 (7)	45 (3)	37 (1)	30 (4)
Trichloroacetic acid	100	75 (5)	74 (3)	65 (3)	51 (4)
Bromodichloroacetic acid	200	45 (2)	39 (3)	33 (6)	28 (3)
Dibromochloroacetic acid	500	37 (2)	22 (2)	37 (1)	33 (6)
Tribromoacetic acid	1000	33 (2)	25 (1)	31 (3)	27 (6)
Monochloroacetic acid	30	35 (8)	–	–	–
Monobromoacetic acid	20	51 (9)	–	–	–
Dichloroacetic acid	30	45 (12)	–	–	–
Bromochloroacetic acid	20	53 (6)	–	–	–
Dibromoacetic acid	10	48 (9)	–	–	–
Trichloroacetic acid	10	69 (8)	–	–	–
Bromodichloroacetic acid	20	56 (7)	–	–	–
Dibromochloroacetic acid	50	45 (5)	–	–	–
Tribromoacetic acid	100	39 (8)	–	–	–

Reprinted from [327] with permission of Elsevier.

Thus, reasoning from the consideration of the above studies, there can be no preference for Oasis HLB over hypercrosslinked adsorbing materials or vice versa. There is no general answer to the question which SPE material is better. The commonly accepted practice is to compare commercially available cartridges rather than the adsorbing materials as such. One easily overlooks the fact that the amount of the material in the cartridge and the size of polymer particles are different, which would imply that different flow rates and volumes of the sample should be applied in the loading procedure. Similarly, elution conditions must also be adjusted to the sizes of the particles and cartridge. Unfortunately, these experimental details are often omitted in publications and the difference in the efficiency of the SPE procedure is often discussed entirely in terms of chemical structure of materials; see, for example [330].

Table 14.8 SPE recoveries for the extraction from 150 mL groundwater spiked at 50 µg/L for the different sorbent materials (pH 2.5 with sulfuric acid, elution with 1 mL water or 5 mM trifluoroacetic acid and 6 mL methanol/acetone)

Sulfonate	Recovery (%)			
	LiChrolut EN (n 3)	HR-P (n 3)	Isolute ENV+ (n 3)	Oasis HLB (n 4)
2-Amino-1,5-naphthalenedisulfonate	Nd	Nd	Nd	Nd
1,3,6-Naphthalenetrisulfonate	Nd	Nd	Nd	Nd
1,3-Benzenedisulfonate	Nd	Nd	Nd	Nd
1,5-Naphthalenedisulfonate	Nd	Nd	Nd	Nd
2,6-Naphthalenedisulfonate	Nd	Nd	Nd	Nd
1-OH-3,6-Naphthalenedisulfonate	Nd	Nd	Nd	Nd
1-Amino-5-naphthalenedisulfonate	12	21	12	Nd
Benzenesulfonate	6	4	3	Nd
1-Amino-4-naphthalenesulfonate	50	37	27	Nd
2-OH-3,6-Naphthalenedisulfonate	Nd	Nd	Nd	Nd
1-OH-6-Amino-3-naphthalenedisulfonate	26	25	24	2
3-Nitrobenzenesulfonate	105	113	78	39
1-Amino-6-naphthalenesulfonate	55	54	36	Nd
4-Methylbenzenesulfonate	71	63	36	5
1-OH-4-Naphthlenesulfonate	88	85	58	30
4-Chlorobenzenesulfonate	113	110	83	51
2-Amino-1-naphthalenesulfonate	86	85	66	64
1-Amino-7-naphthalenesulfonate	69	67	52	24
4-Chloro-3-nitrobenzenesulfonate	91	94	72	97
1-Naphthalenesulfonate	88	92	77	91
2-Naphthalenesulfonate	84	85	71	86
Diphenylamine-4-sulfonate	63	66	54	67

Nd = not detected.
Reprinted from [328] with permission of Elsevier.

Certainly, the difference in the nature of the SPE polymers has to be carefully observed. The improved hydrophilic–lipophilic balance (HLB) of the Oasis packing does not generally improve the adsorption strength of the material. It only improves its wetting ability and makes it less subjected to experimental mistakes like letting the cartridge run dry before the extraction procedure. One should take into account that the hydrophobic polystyrene material easily loses water and that air entering the fine pores of the material will not be replaced back by water during the loading of the cartridge. The better wetted Oasis HLB is less demanding in this respect.

Another basic difference between the two materials is that Oasis HLB is macroporous (80 Å pores) while LiChrolut EN and Isolut ENV+ are basically nanoporous materials with the maximum pore size distribution located at 20–30 Å. For this reason, the latter sorbents perform much better in extracting smaller molecules with molecular masses less than 500 Da, as is the case with the above halogenated acetic acid derivatives [326, 327] or aryl sulfonates [328]. On the contrary, macroporous Oasis HLB is the material of choice when larger analytes are the target of pre-concentration procedures, as was the case with tetracycline and macrolide antibiotics [321–324]. Another vivid example of this kind is pre-concentration of soy isoflavones [331], where Oasis HLB and macroporous Strata X display by far higher recovery values. On the other hand, the size exclusion effect can be exploited purposefully when only smaller molecules have to be retained with the elimination of the major sample matrix. The best example of this type of applications is the analysis of drugs and drug metabolites in whole blood or blood plasma [273–275]. Here the microporous hypercrosslinked polystyrene Purosep-270 is the best possible SPE material since it functions as an RAM.

In a recent review on SPE of polar compounds, Fontanals et al. [332] analyze the use of many other polymeric adsorption materials that contain various polar functional groups introduced into the polymer in the form of a polar co-monomer at the stage of polymerization or at a later stage by an appropriate chemical modification of the final material. The information accumulated by different groups is rather controversial. It leaves the impression that the very popular general notion of better adsorption of polar compounds on SPE materials that incorporate some polar functions is not yet convincingly supported by experimental findings or theoretical considerations. Indeed, the polar groups of a polymer immersed into aqueous media should be strongly hydrated with water molecules, and there are no reasons for them to exchange the latter against organic molecules, the polar groups of which are also hydrated. As to the retention of organic molecules on hydrophobic hypercrosslinked polystyrene from aqueous media, it is governed by the combination of hydrophobic and π–π interactions. A polar group in an aromatic ring of an analyte reduces the strength of hydrophobic interactions, but simultaneously enhances the π–π interactions. For this reason retention of phenol on the aromatic polymer is that much stronger than on the hydrophobic C-18 silica. Isolated polar groups in the analyte molecule are also able to enter attractive π–π interactions with the exposed aromatic rings of the hypercrosslinked polymer, as will be shown in the next section, thus combining with the hydrophobic-type retention of the remaining part of the analyte molecule. This does not mean, however, that there could not emerge combinations where some polar groups of the SPE polymer enter specific dipole interactions or form hydrogen bonds with the analyte molecule, which significantly enhance the retention of that particular polar analyte.

Thus, the existing porous polymers, incorporating or devoid of polar groups, should be considered as complementary, rather than competing, SPE materials, each one having its own optimal application niche.

11. $\pi-\pi$-INTERACTIONS AND SPE FROM NON-AQUEOUS MEDIA

In all the above-discussed examples of using SPE materials for adsorption and pre-concentration of trace organic compounds from aqueous or aqueous-methanolic media, the polymers basically functioned as hydrophobic adsorbing media. Still, hypercrosslinked polystyrene differed markedly from the well-known silica–C-18 reversed phase SPE materials and macroporous styrene–DVB polymers by a much stronger retention power with respect to weakly hydrophobic aromatic compounds (like phenol or catechol) and even polar aliphatic compounds. The reason for the enhanced retention of polar compounds is that they exhibit the ability to enter $\pi-\pi$-type interactions with π-electronic systems of exposed phenyl rings of the hypercrosslinked polymer. *Ab initio* calculations show that the energy of $\pi-\pi$-interactions between a benzene ring and the carbonyl group of formaldehyde, 1.87 kcal/mol, is higher than the interaction energy between two parallel-oriented benzene molecules, 0.49 kcal/mol [333]. The same type of $\pi-\pi$-interactions contributes much to the enhanced retention of PAHs on stationary phases with grafted carbonyl-reach ligands such as poly(L-alanine) [333], of nucleic acid constituents on L-glutamide-derived non-crystalline stationary phases [334], and of aromatic hydrocarbons on poly(4-vinylpyridine) [335]. Since acetonitrile impedes the $\pi-\pi$-type intermolecular interactions, selective adsorption of polar analytes on phenyl-incorporating stationary phases manifests itself most markedly in water and water–methanol mixtures [336, 337]. The same is also valid for the SPE on hypercrosslinked polystyrene: if an organic modifier in the aqueous sample is needed, for example, for the solubility of analytes, it is better to use methanol or other alcohols, while adding acetonitrile would then facilitate desorption of thus accumulated analytes.

The role of π–π-type interactions between hypercrosslinked polystyrene, on the one hand, and compounds containing aromatic, heteroaromatic, carbon–carbon double bonds, and/or carbonyl functional groups, on the other hand, has been discussed in Chapter 13, Section 4 dealing with HPLC on polystyrene-type packings. It has been emphasized there that the selective π–π-interactions are even more effective in nonpolar media such as hexane or other aliphatic hydrocarbons. This property opens an unique possibility to selectively isolate the above π-electron-containing classes of organic compounds from nonpolar media.

This idea was embodied for the first time by extracting selectively four polar furan derivatives, 5-hydroxymethylfurfurol, furfurol, 2-acetyfuran, and 5-methylfurfurol, from transformer oil, which is mostly composed of higher aliphatic hydrocarbons. Being the degradation products of cellulose-type electric insulation in transformers, the furan-derived compounds serve as specific markers for the insulation diagnostics. To study the extraction of these polar analytes, 2 mL pure oil was diluted with 10 mL hexane, spiked with 100 μg/L of each component, and then 2 mL of the final solution was percolated through 20 × 8 mm ID cartridge packed with 500 mg 70–140 μm particles of Purosep 200. The loaded cartridge was washed with hexane and dried in a gentle flow of nitrogen. After eluting with acetonitrile–water 1:1 mixture, the above analytes were quantified by HPLC–UV [338]. As Table 14.9 indicates, Purosep 200 almost completely recovers the analytes with excellent reproducibility. The LOD of each component does not exceed 30 μg/L, which permits determination of the target analytes at 1% level of their allowed tolerance limit.

Similarly, using the same approach, it proved to be possible to control the presence of priority PAHs or pesticides in vegetable oil [339]. The developed method includes mixing an oil sample with hexane, selective SPE of the aromatics with hypercrosslinked polystyrene, washing the cartridge with hexane, eluting polyaromatic hydrocarbons with dichloromethane, and, finally, analysis of the analytes by HPLC with fluorescence detection. Figure 14.2 shows the chromatogram of a 16 PAH test mixture as well as the chromatogram of a real smoked fish sample treated according to the developed procedure. Recoveries of PAHs vary between 80 and 99%. The above suggested pre-concentration of aromatic hydrocarbons from hexane extracts of fatty food products eliminates the most unpleasant stage of previously applied analytical techniques, which was the alkaline hydrolysis of oils and fats of the sample.

As no hydrophobic interactions function in organic solvents, the π–π-interactions are the only mechanism of retaining certain organic analytes

Table 14.9 Mean recoveries for furan derivatives extracted from two samples of transformer oil spiked at 100 µg/L using one and the same cartridge with Purosep 200 ($n = 10$)

Compound	1st oil sample	2nd oil sample
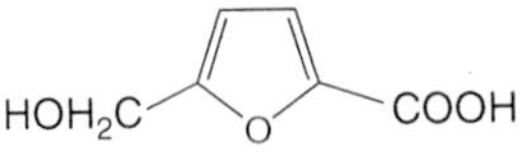 5-hydroxymethylfurfurol	76.5 ± 1.3	79.5 ± 0.4
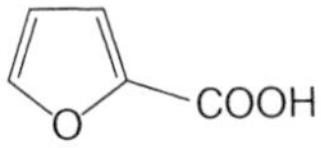 furfurol	99.0 ± 0.5	93.9 ± 0.6
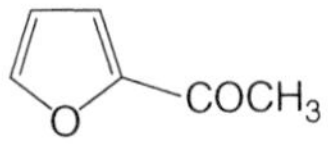 2-acetylfuran	95.2 ± 1.1	95.5 ± 1.2
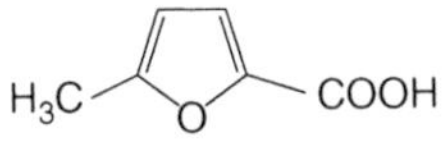 5-methylfurfurol	98.3 ± 0.9	98.4 ± 0.8

Source: After [338].

from these media on hypercrosslinked polystyrene. This opens an entirely new principle of pre-concentrating trace components, which will find an ever-growing application of hypercrosslinked polystyrene SPE materials in the analysis of manifold extracts. Note that the best suited solvents for preparing the extracts should be aliphatic hydrocarbons (as hexane), aliphatic alcohols, and water–alcohol mixed solvents. Desorption of the thus concentrated analytes should be performed with solvents disrupting the π–π-interactions (as dichloromethane or chloroform).

12. PRE-CONCENTRATION OF VOLATILE ORGANIC COMPOUNDS IN AIR

Volatile organic compounds (VOCs) are pollutants commonly found in trace concentrations in indoor and outdoor air as well as – in higher concentrations – in industrial workplaces. They are assumed to be

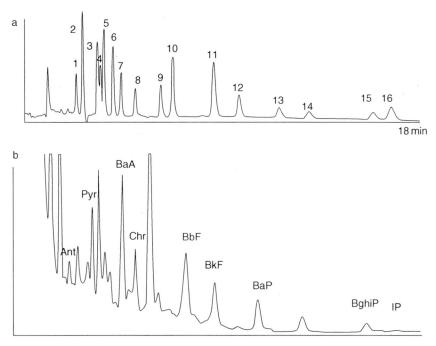

Figure 14.2 Chromatograms of (a) priority 16 PAH test mixture and (b) priority PAH in smoked fish after their SPE pre-concentration. Stationary phase: 250 × 4.6 Wakosil II C-18 AR; mobile phase: acetonitrile–water 85:15; 1 mL/min; 35°C, (1) naphthalene (Na), (2) acenaphthylene (Ap), (3) acenaphthene (Ac), (4) fluorene (F), (5) phenantrene (Phe), (6) anthracene (Ant), (7) fluoranthene (Fl), (8) pyrene (Pyr), (9) benzo[a]anthracene (BaA), (10) chryzene (Chr), (11) benzo[b]fluoranthene (BbF), (12) benzo[k]fluoranthene (BkF), (13) benzo[a]pyrene (BaP), (14) dibenzo[a,h]anthracene (DahA), (15) benzo[ghi]pyrene (BghiP), (16) indeno[1,2,3-cd]pyrene (IP). (After [339].)

responsible for the "sick building syndrome." Several VOCs have been identified as potent carcinogens and mutagens.

Many adsorbing materials have been evaluated for collecting VOCs from air: activated carbons, silica gels, alumina, zeolites, porous styrene–DVB copolymers, and Tenax. The latter is certified material for collecting compounds of medium and low volatility. Its benefit is high thermostability that allows a complete desorption of the sampled compounds by a rapid increase in the temperature of the sorbent. However, its adsorption capacity for small and highly polar molecules is far from ideal. To readily adsorb this kind of compounds and then fully release them on heating to moderate temperatures is the benefit of hypercrosslinked polystyrene.

In order to evaluate the suitability of MN-200 sorbent for sampling VOCs in air, the following parameters were determined for eight typical indoor pollutants: breakthrough volume, percentage recovery, storage stability, and background levels [340]. Tenax GR (matrix of Tenax with 23% graphite) was chosen as the best known reference material for the analysis of VOCs.

Table 14.10 presents results of determining the breakthrough volumes on MN-200 for the eight selected compounds at their infinite dilution in air at 20°C. All values were found to exceed the 1 L/g that was considered to be required for collecting efficiently the trace amounts of vapors (the smaller BTV value for isooctane can be explained by the size exclusion of the bulky molecule from small nanopores of the polymer). Accordingly, all the above compounds could be safely determined by their pre-concentration on MN-200 followed by either thermal desorption at 210°C or extraction with acetone. Recoveries in the thermal desorption mode were all close to 100%, with the exception of isooctane where it dropped to $98.2 \pm 0.2\%$. By comparison, Texan GR gave recovery of 25% for pentane and 35% for dichloromethane under equal experimental conditions. This is logical as the specific surface area of Tenax GR is less than $35 \, m^2/g$, compared to the $1000 \, m^2/g$ that is characteristic of MN-200.

Importantly, recoveries of the VOCs were found to drop by less than 3% when the samples collected on MN-200 were stored for 15 days at room temperature (20–25°C), with the generally acceptable criterion being 10% loss. Compared with Tenax (3% loss for hexane and 8% for benzene),

Table 14.10 Breakthrough volumes on MN-200 for 8 volatile organic compounds measured at 5 elevated temperatures (50–210°C) and then extrapolated to 20°C at "infinite dilution" of the vapors

Compound	BTV (L/g)
Pentane	13
Octane	109
Undecane	329
Methanol	5.8
Dichloromethane	5.1
Toluene	26
Isooctane	3.7
Cyclohexane	188

Source: After [340].

MN-200 can not only collect VOCs efficiently but it can also "hold" them without significant loss until analysis. In addition to the high adsorption capacity, the polymer has the advantage of being easily cleaned and regenerated and allowing desorption of the adsorbed compounds at much lower temperatures.

Most probably, future concentration cartridges will incorporate both Tenax and hypercrosslinked polystyrene SPE material in sequence, in order to combine the positive characteristics of both sorbents and collect the whole range of VOCs, from the heavy ones on Tenax to the very light on MN-200.

CHAPTER 15

Hypercrosslinked Polystyrene as Hemosorbents

1. HEMOPERFUSION VERSUS HEMODIALYSIS IN BLOOD PURIFICATION

The kidneys are important excretory organs whose main functions are to remove both the superfluous water and the biologically useless as well as toxic metabolic products of various nature and molecular weights that accumulate in the blood. Approximately 180 L of blood per day passes through the kidneys, generating about 1.5 L of urine. The latter contains hundreds of organic compounds, the most important of which are the protein digestion and metabolic products, urea, creatinine, uric acid, etc. When the kidneys cannot operate properly, toxic materials accumulate in blood and other physiological fluids and lead to death within 10–12 days.

Replacing the ill kidney with a healthy one by transplant stimulates the rejection mechanisms against the foreign organ in the living body, unless the donor is a near relative. Therefore anti-rejection drugs must be given to the recipient patient, which always have harmful side effects, so that a transplanted kidney cannot generally be expected to keep functioning for more than 5 years.

The only alternative way of removing wastes from the organism is the external purification of circulating blood (or peritoneal fluid, which, however, is much less efficient). The first attempt to remove urea from blood through adsorption dates back to 1948 [341]. This technique, called hemoperfusion, implies the direct contact of blood from patients with a sorbent system. Though capable of eliminating some particular components from the blood, hemoperfusion (which is based on adsorption) does not solve another even more acute problem, that of removing excess water. This can be done through hemodialysis, the common modern way of supporting life in the case of renal failure. By letting blood equilibrate with a special dialysate aqueous solution across a semipermeable polymeric membrane, hemodialysis allows both the excess water and small molecules to migrate down the concentration gradient from the blood into the dialysate fluid.

Comprehensive Analytical Chemistry, Volume 56
ISSN 0166-526X, DOI 10.1016/S0166-526X(10)56015-0

The introduction of hemodialysis into medical practice extends the life expectancy of patients with end-stage renal disease (ESRD) for about 5 years on average. At present there are over 350,000 patients treated for chronic kidney failure in the USA and nearly 1 million worldwide. These numbers increase by about 6% every year. Though Medicare spends annually about US\$45,000 per patient, the morbidity and mortality statistics for this patient population is horrendous. Between 23 and 25% of the US dialysis population die each year, and the average patient spends up to 16 days in the hospital each year being treated for serious, often life-threatening, complications.

Moreover, hemodialysis is a slow process, which keeps the patient connected to the stationary dialysis machine for several hours. This procedure has to be repeated three or four times a week. Beside the high consumption of the apyrogenic physiological dialysate fluid (about 120 L), the technique is expensive as well as unpleasant and inconvenient for the patient. The patient will feel unwell both before and after dialysis. Before dialysis the waste products build up in the body, and after dialysis there is a dramatic distortion of the balance of chemical equilibria and processes in the body due to the rapid removal of about 3 L water and a whole pool of molecules of molecular weight below 500 Da. Among these molecules are all amino acids, nucleotides, some mineral ions, and many other useful components.

The first hemodialysis devices utilized natural cellulose (cuprophan) membranes, which possessed predominantly small pores. These membranes permitted the removal of excess fluid, ions, and small molecules, but prohibited the removal of substances above approximately 1200 Da in size. Larger molecules, such as β_2-microglogulin (β_2M, 11.8 kDa), accumulated in the blood and were thought to contribute to many of the additional health problems and high mortality of patients on dialysis. This idea, coined the "middle molecule hypothesis" by Bapp et al. [342], led to the development of new synthetic polysulfone or polyacrylonitrile dialysis membranes that possessed larger pores and, in combination with equipment to control transmembrane pressure, permitted more efficient elimination of middle molecules.

The use of "high-flux" membranes gained significant support in 1985 when Geyjo et al. [343] conclusively established the link between the accumulation of β_2M and a complication of long-term dialysis called dialysis-related amyloidosis (DRA). As kidney failure progresses, β_2M concentration in the extracellular compartments increases, often to levels

30–60 times of normal. In DRA, insoluble plaques of β_2M and its glyco-silation products, known as amyloid fibrils, cause a progressive crippling arthritis and inflammatory arthropathy of the joints and spine. Patients with DRA often require surgery to correct carpal tunnel syndrome, a common manifestation of DRA, and lifelong medications to ameliorate joint pain and inflammation. Several investigators have shown that the partial removal of β_2M with high-flux membranes prolongs, but does not prevent, the onset of DRA in dialysis patients.

Fresh high-flux dialysis membranes achieve approximately 23–37% reductions in plasma β_2M levels [344]. However, dialyzer reuse signifi-cantly impairs the removal of β_2M [345]. Indeed, it has been recognized that non-specific adsorption of middle molecules on the surface of these synthetic high-flux membranes, rather than diffusion through the mem-brane, can account for significant amounts of the device's clearance [346, 347]. With the surface area of membranes in the dialysis device amounting to less than $2\,m^2$, the adsorption capacity of the device is obviously too small.

A much more efficient approach can be expected with the use of adsorbing materials, for example, activated carbons, because the specific surface area of these materials easily approaches $1000\,m^2/g$ and a hemoper-fusion device can easily incorporate several hundred grams of the adsorbent. Indeed, carbon was shown to remove creatinine, uric acid, guanidine, indoles, phenolic compounds, and organic acids (but not urea) in uremic patients more efficiently than with the dialysis equipment [348]. However, activated carbons are not hemocompatible and cause coagulation of blood. Besides, activated carbons are not resistant to attrition, and, by releasing fins, cause embolism of blood capillaries. A recent review of the use of sorbent hemoperfusion in ESRD [349] describes manifold attempts to eliminate the above problems by introducing a microencapsulation techni-que [350], by which the activated charcoal particles were coated with a polymer membrane, such as cellulose nitrate (collodin), cellulose acetate, methacrylic hydrogel, or crosslinked albumin. Unfortunately, every coating membrane creates significant diffusion barriers to larger toxic molecules and dramatically slows down their sorption kinetics, while not preventing the loss of many biologically important smaller molecules such as amino acids, glucose, hormones, and calcium. [351] Though hemoperfusion on coated carbons reduces concentration of toxic middle molecules, the undesired losses of fibrinogen, fibronectin, platelets, leukocytes, and complement activation are also observed [352].

A special polymeric adsorbing material (BM-01, Kaneka, Japan) has been described [353] for the selective removal of β_2M from the blood of dialysis patients. The adsorbent consists of porous cellulose beads modified with hexadecyl groups that retain β_2M through hydrophobic interactions. The adsorption capacity of this material is 1 mg of β_2M per milliliter of adsorbent. Using a hemoperfusion cartridge containing 350 mL of these cellulose beads in sequence with a high-flux hemodialyzer, several small clinical trials were performed. During 4–5 h of treatment, about 210 mg of β_2M was removed, thus reducing its concentration in the blood by 60–70% of the initial levels [354]. Clinical improvement was partial, but marked in those patients subjected to thrice-weekly treatment with this device for a period of 2 months or more. Subjective improvements were seen in joint mobility, joint pain, and nocturnal awakening. The use of this device is currently restricted to Japan and concerns over its high cost have largely prevented more widespread use [355].

2. HYPERCROSSLINKED POLYMERS FOR THE REMOVAL OF β_2-MICROGLOBULIN

There would seem to be a need for a new cost-effective, selective, and high-capacity adsorbent for the removal of β_2M. Hypercrosslinked networks are undoubtedly prospective materials.

β_2M is a relatively hydrophobic globular protein of 11.8 kDa in molecular weight. Assuming the molecule to be spherical in shape, its diameter can be estimated as about 3.35 nm. Thus, hydrophobic polymers with an enhanced proportion of mesopores, in the range from 4 to 10 nm, are needed. The majority of hypercrosslinked polystyrenes, even those crosslinked with such elongated conformationally rigid bifunctional reagents as p-xylylene dichloride, represent a microporous transparent material with an average pore diameter of 2–3 nm. Some commercially available biporous hypercrosslinked polystyrene materials, such as MN-200, possess both the 1–3 nm micropores and large transport pores of about 80–100 nm in diameter. Adsorbents of this type showed high affinity to β_2M and could be made hemocompatible in several different modes [356]. However, the porous structure of these materials was not considered to be optimal for the adsorption of β_2M.

Polymerization of divinylbenzene (DVB) was thought to offer the opportunity of preparing beaded adsorbing materials with the desired pore sizes. The phase separation is well documented that takes place during the formation of a rigid polymeric network in the presence of an organic diluent that is miscible with the monomers, but causes precipitation of the polymer, resulting in macroporous materials. Macroporous copolymers of styrene with DVB are the best known materials of this kind. On the contrary, largely microporous structures are created if the copolymerization proceeds in the presence of a diluent which is compatible with both the co-monomers and the polymer. No phase separation takes place if the amount of such diluent remains relatively small. If the amount of the latter exceeds the swelling ability of the final rigid network, excess solvent forms a separate phase, thus introducing additional larger pores. From this consideration, the structure of poly-DVB prepared with toluene, a good solvating medium for the final polymer, must be similar to that of microporous or biporous hypercrosslinked networks.

When taking into account the above regularities, it was logical to suppose that the required mesoporous networks could form when the diluent represents an organic medium situated between the thermodynamically good solvents and precipitators for the polymer. Indeed, preparation of hypercrosslinked networks by crosslinking linear polystyrene chains dissolved in cyclohexane, a rather poor solvent for polystyrene even at the temperature of synthesis, 60°C, resulted in basically mesoporous materials having an exceptionally high adsorption capacity toward β_2M. Unfortunately, we have failed thus far to prepare these materials in a beaded form of sufficient mechanical strength.

The above approach of using a diluent of an intermediate thermodynamic quality during the polymerization of DVB has been intensively examined and, indeed, resulted in materials with enhanced proportions of mesopores. In order to create a rigid polymer of desired porosity, DVB (usually more than 30% in its mixture with styrenic co-monomers) must be copolymerized in the presence of a sufficient amount of a poor diluent (usually 100% or more of the volume of the co-monomers). Of crucial importance is the nature of the poor solvent. Besides cyclohexane, mixtures of a thermodynamically good solvent (ethylene dichloride, toluene, etc.) with precipitating media (hexane, octane, isooctane, higher aliphatic alcohols, etc.), taken in an appropriate proportion, can be applied. Microphase separation during the suspension copolymerization of such a mixture should take place when the major part of the co-monomers has converted into polymer.

The total volume of pores and voids in the final polymer approaches the volume of the porogen in the mixture under polymerization, provided that the network structure is rigid enough to prevent the collapse of the bead on removing the diluent after the synthesis. In our experiments, a series of copolymers were obtained with the total porosity varying between 1.0 and 1.7 cm^3/g and an apparent specific surface area of 550–800 m^2/g. Pore size distribution, when measured using the multipoint nitrogen adsorption technique, was found to be sensitive to the composition of the initial mixture under polymerization as well as the polymerization protocol. Typically, a broad pore size distribution was found with the diffuse maximum located between 15 and 25 nm. A sufficiently large portion of pores with smaller diameters was shown by the high adsorption capacity of the material toward smaller toxic molecules, which is generally characteristic of hypercrosslinked polystyrene materials. The typical property of the hypercrosslinked network, namely the swelling in any liquid media, was also strongly expressed. The volume of polymers increased by a factor of at least 1.3 when dry material was wetted with water and 1.5 when wetted with methanol (non-solvents for polystyrene), with a further smaller increase on substituting methanol for a good solvent, such as toluene. Contrary to this, typical macroporous styrene–DVB copolymers are known to maintain their volume constant, both in a dry or wetted state.

The mechanical strength of the mesoporous polymers was found to be good. Each bead of 0.4 mm in diameter could tolerate a load as high as 300–450 g before destruction. This is an important parameter, because crushed particles and fines released from an adsorbent bed could embolize into the patient's circulation. The optimal bead size of the polymeric adsorbent was found to be 0.3–0.8 mm, which prevented high back pressures at moderate to high flow rates of viscous liquids, such as whole blood and plasma, through a 200–300 mL cartridge packed with the polymer.

The important property of the above mesoporous hypercrosslinked polystyrene–poly-DVB-type materials is their high adsorption capacity toward small proteins and lower affinity to large protein molecules, probably due to size exclusion effects. Cytochrome C, which has a molecular weight of 13 kDa, was used as a substitute for β_2M (11.8 kDa) in the adsorption experiments. From a very dilute solution of cytochrome C in a neutral phosphate buffer, the polymer removed 70–98% of the protein. This corresponds to an adsorption capacity of 11–14 mg of protein per milliliter of wet beads, or approximately 30 mg/g of dry polymer. In contrast to cytochrome C, adsorption of larger albumin molecules from relatively concentrated solutions (35 mg/mL) was found to be moderate, about 5–7% of the initial amount [357].

Accordingly, materials that were rated good in screening tests with cytochrome C and albumin removed 95% of β_2M and less than 5% of albumin and other essential proteins during *in vitro* tests with blood or plasma from ESRD patients (Fig. 15.1). Both the larger size of albumin molecules (MW 66 kDa, 6.0 nm in diameter) and their higher hydrophilicity may contribute to the tendency of the mesoporous adsorbent to remove β_2M (MW 11.8 kDa, 3.35 nm in diameter) with a relatively high apparent selectivity compared to albumin. Because of the specificity of the problem under discussion, most important here would be to see the percentage reduction of initial concentrations of proteins of interest, rather than the absolute amounts of these proteins adsorbed. The loss of a few grams of albumin per treatment session can be easily tolerated by the patient.

Actually, with the initial concentrations of albumin and β_2M differing by nearly 3 orders of magnitude (about 40 g/L and 60 mg/L, respectively), the apparent size selectivity of any adsorption process must be principally higher than that of dialysis. Indeed, in the course of a sorption process, the sorption sites that are attainable to albumin or other major plasma components will soon be saturated, while sorption sites for β_2M remain far from saturation till the very end of the hemoperfusion session. Therefore, the ratio of β_2M/albumin adsorbed will steadily improve with the continuation of the procedure. In contrast, the loss of major protein components in a

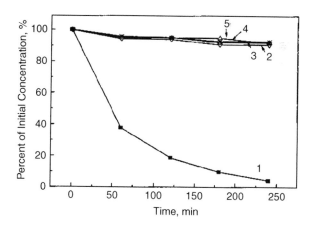

Figure 15.1 Selectivity of protein sorption on the hypercrosslinked mesoporous polymer in dynamic *in vitro* experiment with plasma. Experimental conditions: the polymer to plasma ratio is 1:20 (v/v); 37°C; initial concentrations of proteins are as follows: (1) β_2-microglobulin   6.35 mg/dL, (2) albumin   3.15 g/dL, (3) prealbumin 30 mg/dL, (4) transferrin   317 mg/dL, (5) total protein   6.2 g/dL. *(After [357].)*

dialysis process will continue at a constant rate throughout the entire procedure, whereas removal of $\beta_2 M$ will slow down as its concentration decreases. In other words, with the longer continuation of the procedure, adsorption will predominantly remove the minor component of a mixture, whereas long-term dialysis will predominantly affect the major component. This is not surprising, because dialysis is a steady dynamic process that is entirely governed by the concentration gradients of the components across the membrane (and kinetics of their through-pore diffusion), whereas the adsorption process for the major component stops on approaching the thermodynamic equilibrium for that component.

In closing, we refer to the recent publication [358] in which the synthesis of mesoporous poly-DVB and the characterization of its porous structure were reported. The polymer readily adsorbs another surrogate for $\beta_2 M$, lysozyme of hen egg white with molecular weight of 14 kDa and size of about 4 nm, but exhibits a pretty low affinity to human serum albumin. At present, however, it is hard to evaluate whether there is any prospect of real application of the sorbent because no information is available on its hemocompatibility.

3. BIOCOMPATIBILITY OF HYPERCROSSLINKED POLYSTYRENE: *IN VITRO* AND *IN VIVO* STUDIES

During blood contact with any foreign material, several components in the blood are activated through a variety of enzymatic and cellular processes. In the setting of chronic hemodialysis with up to 15 h of weekly blood–material interactions, the immune system, in particular, appears to be in a constantly activated, inflammatory state. Some have suggested the high incidence of infection and septicemia in these patients is due, in part, to the activated immune system's inability to adequately respond to microbial pathogens. Therefore, the search for more biocompatible materials for hemodialysis and hemoperfusion is of great importance.

Previous experience with polystyrene-type adsorbent materials (Amberlite XAD-4) in clinical hemoperfusion was characterized by significant bioincompatible responses such as complement activation and sharp drop of white blood cell (WBC) and platelet counts (neutropenia and thrombocytopenia). This was believed to be due to the hydrophobic nature of the polystyrene surface. Similar problems were also noted with activated carbon adsorbents. The biocompatibility of both materials needed to be improved by applying more hydrophilic polymers as surface coatings with dip or spray methods [349]. A major drawback of these surface coatings was that they tended to sharply reduce the efficiency of adsorption.

It has been noted that hypercrosslinked polystyrene does not adsorb proteins to the same extent as do conventional polystyrene or macroporous polystyrene [190]. Most probably, the open-work hypercrosslinked material does not expose a dense hydrophobic surface for the proteins to adsorb. The same is true for the marked inertness of both the new hypercrosslinked mesoporous poly-DVB materials and industrial hypercrosslinked sorbent MN-200 toward whole blood cells. Blood (50 mL) spiked with citric acid freely flows through a 5 mL column filled with beads of the above sorbents, contrary to many other coated carbon-type and polymeric hemosorbents tested under identical conditions, where blood coagulation and column clotting were found to occur [359]. Indeed, the new mesoporous poly-DVB sorbent, without any additional modification of the surface, was found to be sufficiently biocompatible and not cause any early coagulation effect in a standard plasma recalcification test.

Still, further enhancement of the hemocompatibility of the polymer that is intended for a prolonged treatment of ESRD patients was thought to be important. Therefore, the outer surface of polymer beads as well as that of larger pores was subjected to chemical modification. It had been previously recognized that the consumption of vinyl groups of DVB in its polymers and copolymers is never complete. Up to 30% of DVB involved in polymerization fails to function as a divinyl crosslinking agent and retains one of its vinyl groups intact. By analogy with hypercrosslinked polystyrene, where the remaining pendent chloromethyl groups largely concentrate on the surface of the beads and their large pores, it is logical to assume that the pendent vinyl groups of DVB also concentrate on the polymer surface. There they fail to find a partner for the addition reaction during the polymerization process. It is convenient to use these surface-exposed vinyl groups to make the surface more hydrophilic and biocompatible. We developed several simple surface modification procedures to accomplish this goal. One approach is based on the oxidation of the vinyl groups to epoxy groups, which can then be hydrolyzed to diol functions or be used for the addition of such hydrophilic species as serine, aspartic acid, ethylene glycol, and polyethylene glycol. Another simple approach to create blood-compatible materials is the grafting of flexible hydrophilic polymer chains by radical graft polymerization of such monomers as 2-hydroxyethyl methacrylate, acrylamide, and *N*-vinylpyrrolidone to the pendent surface-exposed vinyl groups. It is difficult to follow and quantitate any of the above chemical transformations by conventional analytical techniques, unless large amounts of functional polymers are grafted onto the surface. Nevertheless, the success of the hydrophilization is always evident from the behavior of the dry polymer with respect to water [347]. Whereas the untreated dry porous material is hydrophobic and remains floating on the surface of water plus small amount of dioxane or ethanol, the modified material is hydrophilic and easily sinks in the solution. These methods of minor surface modification provide the required hemocompatibility without compromising the adsorption kinetics of the material toward small protein molecules.

The modified polymer beads [347] passed all of the standard battery of biocompatibility tests required by the International Organization for Standardization guidelines (ISO 10993). The tests included *in vitro* coagulation tests (plasma recalcification time), hemolysis study (extraction method), cytotoxicity study using the ISO elution method, etc. In *in vivo* experiments, extracts of the polymer beads did not elicit pyrogenic irritation or sensitization reactions in laboratory animals (acute systematic toxicity study in the mouse, acute intracutaneous reactivity study in the rabbit, rabbit pyrogen study).

As further evidence of the modified polymer's excellent hemocompatibility, *in vivo* trials incorporating the polymer into a hemoperfusion device were conducted at the University of California at Davis. A polycarbonate cartridge containing 100 mL of the polymer was steam autoclaved at 120°C for 45 min and flushed with 1 L of sterile saline prior to use. Several times two healthy canines underwent 5 h of hemoperfusion at a flow rate of 200 mL/min. No adverse effects such as fever or hypotension were noted. Temperature, blood pressure, mixed venous oxygen saturation, and hematocrit were continuously monitored during the procedure and all remained unchanged throughout the procedure.

Neutropenia and thrombocytopenia are commonly observed phenomena during extracorporeal blood circulation and are considered to be the most sensitive markers of hemocompatibility. The graphs of WBC and platelet counts in the above canine experiments showed a characteristic reduction (by 13 and 27%, respectively) at 15–30 min. To generate data for comparison, the same canines were also subjected to 5 h of hemodialysis under identical conditions with a modified cellulose dialyzer. In both the dialyzer and hemoperfusion experiments, WBC counts quickly return to normal and remain unchanged through the remainder of the procedure. Interestingly, platelet counts also returned to normal in the hemoperfusion group but remained low in the dialyzer group. This would suggest superior biocompatibility of the surface-modified mesoporous poly-DVB adsorbent beads. With this respect it is important to note that, contrary to canine blood, no changes occurred in the platelet and leukocyte counts when human volunteer blood passed over the resin *ex vivo* [360] or *in vivo* in man during clinical hemodialysis, as will be shown below.

According to the neutral nature of the polymer, no significant change in the concentration of ions (Na, K, Mg, Ca, Cl, CO_2, PO_4) was observed in the blood of the dogs during a 5-h-long hemoperfusion procedure. Interestingly, blood urea nitrogen does decrease to a noticeable extent during the perfusion. It may well be that our polymer is adsorbing some urea, though the latter is a very hydrophilic and small molecule. Other major blood components examined (glucose, creatinine, cholesterol, albumin, total proteins) in these tests did not change with the exception of triglycerides, the content of which dropped noticeably.

4. CLINICAL STUDIES

The excellent sorption capacity of the hypercrosslinked mesoporous poly-DVB with respect to selective removal of β_2M from its mixtures with albumin and other serum proteins, combined with superior hemocompatibility of the beads' surface modified with poly(*N*-vinyl)pyrrolidone, justified the manufacturing of an experimental batch of the material for initial clinical studies. The polymer was named BetaSorb^{TM} (RenalTech International, USA) and was used in 300 mL cylindrical polysulfone devices that were steam-sterilized and filled with normal saline containing 1000 IU heparin. The device was placed in line with the dialysis circuit, upstream of the dialyzer, in order to not affect the pressure drop across the dialyzer membrane. The blood flow was maintained at the customary value of 400 mL/min, again the optimal flow rate for the dialyzer. The complete setup of the combined hemoperfusion–hemodialysis treatment [361] is displayed in Fig. 15.2.

Initial experiments on two ESRD patients at a hospital in Vicenza (Italy) clearly demonstrated the full safety and efficiency of the Beta-Sorb^{TM} device [361]. During each of the 3-h-long treatments, platelet

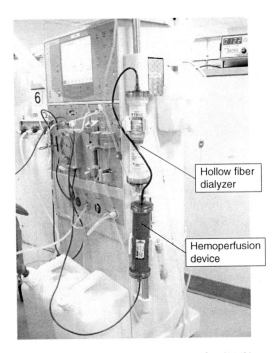

Figure 15.2 *Operational unit for blood purification. (After [361].)*

and leukocyte counts remained stable, as did serum albumin concentrations. In fact, these figures even increased during the session because of the removal of excess water from the circulation through dialysis membranes. On the contrary, $\beta_2 M$ levels dropped by 70 and 80% from their initial levels. Interestingly, these levels slightly rebound in 30 min after the hemoperfusion process. This finding indicates that an active exchange of small protein molecules exists between different compartments of physiological liquids in the organism. Therefore, it is principally possible to remove in one treatment session more of the toxin than initially circulated with blood. It also encourages the continual use of hemoperfusion in regular dialysis procedures with the hope of prevention or even reversal of dialysis-related amyloidosis (DRA). Indeed, after the treatment period of 3 and 6 weeks (the time allotted for the experiment for the two patients), where on the average 140 mg of $\beta_2 M$ was removed per session, a 30% sustained reduction of the pre-dialysis $\beta_2 M$ levels was registered in one patient for a period of 3–4 weeks [362]. During the hemoperfusion session, in addition to $\beta_2 M$, the adsorptive resin was also found to remove other "middle-molecular-weight" toxins, such as tumor necrosis factor-α (TNF-α, 17.2 kDa) and interleukin-1β (IL-1β, 17 kDa) [361], leptin (16 kDa), angiogenin (homologues of degranulation inhibitory protein I, Dip1, 14.4 kDa), and several cytokines [362]. In fact, nearly 70 "classical" and "new" uremic toxins have been found to accumulate in blood of ESRD patients [363]. Among them recent studies have identified a number of proteins, including those larger in size than $\beta_2 M$, which are believed to relate to malnutrition of renal failure patients, affect their immune system, or cause cardiovascular problems. Many of the toxic proteins result from the fact that hemodialysis patients are constantly exposed to a microinflammatory environment. Indeed, exposure to bacterial contaminants from the dialysis water system, poorly biocompatible dialyzer membranes and tubings, and bacterial infections induce the release of proinflammatory cytokines, which generate a chronic inflammatory state and then an acute phase reaction. Only the low-molecular-weight toxins are being removed during regular dialysis, but not the "middle-molecular-weight" toxins with MW > 1500 Da. Adsorption of the latter by the BetaSorb^{TM} device may sufficiently increase the therapeutic effect of a combined hemoperfusion/hemodialysis treatment [363, 364].

Much more extended clinical studies with BetaSorb^{TM} device were performed in the USA. Averaged data for 100 patients are as follows:

Initial concentration of $\beta_2 M$	34 ± 18.5 mg/L
Total reduction of $\beta_2 M$	$70.6 \pm 10.3\%$
Final concentration of $\beta_2 M$ after 30 min rebound	14 mg/L
Total removal of $\beta_2 M$ per session	402 ± 142 mg

No complications related to the combined hemoperfusion/hemodialysis sessions were registered. The treatment was well received by the patients. These results speak for themselves. They prove both the high safety and efficiency of the hemoperfusion procedures on hypercrosslinked poly-DVB.

The long-term consequences of adding the adsorption device to the conventional hemodialysis system in treating renal patients still remain to be evaluated. For many decades, hemodialysis has been the most effective method of supporting patients with renal failure. In the United States alone in the year 2000, over 15 billion dollars was spent on patients with chronic kidney failure. The cost of the dialysis treatment represents only

30% of these expenditures. The remainder is accounted for by the morbidity related to chronic kidney failure, problems such as DRA, infections, nerve dysfunction, and malnutrition. Most probably, providing modern expensive hemodialysis machines with additional disposable cost-efficient hemoperfusion cartridges could contribute much to reducing this remaining 70% portion of the costs and improving the quality of patients living with kidney failure.

5. FURTHER PERSPECTIVES FOR HEMOPERFUSION ON HYPERCROSSLINKED SORBENTS

The finding that hemoperfusion through the polymeric BetaSorb^{TM} column effectively removes several middle-molecular-weight toxic proteins and that such toxins and cytokines are heavily involved in the systemic inflammatory response syndrome (SIRS) have stimulated extensive preclinical studies to demonstrate the wider utility of the hemoperfusion procedure. Winchester et al. [365] list 24 cytokines and other protein and glycoprotein molecules in the molecular weight range of 4–110 kDa that are implicated in SIRS. In addition to ESRD, SIRS is caused by various other health problems, such as acute renal failure (350,000 cases per year in the USA), cardiopulmonary bypass surgery (1.5 million/year), and sepsis.

The leading cause of acute renal failure in the USA is sepsis, with over 750,000 cases per annum and an absolute mortality rate of 28.6%. Severe sepsis, regardless of etiology, is complicated by failure of multiple organs, of which renal failure is seen in 22% of patients (in this setting the mortality is 38.2%). Since sepsis is preceded by the SIRS and accompanying intensive cytokine release, attention has been directed to correcting circulating cytokine levels. In sepsis, in response to endotoxin (lipopolysaccharide component of Gram-negative bacterial cell walls), mononuclear phagocytes are activated resulting in elevated concentrations of several cytokines, including TNF-α, interleukins IL-1β, IL-4, IL-6, IL-8, and IL-10, interferon-γ, and others that have been linked to mortality in patients. The neutral polymer BetaSorb^{TM} was found to be inactive against endotoxin, probably because of the very high molecular weight of the latter. However, the modified adsorptive resin CytoSorb^{TM} was shown to actively remove proinflammatory cytokines, TNF-α, many interleukins, leptin, and others. In preliminary experiments on 40 rats that were subjected to 100%-lethal doses of *Escherichia coli* endotoxin (LPS, 20 mg/kg), a substantially increased survival rate was registered in the group that received hemoadsorption with a 10 mL cartridge for 4 h [365]. Blood analysis showed a definite clearance effect that was most pronounced on IL-6, which was no longer detectable after 60 min, and least pronounced on TNF-α, a third of which was still present after 120 min. Thus, sepsis treatment seems to be one of the most obvious and straightforward application areas of hemoperfusion using hypercrosslinked CytoSorb™ polymers.

In fact, most surgeries involving extracorporeal blood circulation, in particular cardiopulmonary bypass surgery, could strongly benefit from adding blood perfusion through a polymeric adsorbing material. It would remove the unavoidably emerging proinflammatory proteins, cytokines, and cell degradation products from the organism, which would reduce the probability of any post-surgery complication.

The same extracorporeal hemoperfusion procedure should be applied to brain-dead potential organ donors. While being severely damaged in most cases, their bodies produce large amounts of toxins, which soon results in the failure of many organs, thus making them no longer suitable for transplantation. In the USA alone, there are over 80,000 patients on the waiting list for organ transplantation, while only 6000 per year will be treated.

Another important application field for hypercrosslinked polystyrene–type adsorbing materials is hemoperfusion in the case of acute intoxication. In Russia this problem causes more than 50,000 deaths yearly. Removing toxic xenobiotics from blood is the quickest way to avoiding their harmful action in the organism. Activated carbons and hypercrosslinked polymers (unmodified mesoporous poly-DVB and industrial MN-200) remove 1,2-ethylene dichloride and tetrachlorocarbon from aqueous model solutions equally rapidly and completely, but only the polymeric sorbents were found to maintain their high efficiency in the case of sorption of the same chloroorganic toxicants from whole blood [349]. This finding correlates with the fact of rapid adsorption of proteins and blood cells on the surface of carbon beads, which immediately sets barriers to mass exchange with the sorbent interior.

Extension of these tests to include other classes of toxic xenobiotics confirmed excellent activity of hypercrosslinked polystyrene sorbents with respect to overdosage of some drugs such as phenobarbital. Though the sorption efficiency against methanol and phosphor-organics (trichloromethaphos-3) was found to be lower, still, even in these cases, polymeric materials were far superior to carbonaceous hemosorbents [366, 367].

Indeed, in experiments on dogs who received with their food a dose of ethylene dichloride (2000 mg/kg) and then, after 1 h of exposure to the poison, a hemoperfusion therapy using 100 mL cartridges with different adsorbing materials, only hypercrosslinked polymers showed an intensive removal of ethylene dichloride from the blood with no changes in blood formulation (platelets, white and red blood cells, monocytes, lymphocytes). Carbonaceous materials coated with a hemocompatible layer produced hemolysis and severe drops in platelet and WBC counts, with a slow kinetics of detoxification.

In a group of 20 rats that received an absolutely lethal dose (2.5 LD) of carbon tetrachloride into the stomach and then, after 15 min, an oral portion of sorbent MN-200 (1.0 g/kg), 90% of animals were alive after 4 days, while in the reference group with no polymer ingestion no animal survived the second day.

All these experiments demonstrate the high efficiency of hypercrosslinked polystyrene–type adsorbing materials in the capacity of hemosorbents and enterosorbents with broad application perspectives.

Finally, the development and successful testing of a hemosorbent that displays long-term biocompatibility and high sorption capacity toward minor toxic protein components in the blood opens ways for designing a small portable and disposable artificial kidney [368, 356]. One can think of a continuously functioning extracorporeal device (Fig. 15.3) that combines selective adsorption of middle-molecular-weight toxins with ultra-filtration though a membrane that would slowly remove excess water and all dissolved small toxic molecules. Filtration through the membrane could be easily induced by a slight vacuum applied on one side of the membrane, while its other side is swept with blood. Permeability of modern dialyzer membranes is fully sufficient to provide a desired flow of the filtrate under the action of an easily adjustable vacuum generated by a simple squeezed plastic bottle. The latter would simultaneously serve for collecting the filtrate and registering its amount. The blood flow should also pass an adsorbent polymer-filled chamber. When connected with implanted cannula and suitable tubings to the blood circuit of the patient (ESRD patients have a constant access to the blood circuit, anyway), the device could maintain a constant blood flow due to the normal pressure difference between the artery and vein (which is about 5300–6600 Pa).

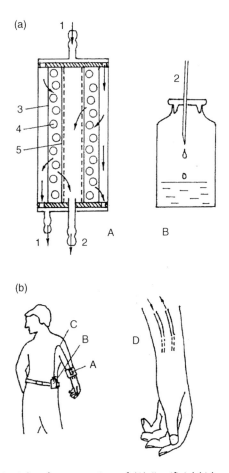

Figure 15.3 (a) Principle of construction of (A) "artificial kidney" and (B) "artificial bladder": (1) flow of blood, (2) flow of ultrafiltrate, (3) plasma-filtrating membrane, (4) adsorbing material, (5) ultrafiltration membrane. (b) Reasonable disposition of the "artificial kidney"; (A) artificial kidney, (B) artificial bladder, (C) artificial ureter, (D) implanted arteries and venous canulas. *(Reprinted from [356] with permission of Elsevier.)*

Carrying such a miniaturized portable artificial kidney could substantially increase the periods between normal hemodialysis procedures, thus making the patient much less dependent on his residency clinic.

CHAPTER *16*

Hypercrosslinked Ion-Exchange Resins

1. ION EXCHANGE CAPACITY AND SWELLING BEHAVIOR OF HYPERCROSSLINKED STRONG ACIDIC ION-EXCHANGE RESINS

Polystyrene networks obtained by post-crosslinking of both linear polystyrene and copolymers of styrene with divinylbenzene (DVB) readily enter a sulfonation reaction with concentrated sulfuric acid even under mild conditions, reflecting enhanced accessibility of the polymer interior to the chemical reagent. The ion exchange capacity of the resulting resins containing no more than 25% crosslinks is close to the theoretical value, 5.43 meq/g, that corresponds to one functional group per phenyl ring of polystyrene. Hypercrosslinked exchangers exhibit a slightly lower exchange capacity, which is still similar to that of many commercial gel-type resins and is fully sufficient for many practical applications (Table 16.1).

All sulfonated resins have high affinity to water. The dependencies of their swelling in water versus the degree of crosslinking are similar to those of swelling of starting networks in toluene. In the case of resins based on linear polystyrene, the plots exhibit a clear-cut minimum located in the vicinity of a 40% degree of crosslinking. The sulfonates manifesting the lowest water regain absorb 0.46 mL/meq water, which is comparable with the swelling of conventional gel-type ion exchanger with 2–3% DVB (Fig. 16.1). Water uptake of hypercrosslinked exchangers based on styrene–1% DVB copolymer decreases until a minimum value at $X = 40\%$ and does not change anymore with further crosslinking. It should be also noted that, due to the rigidity of the network, swelling of hypercrosslinked sulfonates in water equals that of their initial copolymers in toluene. This is different from the situation with conventional gel-type resins, which swell in toluene less than their sulfonation products in water. Obviously water hydrates the HSO_3 groups of the resin stronger than toluene solvates the initial polystyrene matrix.

Comprehensive Analytical Chemistry, Volume 56
ISSN 0166-526X, DOI 10.1016/S0166-526X(10)56016-2

Table 16.1 Ion exchange capacity (meq/g) of the macronet isoporous, hypercrosslinked, and conventional sulfonates of polystyrene [369]

X (%)	Linear polystyrene crosslinked with			St-1%DVB crosslinked with		Gel-type St–DVB resins
	MCDE	XDC	CMDP	CMDP	MCDE	
2.3	–	–	–	5.30	–	5.42
4	5.22	5.25	5.29	4.61	–	–
6	–	–	–	5.31	5.49	5.23
8	–	–	–	5.34	–	5.13
11	5.35	5.28	5.25	5.28	5.49	5.00
25	5.35	5.12	5.02	–	5.23	–
43	5.00	4.25	5.00	–	5.00	–
66	4.78	4.40	4.50	–	4.74	–
100	4.68	4.16	4.26	–	4.41	–

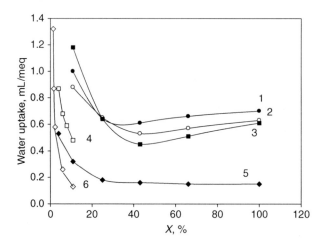

Figure 16.1 Water uptake as a function of the crosslinking degree of macronet isoporous and hypercrosslinked sulfonates based on (1–3) linear polystyrene, styrene copolymers with (4) 0.8 and (5) 1% DVB, and crosslinked with (1, 5) monochlorodimethyl ether, (2, 4) 1,4-bis(chloromethyl)diphenyl, (3) *p*-xylylene dichloride, as well as (6) conventional gel-type sulfonated resins.

A practically important property of Styrosorb hypercrosslinked ion-exchange resins is the stability of their volume, which remains nearly constant regardless of the ionic strength and pH of the outer electrolyte

Table 16.2 Change in volume in various media, relative to the volume in water (%), of sulfonated resins (H$^+$ form) based on different matrices [369]

X (%)	K_2 (mL/meq)^a	KOH 0.1 N	HCl 0.1 N	FeCl$_3$ 5.0 N	Ethanol	Dioxane	Benzene
Linear polystyrene crosslinked with CMDP							
11	0.88	15	5	–	25	64	73
43	0.52	0	0	–	5	14	38
100	0.62	0	0	–	0	0	17
Linear polystyrene crosslinked with XDC							
11	1.17	23	15	–	40	60	64
43	0.62	0	0	–	5	8	25
100	0.50	0	0	–	0	0	15
Styrene–1% DVB copolymer crosslinked with MCDE							
11	0.32	11	12	17	0	–	50
43	0.16	0	0	0	0	–	15
100	0.16	0	0	0	0	–	10
Gel-type styrene–DVB copolymers							
1.4	1.28	18	12	–	30	66	68
2.3	0.58	16	6	–	12	46	62

$^a K_2$ = swelling with water measured by weight.

solution, as well as the nature and charge of the counter ions (Table 16.2). Thus, ion exchanger at the degree of crosslinking $X = 100\%$ do not change their volume when water is replaced with concentrated solutions of HCl, NaOH, FeCl$_3$, ethanol, or dioxane. Even in benzene, the hypercrosslinked strong acidic ion-exchange resins still retain their swollen state (conventional gel-type cation exchanger collapse in benzene to the volume of dry resin).

As can be seen from Table 16.3, the swelling of hypercrosslinked sulfonated ion exchanger in nonpolar benzene, heptane, and ethylene dichloride is only insignificantly smaller than that in water or polar methanol. Their functional sulfonic groups thus remain fully accessible to counter ions and organic molecules in any liquid media, whereas conventional ion exchanger can only be used in water or polar organic solvents. The unique ability of hypercrosslinked sulfonated resins to retain their expanded (swollen) state in nonpolar media opens ways for their successful use as strong acidic catalysts or high-capacity base scavengers in organic synthesis.

Table 16.3 Solvent uptake (mL/g) in various media of strong acidic ion-exchange resins Styrosorb 1 based on linear polystyrene post-crosslinked with XDC

X (%)	EC (meq/g)^a	Water	Ethanol	Benzene	n-Heptane	EDC
11	5.07	7.20	5.58	0.45	0.67	0.40
25	4.98	3.30	3.04	0.49	0.73	0.45
43	4.86	2.23	2.12	0.69	0.94	0.72
66	4.56	1.85	1.85	0.85	1.11	0.88
100	4.32	1.56	1.47	1.14	1.32	1.28

^{a} EC = ion exchange capacity.
Reprinted from [370] with permission of Akademizdattsentr "Nauka" RAN.

2. KINETICS OF ION EXCHANGE ON HYPERCROSSLINKED RESINS

The relatively high water uptake, combined with rather uniform distribution of crosslinks in the polymeric matrix of hypercrosslinked ion exchanger, enhances the permeability of sulfonated resins with regard to large organic ions. When discussing the kinetics of ion exchange on the resins, two parameters should be taken into consideration, namely the diffusion rate of organic ions into resin beads and the accessibility of functional groups to these ions. To characterize the permeability of hypercrosslinked sulfonated polystyrenes, voluminous tetrabutylammonium (TBA) ions of 9.6 Å diameter were chosen as the test ions [371].

TBA ions migrate so fast into the water-swollen particles of Styrosorb 1 resins based on linear polystyrene that it does not seem possible to correctly determine the limiting stage of $H^+-(C_4H_9)_4N^+$ exchange and calculate the coefficients of diffusion. Because of this, the time required to substitute half of the exchange sites ($\tau_{0.5}$) and the time of approaching equilibrium (τ_{eq}) were used to characterize the process.

Usually, the permeability of ion-exchange resins has been related to the number of crosslinks: the higher the degree of crosslinking, the lower the permeability. However, as Fig. 16.2 shows, irrespective of the matrix structure, $\tau_{0.5}$ depends on the swelling capacity of the resins, rather than the density of their network. The ion exchange is really fast both in the weakly crosslinked (1.4–2.3% DVB) standard ion exchanger and in the hypercrosslinked resins based on linear polystyrene: half of the protons are replaced by TBA ions within 1.5 min or less.

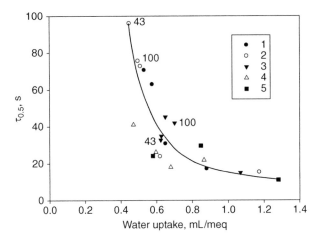

Figure 16.2 Dependence of the time of saturation of 50% functional groups with $(C_4H_9)_4N^+$ ions on the resin water uptake for the sulfonates prepared by crosslinking (1–3) linear polystyrene with (1) 1,4-bis(chloromethyl diphenyl), (2) p-xylylene dichloride, and (3) monochlorodimethyl ether; (4) styrene–1% DVB copolymer crosslinked with 1,4-chloromethyl diphenyl; (5) sulfonates based on conventional styrene–DVB copolymers; numbers denote the degree of crosslinking. *(After [371].)*

Sulfonated resins based on post-crosslinked styrene copolymer with 1% DVB swell in water to a remarkably lower extent. The $H^+–(C_4H_9)_4N^+$ exchange proceeds slower, thus allowing us to calculate the interdiffusion coefficients, $\bar{D}$, at the initial stage of the ion exchange [372]. With decreasing water uptake, the effective interdiffusion coefficients drop sharply (Table 16.4). Still, one can see again that similar $\bar{D}$ values correspond to resins having similar swelling capacity, regardless of their degree of crosslinking.

Thus, at the initial stage of ion exchange, the swelling of the gel-type resins in water appears to be the basic factor determining the rate of the process (to a certain extent, one can consider the homogeneous hypercrosslinked networks to be gel polymers). The differences in the structure of networks begin to tell upon the rate of ion exchange toward the last stages of the process. The highest rate of approaching equilibrium (within 12 min or less) is characteristic of the resins prepared by crosslinking styrene–1% DVB copolymer with 4,4′-bis(chloromethyl)diphenyl to 3–11% degree of crosslinking (Fig. 16.3). Note that the exchange of protons against TBA ions in these samples occurs quicker compared to standard exchangers with the same water regain.

Table 16.4 Swelling in water (K_2, mL/meq) and H^+–$(C_4H_9)_4N^+$ effective interdiffusion coefficients ($\bar{D}$, cm^2/s) for the sulfonated resins based on different matrices

Styrene–1% DVB copolymer crosslinked with MCDE			Gel-type styrene–DVB SA ion-exchange resins		
X (%)	K_2	$\bar{D}$	X (%)	K_2	$\bar{D}$
11	0.32	2.6×10^{-7}	6	0.28	1.8×10^{-7}
25	0.18	6.6×10^{-9}	8	0.26	4.4×10^{-8}
43	0.16	3.2×10^{-9}	12	0.14	2.0×10^{-8}
66	0.16	2.9×10^{-9}	–	–	–
100	0.16	3.2×10^{-9}	–	–	–

Reprinted from [372] with permission of Akademizdattsentr "Nauka" RAN.

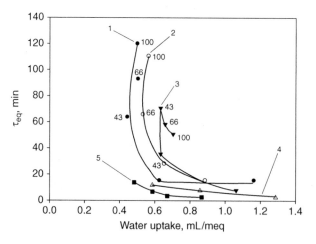

Figure 16.3 Dependence of the time required for establishing $(C_4H_9)_4N^+/H^+$ equilibrium (τ_{eq}) on the resin water uptake for sulfonates prepared by crosslinking (1–3) linear polystyrene with (1) p-xylylene dichloride, (2) 1,4-bis(chloromethyl diphenyl), and (3) monochlorodimethyl ether; (4) styrene–1% DVB copolymer crosslinked with 1,4-bis(chloromethyl diphenyl); (5) sulfonates based on styrene–DVB copolymers; numbers denote the degree of crosslinking. (After [371].)

The time needed for the saturation with TBA ions of hypercrosslinked exchangers Styrosorb 1 crosslinked with 1,4-bis(chloromethyl diphenyl) (CMDP) or xylylene dichloride (XDC) progressively rises with the degree of crosslinking and reaches 2 h for samples with $X = 100\%$. Distinctly, saturation of sulfonates with matrices obtained by crosslinking linear polystyrene with monochlorodimethyl ether (MCDE) is also slow, but it decreases from 70 to 50 min with the degree of crosslinking rising from 43 to 100% (Fig. 16.3). Still, these equilibration times are significantly longer than those required for conventional gel-type ion-exchange resins with comparable water uptake. Obviously, the high elasticity of the slightly crosslinked gel-type networks assists the diffusion of large ions to the exchange sites, while both small dimensions and rigidity of meshes in the hypercrosslinked network hinder the diffusion of large TBA ions to the last spatially restricted functional groups of the matrix.

As a result, some 12–13% of sulfonic groups in the matrix obtained by crosslinking linear polystyrene with MCDE to 100% fail to exchange H^+ ions by large $(C_4H_9)_4N^+$ ions, while the extent of the exchange further drops to nearly 50% in the case of the more rigid matrix obtained by post-crosslinking styrene–1% DVB copolymer. Longer cross-bridging reagents, bis-chloromethylated benzene and diphenyl, result in a matrix with larger and more flexible meshes, which restrict the hydrogen–TBA exchange less severely. To give a comparison, conventional gel-type sulfonated resins do not permit the H^+–$(C_4H_9)_4N^+$ exchange to proceed further than 50% when the DVB content exceeds 10%.

To sum up, ion-exchange resins with a hypercrosslinked matrix are characterized by very fast kinetics of ion exchange at the beginning; here the interdiffusion of ions is mainly governed by the water uptake of the resins. However, the rate and the extent of exchange of small H^+ or Na^+ ions against large $(C_4H_9)_4N^+$ ions drop at the final stage of exchange, particularly for networks with small and rigid meshes.

3. SELECTIVITY OF ION EXCHANGE ON HYPERCROSSLINKED STRONG ACIDIC CATION-EXCHANGE RESINS

It is well known that the ion selectivity of conventional gel-type cation-exchange resins is determined by the radius and hydration extent of the cations. However, taking a specific pair of cations, the ion exchange coefficient does not remain constant, but changes during the exchange process with changing ratio of the two cations residing in the resin phase. This phenomenon is logically explained by the

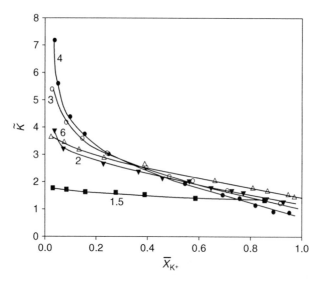

Figure 16.4 Dependence of concentration constants of H^+–K^+ exchange on the extent of saturating functional groups with potassium ions for sulfonates (1–4) prepared by crosslinking linear polystyrene with p-xylylene dichloride to (1) 11, (2) 43, (3) 66, (4) 100% and for conventional resins KRS (Russia) containing (5) 3 and (6) 8% p-DVB. *(After [376].)*

energetic non-equivalence of exchange sites in the resin phase, which, in its turn, results from the heterogeneous distribution of DVB cross-links in the resin matrix [373–375]. Hypercrosslinked polystyrene matrices are assumed to be crosslinked more uniformly, but, nevertheless, the dependence of the ion exchange coefficient on the ratio of H^+ and K^+ ions in the resin phase (Fig. 16.4) was found to be even more strongly pronounced [376]. This unexpected finding can be explained by the rigidity of the highly crosslinked polymer matrix, which substantially reduces the rotation freedom of sulfonated aromatic rings involved in the formation of small meshes. If both the original repeating unit of polystyrene and the cross-bridges introduced during the post-crosslinking reaction receive sulfonic groups, which is indeed the case at high sulfonation extents, then two or even three sulfonic groups may be forced to locate close to each other. This constellation would form a "cage" with a definite size and shape and an enhanced density of negative electrostatic charge, which could favor binding of a particular cation:

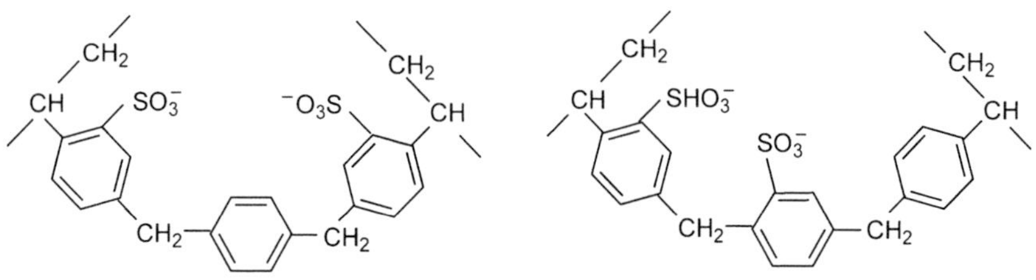

Even a few such local zones with increased affinity to a metal cation can be responsible for the extremely high values of concentration constants ($\tilde{K}$) of H^+–K^+ exchange at low loadings of the resin with potassium ions. On the other hand, after entering the above narrow cage, the strongly bonded cation can block the ion "trap" and prevent sorption of a second metal cation. In this case, the remaining sulfonic neighbor groups have to be neutralized with protons that have no *eigen*-volume in the aqueous phase. Therefore, the last functional groups of the resin retain express affinity to protons and the ion exchange constant drops to an unusually low value (Fig. 16.4).

It follows from the above that sulfonated hypercrosslinked polystyrene resins taken in the H^+ form should be able to eliminate trace metal cations from concentrated mineral acids much more efficiently than conventional cation exchanger [377], which could be exploited in the preparation of high-purity acids.

4. POROSITY OF DRY HYPERCROSSLINKED STRONG ACIDIC ION-EXCHANGE RESINS

When carefully dried, hypercrosslinked sulfonated resins were unexpectedly found to be non-porous materials. Indeed, after introducing 4.4 meq/g of sulfonic functions in a matrix obtained via post-crosslinking linear polystyrene with XDC to 100% and having an apparent specific surface area as large as 1000 m^2/g, one obtains the material with negligible porosity, less than 1 m^2/g. Of course, the voluminous sulfonic groups occupied a certain part of the matrix free volume. But more important could be the additional contraction of the network due to the strong cohesion energy evoked by dipoles of polar functional sulfonic groups. In any case the dry material becomes dense and impermeable even for molecules of nitrogen. Neither does it sorb acetone and hexane vapors, testifying once more to the non-porous structure of the dry sulfonic resin.

On the contrary, the sulfonated resin absorbs vapors of those substances that strongly interact with the HSO_3 groups; these are water and methanol (Fig. 16.5) [13]. If one compares the sorption of water vapors onto the cation-exchange resin Styrosorb 1 and macroporous sulfonic resin KU-23 (Russia), both having similar exchange capacity, it is easy to ascertain that in the range of relative pressures from 0 to 0.7 the adsorption is proportional to the number of exchange sites, and, therefore, the initial sections of the two isotherms coincide. Sorption of methanol vapors is, naturally, lower. The desorption branches of the isotherms for both water and methanol are located above their adsorption branches throughout the entire range of relative pressures, thus pointing to the swelling of the hypercrosslinked resin with vapors of the above sorbates.

Notably, sulfonated hypercrosslinked resins readily swell in liquid methanol or ethanol, and the latter can be then replaced with hexane, without causing any collapse of the network (see Table 16.3). We would like to emphasize here a close analogy in the behavior of dry sulfonated resins with respect to hexane and that of dry initial hypercrosslinked

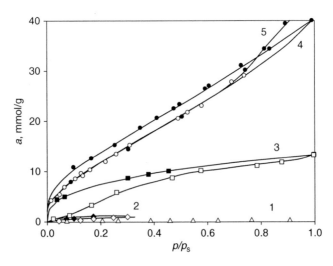

Figure 16.5 Sorption isotherms at 25°C of vapors of (1) n-hexane, (2) acetone, (3) methanol, and (4, 5) water onto (1–4) sulfonated Styrosorb 1 crosslinked with p-xylylene dichloride to 100% and onto (5) cation exchanger KU-23, both taken in H^+ form. *(After [381].)*

polystyrene with respect to water. It is only with the assistance of ethanol that the network of sulfonated resins acquires an extended swollen structure then preserved in hexane, and the network of hydrophobic hypercrosslinked polystyrene becomes swollen and then compatible with water.

It is self-evident that the biporous hypercrosslinked sulfonated resin MN-500 retains its porous structure even in its dry state, since large transport pores always remain open and accessible to vapors of any compounds.

5. ANION-EXCHANGE RESINS

Macronet isoporous polymers with a degree of crosslinking less than 25% readily enter the reaction of chloromethylation. The amount of introduced chlorine, 20–22%, comes close to that corresponding to one $ClCH_2$ group per styrene repeating unit (23% Cl).

The mobile chlorine atoms in the $ClCH_2$ groups can easily be substituted with iodine by boiling the chloromethylated polymer with twofold molar excess of sodium iodide in a mixture of acetone and dioxane (3:1, v/v) [378]. A polymer with an 83% degree of iodomethylation was thus obtained on the base of a macronet isoporous matrix prepared by post-crosslinking of a styrene–0.8% DVB copolymer with 11% additional bridges formed by 4,4′-bis(chloromethyl)diphenyl.

By reacting chloromethylated or iodomethylated products with amines under standard conditions, various macronet isoporous anion-exchange resins with high exchange capacity can be obtained. Interestingly, even a reasonably small amount of rigid post-crosslinks, just 5–10%, when introduced into a swollen network, favorably affect the permeability of final ion exchanger. To demonstrate the difference between conventional styrene–DVB copolymers and macronet isoporous networks, two resins were prepared. The first was based on a styrene–0.8% DVB copolymer and the second on the same material post-crosslinked in swollen state with 5 mol.% of the above CMDP. Both samples were chloromethylated and then aminated with methyl esters of α-amino acids in the presence of sodium iodide as a catalyst (0.5 mol per 1 mol of $ClCH_2$ groups). The ester groups were finally hydrolyzed with sodium hydroxide in a water/dioxane solution at room temperature [379]. Table 16.5 presents the results. As the permeability of the macronet matrix was obviously enhanced, the hydrolysis of ester bonds in the polymer proceeded much faster compared to the hydrolysis rate in the conventional styrene–DVB product.

Table 16.5 Hydrolysis of methyl esters of α-amino acids attached to macronet isoporous polymer (CMDP, 5% – Support I) and styrene–0.8% DVB copolymer (Support II) [379]

Amino acid	Support I		Support II	
	τ (days)	α (%)	τ (days)	α (%)
DL-α-Aminobutyric acid	20	40	40	0
DL-β-Phenyl-α-alanine	20	40	40	0
DL-Norvaline	18	5	40	0
DL-Serine	6	100	40	traces

τ and α = time and the degree of hydrolysis, respectively.

The markedly enhanced swelling of the macronet isoporous networks allowed Davankov to prepare efficient chiral resins containing residues of optically active α-amino acids. The latter were complexed with copper(II) ions to form chiral adsorption centers for enantiomers of amino acids, hydroxy acids, and other compounds capable of coordinating Cu ions. The new technique was termed ligand exchange chromatography and was the first successful separation approach to resolve racemic compounds into individual enantiomers in liquid chromatography [380]. Many attempts by other researchers to repeat these results by starting from conventional styrene–DVB copolymers resulted in chiral resins with low swelling in aqueous media and low chromatographic efficiency.

Attempts can be mentioned of using a hypercrosslinked anion-exchange resin with ethylene diamine functional groups for selective adsorption of carbon dioxide [381]. The tested weak basic anion-exchange resin had a developed porous structure with specific surface area of $700\,m^2/g$, which is the essential condition for adsorption from a gaseous phase. However, no remarkable difference in the adsorption capacity was observed between this Styrosorb resin and macroporous anion-exchange resin AN-221 (Russia), which also contained a similar amount of ethylene diamine functions. Generally, the absorption of carbon dioxide was found to be caused by the formation of carbamate complexes with ethylene diamine functional groups [382–385]. The dynamic capacity of both types of resins amounted to 15–20 mg/g. The decomposition of the carbamate with the release of CO_2 proceeds easily in vacuum at 70°C.

It must be noted that preparation of really hypercrosslinked anion-exchange resins presents specific difficulties that have not yet been circumvented. Chloromethylation of hypercrosslinked networks proceeds with a small yield only, giving no more than 13–15% chlorine in the product, independent of the conditions of chloromethylation. Obviously, in the hypercrosslinked network many phenyl groups and cross-bridges are located at a distance that favors the formation of new bridges with the involvement of newly introduced chloromethyl groups. For that reason, the ion exchange capacity of final aminated hypercrosslinked materials does not exceed 3 meq/g, and, moreover, the matrix becomes crosslinked so densely that adsorption of larger anions is spatially hindered.

6. PROPERTIES OF COMMERCIAL HYPERCROSSLINKED ION-EXCHANGE RESINS

Purolite International (UK) offers a whole series of ion-exchange resins based on a hypercrosslinked polystyrene matrix. A list of these products and their general properties are presented in Table 9.6. In the following, only specific features are dealt with, which distinguish hypercrosslinked materials from conventional ion exchanger.

The expressed non-equivalence of sorption sites in hypercrosslinked ion-exchange resins, resulting in the existence of certain "ion traps" for ions of various sizes, must be a general phenomenon (see Section 3). Indeed, commercial strong acidic cation-exchange resin MN-500 in H^+ form shows good selectivity for alkali and alkali-earth ions [386]. Also, MN-500 in NH_4^+ form can be used for removing trace Na^+ ions from rather concentrated aqueous ammonia solutions.

Tai et al. [387] studied the removal of toxic Cu^{2+}, Zn^{2+}, and Ni^{2+} ions by passing 0.25 mM aqueous solutions of corresponding chloride salts through dry MN-600, the hypercrosslinked weakly acidic cation exchanger. Although 90% breakthrough capacity (0.04–0.08 mmol/g) for the salts proved to be 2–5 times smaller than the total resin capacity for sodium ions, 0.42 meq/g, nevertheless 75, 50, and 40 bed volumes of feed were purified from $CuCl_2$, $ZnCl_2$, and $NiCl_2$, respectively, till the salts emerged at the column outlet.

The carboxylic-type hypercrosslinked resin MN-600 was examined comprehensively by Saha and Streat [388], in comparison with polyacrylic acid C-104E (Purolite), with the aim of evaluating the performance of the resins for trace heavy metal removal. Characterization of these polymers involved scanning electron microscopy, BET and Langmuir surface area measurements, Fourier transform infrared (FTIR) spectroscopy, X-ray photoelectron spectroscopy, elemental analysis, zeta potential measurements, and exchange capacity determination. Naturally, under static conditions MN-600 takes up smaller amounts of copper, nickel, or zinc ions than C-104E, because the latter exhibits a 20-fold larger total exchange capacity, 9.65 versus 0.45 meq/g. However, both resins follow a similar pattern of extracting the above bivalent ions: $Cu^{2+} >> Ni^{2+} >> Zn^{2+}$. From the experimentally determined equilibrium data in ternary $H^+/Zn^{2+}/Ni^{2+}$ or $H^+/Cu^{2+}/Ni^{2+}$ systems and in the quaternary $H^+/Cu^{2+}/Zn^{2+}/Ni^{2+}$ system, ion exchange parameters have been determined in a wide range of initial conditions.

The influence of the texture of strong acidic polystyrene ion-exchange resins on H^+/Cu^{2+}, Zn^{2+}, or Cd^{2+} ion exchange equilibrium was investigated [389]. Three resins were chosen as the representatives of three resin generations of exchangers: gel-type C-100, macroporous C-150, and hypercrosslinked MN-500. The last resin was shown to provide the best separation of a mixture of Cu^{2+}, Zn^{2+}, and Cd^{2+} ions, because it demonstrates the largest difference in ion exchange constants. The ion exchange constants calculated from the Langmuir sorption model and predicted by surface complexation theory (both are in good agreement with experimental isotherms) give the same selectivity series, $Cd^{2+} > Zn^{2+} > Cu^{2+}$ for hypercrosslinked MN-500 and gel-type C-100 resins. For the macroporous exchanger this sequence is different, $Cu^{2+} > Cd^{2+} > Zn^{2+}$. Also, the surface complexation theory predicts higher affinity of MN-500 to all the ions compared to conventional resin C-100.

Apart from the high selectivity for the cadmium sorption, the hypercrosslinked cation exchanger MN-500 is characterized by the highest rate of H^+/Cd^{2+} exchange, while the slowest diffusion was found to be peculiar to the macroporous resin C-150 [390]. This result is quite predictable, because in the open-work-type matrix of hypercrosslinked resins there are no dense domains. On the contrary, the pore walls in the macroporous matrix present a dense polymer phase in which the bivalent metal ions could hardly diffuse.

Despite the relatively low exchange capacity, 0.6–0.8 meq/g, the weak basic anion-exchange resin MN-100 absorbs up to 15 mg/g gold in the form of cyanide ions $Au(CN)_2^-$ from aqueous basic solutions [391]. Note that XAD-2 and MN-200 impregnated with N,N'-bis(2-ethylhexyl)guanidine extract 10 times smaller amounts of gold cyanide under the same conditions. The commercial hypercrosslinked ion exchanger MN-300 with quaternary amine groups takes up an even larger quantity of gold cyanide, >40 mg/g [392]. Along with the above high sorption capacity,

MN-100 and MN-300 exhibit high selectivity of extracting gold ions from solutions containing also cyanides of Fe, Ni, Cu, and Ag. The subsequent elution of the loaded sorbent with aqueous 5 g/L NaCl/ethanol (40%) or 5 g/L NaCN/acetone solutions completely recovers $Au(CN)_2^-$ [391]. The stripping of loaded MN-100 and MN-300 was found to be complete, as well, when using ethanol/water or acetone/water sodium hydroxide or sodium cyanide solutions. All the above promising results indicate that MN-100 represents a potentially useful material for extracting gold from mining leach solutions.

Pan et al. [393] compared the sorption of industrially important aromatic sulfonates – sodium benzenesulfonate (BS), sodium p-toluenesulfonate (TS), and sodium 2-naphthalenesulfonate (NS) – onto hypercrosslinked weak basic anion-exchange resin M-101 and conventional macroporous weak basic exchanger D-301 (China). M-101 was prepared by amination of commercial hypercrosslinked sorbent CHA-101 with dimethylamine. At the optimal pH 3.0 the sorption capacity of M-101 decreases in the row of solutes NS > TS > BS. The distinguishing feature of the hypercrosslinked resin consists in that its capacity for sodium benzenesulfonate is practically independent of the presence in solution of the competing inorganic salt, sodium sulfate, in the range of 5–45 mM/L. This finding suggests that the physical adsorption because of π–π-interactions between the solute and the polymeric matrix of M-101 contributes markedly to the resin total sorption capacity. Compare this with the sorption capacity of D-301 exchanger, which decreases by a factor of 3 with increasing concentration of the inorganic salt.

Among very important parameters of commercial ion-exchange resins, which can determine their applicability in specific industrial sorption technologies, the following can acquire utmost importance: mechanical strength and attrition resistance, volume stability, thermostability, resistance to irradiation and aggressive or oxidative media. Because of the extremely high crosslinking density and rigidity of the matrix, hypercrosslinked resins prove to be superior to conventional gel-type and macroporous resins.

Thus, conventional cation exchanger are known to be intolerant to oxidative processes. As a result of matrix degradation, their exchange capacity decreases, the mechanical strength of beads drops, swelling increases, and a considerable portion of leachables emerges in the effluents. A simple standard test with hydrogen peroxide in the presence of iron ions revealed a fundamental advantage of hypercrosslinked sulfonic resin

Table 16.6   Water uptake (wt%) of resins subjected to oxidation using 1% (w/v) hydrogen peroxide in the presence of 500 ppm iron ions at 40°C

Resin	Days				
	0	1	2	3	5
MN-500, H form	56.9	57.4	58.6	60.2	65.5
5% DVB	65.1	74.3	92.9	Sample dissolved	
8% DVB, gel, H form	56.3	62.2	70.0	80.1	Sample dissolved
8% DVB, gel, Na form	49.6	73.3	90.4	Sample dissolved	
8% DVB	64.5	68.2	75.4	92.9	–
12% DVB	55.9	58.2	60.5	70.0	–

Source: After [386].

MN-500 over conventional resins in the resistance to oxidation (Table 16.6). Whereas treatment of gel-type exchangers with oxidative solutions results in complete dissolution of resins in 3 days, MN-500 adds only 3% of moisture. The high selectivity of MN-500 with respect to trace metal cations in combination with its high oxidative resistance makes this commercial sulfonate very useful for condensate polishing.

Thus, hypercrosslinked ion-exchange resins may be inferior to conventional resins in the total exchange capacity, but overbalance them in mechanical, chemical, thermal, oxidative, and irradiation resistance and volume stability. They represent the material of choice if acid-base catalysis in organic media (including chlorobenzene, hexane, or dichloromethane) is needed. Due to the large adsorption surface and high permeability of the whole network they are also indispensible in removing large organic ions and decolorizing complex mixtures of valuable compounds. The high selectivity toward trace ions is a recommendation for using hypercrosslinked ion-exchange sorbents for deep purification of concentrated acids, on the one hand, and for final deionization of pure water and any solvents, on the other hand.

CHAPTER 17

Other Applications of Hypercrosslinked Polystyrene

1. EXTRACTION OF RHENIUM BY IMPREGNATED HYPERCROSSLINKED SORBENTS

Over the last decade, the consumption of rhenium has been rapidly growing due to its increased use in superalloys in the aircraft and space-craft industries and also as platinum–rhenium catalysts in the production of lead-free high-octane gasoline. Usually, rhenium has been isolated from diluted sulfuric acid solutions of mixed salts obtained on hydrome-tallurgical processing of polymetallic rhenium-containing raw materials. The extraction has been conducted by means of special sorbents that are selective to this metal. However, highly selective resins exhibiting high adsorption potential for rhenium are hardly accessible and so the search for new materials is fully justified. Troshkina et al. [394] suggested a simple route of preparing complex-forming resins by impregnating com-mercial hypercrosslinked sorbent MN-202 or macroporous Polysorb-b (Russia) with chelating agents $(C_4H_9CH(C_2H_5)CH_2OC_2H_4)_3N$ (OKSAM), cyclohexylamine (CHA), and trialkylamine (TAA) produced by Purolite International (UK). When contacting 1 g of MN-TAA, MN-OKSAM, and MN-CHA impregnates with $100\,mL$ of $100\,mg/dm^3$ rhenium sulfate in sulfuric acid solution (pH 2), the authors managed to extract 48, 39, and $34\,mg/g$ Re, correspondingly. Under the same conditions, Polysorb-b impregnated with a comparable amount of TAA extracts only $18\,mg/g$ Re.

2. HETEROGENEOUS MEMBRANES FILLED WITH HYPERCROSSLINKED POLYSTYRENE

Separation of gases and vapors by means of selective permeable membranes belongs to progressive modern technologies that successfully compete with gas separations by adsorption on porous solids. To compare the influence of different polymer fillers on the properties of heterogeneous membranes,

Comprehensive Analytical Chemistry, Volume 56
ISSN 0166-526X, DOI 10.1016/S0166-526X(10)56017-4

Hradil et al. [395, 396] prepared a set of composite membranes containing fine particles of conventional macroporous resins or hypercrosslinked poly-styrene adsorbing materials in films of poly(2,6-dimethyl-1,4-phenylene oxide) as a binder. Hypercrosslinked resins were either a commercial product, Lewatit EP63 (Bayer AG), or were obtained by crosslinking (i) a macroporous styrene–divinylbenzene (DVB) copolymer with carbon tetra-chloride (Hyp–St–DVB) or (ii) a linear polystyrene with monochlorodi-methyl ether. In the latter case the reaction of bridging was conducted in a pretty diluted ethylene dichloride solution at stirring, which resulted in obtaining a particulate (1–5 µm) product.

Examination of thus prepared heterogeneous membranes showed that they are superior in gas permeability to homogeneous ones made of the binder alone. (Permeability is a quantity of gas that passed through the unity of area in unity of time at a given difference in gas partial pressure on two sides of the membrane, multiplied by membrane thick-ness.) As was found, the permeability of heterogeneous membranes estimated by H_2 depends on the type of polymeric filler and increases in the following order:

$$Poly(EDMA) < XAD\text{-}4 < Hyp-St-DVB < Lewatit\ EP63,$$

while the selectivity of H_2/CH_4 separation increases from 3 to 120 in another sequence of the sorbents,

$$XAD\text{-}4 < poly(EDMA) < Lewatit\ EP63 < Hyp-St-DVB,$$

where poly(EDMA) is poly(ethylene dimethacrylate). Here, the selectiv-ity was calculated as the ratio of diffusion flux $(mol\,s^{-1}\,m^{-2})$ of hydrogen to that of methane. The authors explain the different order of sorbents in the above rows by assuming that the selectivity of H_2/CH_4 separation depends on the preferential sorption of the organic component in the polymer phase, whereas the permeability of the membrane mainly depends on the porous structure of sorbent particles. Anyway, mem-branes filled with the hypercrosslinked materials exhibit the highest permeability and the highest selectivity in H_2/CH_4 separation process.

Membranes based on the above poly(phenylene oxide) binder and the hypercrosslinked particulate derived from linear polystyrene exhibit even higher selectivity, 540, in H_2/isobutene separation, most likely due to size exclusion effect [396]. This experimental finding indicates that such type of heterogeneous membranes may successfully be used for effective separation of hydrogen and hydrocarbons or, might be, for the separation of small and large molecules of hydrocarbons. The optimal filling of the membrane with the 2–3 μm particles was around 9%. The membranes containing more than 20 wt% of the adsorbent were not mechanically stable.

Polyimide-type membranes are known to be thermally more stable and mechanically more resistant compared to poly(phenylene oxide) membranes, but possess insufficient permeability for gas molecules. When cast from a polyimide solution in N-methylpyrrolidone in the presence of 5–20% hypercrosslinked polystyrene microparticles, heterogeneous membranes of acceptable strength were obtained [397] having a permeability for gases increasing in the following sequence:

$$CH_4 < N_2 < O_2 < CO_2 < H_2.$$

This sequence is in good correlation with the kinetic diameter of gas molecules. The selectivity (relative to nitrogen) of gas separations increases in the same order, especially at enhanced temperatures. This synchronous rise in permeability and selectivity of the above membranes was attributed to the complex mechanism of separation. It incorporates two processes occurring simultaneously, namely the transport of gas molecules through free volume of porous hypercrosslinked polymeric particles and transportation of gas through the film of binder via gas dissolution mechanism [398].

3. NANOCOMPOSITE CATALYSTS OF ORGANIC REACTIONS

Noble metals are known to actively catalyze reduction and oxidation reactions of many organic compounds. The catalytic activity depends on the metal surface area available to the reacting molecules. From the viewpoint of economy the most beneficial are nano-dispersed metal catalysts. However, nanoparticles tend to agglomerate, so that the catalyst gradually loses its activity. Size stabilization of the catalyst nanoparticles can be achieved by immersing them into a polymeric matrix with an optimized size of cavities that incorporate the metal clusters. Hypercrosslinked polystyrene offers such an opportunity.

By wetting Styrosorb 2 beads with a 10% solution of H_2PtCl_6 acid in tetrahydrofuran or methanol, drying the beads, and treating them with hydrogen (or H_2O_2 in alkaline media), catalytically active nanocomposites were prepared [399]. Measuring the size distribution of the Pt nanoparticles resulted in a mean diameter of 1.3 nm. Interestingly, this size corresponds to a perfect cluster consisting of 55 atoms that are arranged in two layers around one central atom according to the general formula for regular clusters, $1 + \Sigma(10n^2 + 2)$, composed of densely packed identical spheres (where n is the number of shells around the central sphere). However, in catalytic processes the initial size distribution was found to broaden, with 70% of particles remaining less or equal to 2.0 nm in diameter, but some particles growing to 5 nm.

The Pt-containing (2–4% Pt) nanocomposite beads placed in an aqueous basic solution of L-sorbose catalyze oxidation of the latter with O_2 to 2-keto-L-gulonic acid, an intermediate in the production of vitamin C [399]. Within first 100 min at 60–80°C the catalytic activity gradually develops, resulting in a 100% conversion of L-sorbose with the yield of ketogulonic acid up to 98% [400]. With the size of Pt clusters controlled by the size of matrix network meshes, the catalytic properties of the composite material remain stable even after 30–50 reaction cycles [400].

Similarly, in the presence of Pt, Pd, or Ru nanoparticles formed in a hypercrosslinked matrix, lactose is selectively oxidized to lactobionic acid [401] and D-glucose to gluconic acid [402], both presenting intermediates in the synthesis of important biologically active substances.

Industrial MN-270 impregnated with Pt nanoparticles was recently tested as a heterogeneous catalyst for oxidation of phenol dissolved in water [403]. In batch experiments, when strongly stirring and heating to 95°C suspension of the nanocomposite catalyst (containing 5.15×10^{-3} M Pt) in 0.2 M phenol solution (18.8 g/L) through which oxygen was passed with a flow rate of 10^{-2} L/s, 100% conversion of phenol was achieved. At that, 98.5% of phenol was oxidized to CO_2 and H_2O. According to X-ray photoelectron spectroscopy, the nanoparticles incorporated platinum in oxidation states 0, II, and IV. As in the case of L-sorbose oxidation, the first phenol oxidation reaction cycle was accompanied by an increase in the average dimensions of platinum nanoparticles from 2.2 to 3.5 nm and noticeable broadening of particle size distribution. As per the authors' opinion, it is caused by "possible swelling of nanoparticles with the reaction components." If a long-run total conversion of phenol to CO_2 can be really achieved, the developed method may be considered very promising for wastewater purification.

Finally, it should be mentioned that most recently the (-)-cinchonidine-modified platinum-containing MN-270 was reported to exhibit notable enantioselectivity in hydrogenation of ethyl pyruvate to (R)-ethyl lactate, with enantiomeric excess (ee) of the product as high as 75% [404].

4. STORAGE OF HYDROGEN AND METHANE ON HYPERCROSSLINKED POLYSTYRENE

The wide use of hydrogen as a fuel and further development of hydrogen technologies are currently hampered by the absence of a method for hydrogen storage which would be simple, safe, and cost efficient. The 2010 US Department of Energy (DOE) storage target is 6.0 wt% hydrogen at 77 K and 10 bar pressure. The target set for 2015 is 9 wt% or 81 kg H_2 per m^3, which is extremely challenging and so far even seems clearly out of reach, particularly because it is a system target that includes into calculation the weight of the pressure container, valves, cooling systems, etc. In fact, capacities significantly greater than 6.0–9.0 wt% solely on the basis of the storage material are required.

Various porous materials have been tested as adsorptive H_2 media; these are activated carbons, zeolites, metal-organic and covalent organic framework materials, carbon nanotubes, and polymeric resins. However, none of these adsorbents approaches the required threshold thus far. The basic reason for the low H_2 adsorption is the extremely weak interaction of gaseous hydrogen with the adsorbents even at 77 K. Still, the potential of microporous hypercrosslinked polystyrene for adsorbing H_2 is considered to be rather high compared to other polymeric adsorbents, because the former exhibits enhanced electron-donating property, in addition to extremely large specific surface area values. Moreover, polystyrene is built of the lightest possible elements, carbon and hydrogen. Finally, it is possible to suggest technology for preparing monolithic blocks of hypercrosslinked material, most suitable for packing into a pressure container with a space filling higher than 60%, which is the limit for space usage by irregularly shaped or beaded sorbent particles.

Therefore, hypercrosslinked sorbents were among the first to be tested as materials for hydrogen storage. Table 17.1 summarizes some results achieved by now. Among commercial macroporous and hypercrosslinked polystyrene adsorbents, MN-200 demonstrates the best results. The resin is capable of retaining 1.3 wt% H_2 at the temperature of liquid nitrogen and nearly atmospheric pressure (1.2 bar). In search for better adsorbents, Germain et al. [405] prepared two hypercrosslinked polymer

Table 17.1 Hydrogen storage capacities of various macroporous and hypercrosslinked polyaromatic adsorbing materials at 77 K

Resin	Composition	S_{BET}^{N2} (m²/g)	Capacity for H_2 (wt%)	Pressure (bar)	Ref.
Amberlite XAD-4	Styrene–DVB	1060	0.8	1.2	[409]
Amberlite XAD-16	Styrene–DVB	770	0.6	1.2	[409]
Hayesep N	DVB–EDMA	460	0.5	1.2	[409]
Hayesep B	Poly-DVB modified with poly (ethyleneimine)	570	0.5	1.2	[409]
Hayesep S	DVB-4-vinylpyridine	510	0.5	1.2	[409]
Wofatit Y77	Hypercrosslinked PS	940	1.2	1.2	[409]
Lewatit EP63	Hypercrosslinked PS	1206	1.3	1.2	[409]
Lewatit VP OC 1064	Hypercrosslinked PS	810	0.7	1.2	[409]
MN-200	Hypercrosslinked PS	840	1.3	1.2	[409]
MN-100	Hypercrosslinked PS–N$^+$	600	1.1	1.2	[409]
MN-500	Hypercrosslinked PS–SO$_3^-$	370	0.7	1.2	[409]
From macroporous precursor	Hypercrosslinked PS	1300	1.2	1.2	[405]
From gel precursor	Hypercrosslinked PS	1930	1.55 3.8 5.5	1.2 45 80	[405] [406] [406]
From linear precursor	Hypercrosslinked PS	1466	1.2–1.3 2.75 3.04	1 10 15	[407]
From bis-chloromethyl biphenyl	Hypercrosslinked poly(arylene methylene)	1904	3.68	15	[411]

EDMA = ethylene dimethacrylate.

samples by post-crosslinking via Friedel–Crafts reaction of a beaded gel-type copolymer (vinylbenzyl chloride/2% DVB) and a macroporous copolymer (50% vinylbenzyl chloride and 50% DVB polymerized in toluene). The sample based on the gel-type precursor proved to be slightly better than MN-200, while the macroporous polymer was found to operate worse. In spite of very high specific surface area (1930 and 1300 m^2/g), the above two polymers take 1.55 and 1.2 wt% of hydrogen at 77 K and 1.2 bar with the initial heat of adsorption of hydrogen molecules 6.6 and 6.9 kJ/mol, respectively.

It was noted that the storage capacity cannot be directly related to the BET surface area determined via nitrogen adsorption isotherms. Thus, XAD-4 with its surface area of 1060 m^2/g retains less hydrogen, 0.8 wt%, compared to MN-200 (840 m^2/g and 1.3 wt% H$_2$, respectively). On the other hand, assessing the apparent surface area in the nanoporous polymers from the hydrogen adsorption isotherms measured at 77 K and calculated according to Langmuir equation results in values that correlate much better with the hydrogen uptake [406]. Thus Amberlite XAD-4 and Hypersol-Macronet MN-200 display Langmuir (H$_2$ isotherm) surface areas 425 and 576 m^2/g, respectively, which correlate, at least qualitatively, with the better performance of MN-200 [405]. We may note here that the mechanism of adsorption of nitrogen may significantly differ from that of adsorption of hydrogen. At a temperature of 77 K, which is the boiling point of nitrogen (critical temperature is 126 K), nitrogen is sorbed from the vapor phase, while hydrogen presents a gas in a supercritical state because its critical point is much lower, 33 K. Nitrogen vapors easily condense in the micropores and cause a noticeable swelling of hypercrosslinked polystyrene, while hydrogen predominantly adsorbs on available surfaces. Of course, some finest pores could remain inaccessible to N$_2$ molecules, but accommodate much smaller H$_2$ molecules.

In another paper Germain et al. [406] examined in more detail and at higher pressures the gel-type hypercrosslinked polystyrene prepared from the copolymer of vinylbenzyl chloride with 2.5% DVB, having the N$_2$/BET specific surface area of 1930 m^2/g. Its total, completely reversible adsorption of H$_2$ at 77 K and 80 bar was shown to be as high as 5.5 wt%. Importantly, the adsorption capacity drops on introducing nitro groups or, particularly, bromine atoms into the phenyl rings, indicating that the high electron density in the aromatic rings of parent hypercrosslinked polystyrene is a critical parameter for hydrogen storage.

Lee et al. [407] prepared 50–200 μm particles by suspension polymerization of vinylbenzyl chloride followed by post-crosslinking procedure to arrive at a porous material with the specific surface area of 1466 m^2/g. When examining sorption of hydrogen at pressures up to 15 bar, they paid attention to the very reproducible deviation of the desorption branch from the adsorption isotherm. The narrow hysteresis loop spreading throughout the entire range of pressures was explained by the "tortuosity" and broad pore size distribution of the micropores in the polymer network. This explanation may be incorrect. The hysteresis may well result from the incipient swelling of the polymer with hydrogen gas sorbed at low temperature, due to the rearrangement of conformations of strained network meshes on hosting the hydrogen molecules. This phenomenon was earlier discussed for the case of sorption of nitrogen and water vapors at room temperature (Chapter 10, Section 2).

Pastukhov et al. [408] examined the hydrogen uptake at room temperature by commercial hypercrosslinked polystyrene–type sorbents MN-200, MN-270, and MN-500 filled in a 100 mL stainless steel container. The autoclave was pressurized with hydrogen up to 250 atm. Afterward, the shutoff valve was carefully opened to slowly release the gas. The changing weight of the autoclave, hydrogen pressure, and the volume of the gas released were registered. As can be seen from Fig. 1, the total hydrogen uptake of the system decreases almost linearly with pressure. At the initial highest pressure tested, the loading capacity of MN-270 amounts to 3.5 wt%, while MN-200 is capable of retaining even larger quantity of H$_2$, 4.5 wt%. Though the hydrogen sorption capacity achieved thus far with hypercrosslinked polystyrene remains insufficient for practical applications, other microporous aromatic polymers deserve further investigation as light and easily available materials with highly variable open-network structure. Two review papers [409, 410] underscore the great potential of hypercrosslinked polymers for hydrogen storage. Also, it was stated that the goals for storage capacity are unlikely to be met by simply increasing the surface area of materials, and that therefore by introducing new surface chemistry the isosteric heat of adsorption of hydrogen must be enhanced, which for polystyrene-type materials is not higher than 5–9 kJ/ mol.

The group of Cooper [411] by following the protocols by Tsyurupa and Davankov [412] synthesized hypercrosslinked aromatic networks through self-condensation of *ortho-*, *meta-*, or *para*-xylylene dichloride or

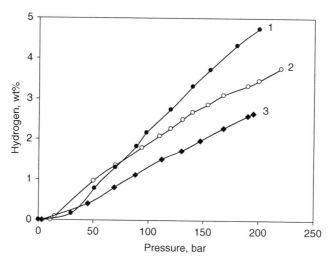

Figure 1   Dependence of hydrogen weight increment on hydrogen pressure at room temperature in an autoclave filled with the commercial hypercrosslinked sorbents (1) MN-200, (2) MN-270, and (3) MN-500. *(After [408].)*

bis(p-chloromethyl)-diphenyl. In the latter case a resin with BET surface area of $1904\,m^2/g$ was produced that displayed hydrogen sorption of 3.68 wt% at 15 bar and 70 K, the highest value reported for hypercrosslinked polyaromatics by that time (2007).

Hypercrosslinked polymers of different chemical natures were introduced by Germain et al. [413] when they treated swollen commercial polyaniline with diiodomethane or paraformaldehyde (see Chapter 9, Section 3.5). Though the materials thus obtained had a reduced specific surface area of slightly over $630\,m^2/g$ and smaller overall hydrogen capacity, they demonstrated an enhanced initial enthalpy of hydrogen adsorption, as high as 9.3 kJ/mol. Polyaniline deserves further attention in view of the recent finding [414] that electrospun microfibers (about 2 µm in diameter) of linear polyaniline slowly but reversibly swell in hydrogen at high pressures, thus allowing storage of 6–10 wt% hydrogen at 80 bar and 125 °C.

The intensive search for other types of microporous materials that started less than a decade ago resulted in the development of polymers with intrinsic microporosity (PIMs; see Chapter 9, Section 3.8) and covalent organic frameworks and metal–organic frameworks (COFs and MOFs; see Chapter 9, Section 3.10) and caused an exponential growth in the number of publications dealing with hydrogen storage. All these new developments seem to be promising, but none of them has showed thus far great advantages over the simplest hypercrosslinked aromatic polymers (see also review [410]). Thus, MOFs have the disadvantage of incorporating heavier elements such as oxygen and metals, but, on the other hand, they offer the possibility of exploiting the enhanced affinity of some metal ions toward hydrogen molecules. In a series of isostructural MOFs [415], the H_2 binding was found to follow the trend of $Zn^{2+} < Mn^{2+} < Mg^{2+} < Co^{2+} < Ni^{2+}$. The initial heats of H_2 adsorption on the Ni^{2+}-containing MOF amount to 12.9 kJ/mol.

COFs, though rather expensive, have the advantage of being amenable to rational design and theoretical description. A Greek group [416] suggested to significantly enhance the hydrogen uptake of COF-105 by introducing three lithium phenolate groups into each triangular hexahydroxytriphenylene building block of the network, the second block being tetrahedral tetra(4-dihydroxyboryl-phenyl)silane [417]. The free space around the lithium atom permits the accommodation of up to five hydrogen molecules. According to Monte Carlo simulations at 77 and 300 K for pressures up to 100 bar, the thus modified COF-105 could reach the DOE target for 6 wt.% hydrogen uptake even at room temperature. If this is really so, lithium functionalization can be undertaken even more easily with simple hypercrosslinked polystyrene where all 8.5 mmol/g aromatic rings are accessible to chemical modification.

Over the last decade the problems of storage of natural gas and its transportation over long distances became equally relevant. Estimation of the suitability of hypercrosslinked adsorbing materials for methane storage was undertaken [408] using the aforementioned technique of weighing the autoclave packed with 100 mL MN-270 and filled with methane to 225 bar at room temperature. The system accommodated 15.5% of methane by weight of MN-270, which is still insufficient for practical storage of natural gas. The hypercrosslinked aromatic polymer prepared by self-condensation of p-xylylene dichloride was reported [418] to retain 8.3% methane at 20 bar (298 K), with the isosteric heat of sorption of ~21 kJ/mol. Thus, while hydrogen storage is still hampered by the very low temperatures involved (77 K), the targets for methane storage (35 wt%) might be reached much faster (see review [419]).

5. CARBONACEOUS SORBENTS BASED ON HYPERCROSSLINKED POLYSTYRENE

Hypercrosslinked polystyrene is a well-suited material to obtaining new carbonaceous products. As discussed in Chapter 7, Section 8.3, pyrolysis of these polymers results in porous carbons with a yield as high as 45–55%, while conventional styrene–DVB copolymers and even poly-DVB undergo complete depolymerization under same conditions. Particularly high carbon yields are characteristic of polymers containing sulfonic groups (MN-500 and MN-270sulf), carboxylic groups (MN-600), or residual chlorine (MN-202). Pyrolysis of the last two industrial resins is especially beneficial, because no hazardous sulfur oxides are evolved. Still, heating the hypercrosslinked polymers to 600°C (or higher) in an inert atmosphere causes a loss of nearly half of the sample weight and a significant reduction in the size of the beads (by a factor of about 2). Rupture of the most strained covalent bonds of the network during its thermodestruction leads to a global rearrangement of hypercrosslinked structure with reduction of the free volume of the polymer. Nevertheless, the overall morphology and porosity of the material remains preserved [408]. Table 17.2 shows parameters of the porous structure of carbonaceous materials, calculated from adsorption isotherms for nitrogen at −196°C according to the micropore volume filling theory [408].

The carbonaceous products are predominantly microporous materials. Their inner surface area is significantly smaller than that of starting polymers and is formed largely by micropores. The contribution of mesopores to the total surface area is small. At the same time, by comparing the solvent uptake of the carbons with their pore volume estimated as a volume of nitrogen absorbed at the relative pressure of 0.95–0.99, one may arrive at the conclusion that samples obtained from the sulfonate MN-500 contain a certain portion of large pores. The largest pore volume is characteristic of the carbon prepared by pyrolysis of MN-500 at 850°C in carbon dioxide atmosphere. CO_2 is known to additionally remove carbon from the material at a high temperature in the form of CO. This process leads to the so-called activated carbons. Indeed, the specific surface area of this product is particularly high, thus pointing to the formation of many new micropores.

The sorption properties of the above carbonaceous materials were examined by means of gas chromatography [420], which allows quantitative characterization of sorbent–sorbate interactions at a small coverage of sorbent surface. Table 17.3 presents the specific retention volumes of various organic compounds measured for different carbons.

Table 17.2 Porous structure of carbonaceous materials

Initial polymer	Pyrolysis t (°C)	S (m²/g)	S_{me} (m²/g)	W_s (cm³/g)	W_{mi} (Cm³/g)	x_o (nm)	Solvent uptake (mL/g)	
							Toluene	Ethanol
MN-270sulf	600	560	45	0.28	0.18	0.33	0.23	0.28
MN-202	600	670	20	0.24	0.21	0.39	0.22	0.29
MN-600	600	750	5	0.24	0.23	0.18	0.28	0.31
MN-500	600	680	50	0.28	0.22	0.23	0.50	0.50
MN-500	900	630	35	0.26	0.21	0.29	0.46	0.47
MN-500 (CO_2)	850	1540	130	0.64	0.52	0.44	1.04	1.06
MN-500-initial	–	290	60	0.29	0.09	0.63	–	–

S_{me} = surface of mesopores; W_s = pore volume identified with the volume of nitrogen measured at the highest relative pressure; W_{mi} = volume of micropores; x_o = half-width of slit-like pores.
Source: After [420].

Table 17.3 Polarizability (α_p) of adsorbate molecules and their specific retention volumes (mL/g) at 260°C

Adsorbate	α_p (Å^3)	C-150 (600 °C)[a]	MN-270sulf (600°C)	MN-500 (600°C)	MN-500 (900°C)	MN-500 (850°C, CO$_2$)
C$_3$H$_8$	6.2	30	35	65	11	107
n-C$_4$H$_{10}$	8.1	140	159	375	70	612
n-C$_5$H$_{12}$	10	710	733	1779	416	3311
CH$_3$OH	3.2	60	8	12	22	21
C$_2$H$_5$OH	5.1	228	17	26	61	56
n-C$_3$H$_7$OH	6.9	549	44	92	133	121
(CH$_3$)$_2$CO	6.3	299	282	484	237	557
CH$_3$NO$_2$	4.8	147	168	208	160	302
CH$_2$Cl$_2$	6.4	137	146	241	206	300
(C$_2$H$_5$)$_2$O	9.0	412	443	1095	588	2068

[a] C-150 is macroporous styrene–divinylbenzene sulfonate produced by Purolite Int. (UK).
Source: After [420].

As can be seen, the retention volumes of adsorbates are basically proportional to their polarizability, which has usually been interpreted as the domination of dispersion forces in the sum of all sorbent–sorbate interactions. The largest retention of alkanes is characteristic of the activated carbon obtained at 850°C in CO_2 atmosphere, while the smallest retention is peculiar to the sample prepared at the same temperature but in the presence of nitrogen. The carbonaceous material based on conventional macroporous sulfonate C-150 retains alcohols much stronger than the other carbons. On the contrary, the sample based on microporous MN-270 sulfonate demonstrates the weakest retention of methanol, ethanol, and propanol. The MN-500-based carbon (600°C) exhibits affinity to polar acetone, nitromethane, and ethyl ester, in this respect being inferior insignificantly to the CO_2-activated carbon. This suggests that some polar functional groups are present on the surface of the sorbents. Yet, in accordance with the data of elemental analysis, all carbonaceous products obtained by pyrolysis of hypercrosslinked polystyrenes contain a few per cent of oxygen-incorporating groups; that is why these materials should be attributed to weakly specific adsorbents.

Asnin at al. [421–423] examined adsorption isotherms of benzene and chlorobenzene vapors on experimental microporous carbon D4609 (Purolite Int.). Of the total $0.65\,cm^3/g$ pore volume accessible to nitrogen, micropores of about 1.5 nm in diameter comprise $0.55\,cm^3/g$. Surprisingly, the limiting amount of adsorbed chlorobenzene was found to exceed that of benzene [423], in spite of the fact that exclusion effects could be expected to result in opposite sorption proportions. Still, the correctness of that finding was corroborated by direct measurements of swelling of the material in the two solvents. While the volume of the carbon remained unchanged on immersing the beads into benzene, an increase in their volume by 3% in chlorobenzene was reliably registered. The microporous structure of the carbon obviously responds to the strong solvation of the condensed aromatic rings of the carbon with electron-deficient chlorobenzene molecules via π–π-type interactions.

It should be mentioned here that interesting paramagnetic properties of materials prepared by pyrolysis of hypercrosslinked polystyrene allowing their use as oxygen sensors have been described in Chapter 7, Section 8.3. Application of microporous carbons in the capacity of stable column packing in preparative separation of aggressive inorganic electrolytes according to size-exclusion chromatography was discussed in Chapter 12, Section 10.1.

In short, pyrolysis of hypercrosslinked polystyrene beads leads to chemically and mechanically robust carbonaceous materials in a beaded form and with a high carbon yield. This could present a convenient way of converting used-up or contaminated (i.e., in blood perfusion procedures) hypercrosslinked adsorbing materials into a product of high value and many application possibilities.

Literature search shows that number of publications dealing with hypercrosslinked polymers grows exponentially. No wonder that during the couple of months were the manuscript was processed by the publisher, several new articles became available. We would like to mention here some most important new contributions to the field.

Reports on *industrial application* of commercially available hypercrosslinked polystyrene resins and attempts to implement the materials into industrial technologies are particularly gratifying. Such an innovation is the purification of HCl gas [424]. In China, about 600,000 tons of chlorobenzene is produced every year by chlorination of benzene in the presence of ferric chloride as a catalyst. This reaction also generates hydrogen chloride in stoichiometric amounts. The HCl gas is contaminated with vapors of both benzene and chlorobenzene in concentrations of about 20–50 and 5–15 g/Nm^3, respectively. While pure hydrogen chloride gas could be used for manufacturing vinyl chloride or other valuable chemicals, the contaminated HCl gas is only useful for the preparation of low-quality technical hydrochloric acid. Therefore, removal of the organics from the dry hydrogen chloride is an important industrial task. Attempts to achieve this by using activated carbon failed because of its low adsorption capacity and sharp rise of bed temperature due to the interaction of HCl with many functional groups on the surface of the material. On the contrary, hypercrosslinked polystyrene, when examined on a pilot-scale adsorption module containing 65 kg of the polymer, showed a good adsorption capacity of over 280 mg/g with respect to the mixed contaminants. Importantly, the regeneration of the polymer could be easily accomplished using water steam of 120 °C. Continuous 6-months runs of the column without any loss of the polymer performance resulted in both recovery of about 95–98% of benzene and chlorobenzene and safe obtaining of pure hydrogen chloride.

In a detailed recent paper Saha et al. [425] have reported about the high potential of commercial non-functional hypercrosslinked polystyrene resins MN-200 and MN-250 in the process of removal from water of two endocrine disrupting substances, 17β-oestradiol and 17α-ethinyl oestradiol. They are present in sewage, some industrial effluents and, therefore, circulate in surface waters. Even at very low concentration levels (ng/L), the contaminants interfere with hormonal stem controlling reproduction system of man, fish and other wildlife species. These compounds have to be carefully removed from drinking water. In this respect, the hypercrosslinked polystyrene sorbents prove to be very efficient. As was found, at the equilibrium concentration of 0.006 µmol/L resins MN-200 and MN-250 are capable of absorbing 1.57 µmol/g of 17β-oestradiol. Sorption capacity of these resins toward 17α-ethinyl oestradiol is lower, around 0.4 µmol/g. Most probably the latter compound with a more bulky structure experiences partial size exclusion from nanopores of the sorbents.

By impregnating commercially available hypercrosslinked polymers of the MN-series with platinic acid in tetrahydrofuran, followed by $NaHCO_3$ treatment, a composite material was obtained with nanodispersed metal (∼1% Pt) in the polymer matrix. The material actively catalysis oxidation of phenol to CO_2 by air oxygen in aqueous systems, significantly outperforming the activity of the conventional Al_2O_3-Pt catalyst or Pt(5%)/activated carbon [426]. Thus, the decontamination of industrial effluents and fine polishing of potable water are the perspective potential area for large-scale application of hypercrosslinked polymers.

Not less successful are *analytical applications* of hypercrosslinked polymers. Thus, three types of micron-sized, high specific surface area hypercrosslinked resins containing hydrophilic moieties have been suggested for solid-phase extraction of very polar compounds and some pharmaceuticals from pure water and complex environmental aqueous matrices. One product, a weak basic hypercrosslinked anion-exchange resin was prepared starting from the monosized microbeaded (6.66 ± 0.18 µm) copolymer of 75% vinylbenzyl chloride and 25% divinylbenzene obtained via precipitation polymerization in acetonitrile [427]. After involving the majority of the chloromethyl groups into post-crosslinking reaction, the residual chlorine of the polymer was substituted with 1,2-ethylene diamine to give a hypercrosslinked anion exchanger with nitrogen content of 0.75 meq/g. 200 mg of the resin were then used for the pre-concentration of acidic pharmaceuticals in a cartridge coupled on-line with the analytical C18 column of the liquid chromatograph. After percolating 50 to 500 mL complex water sample adjusted to pH 7 through the cartridge and washing it with 1 mL methanol to remove all neutral and basic interfering compounds, the acidic analytes (pK_a < 6) were eluted from the cartridge. The optimized eluent contained 20% acetonitrile and the rest of the solution was the NH_4Cl/NH_4OH buffer of pH 9.2. Further optimization of the mobile phase for the C18 column was performed by mixing the effluent from the cartridge with an additional amount of acetonitrile. To enable this, a system of two pumps and two six-port valves was used suggested earlier by Pocurrull et al. [216] (see Chapter 17, Section 2). When using the above technique to analyze samples of river water or sewage treatment plant effluent spiked with 11 pharmaceuticals at the level of 0.4 and 1 μg/L, respectively, selective determination and very good recoveries were achieved for acidic components, naproxen, fenoprofen, ibuprofen and gemfibrozil.

Another material, a microbeaded hypercrosslinked weak cation exchanger, was again prepared by precipitation copolymerization of a mixture of 10% methacrylic acid, 50% vinylbenzyl chloride and 40% DVB in acetonitrile, followed by post-crosslinking procedure [428]. 1000 mL aqueous solution of pH 7 of basic (caffeine, antipyrine, propranolol, carbamazepine) and acidic (acetaminophen, naproxen, diclofenac) pharmaceuticals, spiked at 20 μg/L level, was percolated trough 200 mg cartridge packed with 6 μm particles of the above cation exchanger. After loading, the cartridge was washed with 2 ml 5%NH_4OH in methanol to remove all acidic compounds and neutral interferences. The retained basic components were then eluted from the cartridge using 5 ml of 2% trifluoroacetic acid in methanol and finally analyzed by liquid chromatography in an off-line procedure. The recoveries of the basic pharmaceuticals were found to achieve 80 to 100%. Interestingly, commercial sorbents Strata-X-CW and Oasis-WCX, having similar amounts of carboxylic functions (~0.75 meq/g), retain the tested compounds poorly and lose them already at the washing step. This suggests that carboxylic groups are not the only factor determining the retention of basic pharmaceuticals on hypercrosslinked polystyrene adsorbing materials. The developed protocol ensures the quantitative and selective extraction of low μg/L levels of basic pharmaceuticals present in 500 mL of river water or 250 mL of effluent waste water.

Finally, a neutral hydrophilic product was obtained by post-crosslinking microbeads (5.5 ± 1.8 μm) resulting from the precipitation copolymerization of a mixture of 20% 2-hydroxyethyl methacrylate (HEMA), 40% vinylbenzyl chloride and 40% DVB [429]. 200 mg of the copolymer were placed in a cartridge and loaded with 1000 mL of solution (pH 2.5) spiked at 50 μg/L level of each analyte, oxamyl, methomyl, antipyrine, phenol, guaiacol, 4-nitrophenol, salicylic acid, 2,4-sinitrophenol, (4-chloro-3-methylphenoxy) acetic acid and ibuprofen. After eluting with 7 mL methanol, all analytes were found by gradient elution on a

C18 column with very high recovery rates ranging from 88 to 93%. However, commercially available hypercrosslinked polystyrene-type materials LiChrolut EN, Isolute ENV+ as well as Oasis HLB gave similar results, and recoveries on Isolute ENV+ were even slightly better for the majority of the above analytes, in spite of the 10 to 20 times larger particle size of that SPE sorbent. We have to note here that the popular notion of the polar groups on the surface of hydrophobic sorbents enhancing the adsorption of polar analytes, still lacks any convincing experimental support.

A group of Chinese scientists from Nankai University described the synthesis of hydrophilic hypercrosslinked biporous sorbents derived from two macroporous copolymer-precursors, methyl methacrylate/divinylbenzene and ethylene dimethacrylate/divinylbenzene, by involving pendent double binds of diolefines in the post-crosslinking process [430]. The polymers possess high specific inner surface area and sufficient polarity that, on authors' opinion, must enhance their capability to extract polar phenol from aqueous solutions. Meanwhile, even the best sample from a set of synthesized products adsorbs no more than 200 mg phenol per 1 g of the polymer at the equilibrium concentration of 3.5 g/L. One bed volume of this water-swollen sample purifies merely 15 bed volumes of the 300 mg/L solution sent through the column, which is seriously inferior to the performance of hydrophobic non-functionalized hypercrosslinked polystyrene, e.g. MN-200 (see Chapter 10). Again, we have no reason to suppose any positive contribution from polar ester groups to the sorption of phenol.

Another attempt to introduce hydrophilic groups into hypercrosslinked polystyrene consists of adding a few percent of hydroquinone during the self-crosslinking procedure of a macroporous chloromethylated styrene-6% DVB copolymer swollen in nitrobenzene. The additive reduces the specific surface area of the product, but it still readily adsorbs β-naphthol [431].

More recently, Zeng et al. [432] prepared polar hypercrosslinked polymers by polycondensation of hydroquinone, catechol or bisphenol-A with formaldehyde in a mixture of ethanol and concentrated hydrochloric acid at 120 °C in an autoclave. Depending on the molar ratio of the comonomers and the extent of dilution, resulting polymers have specific inner surface area ranging from 500 to 1000 m^2/g and pore volume as large as 0.4-2.4 cm^3/g. Notably the polymers take up water, ethanol, hexane and toluene nearly to the equal extent, up to 2-5 ml/g. The black colored polymers must have a rigid microporous hypercrosslinked structure differing from hypercrosslinked polystyrene by having many polar phenol-type hydroxyls. Sorption properties of the material still remain to be examined.

With regard to the microanalytical applications of hypercrosslinked polymers the new contribution from the group of Svec and Fréchet [433] must be poined out. The authors subjected to hypercrosslinking procedure a monolithic poly(styrene-*co*-vinylbenzyl chloride-*co*-divinylbenzene) precursor capillary column to arrive at the monolith containing an array of small pores. This monolithic column exhibited a surface area of 663 m^2/g or more than 1 order of magnitude larger than measured for the precursor column. This monolithic column affords good separation of uracil and alkylbenzenes in isocratic mobile phase mode and also proved useful for separations in size exclusion mode. A column efficiency as high as 73000 plates/m was determined for uracil. In contrast, the presence of mesopores in this hypercrosslinked monolithic column had a detrimental effect on the separation of proteins. Finally, an excellent review by Svec must me reffered to, entitled "Porous polymer monoliths: Amazingly wide variety of techniques enabling their preparation" [434] that does not jet contain information about hypercrosslinked monoliths, but sums up experience of many research groups involved into development of polymeric continuous separation media for various chromatography techniques.

Progress in the similarly popular research area, that of design, structure and application of metal–organic frameworks (MOFs), is treated in detail in the most recent monograph by MacGillivray [435].

REFERENCES TO PART THREE

[1] Kel'tsev, N.V. Basic Principles of Adsorption Technique, Khimiya, Moscow, 1976.
[2] Brunauer, S. Adsorption of Gases and Vapors, vol. 1, Gosizdatinlit, Moscow, 1948.
[3] Dubinin, M.M., Zaverina, E.D., Radushkevich, L.V. Zh. Phiz. Khimii 21 (1947) 1351–1362.
[4] Khirsanova, I.F., Soldatov, V.S., Martsinkevich, R.V., Tsyurupa, M.P., Davankov, V.A. Kolloid. Zh. 40 (1978) 1025–1028.
[5] Gregg, S.J., Sing, K.S.W. Adsorption. Surface Area and Porosity, second ed., Mir, Moscow, 1984.
[6] Davankov, V.A., Tsyurupa, M.P. Pure Appl. Chem. 61 (1989) 1881–1888.
[7] Rozenberg, G.I., Shabaeva, A.S., Moryakov, V.S., Musin, T.G., Tsyurupa, M.P., Davankov, V.A. React. Polym. 1 (1983) 175–182.
[8] Liu, P., Long, C., Qian, H.M., Li, Y., Li, A.M., Zhang, Q.X. Chin. Chem. Lett. 20 (2009) 492–495.
[9] Podlesnyuk, V.V., Hradil, J., Kralova, E. React. Funct. Polym. 42 (1999) 181–191.
[10] Belyakova, L.D., Kiselev, A.V., Shevchenko, T.I. Theory and Practice of Adsorption Processes (Russian), No. 18 (1986) 52–55.
[11] Belyakova, L.D., Kiselev, A.V., Platonova, N.P., Shevchenko, T.I. Kolloid. Zh. 42 (1980) 216–222.
[12] Belyakova, L.D., Shevchenko, T.I., Blega, M., Votavova, E., Kalal, Ya. Kolloid. Zh. 49 (1987) 847–852.
[13] Belyakova, L.D., Shevchenko, T.I., Davankov, V.A., Tsyurupa, M.P. Adv. Colloid Interface Sci. 25 (1986) 249–266.
[14] Timofeev, D.P. Kinetics of Adsorption, AN SSSR, Moscow, 1962.
[15] Rudobashta, S.P. Modeling of Mass- and Energy Transfer, Khimiya, Leningrad, 1979.
[16] Shabaeva, A.S. "Absorption of organic solvent vapors onto hypercrosslinked styrene polymers", PhD Thesis, Kazan State Technical University, Kazan, 1986.
[17] Davankov, V.A., Tsyurupa, M.P., Tarabaeva, O.G., Shabaeva, A.S. in: Greig, J.A. (Ed.), Ion Exchange Developments and Applications, Proceedings of IEX'96, SCI, 1996, pp. 209–216.
[18] Shilov, N.A., Lepin', L.K., Voznesenskii, S.A. J. Russ. Phys. Chem. Soc. 61 (1929) 1107–1123.
[19] Romankov, P.G., Rashkovskaya, N.B. Drying in Suspended State Khimiya, Leningrad, 1979.
[20] Belyakova, L.D., Vasilevskaya, O.V., Tsyurupa, M.P., Davankov, V.A. Zh. Fiz. Khimii 69 (1995) 669–700.
[21] Belyakova, L.D., Vasilevskaya, O.V., Tsyurupa, M.P., Davankov, V.A. Zh. Phyz. Khimii 70 (1996) 476–481.
[22] Belyakova, L.D., Kurbanbekov, E., Larionov, O.G., Tsyurupa, M.P., Davankov, V.A. Russ. Chem. Bull. 48 (1999) 1484–1487.
[23] Belyakova, L.D. Collected articles of All-Russian symposium on theory and practice of chromatography and electrophoresis, devoted to 95 university of chromatography discovering by M.S. Tswett, Samara State University, Samara, 1999, pp. 151-162.
[24] Belyakova, L.D., Voloschuk, A.M., Vorob'eva, L.M., Kurbanbekov, E., Larionov, O.G., Tsyurupa, M.P., Pavlova, L.A., Davankov, V.A. Russ. J. Phys. Chem. 76 (2002) 1515–1522.
[25] Belyakova, L.D., Kurbanbekov, E., Larionov, O.G., Tsyurupa, M.P., Davankov, V.A. Kolloid. Zh. 61 (1999) 617–623.
[26] Baya, M.P., Siskos, P.A., Davankov, V.A. J. AOAC Int. 83 (2000) 579–583.
[27] Tikhonov, V.I., Volkov, A.A. Science 296 (2002) 2363–2364.

[28] Farkas, A. Orthohydrogen, Parahydrogen and Heavy Hydrogen, Cambridge University Press, London, 1935.

[29] Veber, S.L., Bagryanskaya, E.G., Chapovsky, P.L. Russ. J. Exp. Theor. Phys. 129 (2006) 86–95.

[30] Tikhonov, V.I., Volkov, A.A. ChemPhysChem 7 (2006) 1026–1027.

[31] Kapralov, P.O., Artemov, V.G., Makurenkov, A.M., Tikhonov, V.I., Volkov, A.A. Russ. J. Phys. Chem. 83 (2009) 663–669.

[32] Lanin, S.N., Bardina, I.K., Zhuikova, O.S., Kovaleva, N.V. Russ. J. Phys. Chem. A 81 (2007) 1–7.

[33] Lanin, S.N., Kovaleva, N.V., Lanina, K.S., Litvincheva, L.A. in: Kurganov, A.A. (Ed.), Chromatography for Prosperity of Russia, Moscow, Granitsa, 2007, pp. 204–214.

[34] Tsyurupa, M.P., Maslova, L.A., Andreeva, A.I., Mrachkovskaya, T.A., Davankov, V.A. React. Polym. 25 (1995) 69–78.

[35] Ghanadzadeh, A., Zeini, A., Kashef, A., Moghadam, M. J. Mol. Liq. 138 (2008) 100–106.

[36] Davankov, V.A., Volynskaya, A.V., Tsyurupa, M.P. Vysokomol. Soedin. (Russia) B22 (1980) 746–748.

[37] Valderrama, C., Cortina, J.L., Farran, A., Gamisans, X., de las Heras, F.X. React. Funct. Polym. 68 (2008) 679–691.

[38] Valderrama, C., Cortina, J.L., Farran, A., Gamisans, X. de las Heras, F.X. React. Funct. Polym. 68 (2008) 718–731.

[39] Huang, J.-H., Huang, K.-L., Liu, S.-Q., Wang, A.-T., Yan, C. Colloid Surfaces A: Physicochem. Eng. Aspects 330 (2008) 55–61.

[40] Kulemin, V.V., Plavnik, G.M., Kareta, V.I., Maslova, L.A., Tsyurupa, M.P., Davankov, V.A. Radiokhimiya 30 (1988) 488–495.

[41] Soldatov, V.S., Pokrovskaya, A.I., Khirsanova, I.F. Isvestia AN BSSR, Ser. Khim. Nauk (3) (1978) 124–126.

[42] Neely, J.W., Isacoff, E.G. Carbonaceous Adsorbents for the Treatment of Ground and Surface Waters, Marcel Dekker, New York & Basel, 1982.

[43] McKay, G. in: McKay, G. (Ed.), Use of Adsorbents for the Removal of Pollutants from Wastewaters, CRC Press, Boca Raton, FL, 1996, pp. 99–132.

[44] Gordienko, S.V., Mikhailov, K.S., Tischenko, A.I., Penkina, V.I., Belikov, V.M. Prikl. Biokhimia. Mikrobiologiya 3 (1967) 437–441.

[45] Dressler, M. J. Chromatogr. 165 (1979) 167–206.

[46] Selemenev, V.F., Nenovinskaya, V.N., Oros, G.Yu., Sholin, A.F. Izv. Vuzov. Pischevaya Tekhnologiya 6 (1981) 48–50.

[47] Tsyurupa, M.P., Davankov, V.A., Krivonosova, L.G., Selemenev, V.F., Chikin, G.A. Prikl. Biokhimiya Mikrobiologiya 21 (1985) 72–77.

[48] Ryskal, V. private communication.

[49] Tsyurupa, M.P., Maslova, L.A., Davankov, V.A., Selemenev, V.F., Parshina, T.A., Papanova, I.V. Patent USSR, 1690338, 1989.

[50] Davankov, V.A., Rogozhin, S.V., Tsyurupa, M.P., Vainerman, E.S., Bakari, T.I. Patent USSR 1077242, 1983.

[51] Tsyurupa, M.P., Khodchenko, E.L., Davankov, V.A. Kolloid. Zh. 5 (1983) 1016–1018.

[52] Kunin, R. in: Dorfner, K., de Gruyter, Walter (Eds.), Ion Exchangers, Berlin & New York, 1991, pp. 659–676.

[53] Tsyurupa, M.P., Ilyin, M.M., Andreeva, A.I., Davankov, V.A. Fresenius J. Anal. Chem. 352 (1995) 672–675.

[54] Streat, M., Sweetland, L.A. React. Funct. Polym. 35 (1997) 99–109.

[55] Azanova, V.V., Hradil, J. React. Funct. Polym. 41 (1999) 163–175.

[56] Oh, C.-G., Ahn, J.-H., Ihm, S.-K. React. Funct. Polym. 57 (2003) 103–114.

[57] Huang, J., Yan, C., Huang, K. J. Colloid Interface Sci. 332 (2009) 60–64.

[58] Pan, B., Du, W., Zhang, W., Zhang, X., Zhang, Q., Pan, B. et al. J. Chen, Environ. Sci. Technol. 41 (2007) 5057–5062.

[59] Zhang, W., Chen, J., Pan, B., Chen, Q., He, M., Zhang, Q. et al. React. Funct. Polym. 66 (2006) 395–401.

[60] Zhang, W., Chen, J., Pan, B., Zhang, Q. React. Funct. Polym. 66 (2006) 485–493.

[61] Valderrama, C., Barios, J.I., Caetano, M., Farran, A., Cortina, J.L. React. Funct. Polym. 70 (2010) 142–150.

[62] Meng, G.-H., Li, A.-M., Yang, W.-B., Liu, F.-Q., Zhang, Q.-X. Chin. J. Polym. Sci. 24 (2006) 585–591.

[63] Boehm, H.P. Carbon 32 (1994) 759–769.

[64] Boehm, H.P., Diehl, E., Heck, W., Sappok, R. Angew. Chem. Int. Ed. 3 (1964) 669–677.

[65] Sun, Y., Chen, J., Li, A., Liu, F., Zhang, Q. React. Funct. Polym. 64 (2005) 63–73.

[66] Bai, X., Liu, B., Yan, J. React. Funct. Polym. 63 (2005) 43–53.

[67] Dale, J.A., Nikitin, N.V., Moore, R., Opperman, D., Crooks, G., Naden, D., Belstein, E., Jenkins, P. in: Greig, J.A. (Ed.), Ion Exchange at the Millenium, Imperial College Press, 2000, pp. 261–268.

[68] Streat, M., Sweetland, L.A. Trans IchemE B, 76 (1998) 127–134.

[69] Streat, M., Sweetland, L.A., Horner, D.J. Trans IchemE B, 76 (1998) 135–141.

[70] Streat, M., Horner, D.J. Trans IchemE B, 78 (2000) 363–382.

[71] Streat, M., Sweetland, L.A., Horner, D.J. Trans IchemE B, 76 (1998) 142–150.

[72] Chang, C.-F., Chang, C.-Y., Hsu, K.-E., Lee, S.-C., Höll, W. J. Hazardous Materials 155 (2008) 295–304.

[73] Dawson-Ekeland, K.R., Stringfield, R.T. Patent USA 5,021,253, 1991.

[74] Chikin, G.A., Myagkoi, O.N. (Eds.), Ion Exchange Methods of Substance Purifications, Voronezh, Voronezh State University, 1984.

[75] Stringfield, R.T., Goltz, H.R., Norman, S.I., Bharwada, U.J., LaBrie, R.L. Patent USA 4,950,332, 1990.

[76] Norman, S.I., Stringfield, R.T., Gopsill, C.C. Patent USA 4,965,083, 1990.

[77] Belfer, S., Daltrophe, N. in: Greig, J.A. (Ed.), Ion Exchange Developments and Applications, SCI, UK, 1996, pp. 538–547.

[78] Pan, B.C., Xiong, Y., Li, A.M., Chen, J.L., Zhang, Q.X., Jin, X.Y. React. Funct. Polym. 53 (2002) 63–72.

[79] Jerabek, K., Hankova, L., Prokop, Z. React. Polym. 23 (1994) 107–112.

[80] Tsyurupa, M.P., Pankratov, E.A., Tsvankin, D.Ya., Zhukov, V.P., Davankov, V.A. Vysokomol. Soedin. A, 27 (1985) 339–345.

[81] Zhai, Z.C., Chen, J.L., Fei, Z.H., Wang, H.L., Li, A.M., Zhang, O.X. React. Funct. Polym. 57 (2003) 93–102.

[82] Li, H., Jiao, J., Xu, M., Shi, Z., He, B. Polymer 45 (2004) 181–188.

[83] Valderrama, C., Gamisans, X., de las Heras, F.X., Cortina, J.L., Farran, A. React. Funct. Polym. 67 (2007) 1515–1529.

[84] Valderrama, C., Cortina, J.L., Farran, A., Gamisans, X., Lao, C. J. Colloid Interface Sci. 310 (2007) 35–46.

[85] Shkutina, I.V., Stoyanova, O.F., Kovaleva, T.A., Selemenev, V.F., Lunina, V.V., Gun'kina, L.A. et al. Sorption Chromatogr. Processes (Russia) 7 (2007) 271–275.

[86] Shkutina, I.V., Stoyanova, O.F., Selemenev, V.F. Zh. Prikl. Khimii (Russia) 74 (2001) 869–871.

[87] Shkutina, I.V., Stoyanova, O.F., Selemenev, V.F. Sorption Chromatogr. Processes (Russia) 6 (2006) 1066–1070.

[88] Grebenkin, A.D., Lukin, A.L., Kotov, V.V. Sorption Chromatogr. Processes (Russia) 6 (2006) 1036–1039.

[89] Safonova, E.F., Selemenev, V.F., Slivkin, A.I., Podryadnaya, E.N., Brezhneva, T.A. Pharm. Chem. J. 38 (2004) 385–387.

[90] Stoyanova, O.F., Shkutina, I.V., Merkulova, Yu.D. Sorption Chromatogr. Processes (Russian) 8 (2008) 620–625.

[91] Sentsov, M.Yu., Stoyanova, O.F., Shkutina, I.V., Selemenev, V.F., Butyrskaya, E.V., Merkulova, Yu.D. Sorption Chromatogr. Processes (Russian) 7 (2007) 845–849.

[92] Zhang, W., Xu, Z., Pan, B., Lv, L., Zhang, Q., Zhang, Q. et al. J. Colloid Interface Sci. 311 (2007) 382–390.

[93] Zhang, W., Hong, C., Pan, B., Xu, Z., Zhang, Q., Zhang, Q. Colloids Surfaces A: Physicochem. Eng. Aspects 331 (2008) 257–262.

[94] Simpson, E.J., Koros, W.J., Schechter, R.S. Ind. Eng. Chem. Res. 35 (1996) 1195–1205.

[95] Simpson, E.J., Koros, W.J., Schechter, R.S. Ind. Eng. Chem. Res. 35 (1996) 4635–4645.

[96] Schroeder, P. J. Phys. Chem. 45 (1903) 75–117.

[97] Liu, F.-Q., Xia, M.-F., Yao, S.-L., Li, A.-M., Wu, H.-S., Chen, J.-L., J. Hazard Mater. 152 (2008) 715–720.

[98] Qiu, Y., Ling, F. Chemosphere 64 (2006) 963–971.

[99] Zheng, K., Pan, B., Zhang, Q., Zhang, W., Pan, B., Han, Y. et al. Sep. Purif. Technol. 57 (2007) 250–256.

[100] Pan, B., Chen, X., Pan, B., Zhang, W., Zhang, X., Zhang, Q. J. Hazard. Mater. 137 (2006) 1236–1240.

[101] Cai, J., Li, A., Shi, H., Fei, Z., Long, C., Zhang, Q. J. Hazard. Mater. 124 (2005) 173–180.

[102] Ming, Z.W., Long, C.J., Cai, P.B., Xing, Z.Q., Zhang, B. J. Hazard. Mater. 128 (2006) 123–129.

[103] Jiang, Z.-M., Li, A.-M., Can, J.-G., Wang, C., Zhang, Q.-X. J. Environ. Sci. 19 (2007) 135–140.

[104] Long, C., Li, A., Wu, H., Zhang, Q. Colloids Surfaces A Physicochem. Eng. Aspects 333 (2009) 150–155.

[105] Hu, Z., Zhang, Q., Chen, J., Wang, L., Anderson, G.K. Chemosphere 38 (1999) 2003–2011.

[106] Li, A., Zhang, Q., Zhang, G., Chen, J., Fei, Z., Liu, F. Chemosphere 47 (2002) 981–989.

[107] Albright, R.L. React. Polym. 4 (1986) 155–174.

[108] Bicak, N., Sherrington, D.C. React. Funct. Polym. 27 (1995) 155–161.

[109] Streat, M., Sweetland, L.A. Trans. IChemE B76 (1998) 115–126.

[110] Tsyurupa, M.P., Tarabaeva, O.G., Pastukhov, A.V., Davankov, V.A. Intern. J. Polym. Mater. 52 (2003) 403–414.

[111] Lenfeld, J., Peška, J., Štamberg, J. React. Polym. 1 (1982) 47–50.

[112] Davankov, A.B., Petrov, G.S., Ogneva, N.E., Laufer, V.M. USSR Patent 100692 (1955), priority 19.04.1950.

[113] Hatch, M.J., Dillon, J.A., Smith, H.B. Ind. Eng. Chem. 49 (1957) 1812–1819.

[114] Dow Chemical Co., Midland, Mich., Tech. Service Bull. 164–62, Ion Retardation.

[115] Hatch, M.J., Dillon, J.A. Ind. Eng. Chem. Process Des. Dev. 2 (1963) 253–263.

[116] Nelson, F., Kraus, K.A. J. Am. Chem. Soc. 80 (1958) 4154–4161.

[117] Helfferich, F., Ion Exchange, McGraw-Hill, New York, 1962, p. 134.

[118] Wheaton, R.M., Bauman, W.C. Ind. Eng. Chem. 45 (1953) 228–233.

[119] Brown, C.J., Sheedy, V., Palaologou, M., Thompson, R. Proceedings of Annual Meeting of Minerals, Metals and Materials Society, Orlando, FL, USA, 1997, TP126.

[120] Brown, C.J., Fletcher, C.J., Streat, M. Ion Exchange for Industry, Ellis Horwood, Chichester, 1988, pp. 392-403.

[121] Götzelmann, W., Hartinger, L., Gülbas, M. Teil 1, Metalloberfläche 41 (1987) 208–212; Teil 2, Metalloberfläche 41 (1987) 315–322.

[122] Soldatov, V.S., Polhovsky, E.M., Sosinovich, Z.I. React. Funct. Polym. 60 (2004) 41–48.
[123] Ferapontov, N.B., Gorshkov, V.I., Parbuzina, L.R., Trobov, H.T., Strusovskaya, N.L. React. Funct. Polym. 41 (1999) 213–225.
[124] Yoza, N. J. Chromatogr. 86 (1973) 325–349.
[125] Yoza, N., Ohashi, S. J. Chromatogr. 41 (1969) 429–437.
[126] Porath, J. Metabolism Clin. Exp. 13 (1964) 1004–1015.
[127] Ueno, Y., Yoza, N., Ohashi, S. J. Chromatogr. 52 (1970) 321–327.
[128] Collins, K.D. Proc. Natl. Acad. Sci. USA Biophysics 92 (1995) 5553–5557.
[129] Collins, K.D. Biophys. Chem. 119 (2006) 271–281.
[130] Grover, P.K., Ryall, R.L. Chem. Rev. 105 (2005) 1–10.
[131] Beckstein, O., Tai, K., Sansom, M.S.P. J. Am. Chem. Soc. 126 (2004) 14694–14695.
[132] Carrillo-Tripp, M., San-Roman, M.L., Hernandez-Cobos, J., Saint-Martin, H., Ortega-Blake, I. Biophys. Chem. 124 (2006) 243–250.
[133] Siwy, Z., Fulinsky, A. Phys. Rev. Lett. 89 (2002) Art. No 198103-198106.
[134] Siwy, Z., Apel, P., Baur, D., Dobrev, D.D., Korchev, Y.E., Neumann, R. et al. Surf. Sci. 532–535 (2003) 1061–1066.
[135] Rona, M., Schmuckler, G. Talanta 20 (1973) 237–240.
[136] Ferapontov, N.B., Gagarin, A.N., Gruzdeva, A.N., Strusovskaya, N.L., Parbuzina, L.R. VIII Regional Conference "Problems of Chemistry and Chemical Technology", Voronezh, 2000, pp. 99–101 (Russ).
[137] Tsyurupa, M.P., Davankov, V.A., Ferapontov, N.B., Gruzdeva, A.N., Strusovskaya, N.L. in Cox, M. (Ed.), Ion Exchange Technology for Today and Tomorrow, Proceedings of IEX2004, Cambridge, July 2004, pp. 339–346.
[138] Tsyurupa, M.P., Davankov, V.A. Dokl. Akad. Nauk RAN, 398 (2004) 198–200.
[139] Davankov, V., Tsyurupa, M. J. Chromatogr. A, 1087 (2005) 3–12.
[140] Davankov, V.A., Tsyurupa, M.P., Alexienko, N.N. J. Chromatogr. A, 1100 (2005) 32–39.
[141] Davankov, V.A., Tsyurupa, M.P., Alexienko, N.N. Mendeleev Commun. (2005), N 5, 192–193.
[142] Ma, J.C., Dougherty, D.A. Chem. Rev. 97 (1997) 1303–1324.
[143] Mecozzi, S., West, A.P., Jr., Dougherty, D.A. J. Am. Chem. Soc. 118 (1996) 2307–2308.
[144] Manalo, G.D., Turse, R., Rieman, W.M. III, Anal. Chim. Acta 21 (1959) 383–391.
[145] Hribar, B., Southall, N.T., Vlachy, V., Dill, K.A. J. Am. Chem. Soc. 124 (2002) 12302–12311.
[146] Washabaugh, M.W., Collins, K.D. J. Biol. Chem. 261 (27), (1986) 12477–12485.
[147] Samoilov, O.Y. Strktura Vodnyh Rastvorov Electrolitov I Gidratacia Ionov, Izdatelstva AN SSSR, Moskva, 1957.
[148] Samoilov, O.Ya. in: Horne, R.A. (Ed.), Water and Aqueous Solutions: Structure, Thermodynamics, and Transport Processes, Wiley-Interscience, New York, 1972, pp. 597–612.
[149] Fabricand, B.P., Goldenberg, S. J. Chem. Phys. 34 (1961) 1624–1628.
[150] Bergstrom, P.A., Lindgren, J., Kristiansson, O. J. Phys. Chem. 95 (1991) 8575–8580.
[151] Licheri, G., Piccaluga, G., Pinna, G. J. Chem. Phys. 64 (1976) 2437–2441.
[152] Biggin, S., Enderby, J.E., Hahn, R.L., Narten, A.H. J. Phys. Chem. 88 (1984) 3634–3638.
[153] Soper, A.K., Weckström, K. Biophys. Chem. 124 (2006) 180–191.
[154] Marcus, Y., Ion Solvation, Wiley, New York, 1985.
[155] Bocris, J.O.M., Saluya, P. J. Phys. Chem. 76 (1972) 2140–2151.
[156] Dunsyuryun, D.H., Karpov, G.V., Morozov, I.I. Chem. Phys. Lett. 242 (1995) 390–394.

[157] Kiriukhin, M.Y., Collins, K.D. Biophys. Chem. 99 (2002) 155–168.
[158] Nightingale, E.R. J. Phys. Chem. 63 (1959) 1381–1387.
[159] Gill, D.S. J. Chem. Soc., Faraday Trans. I77 (1981) 751–758.
[160] Engelhardt, H., Beck, W., Schmitt, T. Capillary Electrophoresis, Vieweg, Braunschweig/Wiesbaden, 1994, p. 8.
[161] Erdey-Grus T., Transport Phenomena in Aqueous Solutions, Akadémiai Kiadó Budapest, 1974.
[162] Rasaiah, J.C., Lynden-Bell, R.M. Philos. Trans. R. Soc. London A, 359 (2001) 1545–1574.
[163] Marcus, Y. Chem Rev. 109 (2009) 1346–1370.
[164] Probst, M.M., Randai, T., Heinzinger, K., Bopp, P., Rode, B.M. J. Phys. Chem. 89 (1985) 753–759.
[165] Chizhik, V.I. Mol. Phys. 90 (1997) 653–660.
[166] Egorov, A.V., Komolkin, A.V., Chizhik, V.I. J. Mol. Liq. 89 (2000) 47–55.
[167] Palka, K., Hawlicka, E. J. Mol. Liq. 122 (2005) 28–31.
[168] Voth, G.A. Acc. Chem. Res. 39 (2006) 143–150.
[169] Tuckerman, M.E., Chandra, A., Marx, D. Acc. Chem. Res. 39 (2006) 151–158.
[170] Davankov, V.A., Tsyurupa, M.P., Blinnikova, Z.K. Russ. J. Phys. Chem. A, 82 (2008) 434–438.
[171] Davankov, V.A., Tsyurupa, M.P., Blinnikova, Z.K. J. Sep. Sci. 32 (2009) 64–73.
[172] Tsyurupa, M.P., Blinnikova, Z.K., Pavlova, L.A., Davankov, V.A. in: Cox, M. (Ed.), Recent Advances in Ion Exchange. Theory and Practice, Proceedings of IEX 2008 (Cambridge), SCI, 2008, 77–84.
[173] Blinnikova, Z.K., Maerle, K.V., Tsyurupa, M.P., Davankov, V.A., Sorption and Chromatogr. Processes (Russia), 9 (2009) 323–331.
[174] Welch, Ch.J. in: Cox, G.B. (Ed.), Perparative Enantioselective Chromatography, Blackwell Publishing, Oxford, 2005
[175] Guioshon, G. J. Chromatogr. A, 1079, (2005) 7–23.
[176] Pastukhov, A.V., Davankov, V.A., Belyakova, L.D., Voloshchuk, A.M. Russ. J. Phys. Chem. 78 (2004) 1992–1998.
[177] Laatikainen, M., Sainio, T., Davankov, V., Tsyurupa, M., Blinnikova, Z., Paatero, E. J. Chromatogr. A, 1149 (2007) 245–253.
[178] Laatikainen, M., Sainio, T., Davankov, V., Tsyurupa, M., Blinnikova, Z., Paatero, E. React. Funct. Polym. 67 (2007) 1589–1598.
[179] Müller-Späth, M., Aumann, L., Melter, L., Ströhlein, G., Morbidelli, M. Biotechnol. Bioeng. 100 (2008) 1166–1177.
[180] Davankov, V.A., Timofeeva, G.I., Ilyin, M.M., Tsyurupa, M.P. J. Polym. Sci. Part A: Polym. Chem. 35 (1997) 3847–3852.
[181] Davankov, V., Tsyurupa, M. J. Chromatogr. A, 1136 (2006) 118–122.
[182] Galia, M., Svec, F., Frechet, J.M.J. J. Polym. Sci.: Polym. Chem. 32 (1994) 2169–2175.
[183] Gawdzik, B., Osypiuk, J. Chromatographia 54 (2001) 595–599.
[184] Halász, I., Martin, K. Ber. Bunsenges. Phys. Chem. 79 (1975) 731–732.
[185] Brandrup, J., Immergut, E.H. Polymer Handbook, Wiley, New York, 1980.
[186] Ells, B., Wang, Y., Cantwell, F.F. J. Chromatogr. A, 835 (1999) 3–18.
[187] Unsal, E., Camli, S.T., Tuncel, M., Senel, S., Tuncel, A. React. Funct. Polym. 61 (2004) 353–368.
[188] www.polymerlabs.com
[189] www.hamiltoncompany.com
[190] Beth, M., Unger, K.K., Tsyurupa, M.P., Davankov, V.A. Chromatographia 36 (1993) 351–355.
[191] Pinkerton, T.C., Miller, T.D., Cook, S.E., Perry, J.A., Rateike, J.D., Szczerba, T.J. Biochromatography 1 (1986) 96–105.

[192] Sutton, R.M.C., Hill, S.J., Jones, P. J. Chromatogr. A, 739 (1996) 81–86.
[193] Kiseleva, M.G., Radchenko, L.V., Nesterenko, P.V. J. Chromatogr. A, 920 (2001) 79–85.
[194] Penner, N.A., Nesterenko, P.N. J. Chromatogr. A, 884 (2000) 41–51.
[195] Sutton, R.M.C., Hill, S.J., Jones, P. J. Chromatogr. A, 789 (1997) 389–394.
[196] Davankov, V.A., Sychov, C.S., Ilyin, M.M., Sochilina, K.O. J. Chromatogr. A, 987 (2003) 67–75.
[197] Sychov, C.S., Ilyin, M.M., Sochilina, K.O., Davankov, V.A. J. Chromatogr. A, 1030 (2004) 17–24.
[198] Penner, N.A., Nesterenko, P.N. Analyst, 125 (2000) 1249–1254.
[199] Penner, N.A., Nesterenko, P.N., Khryaschevsky, A.V., Stranadko, T.N., Shpigun, O.A. Mendeleev Commun. 8, (1998) 24–27.
[200] Saunders, G. Chromatogr. Today 1 (2008) 12–14.
[201] Urban, J., Svec, F., Fréchet, J.M.J. Anal. Chem. 82 (2010) 1621–1623.
[202] Svec, F. J. Chromatogr. A, 1217 (2010) 902–924.
[203] Trammell, B.C., Ma, L., Luo, H., Hillmyer, M.A., Carr, P.W. J. Chromatogr. A, 1060 (2004) 61–76.
[204] Yang, X., Ma, L., Carr, P.W. J. Chromatogr. A, 1079 (2005) 213–220.
[205] Sabik, H., Jeannot, R., Rondeau, B. J. Chromatogr. A, 885 (2000) 217–236.
[206] Planas, C., Puig, A., Rivera, J., Caixach, J. J. Chromatogr. A, 1131 (2006) 242–252.
[207] Kiely, T., Donaldson, D., Grube, A. Pesticide Industry Sale and Usage. 2000 and 2001 Marked Estimates, US Environmental Protection Agency, Office of Prevention. Pesticides and Toxic Substances (7503C), EPA-733-R-04-001, Washington, 2004, p. 27.
[208] Fed. Reg., EPA Method 604, Phenols, Part VIII, 40 CFR Part 136, Environmental Protection Agency, 26 October 1984, pp. 58–66.
[209] Lamprecht, G., Huber, J.F.K. J. Chromatogr. A, 667 (1994) 47–57.
[210] Johansen, S.S., Pawliszyn, J., J. High Resol. Chromatogr. 19 (1996) 627–632.
[211] Huck, C.W., Bom, G.K. J. Chromatogr. A, 885 (2000) 51–72.
[212] Sánchez-Rojas, F., Bosch-Ojeda, C., Cano-Pavón, M. Chromatographia 69 (2009) S79–S94.
[213] Gómez-Ariza, J.L., García-Barrera, T., Lorenzo, F. J. Chromatogr. A, 1049 (2004) 147–153.
[214] Chilla, C., Guillén, D.A., Barroso, C.G., Pérez-Bustamante, J.A. J. Chromatogr. A, 750 (1996) 209–214.
[215] Liska, I. J. Chromatogr. A, 665 (1993) 163–176.
[216] Pocurull, E., Marcé, R.M., Borrull, F. Chromatographia 41 (1995) 521–524.
[217] Thurman, E.M., Mills, M.S., Solid-Phase Extraction. Principles and Practice, John Wiley & Sons, New York, 1998.
[218] Simpson, N.J.K. (Ed.), Solid-Phase Extraction. Principles, Techniques and Applications, Marcel Dekker, New York, 2000.
[219] Sun, J.J., Fritz, J.S. J. Chromatogr. 590 (1992) 197–202.
[220] Puig, D., Barceló, D. Chromatographia 40 (1995) 435–444.
[221] Renner, T., Baumgarten, D., Unger, K.K. Chromatographia 45 (1997) 199–205.
[222] Ruban, V.F., Belen'kii, B.G. Zh. Anal. Khimii 63 (1988) 1307–1312.
[223] Brouwer, E.R., Brinkman, U.A. Th., J. Chromatogr. A, 678 (1994) 223–231.
[224] Pocurull, E., Calull, M., Marcé, R.M., Borrull, F. J. Chromatogr. A, 719 (1996) 105–112.
[225] Pocurull, E., Marcé, R.M., Borrull, F. J. Chromatogr. A, 738 (1996) 1–9.
[226] Coquart, V., Hennion, M.-C. J. Chromatogr. 600 (1992) 195–201.
[227] Gawdzik, B., Gawdzik, J., Czerwińska-Bil, U. J. Chromatogr. 509 (1990) 135–140.
[228] Purosep SPEs – Solid Phase Extractants, Purolite, 1997.
[229] Isolute ENV+, International Sorbent Technology, Catalog.
[230] Puig, D., Barceló, D. J. Chromatogr. A, 733 (1996) 371–381.

[231] Gun'ko, V.M., Turov, V.V., Zarko, V.I., Nychiporuk, Y.M. et al. J. Colloid Interface Sci. 323 (2008) 6–17.
[232] Less, M., Schmidt, T.C., von Löw, E., Stork, G.J. Chromatogr. A, 810 (1998) 173–182.
[233] www.sparkholland.com
[234] Masqué, N., Galià, M., Marcé, R.M., Borrull, F. J. Chromatogr. A, 771 (1997) 55–61.
[235] Bernal, J.L., Nozal, M.J., Toribio, L., Serna, M.L., Borrull, F., Marcé, R.M. et al. Chromatographia 46 (1997) 295–300.
[236] Penner, N.A., Nesterenko, P.N., Khryaschevsky, A.V., Stranadko, T.N., Shpigun, O.A. Mendeleev Commun. (1998) 24–27.
[237] Fontanals, N., Puig, P., Galià, M., Marcé, R.M., Borrull, F. J. Chromatogr. A, 1035 (2004) 281–284.
[238] Fontanals, N., Galià, M., Marcé, R.M., Borrull, F.J. Chromatogr. A, 1030 (2004) 63–68.
[239] Fontanals, N., Galià, M., Cormack, P.A.G., Marcé, R.M., Sherrington, D.C., Borrull, F. J. Chromatogr. A, 1075 (2005) 51–56.
[240] Bagheri, H., Mohammadi, A., Salemi, A. Anal. Chim. Acta 513 (2004) 445–449.
[241] Kvistad, A.M., Lundanes, E., Greibrokk, T. Chromatographia 48 (1998) 707–713.
[242] Rodrígues, I., Mejuto, M.C., Bollaín, M.H., Cela, R. J. Chromatogr. A, 786 (1997) 285–292.
[243] Cheung, J., Wells, R.J. J. Chromatogr. A, 771 (1997) 203–211.
[244] Khryaschevsky, A.V., Podlovchenko, M.B., Nesterenko, P.N., Shpigun, O.A. Bull. Moscow State Univ. 39 (1998) 196–200.
[245] Tsysin, G.I., Kovalev, I.A., Nesterenko, P.N., Penner, N.A., Filippov, O.A. Sep. Purif. Technol. 33 (2003) 11–24.
[246] Cledera-Castro, M.M., Santos-Montes, A.M., Izquierdo-Hornillos, R. LC-GC Europe 19 (2006) 424–431.
[247] Önnerfjord, P., Barceló, D., Emnéus, J., Gorton, L., Marko-Varga, G. J. Chromatogr. A, 737 (1996) 35–45.
[248] Guenu, S., Hennion, M.-C. J. Chromatogr. A, 737 (1996) 15–24.
[249] Curini, R., Gentili, A., Marchese, S., Marino, A., Perret, D. J. Chromatogr. A, 874 (2000) 187–198.
[250] Tolosa, I., Douy, B., Carvalho, F.P. J. Chromatogr. A, 864 (1999) 121–136.
[251] Loos, R., Niessner, R. J. Chromatogr. A, 835 (1999) 217–229.
[252] Laganà, A., D'Ascenzo, G., Fago, G. A. Marino, Chromatographia 46 (1997) 256–264.
[253] Soriano, J.M., Jiménez, B., Redondo, M.J., Moltó, J.C. J. Chromatogr. A, 822 (1998) 67–73.
[254] Aguilar, C., Borrull, F., Marcé, R.M. J. Chromatogr. A, 771 (1997) 221–231.
[255] Junker-Buchheit, A., Witzenbacher, M. J. Chromatogr. A, 737 (1996) 67–74.
[256] Shakulashvili, N., Revia, R., Steiner, F. Engelhardt. H. Chromatographia 60 (2004) 145–150.
[257] Carabias-Martínez, R., Rodríguez-Gonzalo, E., Revilla-Ruiz, P., Domínguez-Álvarez, J. J. Chromatogr. A, 990(2003) 291–302.
[258] Sandin-Espoña, P., González-Blázquez, J.J., Magrans, J.O., García-Baudín, J.M. Chromatographia 55 (2002) 681–686.
[259] Majzik, E.S., Tóth, F., Benke, L., Kiss, Zs. Chromatographia 63 (2006) S105–S109.
[260] Bagheri, H., Saraji, M., Barceló, D. Chromatographia 59 (2004) 283–289.
[261] Hogenboom, A.C., Hofman, M.P., Jolly, D.A., Niessen, W.M.A., Brinkman, U.A. Th., J. Chromatogr. A, 885 (2000) 377–388.
[262] Frassanito, R., Rossi, M., Dragani, L.K., Tallarico, C., Longo, A. J. Chromatogr. A, 795 (1998) 53–60.
[263] Martínez, K., Ferrer, I., Barceló, D. J. Chromatogr. A, 879 (2000) 27–37.
[264] Gimeno, R.A., Aguilar, C., Marcé, R.M., Borrull, F. J. Chromatogr. A, 915 (2001) 139–147.

[265] Fontanals, N., Marcé, R.M., Cormack, P.A.G., Sherrington, D.C., Borrull, F. J. Chromatogr. A, 1191 (2008) 118–124.

[266] Masqué, N., Marcé, R.M., Borrull, F. J. Chromatogr. A, 793 (1998) 257–263.

[267] Castillo, M., Puig, D., Barceló, D. J. Chromatogr. A, 778 (1997) 301–311.

[268] Ferrer, I., Barceló, D. J. Chromatogr. A, 778 (1997) 161–170.

[269] Alonso, M.C., Barceló, D. J. Chromatogr. A, 889 (2000) 231–244.

[270] Sacher, F., Lange, F.T., Brauch, H.-J., Blankenhorn, I. J. Chromatogr. A, 938 (2001) 199–210.

[271] Klenner, S., Bergmann, C., Strube, K., Ternes, W., Spillmann, T. Chromatographia 65 (2007) 733–736.

[272] Shkutina, I.V., Stoyanova, O.F., Selemenev, V.F. Sorption Chromatogr. Processes 8 (2008) 272–276.

[273] Pavlinov, S.A., Piotrovskii, V.K. Khim.-Farm. Zh., 24 (1990) 174–176.

[274] Pavlinov, S.A., Piotrovskii, V.K., Metelitsa, V.I., Filatova, N.P., Bochkareva, E.V. Khim.-Farm. Zh. 24 (1990) 17–19.

[275] Oosterkamp, A.J., Irth, H., Beth, M., Unger, K.K., Tjaden, U.R., van de Greef, J. J. Chromatogr. B, 653 (1994) 55–61.

[276] Karlsson, M., Julander, A., van Bavel, B., Lindström, G. Chromatographia 61 (2005) 67–73.

[277] Cohen, A.S., Jönsson, B.A.G. Chromatographia 65 (2007) 407–412.

[278] Mendaš, G., Drevenkar, V., Zapančič-Kralj, L. J. Chromatogr. A, 918 (2001) 351–359.

[279] Hegstad, S., Lundanes, E., Reistad, R., Haug, L.S., Becher, G. Alexander, J. Chromatographia 52 (2000) 499–504.

[280] Dominguez, C., Guillén, D.A., Barroso, C.G. J. Chromatogr. A, 918 (2001) 303–310.

[281] Jimenez, J.J., Bernal, J.L., del Nozal, Ma.J., Toribio, L., Arias, E. J. Chromatogr. A, 919 (2001) 147–156.

[282] Ferreira, V., Gulleré, L., López, R., Cacho, J. J. Chromatogr. A, 1028 (2004) 339–345.

[283] Insa, S., Anticó, E., Ferreira, V. J. Chromatogr. A, 1089 (2005) 235–242.

[284] Ferreira, V., Jarauta, I., Ortega, L., Cacho, J. J. Chromatogr. A, 1025 (2004) 147–156.

[285] Štajnbacher, D., Zupančič-Kralj, L. J. Chromatogr. A, 1015 (2003) 185–198.

[286] Sychev, C.S., Eller, K.I., Davankov, V.A. Zavodskaya Laboratoriya 70 (2004) 17–19.

[287] Loos, R., Niessner, R. J. Chromatogr. A, 822 (1998) 291–303.

[288] Gimeno, R.A., Marcé, R.M., Borrull, F. Chromatographia 53 (2001) 22–26.

[289] Nageswara Rao, R., Venkateswarlu, N., Khalid, S., Narsimha, R., Sridhar, S. J. Chromatogr. A, 1113 (2006) 20–31.

[290] Galdiga, C.U., Greibrokk, T. J. Chromatogr. A, 793 (1998) 297–306.

[291] Stenholm, A., Drevin, I., Lundgren, M. J. Chromatogr. A, 1070 (2005) 155–161.

[292] Fontanals, N., Cormack, P.A.G., Sherrington, D.C. J. Chromatogr. A, 1215 (2008) 21–29.

[293] Tölgyessy, P., Liška, I. J. Chromatogr. A, 657 (1999) 247–254.

[294] Brossa, L., Pocurull, E., Borrul, F., Marcé, R.M. Chromatographia 59 (2004) 419–423.

[295] Müller, M.B., Zwiener, C., Frimmel, F.H. J. Chromatogr. A, 862 (1999) 137–145.

[296] López de Alda, M. J., Barceló, D. J. Chromatogr. A, 938 (2001) 145–153.

[297] Khryaschevsky, A.V., Nesterenko, P.N., Tikhomirova, T.I., Fadeeva, V.I., Shpigun, O.A. Zh. Anal. Khimii 52 (1997) 485–489.

[298] Fax, E., Marcaillou, C., Mondeguer, F., Bare, R., Hess, P. Harmful Algae 7 (2008) 574–583.

[299] Lacorte, S., Perrot, M.-C., Fraisse, D., Barceló, D. J. Chromatogr. A, 833 (1999) 181–194.

[300] Wissiack, R., Rosenberg, E., Grasserbauer, M. J. Chromatogr. A, 896 (2000) 159–170.

[301] Lacorte, S., Fraisse, D., Barceló, D. J. Chromatogr. A, 857 (1999) 97–106.

[302] Zhu, J., Snow, D.D., Cassada, D.A., Monson, S.J., Spalding, R.F. J. Chromatogr. A, 928 (2001) 177–186.

[303] Jacobsen, A.M., Halling-Sørensen, B., Ingerslev, F., Hansen, S.H. J. Chromatogr. A, 1038 (2004) 157–170.

[304] Yang, S., Cha, J., Carlson, K. J. Chromatogr. A, 1097 (2005) 40–53.

[305] Yang, S., Carlson, K.H. J. Chromatogr. A, 1038 (2004) 141–155.

[306] Zimmer, D., Pickard, V., Czembor, W., Müller, C. J. Chromatogr. A, 854 (1999) 23–35.

[307] Oertel, R., Richter, K., Kirch, W. J. Chromatogr. A, 846 (1999) 217–222.

[308] Kollroser, M., Schober, C. Chromatographia 57 (2003) 133–138.

[309] Carabias-Martínez, R., Rodríguez-Gonzalo, E., Domínguez-Álvarez, J., Hernández-Méndez, J. J. Chromatogr. A, 869 (2000) 451–461.

[310] Grey, L., Nguyen, B., Yang, P. J. Chromatogr. A, 958 (2002) 25–33.

[311] Peruzzi, M., Bartolucci, G., Cioni, F. J. Chromatogr. A, 867 (2000) 169–175.

[312] Dugay, A., Herrenknecht, C., Czok, M., Guyon, F., Pages, N. J. Chromatogr. A, 958 (2002) 1–7.

[313] Smith, C.J., Dooly, G.L., Moldoveanu, S.C. J. Chromatogr. A, 991 (2003) 96–107.

[314] Gilar, M., Belenky, A., Wang, B.H. J. Chromatogr. A, 921 (2001) 3–13.

[315] Väänänen, T., Kuronen, P., Pehu, E. J. Chromatogr. A, 689 (2000) 301–305.

[316] Diez, S., Ortiz, L., Bayona, J.M. Chromatographia 52 (2000) 657–663.

[317] Isobe, T., Shiraishi, H., Yasuda, M., Shinoda, A., Suzuki, H., Morita, M. J. Chromatogr. A, 984 (2003) 195–202.

[318] Dias, N.C., Poole, C.F. Chromatographia 56 (2002) 269–275.

[319] Fine, D.D., Breidenbach, G.P., Price, T.L., Hutchins, S.R. J. Chromatogr. A, 1017 (2003) 167–185.

[320] Morales, S., Cela, R. J. Chromatogr. A, 896 (2000) 95–104.

[321] Hernández, M., Borrull, F., Calull, M. Chromatographia 52 (2000) 279–284.

[322] Reverté, S., Borrull, F., Pocurull, E., Marcé, R.M. J. Chromatogr. A, 1010 (2003) 225–232.

[323] Blackwell, P.A., Lützhøft, H.-C.H., Ma, H.-P., Halling-Sørensen, B., Boxall, A.B.A. J. Chromatogr. A, 1045 (2004) 111–117.

[324] Lindsey, M.E., Meyer, M., Thurman, E.M. Anal. Chem. 73 (2001) 4644–4646.

[325] Lin, W.-C., Chen, H.-C., Ding, W.-H. J. Chromatogr. A, 1065 (2005) 279–285.

[326] Martínez, D., Borrull, F., Calull, M. J. Chromatogr. A, 827 (1998) 105–112.

[327] Loos, R., Barceló, D. J. Chromatogr. A, 938 (2001) 45–55.

[328] Loos, R., Alonso, M.C., Barceló, D. J. Chromatogr. A, 890 (2000) 225–237.

[329] Carabias-Martínez, R., Rodriguez-Gonzalo, E., Herrero-Hernández, E., Román, F.J.S.-S., Flores, M.G.P. J. Chromatogr. A, 950 (2002) 157–166.

[330] Sirvent, G., Hidalgo, M., Salvado, V. J. Sep. Sci. 27 (2004) 613–618.

[331] Rostagno, M.A., Palma, M., Barroso, C.G. J. Chromatogr. A, 1076 (2005) 110–117.

[332] Fontanals, N., Marce, R.M., Borrul, F. J. Chromatogr. A, 1152 (2007) 14–31.

[333] Shundo, A., Sakurai, T., Takafuji, M., Nagaoka, Sh., Ihara, H. J. Chromatogr. A, 1073 (2005) 169–174.

[334] Rahman, M.M., Takafuji, M., Ihara, H. J. Chromatogr. A, 1119 (2006) 105–114.

[335] Ihara, H., Dong, W., Mimaki, T., Nishihara, M., Sakurai, T., Takafuji, M., Nagaoka, S. J. Liq. Chromatogr. 26 (2003) 2473–2485.

[336] Yang, M., Fazio, S., Munch, D., Drumm, P. J. Chromatogr. A, 1097 (2005) 124–129.

[337] Croes, K., Steffens, A., Marchand, D.H., Snyder, L.R. J. Chromatogr. A, 1098 (2005) 123–130.

[338] Proskurina, N.A., Il'in, M.M., Davankov, V.A., Sychev, K.S., Kostikov, S.Yu. Russ. J. Phys. Chem. 81 (2007) 424–427.

[339] Sychov, C.S., Davankov, V.A., Proskurina, N.A., Mikheeva, A.Yu. LC-GC Europe 22 (2009) 20–27.

[340] Baya, M.P., Siskos, P.A., Davankov, V.A. J. AOAC Int. 83 (2000) 579–583.

[341] Muirhead, E.E., Reid, A.F. J. Lab. Clin. Med. 33 (1948) 841–844.

[342] Bapp, A.L., Farrel, P.C., Uveli, D.A., Screibner, B.H. ASAIO Trans. 18 (1972) 98–105.

[343] Gejyo, F., Yamada, T., Odani, S., Nakagava, Y., Arakawa, M., Kunitomo, T. et al. Biochem. Biophys. Res. Commun. 129 (1985) 701–706.

[344] Odell, R.A., Slowiaczek, P., Moran, J.E., Schindhelm, K. Kidney Int. 39 (1991) 909–919.

[345] Leypoldt, J.K., Cheung, A.K., Deeter, R.B. Am. J. Kidney Assoc. 32 (1998) 295–301.

[346] Klinke, B., Rockel, A., Abdelhamid, S., Fiegel, P., Walb, D. Int. J. Artif. Organs 12 (1989) 697–702.

[347] Ronco, C., Heifetz, A., Fox, K., Curtin, C., Brendolan, A., Gastaldon, P. et al. Int. J. Artif. Organs 20 (1997) 136–143.

[348] Yatzidis, H. Proc. Eur. Dial. Transplant Assoc. 1 (1964) 83–86.

[349] Winchester, J.F., Ronco, C. Adv. Ren. Replace. Ther. 9 (2002) 19–25.

[350] Chang, T.M.S. ASAIO Trans. 12 (1966) 13–19.

[351] Kokot, F., Pietrek, J., Seredynski, M. Proc. Eur. Dial. Transplant Assoc. 15 (1978) 604–606.

[352] Winchester, J.F. Hemoperfusion in Replacement of Renal Function by Dialysis, third ed., Dordrecht, Kluwer Academic, 1988, pp. 439–459.

[353] Furuyoshi, S., Kobayashi, A., Tamai, N., Yasuda, A., Tanaka, S., Tani, N. et al. Blood Purif. 9 (1991) 9.

[354] Gejyo, F., Teramura, T., Ei, I., Arakawa, M., Nakazawa, T. Artif. Organs 19 (1995) 1222–1226.

[355] Miyata, T., Jadoul, M., Kurokawa, K., Van Ypersele de Strihou, C. J. Am. Soc. Nephrol. 9 (1998) 1723–1735.

[356] Davankov, V.A., Pavlova, L.A., Tsyurupa, M.P., Tur, D.R. J. Chromatogr. B, 689 (1997) 117–122.

[357] Davankov, V., Pavlova, L., Tsyurupa, M., Brady, J., Balsamo, M., Yousha, E. J. Chromatogr. B, 739 (2000) 73–80.

[358] Malik, D.J., Webb, C., Holdich, R.G., Ramsden, J.J., Warwick, G.L., Roche, I. et al., Sep. Purif. Technol., 66 (2009) 578–585.

[359] Davankov, V.A., Elisarov, D.P., El'kin, A.I., Kataev, S.S., Pavlova, L.A., Terekhin, G.A. et al. Toxicol. Bull., 2002, #3, 2–5.

[360] Bosh, T., Wendler, T., Duhr, C., Brady, J., Samtleben, W. J. Am. Soc. Nephrol. 11 (2000) 257A.

[361] Ronco, C., Brendolan, A., Winchester, J.F., Golds, E., Clemmer, J., Polaschegg, H.D. et al. Blood Purif., 19 (2001) 260–263.

[362] Winchester, J.F., Ronco, C., Brady, J.A., Cowgill, L.D., Salsberg, J., Yousha, E., Choquette, M. et al. Blood Purif. 20 (2002) 81–86.

[363] Winchester, J.F., Ronco, C., Brady, J.A., Golds, E., Clemmer, J., Cowgill, L.D. et al. Blood Purif. 19 (2001) 255–259.

[364] Morena, M.D., Guo, D., Balakrishnan, V.S., Brady, J.A., Winchester, J.F., Jaber, B.L. Kidney Int. 63 (2003) 1150–1154.

[365] Winchester, J.F., Kellum, J.A., Ronco, C., Brady, J.A., Quartararo, P.J., Salsberg, J.A. et al. Blood Purif. 21 (2003) 79–84.

[366] Elisarov, D.P., El'kin, A.I., Davankov, V.A., Kataev, S.S., Pavlova, L.A., Reshetnikov, V.I. et al. Toxicol. Bull. 2003, #2, 18–21.

[367] Elisarov, D.P., El'kin, A.I., Davankov, V.A., Kataev, S.S., Pavlova, L.A., Reshetnikov, V.I. et al. Efferent Therapy 9 (2003) #3, 58–61.

[368] Davankov, V.A. US Patent 5545131, 1996.

[369] Tsyurupa, M.P. "Hypercrosslinked polystyrene: a new type of polymeric networks", Dr. Sci. Thesis INEOS AC RAS, Moscow, 1985.

[370] Martsinkevich, R.V., Tsyurupa, M.P., Davankov, V.A., Soldatov, V.S. Vysokomol. Soedin. A 20 (1978) 1061–1065.

[371] Tsyurupa, M.P., Davankov, V.A., Rogozhin, S.V. J. Polym. Sci. Symp. N 47 (1974) 189–196.

[372] Tsyurupa, M.P., Andreeva, A.I., Davankov, V.A. Vysokomol. Soedin. A 22 (1980) 2523–2527.

[373] Soldatov, V., Högfeldt, E. Ion Exch. Membr. 2 (1974) 13–16.

[374] Soldatov, V.S., Martsinkevich, R.V., Pokrovskaya, A.I. Zh. Fiz. Khimii 43 (1969) 2889.

[375] Soldatov, V.S., Pokrovskaya, A.I., Martsinkevich, R.V. Zh. Fiz. Khimii 46 (1972) 867–870.

[376] Martsinkevich, R.V., Soldatov, V.S., Davankov, V.A., Tsyurupa, M.P., Rogozhin, S.V. Zh. Fiz. Khimii 51 (1977) 1465–1468.

[377] Pokrovskaya, A.I., Soldatov, V.S. in: Thermodynamics of Ion Exchange, Minsk, Nauka I Tekhnika, 1968, pp. 84–91.

[378] Davankov, V.A., Rogozhin, S.V., Korshak, V.V., Tsyurupa, M.P. Izv. AN SSSR, N7 (1967) 1612–1614.

[379] Peslyakas, I.H.I. "Synthesis of dissymmetric sorbents with neutral and hydroxyl-containing amino acid functional groups and their application in ligand exchange chromatography", Ph. D. Thesis, Vilnyus, 1972.

[380] Davankov, V.A., Navratil, J.D., Walton, H.F. Ligand Exchange Chromatography, CRC Press, Boca Raton, FL, 1988.

[381] Belyakova, L.D., Davankov, V.A., Kiselev, A.V., Muttik, G.G., Tsyurupa, M.P. Kolloid. Zh. 40 (1978) 1059–1065.

[382] Kiryutenko, V.M., Kiselev, A.V., Lygin, V.I. Kolloid. Zh. 37 (1975) 382–383.

[383] Avgul, N.N., Belyakova, L.D., Vorob'eva, L.D., Kiselev, A.V., Muttik, G.G. et al. Kolloid. Zh. 36 (1974) 928–930.

[384] Belyakova, L.D., Kiselev, A.V., Lyustgarten, E.I., Muttik, G.G., Shevchenko, T.I. Kolloid. Zh. 37 (1975) 540–548.

[385] Belyakova, L.D., Kiselev, A.V., Lyustgarten, E.I., Pashkov, A.B., Platonova, N.P., Shevchenko, T.I. Dokl. AN SSSR 213 (1973) 1311–1313.

[386] Dale, J.A., Irving, J. in: Greig, J.A. (Ed.), Ion Exchange Development and Application, Proceedings of IEX 96, SCI, UK, 1996, 193–200.

[387] Tai, M.H., Saha, B., Streat, M. React. Funct. Polym. 41 (1999) 149–161.

[388] Saha, B., Sreat, M. Ind. Eng. Chem. Res. 44 (2005) 8671–8681.

[389] Oancea, A.M.S., Drinkal, C., Höll, W.H. React. Funct. Polym. 68 (2008) 492–506.

[390] Oancea, A.M.S., Radulescu, M., Oancea, D., Pincovschi, E. Ind. Eng. Chem. Res. 45 (2006) 9096–9106.

[391] Meinhardt, E., Kautzmann, R.M., Ruiz, N., Sastrea, M., Sampaio, M., Cortina, J.L. in: Williams, P.A. and Dyer, A. (Eds.) Advances in Ion Exchange for Industry and Research, RSC, UK, 1999, 338–345.

[392] Cortina, J.L., Kautzmann, R.M., Gliese, R., Sampaio, C.H. React. Funct. Polym. 60 (2004) 97–107.

[393] Pan, B., Zhang, Q., Meng, F., Li, X., Zhang, X., Zheng, J. et al. Environ. Sci. Technol. 39 (2005) 3308–3313.

[394] Troshkina, I.D., Serbin, A.M., Khaing, Z.N., Ushanova, O.N., Demin, Yu.V., Chekmarev, A.M. Sorption Chromatogr. Processes (Russia), 6 (2006) 1022–1027.

[395] Hradil, J., Krystl, V., Hrabánek, P., Bernauer, B., Kočiřik, M. React. Funct. Polym. 61 (2004) 303–313.

[396] Hradil, J., Krystl, V., Hrabánek, P., Bernauer, B., Kočiřik, M. React. Funct. Polym. 65 (2005) 57–68.

[397] Hradil, J., Sysel, P., Brozová, L., Kovářová, J., Kotek, J. React. Funct. Polym. 67 (2007) 432–441.

[398] Villaluenga, J.P.G., Seoane, B., Hradil, J., Sysel, P. J. Membr. Sci. 305 (2007) 160–168.

[399] Sidorov, S.N., Volkov, I.V., Davankov, V.A., Tsyurupa, M.P., Valetsky, P.M., Bronstein, L.M. et al. J. Am. Chem. Soc. 123 (2001) 10502–10510.
[400] Sulman, E.M., Lakina, N.V., Matveeva, V.G., Valetsly, P.M., Bronstein, L.M., Davankov, V.A. et al. Patent RF 2170227, 10.07.2001.
[401] Sulman, E.M., Marveeva, V.G., Bronstein, L.M., Sulman, M.G., Doluda, V.D., Tokarev, A.V. et al. Stud. Surf. Sci. Catal. 162 (2006) 119–121.
[402] Sulman, E., Doluda, V., Dzwigaj, S., Marceau, E., Kustov, L., Tkachenko, O. et al. J. Mol. Catal. A: Chem. 278 (2007) 112–119.
[403] Doluda, V.Yu., Sulman, E.M., Matveeva, V.G., Sulman, M.G., Lakina, N.V., Sidorov, A.I. et al. Chem. Eng. J. 134 (2007) 256–261.
[404] Bykov, A., Matveeva, V., Sulman, M., Valetsky, P., Tkachenko, O., Kustov, L. et al. Catal. Today 140 (2009) 64–69.
[405] Germain, J., Hradil, J., Fréshet, J.M.J., Svec, F. Chem. Mater. 18 (2006) 4430–4435.
[406] Germain, J., Fréchet, J.M.J., Svec, F. Polym. Mater.: Sci. Eng. 97 (2007) 272–273.
[407] Lee, J.-Y., Wood, C.D., Bradshaw, D., Rosseinsky, M.J., Cooper, A.I. Chem. Commun. 25 (2006) 2670–2672.
[408] Pastukhov, A.V., Aleksienko, N.N., Tsyurupa, M.P., Davankov, V.A., Voloshchuk, A.M. Russ. J. Phys. Chem. 79 (2005) 1371–1379.
[409] Zhang, X., Shen, S., Fan, L., J. Mater. Sci. 42 (2007) 7621–7629.
[410] Thomas, A., Kuhn, P., Weber, J., Titirici, M.-M., Antonietti, M. Macromol. Rapid Commun. 30 (2009) 221–236.
[411] Wood, C.D., Tan, B., Trewin, A., Niu, H., Bradshaw, D., Rosseinsky, M.J. et al. Chem. Mater. 18 (2007) 2034–2048.
[412] Tsyurupa, M.P., Davankov, V.A. React. Funct. Polym. 53 (2002) 193–203.
[413] Germain, J., Fréchet, J.M.J., Svec, F., J. Mater. Chem. 17 (2007) 4989–4997.
[414] Srinivasan, S.S., Ratnadurai, R., Niemann, M.U., Phani, A.R., Goswami, D.Y., Stefanakos, E.K. Int. J. Hydrogen Energy 35 (2010) 225–230.
[415] Zhou, W., Wu, H., Yildirim, T. J. Am. Chem. Soc. 130 (2008) 15268–15269.
[416] Klontzas, E., Tylianakis, E., Froudakis, G.E. J. Phys. Chem. C, 113 (2009) 21253–21257.
[417] Kaderi, El-H.M., Hunt, J.R., Mendoza-Cortés, J.L., Côté, A.P., Taylor, R.E., O'Keeffe, W. et al. Science 316 (2007) 268–272.
[418] Wood, C.D., Tan, B., Trewin, A., Su, F., Rosseinsky, M.J., Bradshaw, D. et al. Adv. Mater. 20 (2008) 1916–1921.
[419] Morris, R.E., Wheatley, P.S. Angew. Chem. Int. Ed. 47 (2008) 4966–4981.
[420] Aleksienko, N.N., Pastukhov, A.V., Davankov, V.A., Belyakova, L.D., A.M. Voloshchuk, Russ. J. Phys. Chem. 78 (2004) 1992–1998.
[421] Asnin, L.D., Davankov, V.A., Pastukhov, A.V. Russ. J. Phys. Chem. A, 82 (13) (2008) 2313–2317.
[422] Asnin, L.D., Davankov, V.A., Pastukhov, A.V., Kachmarski, K. Russ. J. Phys. Chem. 83 (2009) 1204–1207.
[423] Asnin, L.D., Davankov, V.A., Pastukhov, A.V., Shchurov, Y.A. Izv. Acad. Nauk Russ. Ser. Khim. (11) (2009) 2151–2155.
[424] Long, Ch., Li, Q., Li, Y., Liu, Y., Li, A., Zhang, Q. Chem. Eng. J. 160 (2010) 723–728.
[425] Saha, B., Karounou, E., Streat, M. React. Funct. Polym. (2010), doi:10.1016/j.reactfunctpolym.2010.04.004.
[426] Sulman, E.M., Matveeva, V.G., Dolida, V.Y., Sidorov, A.I., Lakina, N.V., Bykov, A.V., Sulman, M.G., Valetsky, P.M., Kustov, L.M., Tkachenko, O.P., Stein, B.D., Bronstein, L.M. Appl. Catal. B, 94 (2010) 200–210.
[427] Fontanals, N., Cormack, P.A.G., Sherrington, D.C., Marc, R.M.é, Borrull, F. J. Chromatogr. A, 1217 (2010) 2855–2861.

[428] Bratkowska, D., Marc, R.M.é, Cormack, P.A.G., Sherrington, D.C., Borrull, F., Fontanals, N. J. Chromatogr. A, 1217 (2010) 1575–1582.

[429] Bratkowska, D., Fontanals, N., Borrull, F., Cormack, P.A.G., Sherrington, D.C., Marcé, R.M. J. Chromatogr. A, 1217 (2010) 3238–3243.

[430] Zeng, X., Yu, T., Wang, P., Yuan, R., Wen, Q., Fan, Y., Wang, C., Shi, R. J. Hazard. Mater. 177 (2010) 773–780.

[431] He, C., Huang, J., Yan, C., Liu, J., Deng, L., Huang, K. J. Hazard. Mater. 180 (2010) 634–639.

[432] Zeng, S.-Z., Guo, L., He, Q., Chen, Y., Jiang, P., Shi, J. Microporous Mesoporous Mater. 131 (2010) 141–147.

[433] Urban, J., Svec, F., Fréchet, J.M.J. Anal. Chem. 2010, 82, 1621–1623.

[434] Svec, F. J. Chromatogr. A, 1217 (2010) 902–924.

[435] MacGillivray, L., Metal-Organic Frameworks. Design and Application, Wiley-VCH, Weinheim, Germany, 2010, 350 pages.

CONCLUSION

Concluding the book, we state once again that there is every good reason to consider hypercrosslinked polystyrene as the first representative of a whole new class of polymeric materials with an intrinsic nanoporosity. They are characterized by a rigid open-framework structure that remains accessible to small molecules both in the dry and solvated state. Because of that structure, on the one hand, extremely large apparent specific surface areas are measured for hypercrosslinked materials by conventional gas adsorption techniques, and, on the other hand, the materials actively respond to mechanical forces and to wetting with any liquid media, irrespective of their nature and thermodynamic quality.

During the last decade, in addition to polystyrene, many other polymeric materials were provided with the hypercrosslinked network structure, including hydrophilic polymers, charged polymers, and polysilsesquioxanes. Even nanoporous carbons prepared by pyrolysis of hypercrosslinked polystyrene were shown to exhibit some properties peculiar to typical hypercrosslinked networks, such as swelling in chlorobenzene [1]. Much attention has been paid in recent years to metal-organic frameworks. Considered to be inorganic salts or crystalline coordination compounds, they behave like "hypercrosslinked" networks if provided with a rigid spacious molecular construction. They even show a definite flexibility of their crystal structure and allow passage into their interior of molecules exceeding the size of channels and openings within their crystal lattice [2].

The global scientific community now recognizes the obvious novelty and usefulness of hypercrosslinked networks and accepts our explanation of their properties, above all else their unique ability to swell in non-solvents. The term "hypercrosslinked," which we introduced in the early 1980s, has established itself in the scientific literature; see, for example, [3–5]. Hypercrosslinked networks prepared by intensive post-crosslinking of polymeric chains with rigid links in the presence of excess solvating liquid are also referred to as "Davankov-type" networks [6–9]. By analogy to "hypercrosslinked," a similar term, "hyperbranched," was coined by Webster [10] to denote the dendritic-type structure of statistically branched macromolecular species (see [11]). Though the branching points of the latter have much in common with network nodes, the two polymer types are topologically different and hyperbranched polymers remain molecularly dispersed globular species that do not form bulk monolithic materials.

Interestingly, in spite of the numerous publications that have appeared in the literature concerning the practical applications of hypercrosslinked polymers, no theoretical treatises of the new class of polymers exist to date. There are also very few papers reviewing information on hypercrosslinked polymers [12], apart from some reviews by the present authors. Therefore, we hope that this book will be found useful by specialists working in various fields of research and practice.

REFERENCES

[1] Asnin, L.D., Davankov, V.A., Pastukhov, A.V., Shurov, Y.A. Russ. J. Phys. Chem. 83 (2009) 1204–1207.

[2] Ziv, A., Grego, A., Kopilevich, S., Zeiri, L., Miro, P., Bo, C., Müller A., Weinstock, I.A. J. Am. Chem. Soc., 131 (2009) 6380–6382.

[3] Hradil, J., Krystl, V., Harbanel, P., Bernawer, B., Kocirik, M. React. Funct. Polym. 65 (2005) 57–68.

[4] Oh, C.-G., Ahn, J.-H., Ihm, S.-K. React. Funct. Polym. 57 (2003) 103–111.

[5] Valderrama, C., Cortina, J.L., Faran, A., Gramisans, X., Lao, C. J. Colloid Interface Sci. 310 (2007) 35–46.

[6] Fontanals, N., Cortés, J., Galià, M., Marcé, R.M., Cormack, P.A.G., Borrull, F., Sherrington, D.C. J. Polym. Sci.: Part A, Polym. Chem. 43 (2005) 1718–1728.

[7] Ahn, J.-H., Jang, J.-E., Oh, C.-G., Ihm, S.-K., Cortez, J., Sherrington, D.C. Macromol. 39 (2006) 627–632.

[8] Wood, C.D., Tan, B., Trewin, A., Niu, H., Bradshaw, D., Rosseinsky, M.J., Khimyak, Y.Z., Campbell, N.L., Kirk, R., Stökel E., Cooper, A.I. Chem. Mater. 18 (2007) 2034–2048.

[9] Weder, C. Angew. Chem. Int. Ed. 47 (2008) 448–450.

[10] Kim, Y.H., Webster, O.W. J. Am. Chem. Soc. 112 (1990) 4592–4593.

[11] Seiler, M. Fluid Phase Equilib. 241 (2006) 155–174.

[12] Zhang, X., Shen, S., Fan, L. J. Mater. Sci. 42 (2007) 7621–7629.

Reader, I wish that, as we parted –
whoever you may be, a friend,
a foe – our mood should be warm-hearted.
Goodbye, for now we make an end.
Whatever in this rough confection
you thought – tumultuous recollection,
a rest from toil and all its aches,
or just grammatical mistakes,
a vivid brush, a witty rattle –
God grant that from this little book
for heart's delight, or fun, you took,
for dreams, or journalistic battle,
Got grant you took at least a grain.
On this we'll part; goodbye again!

Pushkin, "Eugen Onegin"
Translation by Charles H. Johnston
Copyright © Charles Johnston, 1977, 1979

INDEX

A

Acenaphthene, 553
Acetone
  sorption isotherm for, 382f
  volume of absorbed, 383t
Acid retardation, 475–480, 476f–479f
  LiCl/HCl in test for, 477–479, 478f
  mechanism of, 447
  NaCl/HCl in test for, 475–476, 476f
  process in ion exchangers of, 448
Activated carbons
  advantages of, 371
  shortcomings of, 372
Activity coefficients of electrolytes,
  446–449
Adsorption isotherms
  BDDT classification for, 374
  BET equation for, 374–377
  hysteresis loop in, 77–81
  Kelvin equation for, 76–77
  sorption isotherms for nitrogen, 374,
    375f, 376f, 379, 380
  specific surface area calculations from, 377
Aerosol OT (AOT)
  formation of gigaporous texture with,
    119–123, 121f–124f
  molar ratio to water of, 119–120
  specific inner surface area as function of
    molar ratio to water, 120, 121f
  pore diameter and volume as function of
    molar ratio to water, 120, 121f
Ag^+, 443, 444f
  radius for, 462t
Al^{3+}, radius for, 460t, 462t
Aldehydes, wine quality with, 546
Aliphatic alkylamines, 551
Amberchrom GC-161, pre-concentration
  of phenolic compounds with, 530, 532
Amberlite XAD-2
  cephalosporin C sorption with, 436
  phenols sorption with, 427
Amberlite XAD-4, 418
  caffeine extraction with, 432, 432t
  cephalosporin C sorption with, 436

chlorogenic acid removal with, 432, 432t
furfural sorption with, 436
gasoline sorption, 423, 423t
hydrogen storage capacity of, 606t
hypercrosslinked sorbents v., 439–440
lecithin sorption with, 422, 422t
lipophilic marine toxins accumulated
  by, 552
phenols sorption with, 424, 424f, 426f, 427
pre-concentration of phenolic
  compounds with, 531, 532, 533t
sorption isotherms for *n*-valeric acid
  onto, 417, 417f, 418t
Amberlite XAD-7
  deswelling, 291–293, 292f
Amberlite XAD-16
  caffeine extraction with, 432, 432t
  cephalosporin C sorption with, 436
  citrus juice de-bittering with, 434
  hydrogen storage capacity of, 606t
Amberlite XAD-180, cephalosporin C
  sorption with, 436
α-Amino acids, 418–419, 421
  hydrolysis of methyl esters of, 596t
Aminophenols, separation of
  dihydroxybenzenes and
  phenylenediamine in RP mode,
  515, 516
Aminosine, 438
Ammonium acetate, resolution of, 500, 501f
Amperozide, 509, 509
Anion exchange resins Styrosorbs,
  595–597, 596t
Anions, 448–449
Anthraquinone sulfonates, 547, 547t
AOT, *See* Aerosol OT (AOT) Aromatic
  amines, 550t
Aromatic compounds, separation in RP
  mode, 515, 516f, 519f
Aromatic sulfonates
  anthraquinone sulfonates, 547, 547t
  benzene sulfonates, 547, 547t
  naphthalene sulfonates, 547, 547t
  stilbene sulfonates, 547, 547t

M